7th International Conference on Recrystallization and Grain Growth 2019

Journal of Physics: Conference Series Volume 1270

Ghent, Belgium
4 - 9 August 2019

Editors:

Leo A.I. Kestens
Hadi Pirgazi
Tuan Nguyen Minh
Roumen H. Petrov

ISBN: 978-1-5108-9560-7
ISSN: 1742-6588

Printed from e-media with permission by:

Curran Associates, Inc.
57 Morehouse Lane
Red Hook, NY 12571

Some format issues inherent in the e-media version may also appear in this print version.

Institute of Physics
Dirac House, Temple Back
Bristol BS1 6BE UK

Phone: 44 1 17 929 7481
Fax: 44 1 17 920 0979

techtracking@iop.org

Additional copies of this publication are available from:

Curran Associates, Inc.
57 Morehouse Lane
Red Hook, NY 12571 USA
Phone: 845-758-0400
Fax: 845-758-2633
Email: curran@proceedings.com
Web: www.proceedings.com

TABLE OF CONTENTS

ELECTRICAL STEELS

CHARACTERIZATION

STEELS

SEVERE PLASTIC DEFORMATION

PROCESSING AND APPLICATION

THEORY AND SIMULATION

DYNAMIC RECRYSTALLIZATION

SUPERALLOYS

IOP Conf. Series: Journal of Physics: Conf. Series **1270** (2019) 011001 doi:10.1088/1742-6596/1270/1/011001

7th International Conference on Recrystallization and Grain Growth

August 4 – 9, 2019

Ghent, Belgium

Chair Organising Committee

Prof. Leo A.I. Kestens

Full Professor
Metal Science and Technology
Ghent University
Department of Electromechanical Systems and Materials

Vice-Chair Organising Committee

Dr. Hadi Pirgazi

Doctor Assistant
Metal Science and Technology
Ghent University
Department of Electromechanical Systems and Materials

Editors

Leo A.I. Kestens

Hadi Pirgazi

Tuan Nguyen Minh

Roumen H. Petrov

This supplement contains the Full Papers of Keynote & Invited Talks and Oral & Poster Presentations presented on the occasion of ReX&GG 2019.

Preface

The Seventh International Conference on Recrystallization and Grain Growth is the continuation of a conference series that was initiated in 1990 with *Recrystallization '90*, organized in Wollongong, followed in 1991 by '*Grain Growth in Polycrystalline Materials*", organized in Rome. Since 2001, both conference series were merged with the organization of the '*1st International Conference on Recrystallization and Grain Growth*' in Aachen, which has ever since been a regularly recurring tri-annual event.

During the three decades that have passed since its launch, this conference series has shaped an active and versatile research community with active participation from fundamental and applied research groups and industrial R&D organizations. Recrystallization and grain growth are defining factors in shaping the microstructure of both structural and functional polycrystalline materials, metals and non-metals alike, and therefore of paramount importance for a wide variety of microstructurally sensitive material properties. Thus, an intensive and mutually fruitful collaboration has emerged between applied R&D groups, often associated with industrial organizations, and more fundamentally oriented scientists. Other scientific fields, such as geology, mineralogy and archeology have increasingly studied microstructures that ensue from thermally activated processes such as recrystallization and grain growth, which may provide clues as to how materials were created and what circumstances prevailed during their history.

A more detailed understanding of the basic mechanisms of recrystallization and grain growth also contributes to the development of new materials with improved and more versatile functional properties. For newly emerging processing technologies, such as additive manufacturing, a better knowledge on recrystallization and grain growth, particularly in far out-of-equilibrium conditions, is of crucial importance for microstructural control of 3D printed metal objects and hence for the successful introduction of additive manufacturing to all its potential markets.

Microstructures of polycrystalline materials, and by consequence microstructural control, are intricately complex. Better characterization techniques to probe deeper and more extensively into the world of crystals, crystal orientations, grain boundaries, interfaces, dislocations and various other types of lattice defects, continue to have large impact on the field, with the large scale application of orientation contrast microscopy and 3D characterization techniques as two of the most emblematic examples.

The data sets generated by advanced characterization techniques are ideally suited to be explored by data analytics and machine learning, which may give rise to the observation of specific patterns or correlations that can be confronted with newly developed theoretical insights.

IOP Conf. Series: Journal of Physics: Conf. Series **1270** (2019) 011002 doi:10.1088/1742-6596/1270/1/011002

All the aforementioned aspects of the field are well represented at the 7th Recrystallization and Grain Growth conference, organized at Ghent University from August 4th through 9th, 2019, with over 250 delegates attending from more than thirty countries. We are confident that these conference proceedings represent a comprehensive update on the current state-of-the-art in the endlessly fascinating field of Recrystallization and Grain Growth.

Leo A.I. Kestens
Hadi Pirgazi
Tuan Nguyen Minh
Roumen H. Petrov

IOP Conf. Series: Journal of Physics: Conf. Series **1270** (2019) 011002 doi:10.1088/1742-6596/1270/1/011002

Tribute to Julian Driver

Roger Doherty

Professor Emeritus of Materials Science and Engineering, Drexel University

In May 2018, the world of Physical and Mechanical Metallurgy and in particular our field of Thermo-Mechanical Processing lost a major figure with the untimely death, at 72, of my friend and scientific colleague, Julian Driver. I first met Julian at Alcoa Technical Center 30 years ago when we were both consulting for Alcoa on the processing of aluminum alloys for beer cans. Subsequently, we met frequently at Recrystallization and Texture conferences and best of all, in 2003, I spent an enjoyable and scientifically very fruitful six months with Julian and Claire Maurice at L'Ecole des Mines in St Etienne, France. While there, a major collaboration occurred between us when I discovered that both Julian and, quite separately, my ex-student and current colleague Indradev Samajdar (together with Bert Verlinden), were, independently from each other, planning a much- needed text book on Thermo-Mechanical Processing. I was then able to bring these two groups together and help plan what the joint book would cover. "Thermo-Mechanical Processing of Metallic Materials", which was edited by my mentor and close friend, the late Robert Cahn, was published by Elsevier in 2007.

While Julian and I shared a mutual interest in the field of Thermo-Mechanical Processing for grain structure control, we both reached the central topic of recrystallization from very different directions. As I will subsequently describe, Julian's research started from microstructure in studies of carefully controlled plastic deformation, initially with cyclic strain and later with monotonic strain often in single crystals. Having developed important insights into the resulting microstructures with their many heterogeneities, Julian was able, later in his career, to understand how such deformed metals underwent recrystallization on annealing. I, on the other hand, starting with my doctoral research in 1960-63, attempted to characterize and understand the recrystallization process. It was only later in my career, first in the 1970s, and later in the 1990s - realizing that understanding recrystallization required understanding the microstructures of plastically deformed metals - that I tried to make some contributions to that field.

For both of us industrial consultancy was a vital part of our research. We both had found that industry has many fascinating metallurgical problems arising from the metals industry's extensive technological but largely empirical experience. Successful Industrial consultancy is a two-way process in which academic scientists can greatly benefit from exposure to industrial problems while, hopefully, their resulting scientific studies can sometimes give improved processing and better final properties.

Before I try to describe Julian's scientific career and his achievements I would like to give what I think of as a relevant historical insight into our field. Professor Fred Rhines, who was an academic in the USA during the relevant period, once remarked to me: "During the 1950's, academic metallurgy was totally transformed but the industrial alloys in use changed by only about 10%". What might be the reason for this contrast? By the start of that decade, industrial research and development based on enlightened empiricism, guided by optical microscopy and X-ray diffraction, had been able to develop the successful industrial alloys then in use. This had been achieved without real knowledge of the fundamental origin of either mechanical properties or microstructural processing. However, academic metallurgy during that decade was totally transformed scientifically by input from other fields: solid

IOP Conf. Series: Journal of Physics: Conf. Series **1270** (2019) 011002 doi:10.1088/1742-6596/1270/1/011002

state physics, physical chemistry and mechanics. The vital new science included first the invention of edge dislocations in the 1930s and later their acceptance by metallurgists. In the UK, Alan Cottrell, then at Birmingham University, brought the concept of dislocations to the metallurgical community. Then there was the insight into homogeneous and heterogeneous nucleation initially of solidification and then of solid state phase changes. Alan's short textbook "Theoretical Structural Metallurgy" also educated the metallurgical community in the scientific basis for microstructure development. David Turnbull at the GE research lab in the US had brought the critical initial quantitative insights from physical chemistry to metallurgy. Finally, there was the development of fracture mechanics showing how metals could be both strong and resistant to fracture. Furthermore, by about 1955, came the arrival of transmission electron microscopy which allowed direct observations of the resulting microstructures. After 1960, there were the many new developments in metallography, particularly scanning electron microscopy and, after 1980, EBSD. Julian's studies using EBSD were of particular significance. These scientific and technological developments created an environment giving our generation of research metallurgists, including myself from 1960 and Julian a few years later, a marvelous opportunity to build our careers in the rapidly changing field of Physical Metallurgy.

In the various and different research fields that he studied, Julian interacted with industrial problems. These interactions facilitated many research achievements and made them industrially useful. Academic insights in our field rarely lead directly to new industrial alloys – but they can and do facilitate improvements of industrial processing which yield better alloys. The aluminum can is a clear example of that. Industry invented the process and also the can body and can lid alloys, but later metallurgical science helped significantly improve what has been described as "A pressure vessel for 10 cents". For example, cube texture that gives 0 and 90° "ears" on can drawing is formed during recrystallization prior to the final cold rolling for increased strength. That rolling gives compensating 45° deformation texture "ears". This was an empirical technical triumph but academic research finally gave the solution (or at least its current version) to the long standing scientific problem of the origin of the cube texture in recrystallized hot rolled commercial aluminum alloys. Cube texture had been originally found even more strongly in heavily cold rolled pure copper. Academic research now enables prediction of the strength of that texture. Julian, together with his students, was a major contributor to that research topic.

Julian grew up in the North of England. This was always rather clear from his northern accent in both his spoken English, and as my wife tells me, in his spoken French - in which he was of course completely fluent. His undergraduate education in the early 1960s was at the University of Sheffield with its close links to the UK steel industry which had many steel mills and research centers in that city in the 1960s. Sadly, there are very few of either now. His PhD studies were in one of the major metallurgical research centers in the world: Cambridge University. In 1960 it housed two major international scientists: Alan Cottrell, who was until 1965 the head of the Metallurgy Department, and in the Physics Department, Peter Hirsch, pioneer in TEM diffraction theory. However, by the time Julian arrived at Cambridge both had moved on. Alan became the Scientific Advisor to the new Labour Government and Peter was appointed Head of the Department of Metallurgy at Oxford University. Julian became a student of the new Dept. Head at Cambridge, Robert Honeycombe, and - as I was astonished to discover while preparing this tribute - Julian's PhD topic was on high temperature fatigue of austenitic steels. Until recently, I was also unaware that up to the late 1980s, Julian continued to publish extensively in the field of cyclic deformation. After Cambridge, he spent a post-doc period in the University of Newcastle with Professor Jack, working on surface nitriding of various steels. He also continued to contribute to that research field while in France where he also studied a topic that I did not even know even existed! This was surface hardening of steel using boron. In 1974, just as the UK joined the Common Market, he made an unusual transition for a successful young British metallurgist: he moved to France to a research and teaching post at L'Ecole des Mines in St Etienne. Roland Fortunier recently told me a nice story about Julian's arrival there. Apparently, and not unreasonably, a foreign appointee was required to pass an exam in French. Julian, not too surprisingly for an English- educated scientist, failed the French exam. Fortunately, the school was so pleased to have him as a teacher and researcher that they sensibly ignored the language rule. Clearly any initial difficulties with the French language did not hinder his success at the school. Before he retired he had become the deputy director of the School's 100 strong Materials and Structures Center in charge of strategy and international relations.

His move to France was, I think, not unconnected with the fact that his first wife was a teacher in St Etienne and her family, I understood, owned a vineyard. Julian once described to me some of the interesting early times he had helping with vine pruning in very cold conditions in the early spring. Those experiences did not seem, however, to limit his enthusiasm for good wine. I greatly enjoyed a marvelous day in early 2003 where Julian took my wife and me on an extensive wine-tasting and wine-buying tour of some excellent vineyards. The resulting wine purchases were an added bonus for our time in St. Etienne. A few bottles even made it back to the USA, imported in our car buried under several centimeters of papers and reports on recrystallization and grain growth.

It was after Julian arrived in L'Ecole des Mines in 1974 that his research expanded to his pioneering work on metal plasticity. I regret not discussing with Julian how and why his research developed in that way. However, I was grateful to Roland Fortunier, one of Julian's early research students in plasticity, for Roland's insight that the presence in L'Ecole des Mines of Professor Claude Goux, with his research on metallic single and bicrystals, was very significant in Julian's research choice. Julian's early research in France was initially on cyclic deformation in various metallic single and bicrystals, though there were also some more applied fatigue studies on polycrystalline alloys. Julian was also working on boronizing of steel. My knowledge of these aspects of Julian's research is from his publication list that I only recently received thanks to another of Julian's PhD students, Romain Quey.

By 1976 Julian's work extended to conventional plasticity with studies relating first to the Bauschinger effect. This was an interesting transition since it involved a single strain reversal. In 1979 came the first of his many classic plasticity studies of monotonically deformed single and bicrystals. The first publication was "The Plane Strain compression of Aluminium Bicrystals" with Serpoul. That work was followed in 1982, with his student Skalli and industrial collaborator Wintenberger, by "Theoretical and Experimental Study of Aluminum Crystals subjected to High Strain" The title indicated what would become a major feature of Julian's work: combining experiments with theoretical analysis and numerical modelling. After 1982, there were still papers on fatigue and boronizing, nitriding and other industrial topics such as friction stir welding. However, more and more of his research group's work was on metal plasticity, in both single and polycrystals. This soon included grain rotation, which was the start of his many contributions to deformation texture, the origin and importance of heterogeneities in the deformed microstructure, such as shear bands, transition bands and variation of stored energies in regions of different orientation. All of these are vital aspects of the deformed microstructure for subsequent recrystallization. In 1985 came the first studies on crystal rotation and the development of deformation textures. Finally, in 1991 came the first of his many publications on recrystallization - the first one, with Roland Fortunier, on plane strain compressed aluminum crystals. It is striking how very fundamental were many of these studies, reflecting Julian's interest in the science of all the topics he investigated; though as I have previously noted he combined his passion for science with the interest and ability to apply his science to the complex problems of metallurgical engineering

In the following years continuing up to and beyond Julian' s retirement in 2012, with his final publication in 2018, there was a marvelous body of research, both with his students and with many other international collaborators in the fields of metal plasticity and recrystallization. Of course, perhaps the most significant and extensive scientific collaboration was with the PhD student who later became his second wife and the mother of their two children, Claire Maurice. One of potentially the most significant of their collaborations was their recent development of the technique of High Resolutions EBSD. HR-EBSD had been initially developed by Angus Wilkinson and David Dingley but was modified by Claire Maurice and Julian Driver. This new technique, which I am still trying to understand fully, allows at a tens of nanometer length scale, the possibility of measuring residual elastic strains and thus stresses in the surface microstructure. This method excludes hydrostatic strain, since the lattice parameter cannot be measured by EBSD to sufficient accuracy for strain measurement. However, strains that distort the crystal symmetry, away from exact cube symmetry for cubic metals, allow for revolutionary new microstructural insights. Combining High Resolution EBSD with X-ray strain measurements which can measure lattice parameters with sufficient accuracy for strain measurement but at a coarser length scale and extending below the metal surface, seems to provide for a potential new microstructural revolution. Of course at a free surface, the stress state cannot be hydrostatic since a free surface will not support any residual normal stress.

During my time in St Etienne, I was fortunate to be invited to play a very minor part in what I regard as a really major scientific contribution. This was the PhD study by Gael Coutourier who was advised by Claire and also by Roland Fourtunier, with typically helpful input from Julian. Gael analyzed, using an elegant finite element method, the problem of "Smith-Zener" drag on grain boundary migration by incoherent spherical particles. In the classic 1948 Smith-Zener analysis, they estimated the maximum retarding force that an individual particle can exert on a migrating grain boundary. This arose from the interfacial tensions causing the boundary as it passes the center of the particle to become locally significantly distorted. The resulting pressure drag on the boundary was this particle force multiplied by the estimated number density of particles on the boundary. For that estimate, Smith and Zener assumed that the grain boundary was smoothly curved. That assumption is of course incompatible with their earlier assumption of a highly distorted boundary morphology! In the many years between Smith's publication of the analysis and Gael's thesis in 2003 there had been many hundreds of papers on the analysis together with a similar number of experimental studies. Gael's work solved the implicit Smith Zener problem: his analysis showed that there were about 4 times the area density of particles on the distorted boundary than had been predicted by Smith-Zener for an undistorted boundary. However, the average force per particle was only about 50% of the Smith Zener maximum force model. So Gael showed that the Smith- Zener first order model was qualitatively correct, though it underestimated the particle drag by a factor of only about 2 times. Not bad for a 60-year-old model that used only 6 lines of simple algebra! I have described this study as my way of illustrating the quality of research undertaken and the quality of the research collaborators who had been Julian's PhD students at L'Ecole des Mines. That research quality greatly reflects Julian's contribution to the lab where he worked for nearly 40 years.

I would like to end by quoting from the tribute to Julian published by five of his previous PhD students in June, 2018 in the French Metallurgy Society Bulletin, for which I was given in an English Translation

I give it here in slightly more conventional English:

"Those of us who had the privilege of knowing Julian remember his characteristic sense of humor, his humility and the enthusiasm with which he transmitted his passion for Science. He allowed his PhD students great freedom of action. Julian liked to transmit his ideas and advice, and to offer help to his close and less close colleagues, some of whom became friends. Every one of them will remember the very open discussions that they had with Julian."

That seems to me to describe my friend almost perfectly. He is greatly missed by all of us fortunate enough to have known and worked with him

Thank you for this opportunity to celebrate my friend Julian Driver's scientific life and contributions.

Roger Doherty
August 2019

Tribute to Lasar Shvindlerman

Günter Gottstein

Distinguished Senior Professor of Physical Metallurgy and Metal Physics,
RWTH Aachen University, Germany

A true scholar and dedicated scientist, with a smile on his face and always ready for a joke and a new idea of what and how to investigate, that is how we will always remember Lasar S. Shvindlerman who passed away suddenly and entirely unexpectedly on March 3, 2018. His seminal contributions to the fields of thermodynamics and kinetics of interfaces as well as grain growth have shaped our current understanding of thermally induced microstructure evolution and will remain unforgotten.

Lasar was born on 20 September 1935 in Gitomir, Ukraine, which was then part of the Soviet Union, his father a Red Army officer and his mother a pharmacist. World War II forced him together with his mother and grandmother to flee to Tashkent, Uzbekistan, while his father served as military pilot.

After World War II, the family moved first to Vienna and then relocated to Kiev, where Lasar graduated from high school and subsequently studied metallurgical engineering at the Kiev Polytechnic Institute. After obtaining his engineering degree, he started work in the steel industry. In order to continue his academic education he moved to Moscow for doctoral studies at the Moscow Institute for Steel and Alloys MISIS. There he met Boris S. Bokstein, who became a life-long close friend and coauthor of many scientific papers and books.

After earning his PhD degree, Lasar began his scientific career 1970 at the Institute of Solid State Physics of the Soviet Academy of Sciences in Chernogolovka where he headed the Group of Internal Interfaces in Metals, which later was promoted to the rank of a laboratory.

During that time he dedicated himself to the investigation of specific grain boundaries as he realized that there is no such thing like an average grain boundary or, as Lasar put it, 'there is no sense of the average body temperature of patients in a hospital'. His approach was to determine steady state properties since only they can be considered as true state properties of interfaces in metals and can be measured accurately and reproducibly. For this he focused on studies of individual, well-characterized grain and interphase boundaries in specially prepared bi-crystalline samples with half-loop or quarter-loop grain boundary geometry. His group generated probably the largest and most valuable set of data on grain boundary mobilities in metals, in particular in aluminum and zinc, analyzed it with respect to low Σ CSL boundaries and studied the influence of impurities on grain boundary motion. I myself became aware of the data only when Lasar's colleague Ch. Kopetzky visited us in Aachen in the 1970ies to report on their research. I could not believe the quantity and precision of the reported data, but my mentor Kurt Lücke, who had visited the group in Moscow, convinced me that a large group of scientists in Lasar's laboratory was engaged in the measurements of grain boundary motion.

With the breakdown of the Soviet Union, Lasar immediately took the chance to actively establish contacts with colleagues in the West, in particular with W. Gust in Stuttgart and R. Faulkner in Loveborough. In 1990 he came to see me in Aachen, and we immediately felt that our interest in grain boundaries and our complementary approaches made it desirable to team up, and we decided to establish a joint 'interface dynamics lab' at my institute (IMM). We convinced Dmitri Molodov to head the laboratory, a former student of Lasar and now a prominent scientist in the field of grain boundaries. This was the beginning of a most successful cooperation which generated a large body of research papers and also our joint book 'Grain Boundary Migration in Metals: Thermodynamics, Kinetics, Applications'. Besides our scientific cooperation we also became close friends, actually Lasar became my best friend, which was certainly unusual at that age.

In the course of our cooperation on all aspects of grain boundary migration we decided to extend our research also to other interfacial features, in particular grain boundary ridges, faceted boundaries, and last but not least, triple and quadruple junctions. By theory and experiment we were able to show that these interfacial constituents, especially the grain boundary junctions can exert a strong impact on microstructure evolution during grain growth and can be successfully utilized for grain boundary engineering.

(picture from obituary written by Eugen Rabkin)

Beginning 1994, Lasar spent most of his time at RWTH Aachen University, where he became a guest professor and lectured with pleasure even at an age of 82. His scientific achievements did not remain unnoticed by the scientific community. He received many awards in recognition of his contributions to physical metallurgy, including the Werner Köster Award of DGM, the Max-Planck-Prize of the Max-Planck Society and the Humboldt foundation, the Alexander-von-Humboldt Research Award, the Honorary G. Kurdyumov Medal, and in 2013 we were jointly honored with the Cyril Stanley Smith Award bestowed by the ReX&GG International Committee. He also served as editor of '*Materials letters*'.

Lasar loved to do research and to engage in scientific discussions. In fact, he came to my office virtually every day with a new idea on what to urgently investigate, and we practically always disagreed initially on the approach, which led to lengthy and in most cases also not very quiet discussions. When we travelled on the train we usually had a compartment for ourselves since other travelers could not stand our hot-tempered and loud discussions for more than five minutes. However, despite all disagreement in the beginning we eventually would come to a common conclusion, and I definitely learned something every time. Scientific discussion was very important to him. Therefore, he became the driving force for the 'Vaalsbroek Meetings on Interfaces' in the first decade of this

IOP Conf. Series: Journal of Physics: Conf. Series **1270** (2019) 011002 doi:10.1088/1742-6596/1270/1/011002

century. These meetings had a quite different format than other conferences. After very short introductory remarks of the speaker most of the time was dedicated to an in-depth discussion of the issue under consideration. Lasar used to refer to these meetings - with some intended exaggeration - as 'Solvay' conferences on interfaces.

He loved to interact with and teach talented students. In fact, he did not only shape their scientific and conceptual thinking but also the personal attitude of his students. He always was keen on keeping in contact with former students and, this way, established an academic family. As his former student and successor as head of the laboratory in Chernogolovka, Boris Straumal, put it: 'W*e are glad that we had a chance to meet Lasar Shvindlerman in our life, to be together so many years, to observe his smile, to laugh at his jokes, to hear his famous table speeches, to work and debate with him'*.

Moreover, he established an extended active international network with famous scientists in the field of interface science, many of which came to see and work with him at IMM in Aachen. ReX&GG was his most favorable conference. He served as member of the international committee of ReX &GG from the day it was founded.

The scientific community of ReX&GG lost one of his most prominent members. We miss Lasar very much, his friendly and helpful personality, his modest human attitude, his wide-spread interest and knowledge, whether murders of Agatha Christie, Japanese fairy tales or history of science, leave alone his outstanding scientific expertise. He shaped our current understanding of interface thermodynamics and kinetics with his seminal contributions to the field. In the last year of his life he dreamt of writing a book on 'The ten most wonderful experiments in Materials Science'. Regretfully his life suddenly ended before his dream became true.

We miss Lasar Shvindlerman as colleague and friend, but his unique personality and his outstanding scientific contributions will always be remembered.

Günter Gottstein
August 2019

IOP Conf. Series: Journal of Physics: Conf. Series **1270** (2019) 011003 doi:10.1088/1742-6596/1270/1/011003

7th International Conference on Recrystallization and Grain Growth

August 4 – 9, 2019

Ghent, Belgium

ORGANISING COMMITTEE

Leo Kestens, UGent, Belgium (Chair)
Hadi Pirgazi, UGent, Belgium (Vice-Chair)
Monireh Azimi, UGent/TU-Delft, Belgium
Felipe Castro Cerda, U de Santiago de Chile, Chile
Stefaan Cottenier, UGent, Belgium
Christophe Detavernier, UGent, Belgium
Jesus Galan Lopez, TU-Delft, The Netherlands
Sepideh Ghodrat, TU-Delft, The Netherlands
Edwin Alexis Lopez Covaleda, UGent, Belgium
Tuan Nguyen Minh, UGent, Belgium
Roumen Petrov, UGent, Belgium
Jilt Sietsma, TU-Delft, The Netherlands
Manuel Sintubin, KULeuven, Belgium
Roger Van Hecke, UGent, Belgium
Ilse Vercruysse, UGent, Belgium
Bert Verlinden, KULeuven, Belgium

INTERNATIONAL COMMITTEE

Giuseppe Carlo Abbruzzese, Italy (Chair)
Nathalie Bozzolo, France
Michael Ferry, Australia
Niels Hansen, Denmark
Elizabeth Holm, USA
W. Bevis Hutchinson, Sweden
Yuichi Ikuhara, Japan
Dorte Juul Jensen, Denmark
John J. Jonas, Canada
Rustam Kaibyshev, Russia
Suk-Joong Kang, Korea
Leo Kestens, Belgium
Sung-Joon Kim, Korea
Qing Liu, China
Weimin Mao, China
Knut Marthinsen, Norway
Claire Maurice, France
Dimitri Molodov, Germany
Eric Palmiere, United Kingdom
Elena Pereloma, Australia
Paulo Rios, Brazil
A.D. (Tony) Rollett, USA
David J. Srolovitz, USA
Nobuhiro Tsuji, Japan
Kohsaku Ushioda, Japan

IOP Conf. Series: Journal of Physics: Conf. Series **1270** (2019) 011004 doi:10.1088/1742-6596/1270/1/011004

7th International Conference on Recrystallization and Grain Growth

August 4 – 9, 2019

Ghent, Belgium

The Organising Committee thanks the following companies for their support:

Silver Sponsors

JEOL Solutions for Innovation

ocas

Bronze Sponsors

acutance SCIENTIFIC TSL Solutions

TESCAN PERFORMANCE IN NANOSPACE

xnovotech

ThermoFisher SCIENTIFIC

Supporter

metals an Open Access Journal by MDPI

IOP Conf. Series: Journal of Physics: Conf. Series **1270** (2019) 011005 doi:10.1088/1742-6596/1270/1/011005

Peer review statement

All papers published in this volume of *Journal of Physics: Conference Series* have been peer reviewed through processes administered by the proceedings Editors. Reviews were conducted by expert referees to the professional and scientific standards expected of a proceedings journal published by IOP Publishing.

IOP Conf. Series: Journal of Physics: Conf. Series **1270** (2019) 012001 doi:10.1088/1742-6596/1270/1/012001

Quasi *in-situ* EBSD characterization of the evolution of microtexture and microstructure in non-oriented electrical steels during cold rolling and annealing

Y. He[1, *], M. Mehdi[1, 2], M. Sanjari[1, 3], E. J. Hilinski[4], L. A. I. Kestens[5], and A. Edrisy[2]

[1]CanmetMATERIALS, Natural Resources Canada, Hamilton, Ontario, Canada
[2]Department of Mechanical, Automotive, and Materials Engineering, University of Windsor, Windsor, Ontario, Canada
[3]NanoPhyll Inc., Hamilton, Ontario, Canada
[4]Tempel Steel Co., Chicago, Illinois, USA
[5]Department of Materials Science and Engineering, Ghent University, Ghent, Belgium

*Corresponding author: youliang.he@canada.ca

Abstract. Non-oriented electrical steel sheets are extensively used to manufacture core laminations for electric motors. The microstructure and crystallographic texture of the final thin sheets have a significant effect on the magnetic quality of the lamination core. Controlling the development of microstructure and texture during thermomechanical processing of these steels is of great practical importance. Although the final microstructure and texture of the steel sheets are affected by all the processing procedures employed, cold rolling and final annealing have the most direct effect since these are usually the final steps to process the steel. Although much effort has been made to study the evolution of microstructure and texture in non-oriented electrical steels, the formation mechanisms of the final microstructure and texture during annealing (especially grain growth) are still not completely understood, which is partially due to the fact that it is very difficult to *directly* investigate these processes using conventional characterization techniques. In this research, a quasi *in-situ* electron backscatter diffraction (EBSD) technique was presented, which enabled the tracking of the changes of the morphology and orientation of individual grains during cold rolling and the subsequent annealing process. The experiments revealed the rotations of individual grains under plane strain compression, and tracked the formation of nuclei from the deformed matrix and the growth of new grains during the recrystallization process. The results provided valuable insights into the evolution of microstructure and microtexture during cold rolling and annealing of non-oriented electrical steels, and helped understand the mechanisms that govern the nucleation and grain growth during recrystallization.

1. Introduction

Non-oriented electrical steels are the most widely used soft magnetic material for the manufacturing of magnetic cores for electric motors and generators. Due to their practical importance, these steels have been investigated for decades to improve their magnetic and other properties [1-5]. The evolution of microstructure and texture during thermomechanical processing of electrical steels has long been a topic of many researches [6-10], and the numerous experiments and simulations carried out in the past have greatly helped scientists and engineers understand the mechanisms governing the formation of the final microstructure and texture in the steel sheets. Based on these studies, the processing parameters have been optimized, and better microstructure and texture have been produced, which improved the magnetic properties of the final laminates. Nevertheless, there are still mechanisms that are not completely understood, which are of both scientific and practical importance. For example, although advanced crystal plasticity models [11, 12] have been developed to simulate the formation of texture during plastic deformation, it was very difficult to directly verify these simulations since the incremental deformation of the individual grains could rarely be traced experimentally. Furthermore, the mechanisms governing the nucleation and growth of new grains during recrystallization are still not completely understood, and there is still debate on whether it is *oriented nucleation* or *oriented growth* that determines the final texture [13].

The existence of these discrepancies was partially due to the lack of appropriate experimental techniques that can directly track the evolution of microtexture and microstructure during the deformation and annealing processes. Although advanced diffraction facilities (e.g. neutron and synchrotron) have been frequently employed to investigate the evolution of deformation texture *in situ* (usually only on tension, compression or torsion, though), it was very difficult to apply these techniques in a rolling mill to investigate the rolling process. Heating stages have also been integrated into conventional scanning electron microscopes (SEM) and transmission electron microscopes (TEM) to enable direct observation of the recovery and recrystallization of metals during annealing, but these facilities usually require large capital investment, and there are limitations as well. For example, the maximum temperature they could reach was usually relatively low, so only a few materials (e.g. those with relatively low melting points: aluminum, zinc, magnesium, etc.) could be studied. This paper presents a simple *quasi in-situ* technology based on conventional electron backscatter diffraction (EBSD) techniques, which can be utilized to track the morphology and orientation changes of individual crystals in electrical steels during conventional rolling and annealing. This can help us better understand the evolution of microtexture and microstructure during these thermomechanical processes.

2. Materials and Experimental Procedure

Two non-oriented electrical steels were utilized for this study, one containing 2.8 wt% of silicon (High Silicon) and the other containing 0.88 wt% silicon (Low Silicon). The chemical compositions of the two steels are listed in Table 1. The steels were melted in a vacuum induction furnace and cast into ingots with a 200 mm × 200 mm cross section. The ingots were reheated to 1038°C and hot rolled to a thickness of about 2.5 mm in a reversing rolling mill. The hot-rolled plates were then annealed at 840°C in a dry 100% hydrogen atmosphere for 60 hours before cold rolling.

Table 1. Chemical compositions of the steels used for this study.

Material	C	Mn	S	P	Si	Al	Fe
Low Silicon	0.002	0.31	0.001	0.01	0.88	0.46	Balance
High Silicon	0.003	0.30	0.001	0.01	2.80	0.52	Balance

To track the deformation of individual grains under plane strain compression (rolling), a rectangular plate of 100 mm × 50 mm × 2.5 mm was cut from the hot-rolled and annealed steel plate (High Silicon) along the rolling direction, and a smaller rectangular sample of 25 mm × 10 mm × 2.5 mm was cut from the plate, as shown in Figure 1a. One surface of the sample was at the rolling direction-normal direction (RD-ND) plane on the center line, and this surface was polished using conventional metallography procedures. The surface was further polished using a colloidal silica suspension (0.05 μm), and the microstructure and microtexture of a designated area (marked by microhardness indents) were characterized by EBSD, which served as the initial microstructure and texture before rolling. The small rectangular sample was then inserted back to the steel plate and cold rolled to pre-defined thickness reduction rates. Because the small rectangular sample was constrained within the larger plate, bulging of the RD-ND plane at the center line was essentially prevented, and the deformation was assumed to be plane-strain compression. After each thickness reduction, the surface was slightly polished using a colloidal silicon suspension, and the same area was examined under EBSD to record the changes of microstructure and microtexture so that the rotations of individual grains can be tracked during each increment of deformation, which was otherwise very difficult to be realized using conventional methods.

To track the nucleation and grain growth of the deformed material during annealing, the Low Silicon steel (0.88 wt% Si) was incline cold rolled at an angle of 45° to the hot rolling direction [10] to a thickness of 0.5 mm. An area on the deformed sample was first marked using microhardness indents and examined under EBSD (Figure 1b), which recorded the microstructure and microtexture of the material before annealing (after cold rolling). The sample was then annealed at a fixed temperature (650°C) for different times to track the nucleation and grain growth of individual grains during recrystallization. After each annealing time, the sample was quenched in cold water to freeze the microstructure. The previously examined surface was slightly polished using a colloidal silicon suspension to prepare the surface for EBSD scan. Usually about 1-2 μm of the surface will be removed by this polishing step. To effectively track the start of recrystallization, the sample was examined at short intervals at the beginning, since the recrystallization occurs very fast from the deformed microstructure. The sample was annealed with extended holding times once all the grains are recrystallized, since the growth of the recrystallized grains within a matrix consisting of all new grains is much slower than the recrystallization from a deformed matrix. In this way both the initial nucleation and the grain growth processes can be traced.

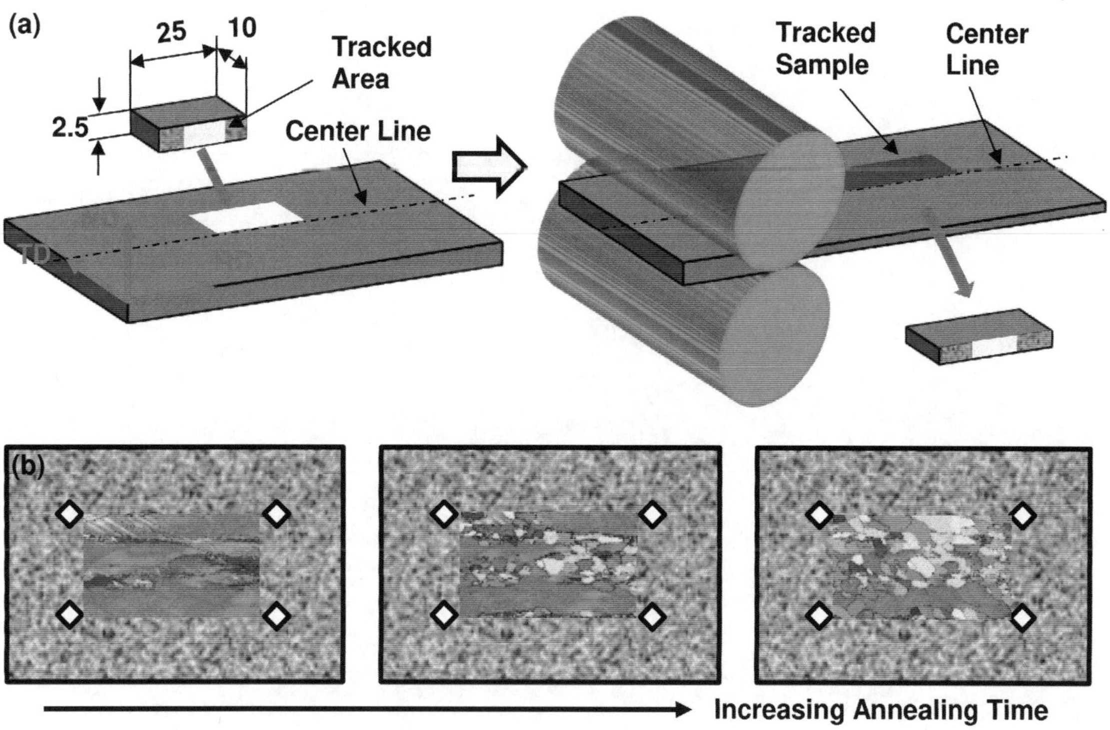

Figure 1. Schematic illustration of the quasi *in-situ* EBSD characterization technique: (a) tracking the cold rolling process, (b) tracking the annealing process.

IOP Conf. Series: Journal of Physics: Conf. Series **1270** (2019) 012001 doi:10.1088/1742-6596/1270/1/012001

3. Results and Discussion

3.1. Tracking cold rolling

An example of quasi *in-situ* EBSD characterization of the cold rolling of a 2.8 wt% Si non-oriented electrical steel is shown in Figure 2. An initial EBSD scan of the hot-rolled and annealed sample was made to reveal the microstructure and microtexture before rolling. The scanned area consists of about 100 grains on the RD-ND section at the center line (Figure 2a). After 8% thickness reduction, essentially all these grains can still be identified, and the orientation change of each grain can be clearly noticed from the variations in color within the individual grains (Figure 2b).

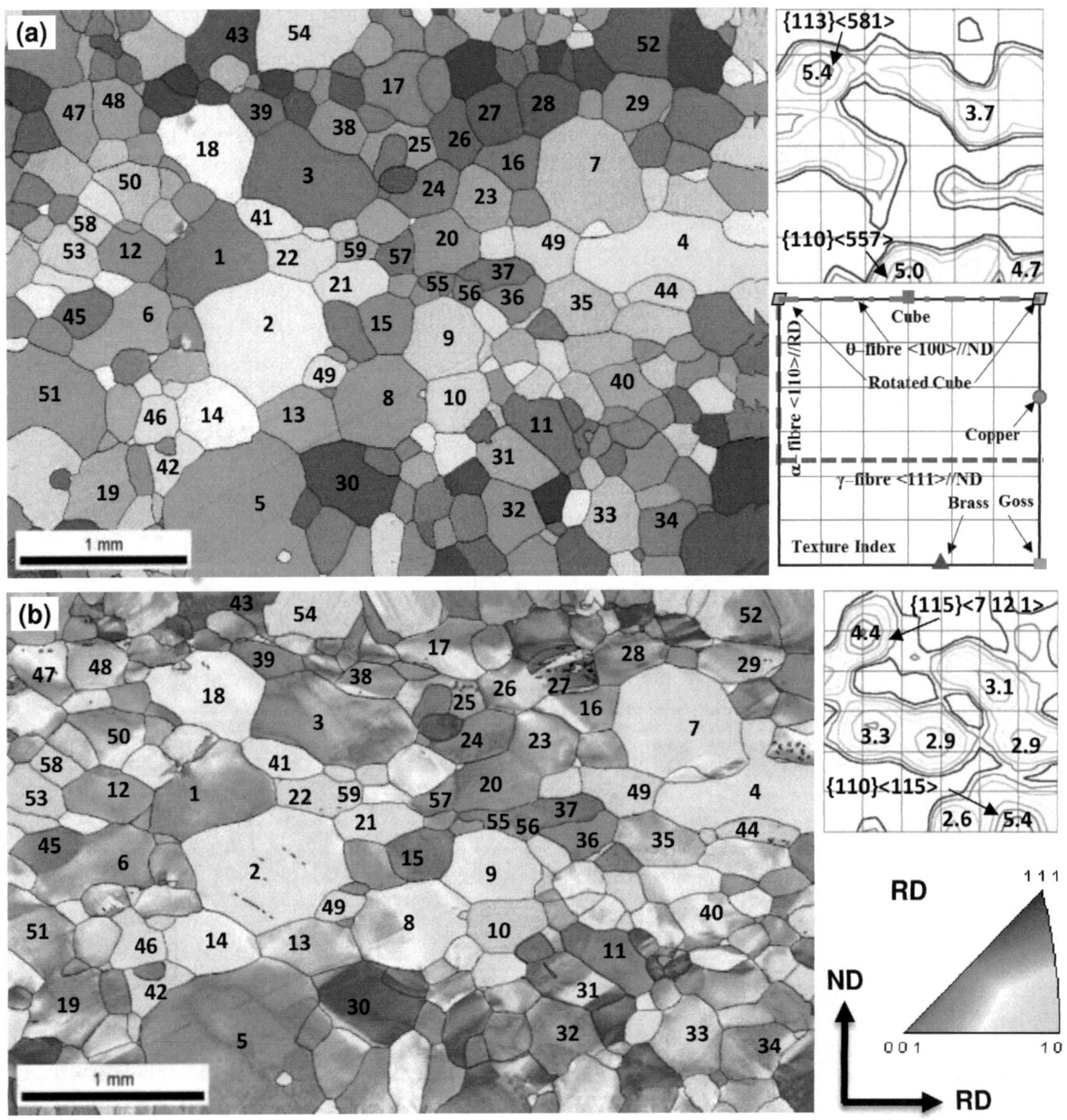

Figure 2. An example of tracking the individual grains during plane-strain compression: (a) EBSD inverse pole figure (IPF) map before deformation, (b) IPF map of the same area after 8% thickness reduction. The textures shown to the right are $\varphi_2 = 45°$ sections of the orientation distribution functions (ODFs) based on Bunge's notation.

Although with only a small thickness reduction (8%), shear bands are already visible in some of the grains (e.g. 2, 22 and 27). It is also noticed that grain subdivision occurred in most of the grains, as can be seen from the color differences within the deformed grains. However, some grains did not show apparent shape or color change, e.g. grains 10, 53 and 58. Apparently, the rotations of these grains are more difficult than the others under plane-strain compression conditions. Some originally separated grains (with distinguished grain boundaries) are merged after deformation, and the grain boundaries may disappear due to the merging. For example, grains 22/59 and 23/20 are merged into two single grains (without apparent boundaries) after deformation. Other examples include grains 37, 55 and 56, which also merge into a single grain after deformation. Thus it is seen that, by quasi *in-situ* EBSD tracking of the individual grains during rolling, it was able to observe some interesting phenomena that might have been overlooked during conventional experiments, where tracing the deformation of the *same grains* was not possible. The orientation data obtained from the tracing of individual grains can be conveniently used to verify the theories used to predict deformation texture [14]. The overall texture of the sample also shows apparent changes, i.e. the initially strong {113}<581>, {110}<557> and Goss textures are weakened and rotated to {115}<7 12 1> and {110}<115>, and the γ-fibre texture was considerably strengthened, even if the reduction rate is only 8%.

Another example of deformation tracking is shown in Figure 3, where a few grains with near {111}<121> orientations (γ-fibre) are isolated from the deformed matrix, and their deformation is tracked from 10% to 20% reductions. It is seen that even though the shapes of the grains are considerably changed, the orientations of these grains are essentially maintained the same as the undeformed ones. The γ-fibre grains have relatively high Taylor factors, and thus have high resistance to plane strain compression. During deformation, these grains tend to retain their orientations. However, apparent orientation variation was noticed in each grain, which contributes to the increase of the orientation spread and the rotation of the original {111}<121> orientations to other orientations (as indicated by arrows). This also results in the decrease of the intensity of the original grains.

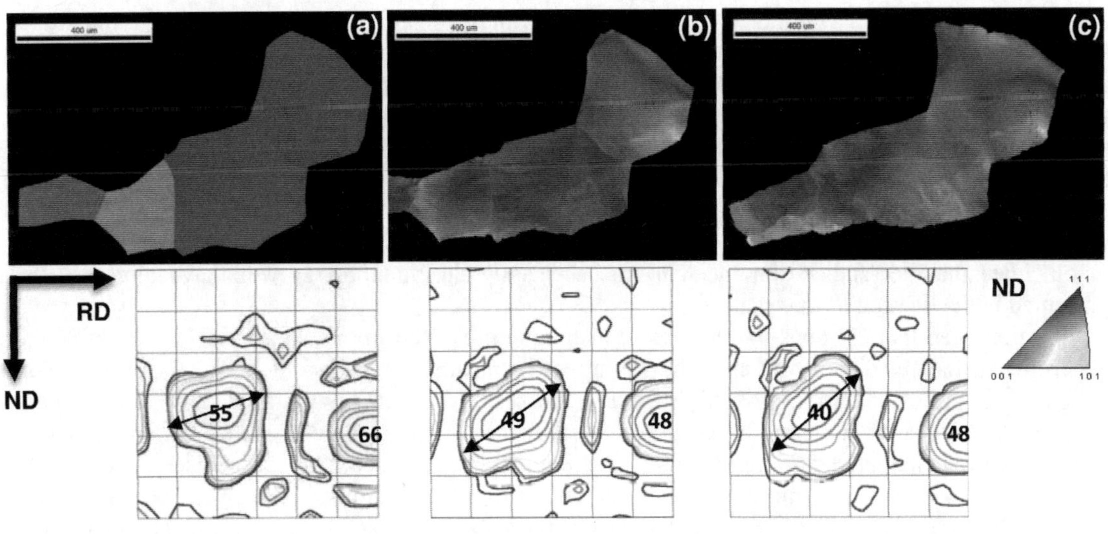

Figure 3. An example to track the deformation of near {111}<121> grains during cold rolling: (a) before rolling, (b) 10% reduction, (c) 20% reduction. Textures are shown on φ_2 =45° sections of the ODFs (Bunge notation).

3.2. *Tracking recrystallization*

Figure 4 shows an example of tracking the recrystallization of a cold-rolled 0.88 wt% Si steel after inclined cold rolling. Figure 4a illustrates the early stages of annealing when new grains nucleate from the deformed matrix, and grow gradually while other nuclei are formed in the meantime. It is seen that, after cold rolling, the steel is mainly composed of elongated <111>//ND and <001>//ND grains with numerous shear/deformation bands within these grains. At the beginning of recrystallization (3 minutes), nucleation preferably starts from the deformed <111>//ND grains, either at the grain boundaries (marked with rectangles) or within the deformed grain along the shear bands (marked with ovals). Only one grain (marked with a rhombus) is noticed to nucleate within a deformed <001>//ND grain. When the annealing time is increased to 6 minutes, some of the new grains grow considerably (e.g. regions 1, 4 and 3), while others only grow slightly (e.g. regions 2, 7 and 6). New grains are also seen nucleated from the deformed matrix, e.g. regions 8, 9 and 10. At this moment, the growing grains are usually constrained within the elongated grains from which they nucleate, and they do not cross the boundaries of those deformed grains. After 10 minutes, all the deformed grains were replaced by recrystallized grains, due to the continued growth of the already recrystallized grains, and due to the nucleation and growth of new grains from the remaining deformed matrix.

Once the recrystallization is complete, i.e. the deformed matrix has been completely consumed and replaced by new grains, further annealing will result in grain growth within the all-recrystallized microstructure, which is much slower than the nucleation and growth of grains from the deformed matrix. There is no nucleation after this, and the subsequent grain growth will determine the final microstructure and texture. Figure 4b shows an example of tracing grain growth during later stages of annealing, where the growth of some grains is tracked during annealing from 54 minutes to 329 minutes. At the lower left corner, the large, red "R" grain (with a "C" shape) is seen gradually embracing the smaller light-green "G" grain inside it, and finally consuming the "G" grain after 149 minutes, to form an equiaxed grain with an almost circular shape. The "B" grain near the top has experienced a different path: initially (54 minutes) a small grain (pink) is seen trapped inside this grain, which is gradually "eaten" by the "B" grain to form two small islands (74 minutes), and are completely consumed after 104 minutes. The size of the "B" grain itself essentially does not change, though, from 54 to 104 minutes, i.e. it does not grow except that the grain inside it has been consumed. In the meantime, all the neighboring small grains around the "B" grain are gradually consumed by the larger growing grains adjacent to them. After 149 minutes, the "B" grain becomes a small grain as all the neighboring grains have grown to much larger than it, and it starts to be consumed by other grains (209 minutes). After 329 minutes, it essentially disappears. These direct observations clearly confirmed that during grain growth, the larger grains had a size advantage, and will eventually grow by consuming their smaller neighbours.

The above observation directly illustrates the morphology changes of individual grains during later stages of annealing. It should be mentioned that, the orientation data obtained from the EBSD scans can be utilized to evaluate the misorientation and mobility of the grain boundaries of each growing grain, which can help unveil the mechanisms governing the growth process. A few examples using such data to elucidate the recrystallization mechanisms, especially during grain growth, have already been presented in the literature [15, 16].

It is thus seen that the quasi *in-situ* EBSD technique presented in this paper provides an effective experimental tool to directly observe the rolling and annealing of electrical steels, which can give insights into the mechanisms governing these processes. However, it should be noticed that, these tracking techniques require careful design of the experiments, and also need good sample preparation skills for EBSD characterization. For rolling tracking, the sample should be cut precisely, and be able to be inserted back into the sample holder so that bulging of the examined surface can be minimized. The sample holder should be the same material (with the same processing history) as the sample to avoid the difference in deformation between the sample and the sample holder. The amount of thickness reduction that can be reached before distorting the examined surface is highly dependent on how the experiments are conducted. Usually, a total of 25-30% thickness reduction can be readily achieved. With more than 30% reduction, the examined surface becomes largely elongated and the bulging of the plane may prevent further tracing of the same area.

For the tracking of recrystallization, there is no bulging issue of the plane to be examined. However, the heating and quenching of the sample during the experiment, i.e. interrupting the annealing process, may result in different behaviors from the real annealing process (continuous). Nevertheless, the results shown in this study did provide some insights into the nucleation and grain growth processes. This technique can be used as an effective experimental tool to investigate the recrystallization of non-oriented electrical steels.

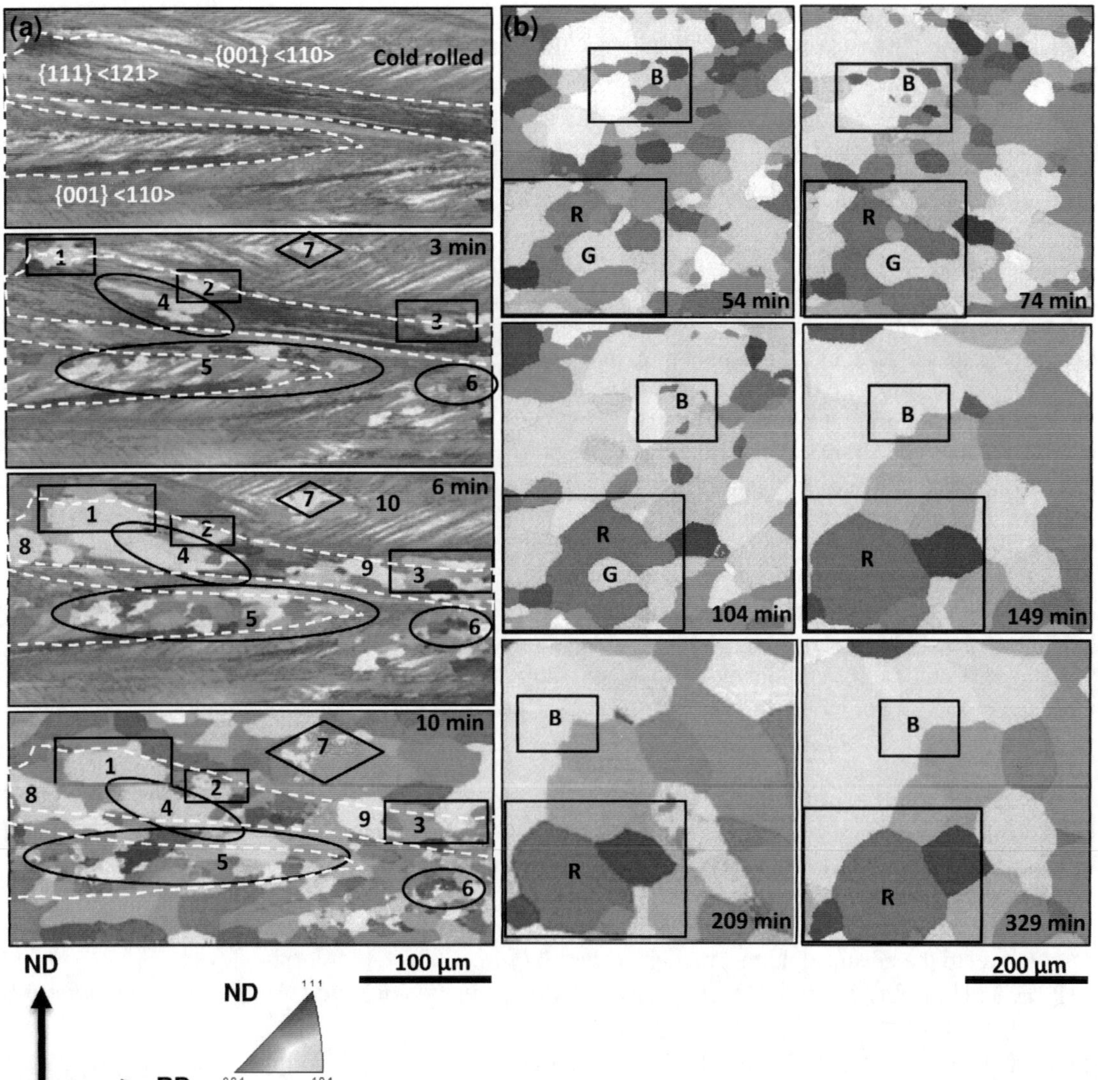

Figure 4. Quasi *in-situ* tracking the evolution of recrystallization during annealing of a cold-rolled 0.88 wt% Si electrical steel: (a) early stage of annealing with nucleation and grain growth from deformed matrix, (b) later stages with grain growth in all-recrystallized microstructure.

4. Summary and Conclusions

In this paper, a quasi *in-situ* EBSD characterization technique was presented as an effective experimental tool to investigate the cold rolling and annealing of non-oriented electrical steels.

The changes of morphology and orientation of individual grains during plane-strain compression can be clearly followed at specific thickness reductions. The orientation data can be used to verify crystal plasticity simulations that involve the effect of grain boundaries. Some interesting phenomena, such as the formation of shear bands, merging of separate grains, and the retention of some grain orientations during deformation can be directly observed.

The tracking of the annealing process can be used to investigate the preferred nucleation from the deformed matrix, and the growth of the nuclei while new grains are created in the meantime. The tracing of the later grain growth process clearly showed how the small grains were consumed by the large grains during extended annealing holding. The information on grain misorientations can be used to evaluate the mobility of the grain boundaries, and thus understand the formation mechanisms of the final texture.

References

[1] C. W. Chen, *Magnetism and Metallurgy of Soft Magnetic Materials*, North-Holland, Amsterdam, The Netherlands, 1977.

[2] G. Lyudkovsky, P. K. Rastogi, M. Bala, "Nonoriented Electrical Steels", *JOM*, 1986, 38: 18-26.

[3] T. Shimazu, M. Shiozaki, K. Kawasaki, Effect of temper rolling on texture formation of semi-processed non-oriented steel", *J. Magn. Magn. Mater.*, 1994, 133: 147-149.

[4] L. Kestens, S. Jacobs, "Texture control during the manufacturing of nonoriented electrical steels", *Texture, Stress and Microstructure*, 2008, Article ID 173083, 1-9.

[5] F. J. G. Landgraf, "Nonoriented Electrical Steels", *JOM*, 2012, 64: 764-771.

[6] T. Tomida, T. Tanaka, "Development of (100) texture in silicon steel sheets by removal of manganese and decarburization", *ISIJ International*, 1995, 35: 548-556

[7] M. Takashima, M. Komatsubara, N. Morito, "{001}<210> Texture Development by Two-stage Cold Rolling Method in Non-oriented Electrical Steel, *ISIJ International*, 1997, 37: 1263-1268.

[8] T. Nakayama, N. Honjou, T. Minaga, H. Yashiki, "Effects of manganese and sulfur contents and slab reheating temperatures on the magnetic properties of non-oriented semi-processed electrical steel sheet", Journal of Magnetism and Magnetic Materials, 2001,234: 55-61.

[9] M. de Campos, F. J. G. Landgraf, I. G. S. Falleiros, G. C. Fronzaglia, "Texture Evolution during the Processing of Electrical Steels with 0.5% Si and 1.25 Si", *ISIJ International*, 2004, 44: 1733-1737.

[10] Y. He, E. Hilinski, J. Li, "Texture Evolution of a Non-oriented Electrical Steel Cold Rolled at Directions Different from the Hot Rolling Direction", *Metall. Mater. Trans. A*, 2015, 46: 5350-5365.

[11] R. Lebensohn, C. N. Tome, "A self-consistent viscoplastic model: prediction of rolling textures of anisotropic polycrystals", *Mater. Sci. Eng. A*, 1994, 175: 71-82.

[12] P. van Houtte, S. Li, M. Seefeildt, L. Delannay, "Deformation texture prediction: from the Taylor model to the advanced Lamel model", *Inter. J. Plast.*, 2005, 21: 589-624.

[13] F. J. Humphreys, M.Hatherly, "*Recrystallization and Related Annealing Phenomena*", Elsevier, Oxford, UK, 2012.

[14] M. Mehdi, Y. He, E. J. Hilinski, L. A. I. Kestens, A. Edrisy. "The origin of the Goss orientation in non-oriented electrical steel and the evolution of the Goss texture during thermomechanical processing", *Steel Research International*, 2019, 1800582: 1-12.

[15] M. Sanjari, Y. He, E. J. Hilinski, S. Yue, L. A. I. Kestens, "Development of the {113}<uvw> texture during the annealing of a skew cold rolled non-oriented electrical steel", *Scripta Materialia*, 2016, 124: 179–183.

[16] M. Sanjari, Y. He, E. J. Hilinski, S. Yue, L. A. I. Kestens, "Texture Evolution during Skew Cold Rolling and Annealing of a Non-oriented Electrical Steel Containing 0.9 wt% Silicon", *J. Mater. Sci.*, 2017, 52: 3281-3300.

Effect of flattening reduction rate of primary recrystallization microstructure and rare earth in copper-containing grain-oriented silicon steel on secondary recrystallization

Zi-li Jin[1,2], Hui-ping Ren [1], Zhi-nan Zhang[1,2], Zhong-wang Wu[1]

[1]School of Material and Metallurgy, Inner Mongolia University of Science and Technology, Baotou 014010, China
[2]Key Laboratory of New Metallic Materials in Inner Mongolia, Baotou 014010, China

E-mail: jinzl6310@126.com

Abstract. The results show that the content of rare earth and the flattening reduction rate after the initial recrystallization annealing have significant effects on the secondary recrystallization process and grain growth behavior in the copper oriented silicon steel produced by the first cold rolling method. After the initial recrystallization of the oriented silicon steel containing copper, the flattening with a reduction ratio of about 9%, relatively coarse recrystallized grains are obtained after high temperature annealing. About 8 ppm of lanthanum and 3 ppm of cerium were added to the oriented silicon steel containing copper, the temperature at which the secondary recrystallization abnormally grew during the high temperature annealing was increased by 30 ℃ ~50 ℃ . After the primary recrystallization annealing, the secondary recrystallized structure of copper-oriented silicon steel containing rare earth is relatively coarse after being flattened.

1. Introduction

Oriented silicon steel is a steel product with long and complex production process and huge energy consumption. The conventional grain-oriented (CGO) silicon steel containing copper is a type of grain-oriented steel with low cost and high productivity [1]. The precipitation temperature of MnS is 1100℃ and Cu2S is 1000℃， It is helpful to reduce the heating temperature of the billet . Rare earth elements have unique electronic shell and strong metal activity and adding rare earth La or Ce to steel can purify grain boundary, refine the solidification structure of the steel and change the morphology and distribution of inclusions. Rare earth can inhibit the precipitation of in austenite, increase the incubation period of precipitation and decrease the precipitation rate [2~4]. Rare earth is easy to combine with oxygen and sulfur in steel, and the effect of rare earth in steel on the solid solution and precipitation of the second phase is obvious with the change of temperature [3,4]， Two-stage cold rolling technology is generally adopted in the production of grain-oriented electrical steels. It consists of first cold rolling, intermediate annealing, second cold rolling by 50%, decarburizing annealing, and final secondary recrystallization annealing[5]. Primary recrystallization textures influence secondary recrystallization, microstructures and textures. Regarding the development of Goss texture, which is most important for magnetic properties, in secondary recrystallization, grain boundary conditions were suggested to be important by Shimizu etal[6].

In the present work, in order to simplify the production process, the Fe-3%Si steel containing Cu and trace rare earth is taken as the research object, the main cold rolling deformation is completed by one-time cold rolling technology, and the flattening process is added after decarburization annealing of the steel sheet. According to the principle of abnormal growth of recrystallized grains during recrystallization annealing of cold deformed metals under critical deformation conditions, secondary recrystallization of oriented silicon steel is promoted to occur in advance and coarser recrystallized grains are obtained. The effects of flattening reduction rate of primary recrystallization annealed structure and rare earth on the behavior of secondary recrystallization in Silicon Steel were studied. Also focus is to grasp the relationship between the flattening reduction rate and the secondary recrystallization temperature and microstructure evolution of the oriented silicon steel during high temperature annealing.

2. Experimental procedure

The experimental steel are 2.33 mm thick oriented silicon steel hot rolled strips produced by China Baotou iron & steel company. The experimental steel compositions are shown in Table 1, One contains rare earth and other not. After acid pickling in aqueous hydrochloric acid solution, the hot-rolled plate was rolled to 0.35mm thick by a four-high cold rolling mill with a support roll of Φ500 x 450 mm, and a work roll of Φ200×470 mm. Then, the cold-rolled specimens were cut into 80 mm x 25 mm coupons with longitudinal axis in both rolling and transverse directions. Finally, the specimens were annealed in a continuous annealing furnace set at 830℃ for 7min with an atmosphere of 75%N2 and 25% H2. Then, the specimens were cooled in air. After the initial recrystallization annealing, the carbon content was determined by a carbon and sulfur analyzer, which showed that the carbon content of specimen 1 was 0.00402% and that of specimen 2 was 0.00345%, the results indicated that decarburization annealing achieves the expected goal successfully, specimen 1 and 2 according to Table 1 were rolled 3%, 6% and 9% by Φ130 mm×250mm cold rolling two-high mill respectively, and the final thickness of these samples are 0.34 mm, 0.33 mm and 0.32 mm. The flatted specimen were coated with MgO, and annealed again to conduct the secondary recrystallization annealing, according to the high temperature annealing process shown in Fig. 1. Specimens were taken out once the temperature reaches to 800℃, 850℃, 900℃, 950℃, 1000℃ and 1150℃ using "disruption" sampling method. The effect of rare earth and percentage of reduction on the microstructure of the specimens were investigated and tracked in the high temperature annealing process.

Table 1. The composition of the experimental steel (wt%).

Specimen	C	Mn	Si	P	S	Cu	Al	N	La	Ce
1	0.030	0.21	3.15	0.009	0.014	0.50	0.020	0.0128	0.0008	0.0003
2	0.028	0.23	3.20	0.009	0.009	0.49	0.019	0.0150	0	0

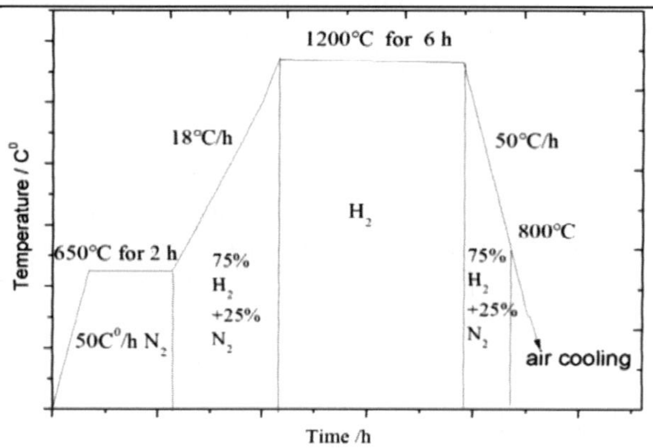

Figure 1. High temperature annealing process.

3. Results and discussion

Fig. 2 and 3 displayed the microstructures of specimens 1 and 2 with 3% flattening-reduction after high temperature annealing, according to Fig. 1. For specimen 1, according to Fig. 2, there is no abnormal growth grains before 930℃. At temperature between 850℃ and 930℃, the grain size of the samples are uniform, while at 800℃~850℃, the mean rain size of the sample is between 7.2mm and 8.3mm. Therefore, the starting temperature of the abnormal growth during secondary recrystallization should be 930℃~950℃. For specimen 2, at the tracking temperature between 800℃~850℃, the grains are uniform. From Fig. 3(c), some abnormal grains were observed at 900℃, which indicated the beginning of secondary recrystallization. At 950℃, the secondary recrystallization has completed and the mean grain size increased by 1~3 mm at 950℃~1150℃.

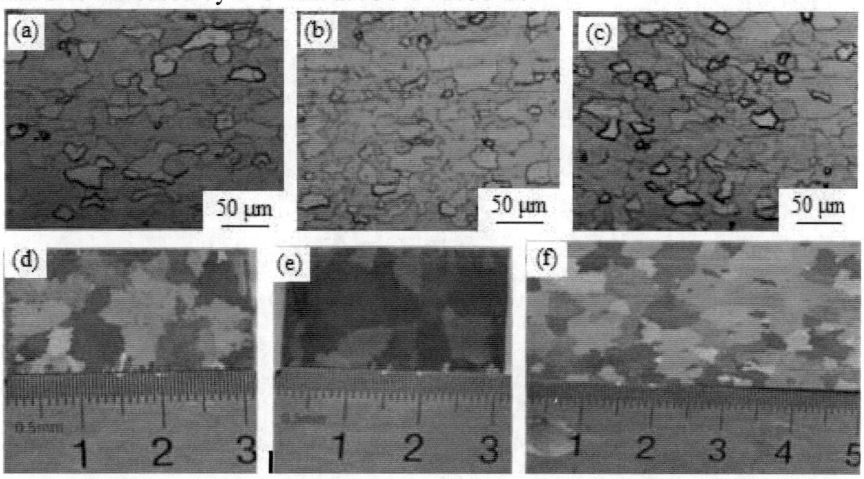

Figure 2. Microstructure of specimen 1 with 3% flattening reduction rate after high temperature annealing. (a) 850℃; (b) 900℃; (c) 930℃; (d) 950℃; (e) 1000℃; (f) 1150℃.

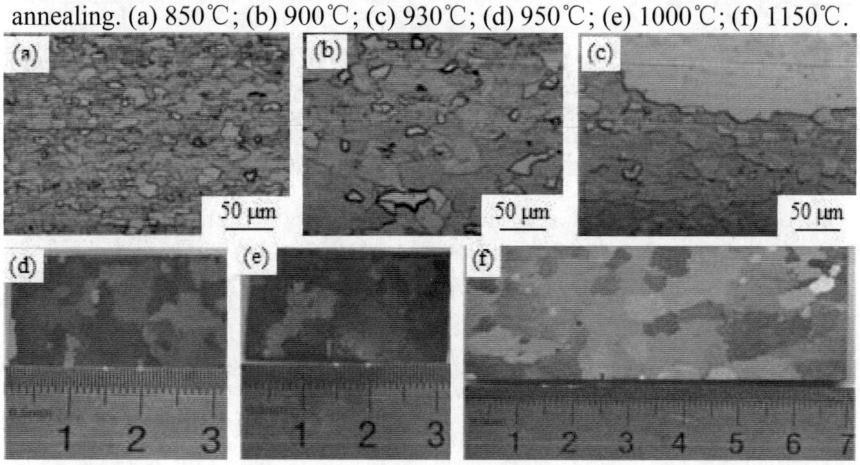

Figure 3. Microstructure of specimen 2 with 3% flattening reduction rate after high temperature annealing. (a) 800℃; (b) 850℃; (c) 900℃; (d) 950℃; (e) 1000℃; (f)1150℃.

Fig. 4 and 5 show the microstructures after 6% flattening reduction and annealing at different temperatures. Fig. 6 and 7 show the microstructures after 9% flattening reduction and annealing at different temperatures. By comparing the microstructures after 6% and 9% reductions, it was found that the grain size after 950℃~1150℃ annealing and cold rolling increased as compared to 3% reduction, and the average grain size increased by 1~3mm. The grain growth in the secondary recrystallization should start at about 930℃~950℃, and the abnormal grain growth of 9% reduction started at 930℃, as shown in the Fig. 6(c). After 6% and 9% reduction, specimen 2 presented abnormal grain growth at the temperature of 900℃, the specimen with 9% reduction shows more apparent abnormal growth. As can be seen in Fig. 7(c), the abnormal growth could be observed in several areas in specimen 2.

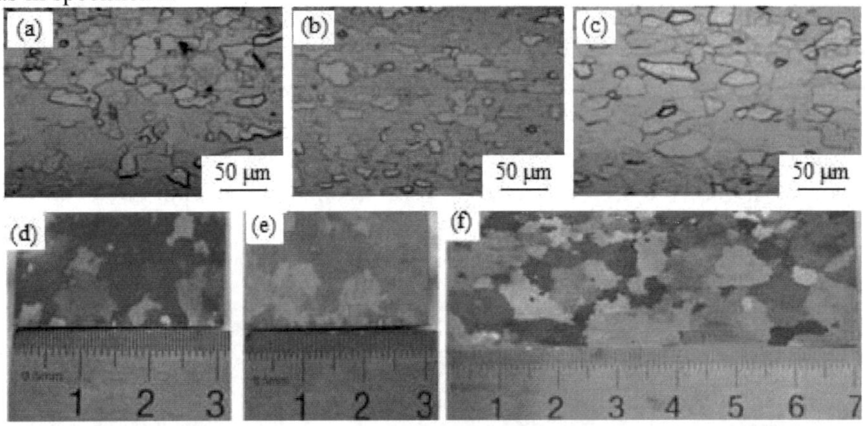

Figure 4. Microstructure of specimen 1 with 6% flattening reduction rate after high temperature annealing. (a) 850℃; (b) 900℃; (c) 930℃; (d) 950℃; (e) 1000℃; (f) 1150℃.

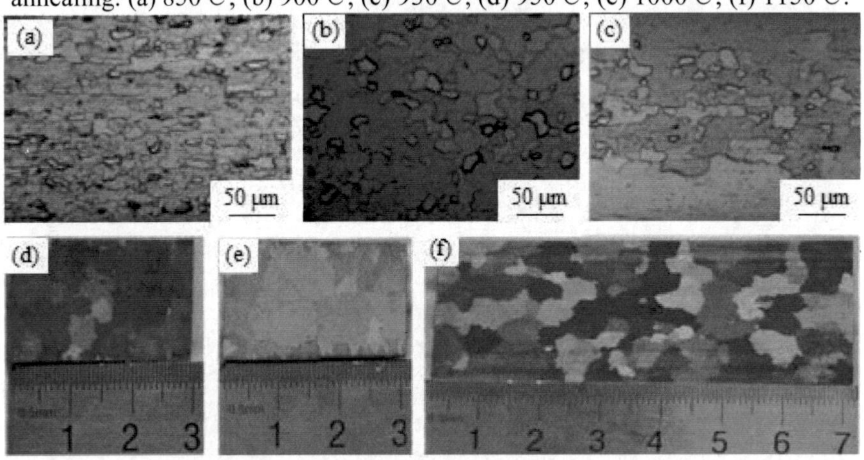

Figure 5. Microstructure of specimen 2 with 6% flattening reduction rate after high temperature annealing. (a) 800℃; (b) 850℃; (c) 900℃; (d) 950℃; (e) 1000℃; (f) 1150℃.

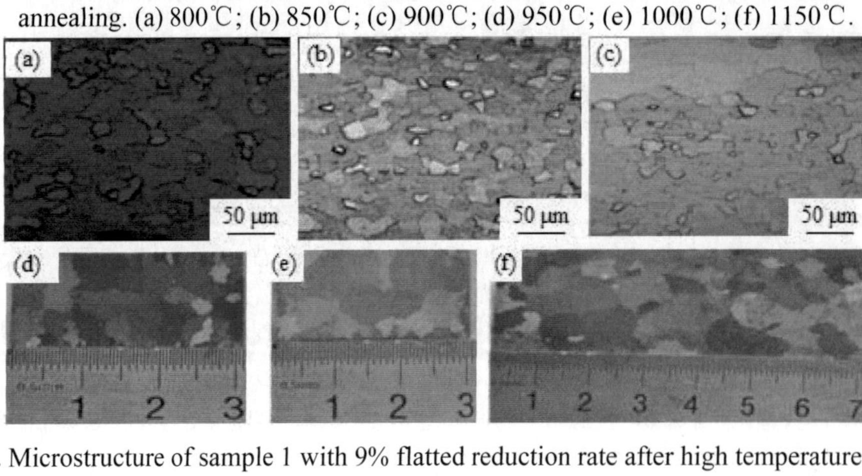

Figure 6. Microstructure of sample 1 with 9% flatted reduction rate after high temperature annealing process. (a) 850℃; (b) 900℃; (c)930℃; (d) 950℃; (e) 1000℃; (f) 1150℃.

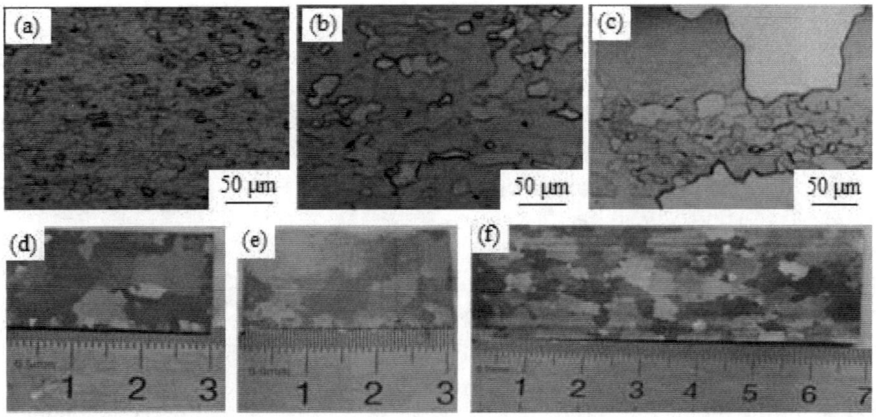

Figure 7. Microstructure of sample 2 with 9% flattening reduction rate after high temperature annealing. (a) 800℃; (b) 850℃; (c) 900℃; (d) 950℃; (e)1000℃; (f) 1150℃.

EBSD was employed to analyze the micro-orientations of selected areas of specimens 1 and 2, which were annealed at 930℃ and 900℃ after 3%, 6% and 9% flattening reductions. As can be seen in Fig. 8(a), (b) and (c), after 930℃ annealing, specimen 1 with the addition of Rare Earth showed large-size grains with a strong Goss orientation of {110}<001>. However, specimens with 3% and 6% reduction don't form an obvious Goss orientation. Under flattening reduction of 3%, 6% and 9%, specimen 2 without Rare Earth could form large-size grains of Goss orientation at the edge of samples.

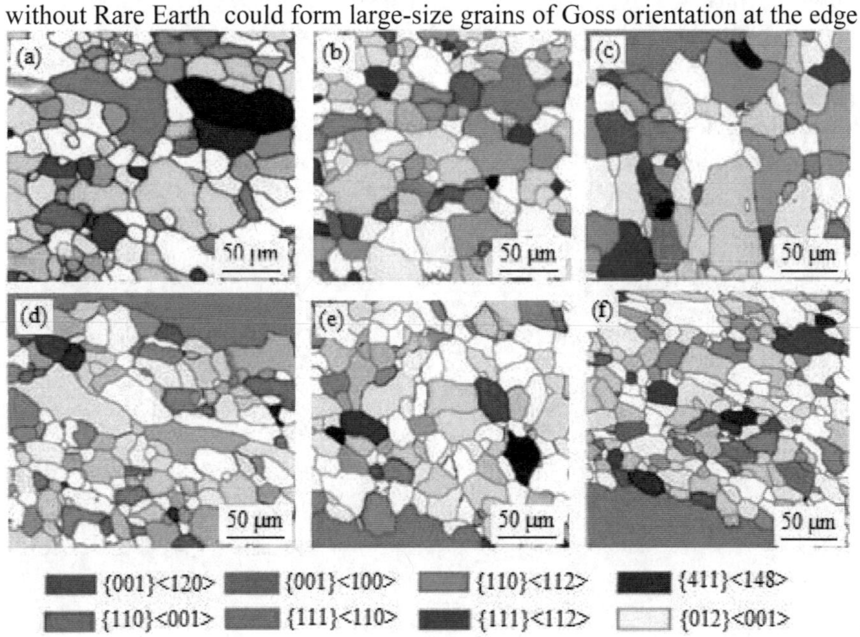

Figure 8. EBSD figures of samples with different compositions and flatted-reductions after hightemperature annealing process.
Sample 1 at 930℃ with (a) 3% (b) 6% (c) 9% flatted reduction rate;
Sample 2 at 900℃ with (e) 3% (f) 6% (g) 9% flatted reduction rate.

Through the above analysis, it was found that the flattening process could further promote the secondary recrystallization after the primary recrystallization of oriented silicon steel, once the flattening reduction rate increased to a certain extent. The main reason is that the flattening increased the storage energy of cold deformation and the recrystallization dynamics. Given 9% of flattening reduction rate, specimens 1 and 2 could form an amount of abnormal grain growth during the high-temperature annealing, which was generally observed in the annealing of cold-rolling metal with the critical deformation. However, there was not apparent impact on the grain growth and the homogeneity of annealing microstructure of samples. The addition of rare earth could raise the critical temperature of abnormal grain growth by a range of 30℃~50℃ during the secondary recrystallization of high-temperature annealing. Additionally, the grain growth was also effectively inhibited during the high-temperature annealing by the addition of rare earth.

4. Conclusions

The addition of La and Ce could postpone the secondary recrystallization and raise the critical temperature of abnormal grain growth by a range of 30℃~50℃. Additionally, the addition of La and Ce can hinder the growth of secondary recrystallized grains during high temperature annealing. The flattening reduction after the initial recrystallization annealing could promote the secondary recrystallization during high temperature annealing. However, the flattening reduction has no apparent impact on the grain growth during high temperature annealing.

References

[1] Yang P,Wang Z, Liu G, Shen X, Bo D, Mao W and Tang S 2018 *Chin. J. Eng.* **40** 200
[2] Fang Q, Li C, Bai H, Ren H and Jin Z 2015 *Chin. Rare Earths.* **36** 32
[3] Lin Q, Guo F and Zhu X 2006 *J. Chin. Rare Earth Soc.* **24** 427
[4] Jiang M, Wang R and Li C 2013 *Chin. Rare Earths.* **24**1
[5] Misshra S, Darrman C and Lucke K 1984 *Acta. Mater.* **32** 2185
[6] Shimizu R, Harase J and Dingley D J 1990 *Acta Mater.* **38** 97

Acknowledgments

The authors would like to acknowledge the financial support given by the National Natural Science Foundation of China (Grant**No**. 51761033).

Investigation on the primary and secondary recrystallization textures of rare earth micro-alloyed grain-oriented silicon steel

Zhongwang Wu[1], Zili Jin[1] , Huiping Ren[2], Ruixiang Liu[1], Tao Li[1], Xin Zhang[1]
[1]Materials and Metallurgy School, Inner Mongolia University of Science and Technology, Baotou 014010, China.
[2]Key Lab of Integrated Exploitation of Baiyun Obo Multi-Metal Resources, Baotou 014010, China.

E-mail: wuzw79@163.com

Abstact. The primary and secondary recrystallization textures of rare earth micro-alloyed grain-oriented electrical steel was examined by using electron backscattered diffraction (EBSD) techniques. The results show that the primary recrystallization texture comprises {111}<112>, {111}<110>, {411}<148>, and {012}<001> components. The benefiting texture components of primary recrystallization increase with increasing rare earth addition after decarburizing annealing. During secondary recrystallization, the proportion of $\Sigma 9$ boundaries are reduced. The sample containing 5ppm rare earth starts secondary recrystallization at 900°C, and the sample containing of 17ppm rare earth starts secondary recrystallization at 950°C. The addition of rare earth can delay the process of secondary recrystallization.
Key words: rare earth, grain-oriented silicon steel, secondary recrystallization, texture

1. Introduction

Grain-oriented silicon steel uses a secondary recrystallization phenomenon to obtain a single {110}<001> texture. In order to develop a perfect secondary recrystallized structure, fine dispersed precipitates are often used as inhibitors. The inhibitor inhibits the normal growth of the primary recrystallized grains by pinning, and causes grains having the {110}<001> orientation to grow abnormally during secondary recrystallization. Therefore, texture is one of the key contents of research on grain-oriented silicon steel [1-3]. Adding trace rare earth elements to steel can improve the strength, toughness and corrosion resistance of steel [4-7]. Therefore, the development of high quality rare earth microalloyed steel is one of the methods to improve the performance of steel. The purpose of this paper is to investigate the primary and secondary recrystallization textures of rare earth micro-alloyed grain-oriented silicon steel.

2. Experimental Methods

The composition of the experimental steel is shown in Table 1. The manufacturing process of the grain-oriented silicon steel is as follows: smelting→ forging→ slab heating at 1250°C→ hot rolling→ pickling→ first cold rolling→ decarburization annealing→secondary cold rolling→ magnesium oxide coating → high temperature annealing. The decarburization annealing is performed in a wet H_2 and N_2 mixed atmosphere at a temperature of 820 °C for 4 minutes. The high temperature annealing process is conducted at 1200 °C at a heating rate of 15 °C/h in a 25% H_2 + 75% N_2 atmosphere, and then purified at 1200°C×6h in 100 % high purity H_2 , and finally cooled. The microstructure of grain-oriented silicon steel was observed by a Zeiss optical microscope. The primary and secondary recrystallization textures of the rare earth micro-alloyed grain-oriented electrical steel was examined by using electron backscattered diffraction (EBSD).

Table 1 The composition of the experimental steel (wt%)

specimen	C	Si	Mn	P	S	Al	Cu	Nb	Ce
No. 1	0.037	2.99	0.25	0.008	0.020	0.016	0.524	0.008	0.0005
No. 2	0.038	3.04	0.25	0.007	0.016	0.021	0.524	0.008	0.0017

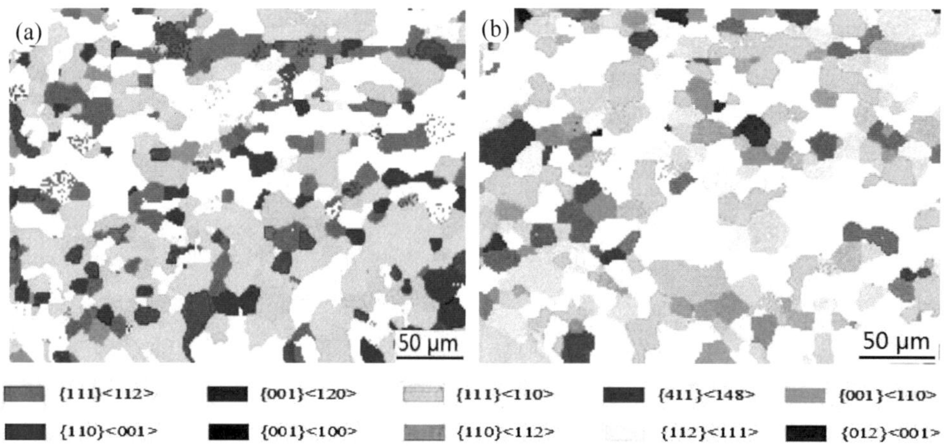

Figure 1. EBSD grain unique color maps of the specimens obtained after decarburization annealing at 820℃
(a) sample No. 1; (b) sample No. 2

3. Results and Discussion

Fig. 1 shows the EBSD grain unique color maps of the specimens obtained after decarburization annealing at 820°C . The typical primary recrystallization texture components are {110}<112>, {111}<110>, {411}<148>, {012}<001>, {001}<100>. The content of Goss ({110}<001>) in primary recrystallization texture is very small, mainly distributed around {111}<112> and {111}<110>. During the cold rolling process, the Goss residual crystal exists in the form of cell structure in transition zones in {111}<112> grains. During decarburization annealing, the Goss texture preferentially nucleates on the {111}<112> grain boundary. The {111}<112> texture not only provides a nucleation site for the Goss texture, but also provides a matrix for the abnormal growth of the Goss texture. The {111}<112> texture in the γ fibre and the {411}<148> texture in the α fibre have close to the Σ9 CLS boundaries with respect to the Goss texture, which is beneficial to the abnormal growth of Goss grains during high temperature annealing. It is also found that {012}<001> texture is a texture that is conducive to the abnormal growth of Goss grains during high temperature annealing.

The area fractions of primary orientations in the primary recrystallized structure are listed in Table 2. The fractions of the Goss texture of the two samples are not very high, and the fraction of sample No. 2 is higher than that of sample No. 1. The sum of the four favorable textures {111}<112>, {111}<110>, {411}<148>, {012}<001> of sample No. 2 is higher than that of sample No. 1. Moreover, the fraction of {110}<112> texture in sample No. 2 was lower than that of sample No. 1. {110}<112> texture severely affects the formation of Goss texture during the high temperature annealing of oriented silicon steel, so it is of great significance to inhibit the texture of {110}<112> during the production of oriented silicon steel. It can be seen from the above analysis that the addition of rare earths is more conducive to the abnormal growth of the Goss grains during the high temperature annealing of the grain-oriented silicon steel.

Table 2 Area fractions of primary orientations in the primary recrystallized structure (%)

specimen	{001}<110>	{111}<112>	{111}<110>	{001}<120>	{110}<001>	{001}<100>	{110}<112>	{112}<111>	{012}<001>	{411}<148>
No. 1	1.3	0.854	13.5	4.16	2.32	7.94	20.2	1.39	8.73	10.9
No. 2	1.13	0.793	11	5.6	4.26	8.76	17.3	3.36	12.1	14.3

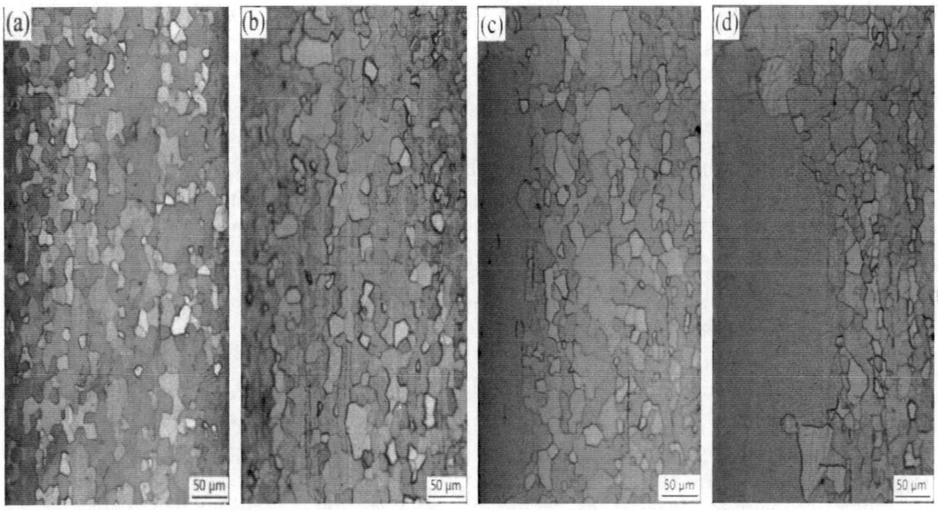

Figure 2. Microstructure of sample No. 1 at different temperature during annealing
(a) 650 °C; (b) 850 °C; (c) 900 °C; (d) 950 °C;

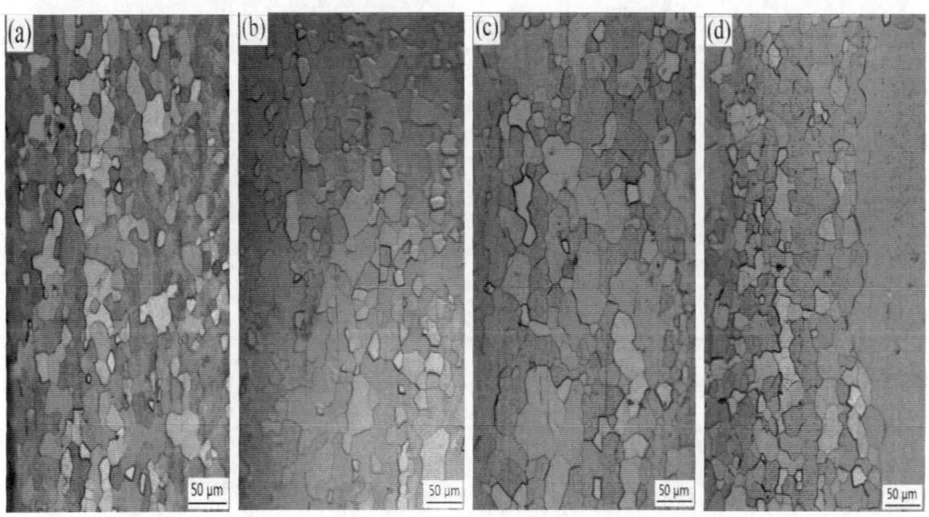

Figure 3. Microstructure of sample No. 2 at different temperature during annealing
(a) 650 °C; (b) 850 °C; (c) 900 °C; (d) 950 °C

Fig. 2 shows the microstructure of sample No. 1 during annealing at 650 °C, 850 °C, 900 °C, 950 °C, respectively. It can be seen that sample No. 1 did not undergo secondary recrystallization at 650 °C and 850 °C, and the crystal grains were fine and uniform. When the temperature is raised to 900 °C, a large grain appears on the edge of the sample, indicating that secondary recrystallization has occurred. As the temperature increases, the abnormally grown Goss grains continuously engulf the small grains around them, and the abnormally grown grains are uneven and the grain boundaries are extremely irregular.

Fig. 3 shows the microstructure of sample No. 2 during annealing at 650 °C, 850 °C, 900 °C, 950 °C, respectively. It can be seen that sample No. 2 did not undergo secondary recrystallization at 650 °C, 850 °C and 900 °C. A large grain appeared on the edge of the sample at 950 °C , which proved that it began to undergo secondary recrystallization. Comparing the secondary recrystallization starting temperature of No. 1 and No. 2 samples, it can be seen that the addition of rare earth to the oriented silicon steel will delay the occurrence of secondary recrystallization.

The EBSD grain unique color maps of sample No. 1 at 900 °C and sample No. 2 at 950 °C during annealing is presented in Fig. 4. As can be seen, the Goss grains are elongated along the RD, and growth of the Goss grain along the ND is being blocked by many equiaxed grains. The sizes of most equiaxed grain are 20-100 μm. The equiaxed grains texture components are {110}<112>, {111}<110>, {011}<120>, {411}<148>,{012}<011>. The secondary recrystallization is induced by special grain boundaries between {111}<112> and Goss grains. Goss grains prefer to swallow {111}<112> primary grains but have difficulty consuming {411}<148> primary grains during high temperature annealing, despite their similar Σ9 relationships [8].

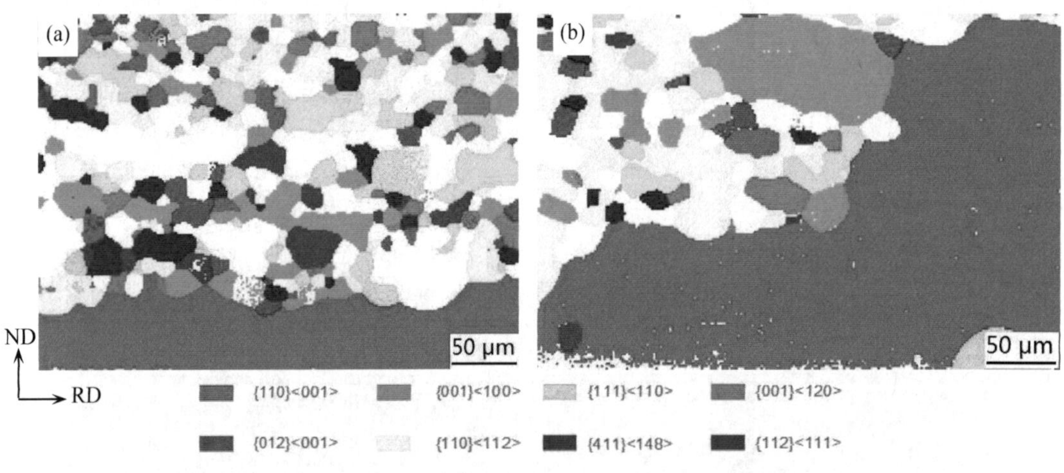

Figure 4. EBSD grain unique color maps of sample No. 1 at 900 °C (a) and sample No. 2 at 950 °C (b) during annealing

Table 3 Proportions of CSL boundaries at different temperatures during annealing (%)

Temperature/℃	Sample No. 1				Sample No. 2			
	Σ3	Σ5	Σ7	Σ9	Σ3	Σ5	Σ7	Σ9
850	4.25	0.49	0.33	1.14	3.75	0.45	0.72	0.52
900	3.55	0.48	0.27	0.39	5.07	0.36	0.44	0.62
950	3.02	0.59	0.31	0.79	4.12	0.28	0.40	0.40

Table 3 shows the proportions of CSL boundaries at different temperatures during annealing. No. 1 sample undergoes secondary recrystallization at 900 °C, and the proportion of the Σ9 grain boundary which is favorable for the abnormal growth of Goss grains, is significantly reduced. Sample No. 2 undergoes secondary recrystallization at 950 °C， and the proportion of Σ9 grain boundary was also significantly reduced. It is indicated that Σ9 grain boundaries contribute to the growth of Goss grains, and they are preferentially consumed. The coincidence site lattice theory holds that the energy of CSL grain boundaries is lower than those of other large angle boundaries. During the temperature rise of high temperature annealing, the pinning force of the CSL grain boundary by the inhibitor is smaller than those of other large angle grain boundaries. When the inhibitor begins to coarsen and decompose, these grain boundaries will be out of pinning earlier, which will help the Goss grains to grow abnormally, especially the Σ9 grain boundary plays a very important role in the process of Goss grain growth.

4. Conclusion

The primary recrystallization texture of rare earth micro-alloyed grain-oriented silicon steel comprises {111}<112>, {111}<110>, {411}<148>, and {012}<001> components. Adding rare earth can improve favorable texture components of primary recrystallization texture. The sample containing of 5ppm rare earth starts secondary recrystallization at 900°C, and the sample containing of 17ppm rare earth starts secondary recrystallization at 950°C. The addition of rare earth can delay the process of secondary recrystallization.

Acknowledgments

The authors would like to acknowledge the financial support given by the National Natural Science Foundation of China (GrantNo. 51761033) and the Inner Mongolia University of Science and Technology Innovation Foundation of China (GrantNo. 2015QDL23, 2014QDL026)

References

[1] Lu X, Fang F , Zhang Y X, Wang Y, Yuan G, Misra R D K, Zhang W N and Wang G D 2018 *Materials Characterization* **142** 68
[2] Wang Y, Xu Y B, Zhang Y X, Fang F, Lu X, Liu H T, and Wang G D 2015 *Journal of Magnetism and Magnetic Materials* **379** 161
[3] Citrawati F, Quadir M Z and Munroe P R 2017 *Procedia Engineering* **184** 750
[4] Xing X, Han Z, Wang H and Lu P 2015 *Journal of Rare Earths* **33** 1122
[5] Sun Z, Zhang C S andYan M F 2014 *Materials & Design* **55** 128
[6] Ahn J H， Jung H D, Im J H, et al. 2016 *Materials Science and Engineering: A* **658A** 255
[7] Song S, Sun H and Wang M, Jung K H and Moon B M 2015 *Journal of Rare Earths* **33** 1204
[8] LIU Gong-tao , LIU Zhi-qiao , YANG Ping and Wei-min MAO 2016 *Journal of Iron and Steel Research*, International **23** 1234

Secondary recrystallization induced by composite precipitates in magnetostrictive Fe-Ga alloy thin sheet

Zhenghua He[1], Yuhui Sha[2], Yongkuang Gao, Songtao Chang, Fan Lei, Quan Fu, Fang Zhang and Liang Zuo

Key Laboratory for Anisotropy and Texture of Materials (Ministry of Education), Northeastern University, Shenyang, China

Corresponding author: [1]hezhenghua@mail.neu.edu.cn; [2]yhsha@mail.neu.edu.cn

Abstract. Secondary recrystallization of Goss ({110}<001>) texture is successfully produced by using nanometer-sized composite inhibitor in primary cold rolled $Fe_{81}Ga_{19}$ alloy sheet. Nanometer-sized sulfides consisting of MnS and Cu_2S are dispersedly precipitated during hot rolling and normalization as inhibitors for secondary recrystallization. Strong γ texture with peak at {111}<112> and weak Goss texture distributed through the sheet thickness in primarily annealed sheets can provide favorable environment for the secondary recrystallization. The higher onset and finishing temperatures of secondary recrystallization confirm the stronger inhibition and driving force originated from the composite inhibitors. Centimeter-sized Goss grains and large magnetostriction coefficient are obtained by primary cold rolling methods without surface energy effect. The result indicates that complete secondary recrystallization in Fe-Ga alloy can be realized by precisely controlling inhibitor characteristic and primary recrystallization texture.

1. Introduction

Fe-Ga alloy is an advanced magnetostrictive material for actuators and sensors in terms of combining excellent mechanical and magnetostrictive properties [1-3]. It is desirable to produce η (<100>//RD) textured Fe-Ga sheets due to the largest magnetostriction coefficient along <100> direction and lower eddy current loss in high frequency use. A series of work has been conducted to promote η texture in Fe-Ga sheet through secondary recrystallization by conventional rolling and annealing process [4-7]. According to current knowledge, secondary recrystallization of η texture requires two necessary conditions: (1) enough pinning force for matrix grains normally originating from precipitate particles (called as inhibitors); (2) preferred growth for η grains by grain boundary mobility advantage or additional driving force from surface energy difference or gradually weaken inhibition.

It is concluded that current methods to promote secondary recrystallization in Fe-Ga alloy mainly include: (a) NbC particles with size of 1-5 μm were added to the molten $Fe_{81}Ga_{19}$ alloy to inhibit the normal grain growth (NGG) [4,8], while the abnormal grain growth (AGG) of Goss ({110}<001>) texture was induced by surface energy effect from H_2S atmosphere [5]. Due to the weak stability and pinning of the coarse NbC particles, complete secondary recrystallization has been achieved by addition of 1~2.5 at.% NbC particles combined with the high temperature annealing of 1200~1250 °C for 2-4 hours [9]. (b) Micronmeter- or nanometer-sized NbC particles were precipitated in rolling and annealing process of $Fe_{83}Ga_{17}$ and $Fe_{82}Ga_{4.5}Al_{13.5}$ alloy sheets, the AGG of Goss texture was induced by surface energy effect under sulfur atmosphere combined with strengthen of Goss texture in primary recrystallization prepared by initial sharp <100>-oriented columnar grains and intermediate annealing [6,10,11]. So far, strong Goss and large magnetostriction coefficient has been achieved by secondary recrystallization with the combination of inhibitor and surface energy modification during high temperature annealing. However, secondary recrystallization texture in Fe-Ga sheets has not been well regulated so that the costly application of surface energy effect and unconventional processing method are necessary. In our previous work [12], nanometer-sized sulfides were dispersedly precipitated during hot rolling, the AGG of Goss texture was produced well-matched pinning effect and driving force by precise control of inhibitor and primary recrystallization texture in two-stage cold rolled $Fe_{81}Ga_{19}$ alloy sheets.

In the grain oriented silicon steel, a few number of Goss grains in primary recrystallization are obtained by primary cold rolling with a large reduction ratio, while the nanometer-sized inhibitors, consisting of MnS, Cu_2S, AlN particles, mainly precipitate at hot rolling and normalization, which provides strong driving force for secondary recrystallization. Finally, sharp Goss texture with lower deviation is produced by prefect-matched pinning force and driving force of secondary recrystallization [13-15]. Primary cold rolling with large reduction ratio significantly reduces the quantity and deviation of Goss nuclei in primary recrystallization, and then provides sufficient growth space for the secondary recrystallization of exact Goss texture. However, it requires stronger inhibition accompanied with more precise matching of precipitation regression process with intrinsic preferred mobility of Goss grain in higher secondary recrystallization temperature. In this paper, nanometer-sized composite inhibitor was precipitated to induce secondary recrystallization of Goss texture in $Fe_{81}Ga_{19}$ alloy sheet prepared by primary cold rolling with a large reduction ratio, the relationship between inhibitor characteristics and secondary recrystallization was studied.

2. Material and methods

The $Fe_{81}Ga_{19}$ alloy ingots with nominal chemical composition (in weight %) of 0.03% C, 22.65 % Ga, 0.25% Mn, 0.15% Cu, 0.01% S and balance Fe were prepared from pure Fe (99.9%) and Ga (99.99%) , pure Cu (99.99%) and master alloy of Fe-C and Fe-Mn by induction melting. The ingots were hot rolled to 2.3 mm at 1200 °C with finishing temperature of 850 °C. Afterwards, hot bands were normalized annealed at 1000°C and then cold rolled to 0.45 mm at 200 °C with 80% reduction. The cold rolled sheets were annealed at 850 °C for 10 min in a wet atmosphere of 75% H_2 and 25% N_2 for primary recrystallization, then heated to 1100 °C at a rate of 20 °C/h in an atmosphere of 75% N_2 and 25% H_2, and finally annealed at 1200 °C for 1 h in 100% H_2 for purification.

The orientation distribution functions (ODFs) of the specimens were measured and calculated at different thickness layers based on X-ray diffraction. The measured layer is defined as the parameter $S=2a/d$, where a and d are the distances from the center and sheet thickness, respectively. Electron backscattered diffraction (EBSD) analysis was also performed to investigate the texture evolution of rolling and annealing process. The precipitates were prepared by carbon extraction-replica technique and studied via transmission electron microscope (TEM) equipped with an energy dispersive X-ray spectroscopy (EDS). Magnetostriction values under applied field from 0 to ± 3000 Oe were measured by strain gauges with gauge area of 4 mm × 6 mm (base area of 6.3 mm × 10.5 mm) positioned along the rolling direction. The saturated magnetostriction coefficient was calculated by $(3/2)\lambda_s=\lambda_{//}-\lambda_{\perp}$, where $\lambda_{//}$ and $\lambda_{\perp}$ are the maximum magnetostriction coefficients under the magnetic field applied parallel and perpendicular to the rolling direction (RD), respectively.

3. Results and Discussion

Figure 1 shows the feature of precipitates in normalized bands and primary recrystallized sheets. The precipitates are determined as MnS and Cu_2S by EDS analysis. A number of spherical, rectangular and lamellar precipitates with size of 10~30 nm are observed in normalized bands. After cold rolling and primary recrystallization annealing, fine precipitates of MnS and Cu_2S are still dispersed in the matrix with size of 20~60 nm, which is critical for efficient pinning effect on primary recrystallized grains [13-15]. The nanometer-sized precipitates can be attributed to the large nucleation rate due to high density of dislocations as preferred nucleation sites by heavy hot rolling and subsequent rapid cooling together with the regulated solubility products of MnS and Cu_2S precipitates by composition design [14,16].

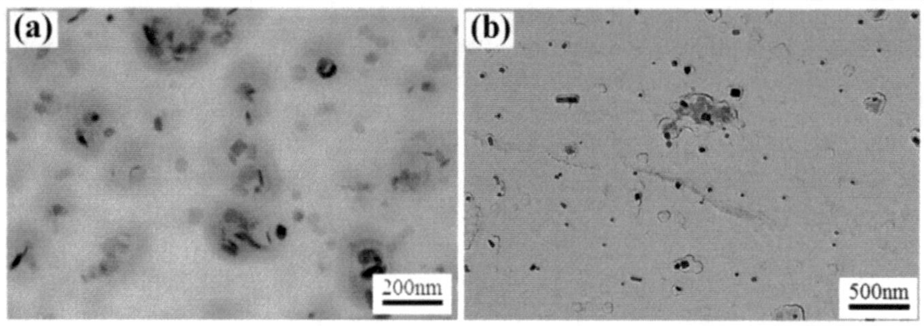

Figure 1. TEM micrograph and EDS of precipitates in (a) normalized band and (b) primary recrystallized $Fe_{81}Ga_{19}$ thin sheet.

Figure 2 shows microstructure and texture of hot-rolled, cold rolled and primarily recrystallized sheets. The hot rolling microstructure and texture (figure 2a,b) exhibit a through-thickness gradient, where fragmented grains oriented by strong Goss and weak brass ({110}<112>) texture occupy subsurface layer and (*S*=0.5) and severely elongated grains with strong α fiber (<110>//RD) with peak at {001}<110> and weak γ fiber (<111>//ND) distributed in center layer (*S*=0). The microstructure of cold rolled sheet is composed of elongated grains with less in-grain shear bands through sheet thickness (figure 2c). The first cold rolling texture is characterized by strong α fiber with peak at {112}<110> and γ fiber (<111>//ND) with peak at {111}<110> at both layers (figure 2d), which is similar to the rolling textures in electrical and ultra-low-carbon steel[17-19]. The primary recrystallized microstructure consists of nearly homogeneous grains (18~21 μm) through thickness. The primary recrystallization texture is characterized by strong γ fiber with a peak at {111}<112> and weak Goss, distinct from the reported primary recrystallization texture in Fe-Ga sheet [11,20,21]. Strong {111}<112> can promote the abnormal growth of secondary recrystallized Goss grains in terms of high grain boundary mobility as in grain-oriented silicon steel [22-24].

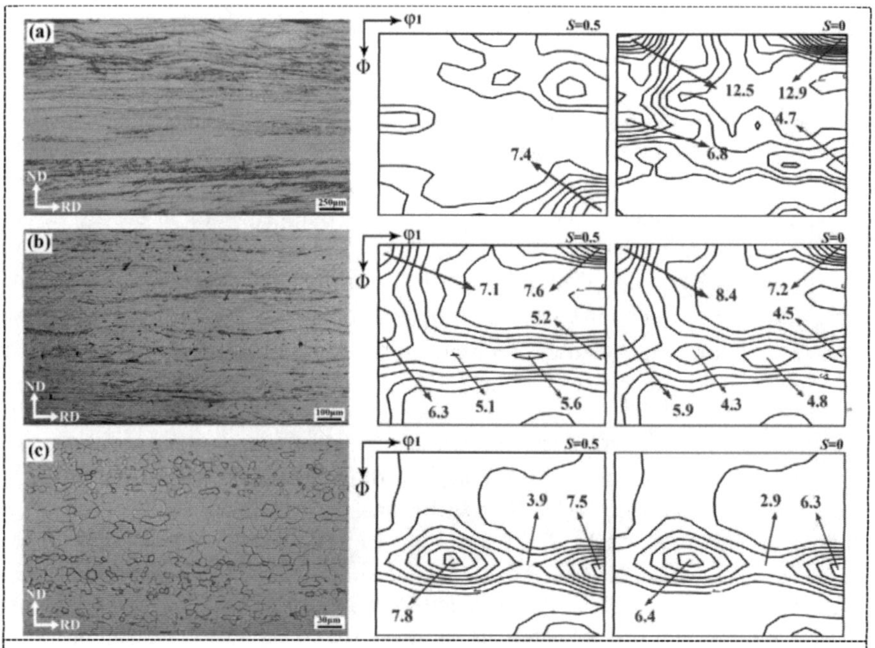

Figure 2. Microstructure and constant φ_2=45° sections of ODFs of (a) hot band, (b) cold rolled and (c) primarily annealed $Fe_{81}Ga_{19}$ strip.

Figure 3 shows microstructure evolution during high temperature annealing by interrupted experiment method. Average grain size increases slowly from 20 μm at 800 °C to 26 μm at 950 °C, indicating that normal grain growth is effectively inhibited by nanometer-sized precipitates. Secondary recrystallization occurs at 1000 °C and nearly completes at 1100 °C. This temperature range is much higher and narrower than Fe-Ga-NbC sheet with the application of surface energy [6,12]. Average grain size of matrix grains further increases from 31 μm to 43 μm as temperature arise from 1000 °C to 1050 °C(figure 4), which means the pinning gradually decrease due to the coarsen of nanometer-sized precipitates as temperature exceeds 1000 °C.

According to EBSD analysis on interrupted samples (figure 5), the abnormally grown grains are identified as Goss orientation by EBSD. The deviation angle increases from 5~8° at 1000 °C with grain size of ~3 mm to 8~15° at 1200 °C with grain size of ~10 mm. This result suggests that the complete secondary recrystallization of sharp Goss texture can be induced in primary cold rolled Fe-Ga sheets by controlled precipitation of nano-sized inhibitor without the application of surface energy effect.

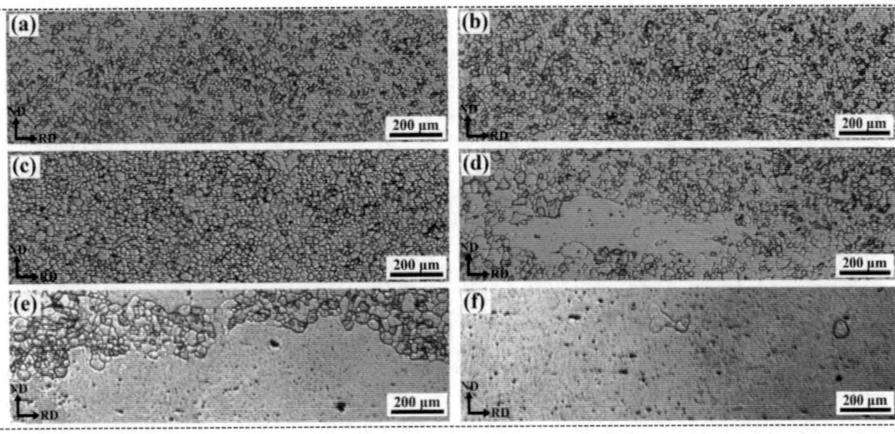

Figure 3. Microstructure evolution during final annealing process of $Fe_{81}Ga_{19}$ thin sheet as heated to (a) 900 °C, (b)950, (c) 975 °C, (d) 1000 °C, (e) 1050 °C and (f) 1100 °C.

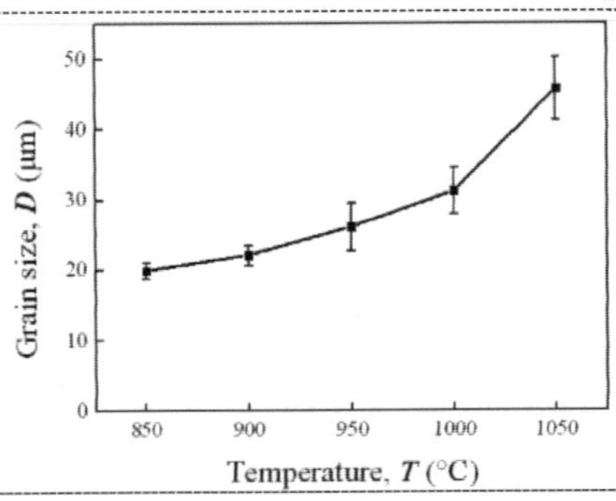

Figure 4. The average grain size with annealing temperature in $Fe_{81}Ga_{19}$ thin sheet.

The magnetostriction coefficient of annealed specimens interrupted at different temperatures is shown in figure 6. The occurrence of abnormal grain growth near 1000 °C leads to an evidently increased magnetostriction coefficient of 95±19 ppm relative to 69±6 ppm at 950 °C. The magnetostriction coefficient reaches up to 164±21 ppm when centimeter-sized Goss grains develop at 1000 °C. The magnetostriction coefficient as high as 210±25 ppm is further obtained after annealing at 1200 °C, corresponding to complete secondary recrystallization of Goss grains. This value is much larger than those reported on Fe-Ga alloys with the help of micronmeter-sized inhibitors [4,10], but lower than the cases with both inhibitors and surface energy [5,6,9]. This is attributed to the larger deviation angle after secondary recrystallization.

IOP Conf. Series: Journal of Physics: Conf. Series **1270** (2019) 012004 doi:10.1088/1742-6596/1270/1/012004

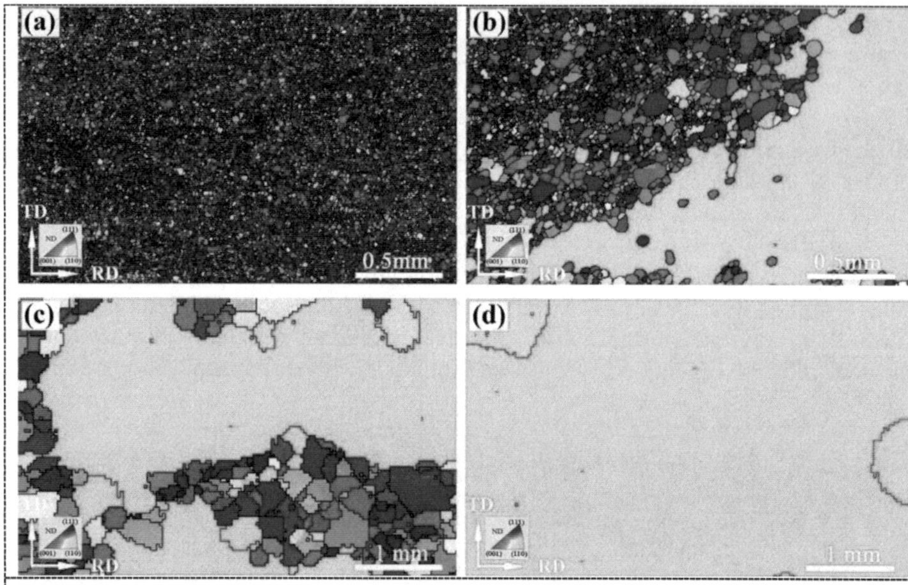

Figure 5. Orientation image maps of $Fe_{81}Ga_{19}$ thin sheet extracted at (a) 950 °C, (b) 1000 °C, (c) 1050 °C and (d) 1200 °C.

The secondary recrystallization within a relatively higher and narrower temperature range indicates that the selected inhibitor and its morphology during hot rolling and normalization can be used for the secondary recrystallization of Goss texture in $Fe_{81}Ga_{19}$ alloy. However, large deviation of Goss texture after secondary recrystallization implies the inhibitor force and its thermal stability of the inhibitor system cannot match well with the AGG of accurate Goss grains. The rapidly coarsening of the nanometer-sized sulfides at annealing temperature higher than 1000°C reduces the driving force of preferred growth of accurate Goss grains. Accordingly, it is concluded that the critical requirements for secondary recrystallization process without the application of surface energy effect in Fe-Ga alloy includes stronger inhibition to primary grains, gradually dissipated inhibition force and precise matching with intrinsic preferred mobility of Goss grain during high temperature annealing.

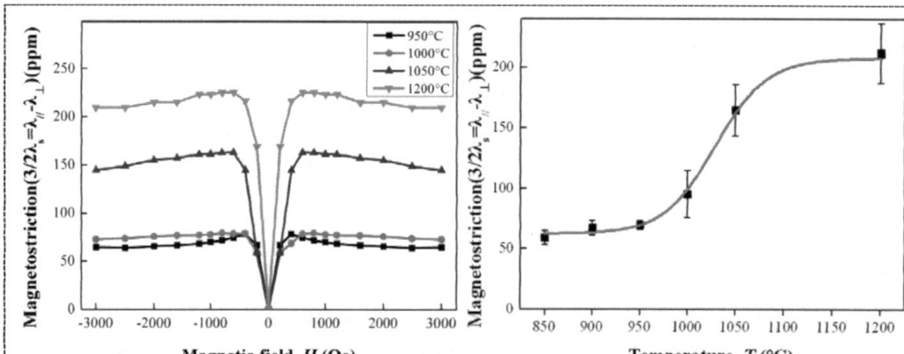

Figure 6. Variation of magnetostriction coefficients ($3/2\lambda_s$) with magnetic field and annealing temperature of $Fe_{81}Ga_{19}$ thin sheet.

4. Conclusion

In the present study, the composite inhibitor composed of nanometer-sized sulfide particles is dispersedly precipitated in hot rolling and normalization annealing. The strong inhibiting force during high temperature annealing promotes Goss grains to grow abnormally by consuming primary recrystallized matrix grains. Centimeter-sized secondary recrystallization Goss grains and large magnetostriction coefficient as high as 235 ppm is obtained without surface energy effect in primary cold rolled $Fe_{81}Ga_{19}$ alloy sheet. The result indicates that complete secondary recrystallization can be realized based on the precise design and control of inhibitor and primary recrystallization microstructure/texture in Fe-Ga alloy by conventional rolling and annealing methods.

Acknowledgments
This work is financially supported by National Natural Science Foundation of China (Grant No. 51671049); Natural Science Foundation of Inner Mongolia (Grant No. 2018ZD10) and the Postdoctoral Science Foundation of China (Grant No. 2018M640257), and State Key Laboratory of Baiyunobo Rare Earth Resource Researches and Comprehensive Utilization (Grant No. 2017Z1953).

References
[1] Clark A E, Restorff J B, Wun-Fogle M, Lograsso T A and Schlagel D L 2000 *IEEE Trans. Magn.* **36** 3238
[2] Guruswamy S, Srisukhumbowornchai N, Clark A E, Restorff J B and Wun-Fogle M 2000 *Scr. Mater.* **43** 239
[3] Kellogg R A, Russell A M, Lograsso T A, Flatau A B, Clark A E and Wun-Fogle M 2004 *Acta Mater.* **52** 5043
[4] Na S M, Yoo J H and Flatau A B 2009 *IEEE Trans. Magn.* **45** 4132
[5] Na S M and Flatau A B 2012 *Scr. Mater.* **66** 307
[6] Yuan C, Li J H, Zhang W L, Bao X Q and Gao X X 2015 *J. Magn. Magn. Mater.* **391** 145
[7] Li J H, Yuan C, Qi Q L, Bao X Q and Gao X X 2017 *Metals* **7** 36
[8] Na S M and Flatau A B 2012 *Smart. Mater. Struct.* **21** 055024
[9] Na S M, Atwater K M and Flatau A B 2015 *Scr. Mater.* **100** 1
[10] Yuan C, Li J H, Zhang W L, Bao X G and Gao X X 2015 *J. Magn. Magn. Mater.* **374** 459
[11] Liu Y Y, Li J H and Gao X X 2017 *J. Magn. Magn. Mater.* **435** 194
[12] He Z, Hao H, Sha Y, Li W, Zhang F and Zuo L 2019 *J. Magn. Magn. Mater.* **478** 109
[13] Mishra S and Kumar V 1995 *Mater. Sci. Eng. B* **32** 177
[14] Jenkins K and Lindenmo M 2008 *J. Magn. Magn. Mater.* **320** 2423
[15] Xia Z S, Kang Y L and Wang Q L 2008 *J. Magn. Magn. Mater.* **320** 3229
[16] Guo W, Meng L, Wang H, Yan G and Mao W 2015 *Front. Mater. Sci.* **10** 66
[17] Park J T and Szpunar J A 2003 *Acta Mater.* **51** 3037
[18] Verbeken K, Kestens L and Jonas J J 2003 *Scr. Mater.* **48** 1457
[19] Sha Y H, Sun C, Zhang F, Patel D, Chen X, Kalidindi S R and Zuo L 2014 *Acta Mater.* **76** 106
[20] Na S M and Flatau A B 2014 *J. Mater. Sci.* **49** 7697
[21] He Z H, Sha Y H, Fu Q, Lei F, Zhang F and Zuo L 2016 *J. Appl. Phys.* **119** 123904
[22] Harase J, Shimizu R and Takahashi N 1990 *Acta Metall. Mater.* **38** 1849
[23] Hayakawa Y and Szpunar J A 1997 *Acta Mater.* **45** 4713
[24] Hayakawa Y and Kurosawa M 2002 *Acta Mater.* **50** 4527

Cold Rolling Reduction Dependence of the Recrystallization Texture of 3% Si Non-oriented Electrical Steel

Lee Seil[1,a] and De Cooman Burno C.[2]

[1]POSCO Technical Research Lab., Pohang, South Korea
[2]Graduate Institute of Ferrous Technology POSTECH, Pohang, South Korea
[a]apull@posco.com

Abstract. The effect of the amount of cold reduction on the crystallographic texture, the microstructure, and the magnetic properties of cold rolled and annealed 3% Si non-oriented (NO) Si steel has been studied. The results indicate that, at constant final thickness, the combination of initial hot-rolled sheet thickness and the cold rolling reduction had a significant influence on the B50 magnetic flux density. The initial thickness of the hot-rolled sheet did not affect the crystallographic texture of the as-hot rolled sheet and the annealed hot-rolled sheet. At constant final thickness, the combination of a larger hot-rolled sheet thickness and a higher cold rolling reduction resulted in the decrease of the B50 magnetic flux density. This was due to an increase of the {411}<148> and {111}<112> texture components intensity and a decrease of the {110}<001> texture component intensity.

1. Introduction

Non-oriented (NO) electrical steel sheet is a magnetic material used as a core material for electrical motors and generators. Modern electrical motors and generators require a combination of a high power density and low energy losses. The energy losses in NO steels (core losses) are consists of three components; hysteresis losses, eddy current losses and the anomalous losses. Addition of Si and Al alloy contents is the most effective method to reduce core losses because an eddy current losses are decreased by an increasing electrical resistivity. There are two more methods to reduce eddy current losses, such as (a) the reduction of the sheet thickness [1], and (b) the optimization of the crystallographic texture [2].

However, magnetic flux density property, which is linked to the power density, is decreased by the alloy additions and thickness reduction. The crystallographic texture improvement is the only method to compensate the decrease of a magnetic flux density property of a NO steel sheet. The improvement of the crystallographic texture of the NO steel sheet makes it, in principle, possible to simultaneously reduce the core losses and increase the magnetic flux density. It is well known that the magnetic properties of α-Fe are anisotropic: while the <100> axes are axes of easy magnetization, the <111> axes are axes of hard magnetization. When a NO electrical steel sheet has a large fraction of grains with one or more <100> axes in the sheet plane, the magnetization parallel to the sheet plane will be easier, the magnetic flux density will increase, and the losses will be reduced.

Current technology allows for the production of NO steel with a thickness in the range of 1.0mm to 0.10mm, corresponding to a cold rolling reduction in the range of 70% to 95%. An increase of the cold reduction could result in the development of crystallographic textures during recrystallization annealing that are suitable for NO electrical steel.

3% Si NO steel differs from ULC steels in two aspects. First, the alloy content of NO steel is significantly higher, and the addition of Si and Al suppress the phase transformation during cooling after the hot rolling. This affects the texture of the hot rolled sheet prior to the cold rolling process. Instead of a transformation texture, a mixed texture of ferrite plane strain and shear texture components are obtained. Hot rolled 3% Si steel is often given an intermediate annealing, which increases the average grain size to a few hundred microns. These coarse grains typically develop a weaker γ-fiber intensity.

2. Experimental procedure

The composition of the NO electrical steel used in the present study was Fe-3.5% (Si+Al)+0.3%Mn-0.01%P-0.004%C-0.002%N-0.001%S (in wt.%). The steel was hot rolled to a thickness of 2.9 mm, 2.5 mm, 2.0 mm, 1.6 mm and 1.4 mm in a 5 stand hot rolling mill. The reduction applied in each stand was in the range of 20% to 50%. The hot rolled sheets, sized 300 mm x 85 mm, were given an intermediate anneal prior to cold rolling. The sheet was heated to the annealing temperature using a heating rate of +20 °C/s, held isothermally over 1000°C to obtain similar grain size in all thickness samples and subsequently cooled in air. The annealing furnace atmosphere was N_2. The samples were pickled and cold rolled to the thickness of 0.5 mm, 0.35 mm, and 0.23 mm. The cold rolling reduction therefore varied in the range of 65% (1.60 mm → 0.50 mm) to 92% (2.90 mm → 0.23 mm). The annealing of the cold rolled sheet was carried out in a continuous annealing simulator. The annealing atmosphere was a 50% H_2 + 50% N_2 gas mixture. The annealing time was 90 seconds. The annealing temperature was about 1000 °C for samples.

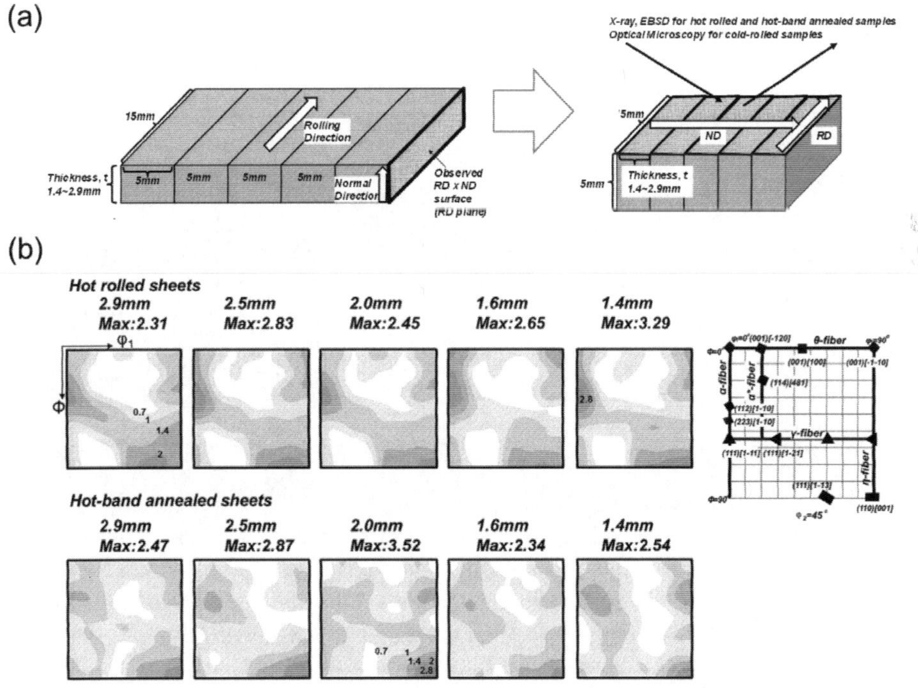

Figure 1. (a) Schematic illustration X-ray analysis. (b) φ_2=45° ODF section for the hot rolled sheets, and the hot-rolled and annealed sheets.

Large area texture measurements on the hot rolled sheet samples and hot-rolled and annealed sheet were carried out using a Bruker D8 X-ray diffractometer (XRD) equipped with a Co target source operated at a voltage of 35kV and a current of 45mA. The XRD measurements were taken from sheet samples as illustrated schematically in figure 1 (a). The orientation distribution functions (ODFs) were calculated from the {110}, {200}, and {211} incomplete pole figures using the series expansion method implemented in the HTEXTools Version 3 software.

The microstructures and texture of the samples were characterized by light-optical metallography (LOM) and electron backscattering diffraction (EBSD) measurements. The EBSD measurements were carried out on a JEOL JSM7001F FE-SEM equipped with a TSL Hikari EBSD set up, and the data were analyzed with the TSL OIM v7.0 software. The EBSD measurements were applied on longitudinal RD plane sections as defined by the rolling and the normal direction.

The microstructure of the hot-rolled, hot-rolled and annealed, and cold-rolled steels was obtained by LOM. The measurements were carried out on the longitudinal sections, defined by the rolling direction and the normal direction. The samples were polished and etched in a Nital solution (5 vol% Nitric acid and 95 vol% ethanol).

The magnetic properties of the sheet samples were measured in a BROCKHAUS Messtechnik MPG 200 single sheet tester with a 60 mm x 60 mm yoke. The magnetic flux density properties of samples at 5000 A/m (B50) were measured along both the rolling and transverse directions.

3. Results and discussion

The φ2=45° section of the ODFs of the hot rolled and hot-rolled and annealed sheet is shown in figure 1 (b). The texture of hot rolled sheet was composed of the {110}<001>, {110}<112>, {100}<011>, {112}<110> and {112}<111> texture components. The texture of hot-rolled and annealed sheets consisted mostly of {110}<001>, {112}<110> and {112}<111> texture components. The intensities of the various texture components did clearly not differ between the hot rolled sheets and annealed sheets. This implies that differences in hot rolling reduction do not significantly affect the texture after hot rolling and after the annealing.

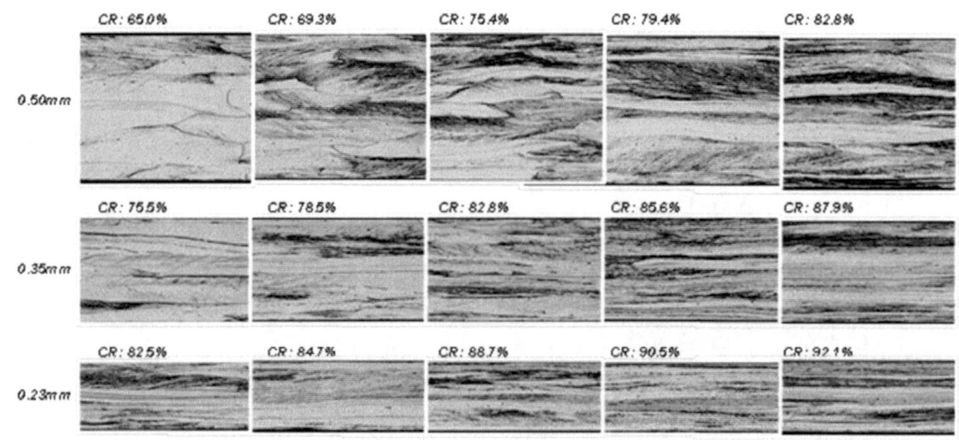

Figure 2. LOM micrographs of the cold rolled steel (Nital etch)

Fifteen samples with different cold rolling reductions were obtained and shown in figure 2. The sheets were cold rolled to a final thickness of 0.23 mm, 0.35 mm, and 0.5 mm. In the resulting set of samples, some of the cold rolled samples had a similar cold rolling reduction but a different final thickness. For example, the samples which were given a 82.8% were reduced in thickness were obtained from the samples having thicknesses of 2.9 mm, 2.0 mm, 1.4mm and cold rolled to 0.5 mm, 0.35 mm, 0.23mm, respectively. In general, the microstructure consisted of a dark etched region and a light etched region. The samples which were given a cold rolling reduction less than 79%, had in-grain deformation bands which extended from grain boundaries into the grain matrix. This microstructure resulted in dark etched regions. The deformation bands were more frequently observed when the cold rolling reduction increased from 65 % to 75%. In samples which were given a more severe reduction, i.e. >79%, no deformation bands were observed. With increasing cold rolling reduction, the grains developed a pan-caked shape when the cold reduction was more than 90%, i.e. all the grains were flattened parallel to the rolling plane. It is clear that the microstructure is determined by the cold reduction ratio rather than the difference in the final cold rolled sheet thickness.

Figure 3 reviews the texture of the samples after the complete 90s annealing treatment. The recrystallized grain size was typically larger than 70 μm. The texture was measured by analysing a 5mm x 5mm area of each sample by means of the EBSD technique. The {111}<112> texture component, the α*-fiber, and the {114}<148> texture component developed with increasing cold reduction. Texture components close to the {110} <001> orientation developed when the cold rolling reduction was less than 80%. The {110}<001> texture component was not observed in the samples which were given a cold rolling reduction larger than 80%. The α-fiber {112} <011>, {111} <011> and {100} <011> texture components, which were present after cold rolling, did not develop during annealing. The maximum texture intensity value increased with increasing cold rolling reduction, from 3.2 at 65%, to 6.4 at 92%. The maximum texture intensity after recrystallization annealing was observed near the {411}<148> orientation for the cold rolled sample which was given a 92.1% rolling reduction.

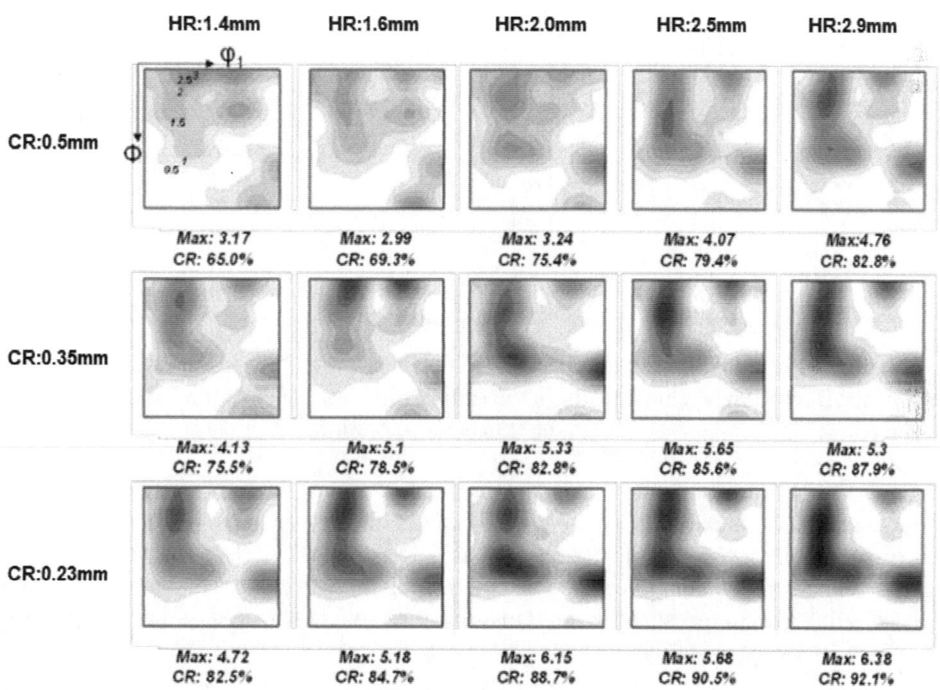

Figure 3. φ_2=45° ODF sections, obtained from EBSD data, after a full recrystallization anneal.

Figure 4 shows that the volume fraction of grains with the {411}<148> orientation increased significantly with increasing reduction. The intensity of the {411}<148> component increased continuously during annealing. This was especially pronounced for the samples with a 0.23mm final thickness. The fraction of {411}<148> oriented grains continued to increase during the grain growth stage following the recrystallization. The volume fraction of {411}<148>-oriented grains showed a clear tendency to increase for higher annealing temperatures and for larger grain sizes. While the {111}<112> orientation remained an important texture component after annealing, the volume fraction of the grains with this orientation was approximately half the volume fraction of the {411}<148>-oriented grains after annealing. The volume fraction of {111}<112>-oriented grains decreased with increasing annealing temperature.

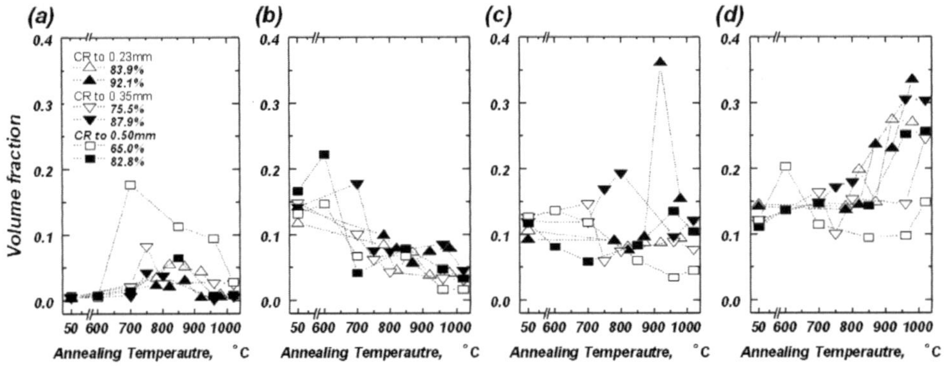

Figure 4. Annealing temperature dependence of the volume fraction of the (a) {110}<001>, (b) {111}<110>, (c)<111>{112}, and (d) {411}<148>-oriented grains.

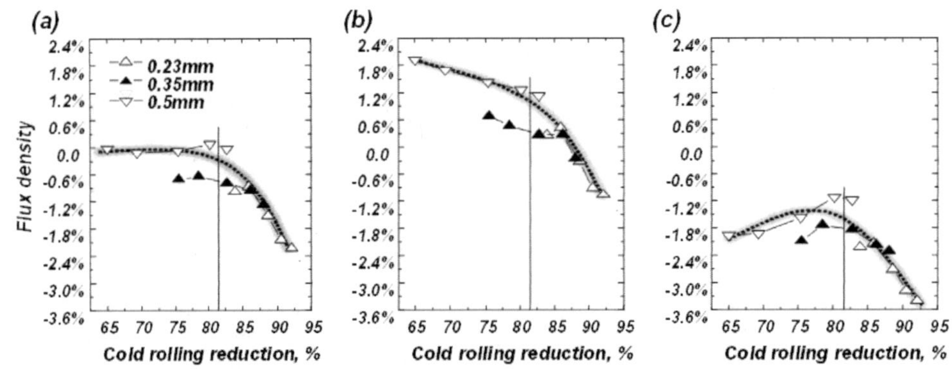

Figure 5. Effect of the cold reduction on (a) the B_{50} magnetic flux density values of averaged values, (b) the B_{50} magnetic flux density in the rolling direction, and (c) the B_{50} magnetic flux density in the transverse direction.

Figure 5 shows the cold rolling reduction dependence of the magnetic flux density B_{50} values. Regardless of the final thickness of the cold rolled sheets, the averaged magnetic flux density decreased with increasing reduction when the reduction was larger than 82%. The B_{50} values measured in the rolling direction decreased continuously with cold rolling reduction for reductions in the range of 65% to 92%. The value of B50 measured in the transverse direction first increased for reductions up to 82%, and decreased for larger reductions.

4. Conclusion

The effect of the cold reduction on the development of the crystallographic texture, the microstructural evolution, and the magnetic properties of cold rolled and recrystallization annealed 3% Si NO steel was investigated by OM, XRD and EBSD measurements. The texture of the hot-rolled sheet and the annealed hot-rolled 3% Si NO sheet is not affected by the hot-rolling reduction. The volume fraction of {411}<148>-oriented grains after an annealing increases with increasing cold rolling reduction. The magnetic flux density decrease is significant for cold rolling reductions exceeding 82%.

References

[1] Broddefalk and M.Lindenmo 2006 *J.M.M.M* **304** 586-588
[2] M.Nakano, T.Okamoto, H.Fukunaga, Y.Yamashiro, K.Ishiyama and K.I.Arai 1999 *J.M.M.M* **196-197** 341-343

Effect of hot band annealing on texture and grain size distribution of primary recrystallization in grain-oriented silicon steel

K J Ko[1,1] and J T Park[1,2]

[1]Steel Product Research Laboratories, POSCO, Pohang, 37859, Korea.

[1]kjko04@posco.com, [2]jtpark20@posco.com

Abstract. During final high-temperature annealing in grain-oriented silicon steel, a few Goss grains grow exclusively fast and consume the matrix grains, which are related to the quality of the final product. The optimization of texture and microstructure in primary recrystallization is crucial to achieve the sharp Goss texture in grain-oriented silicon steel. The driving force for this phenomenon is the reduction of grain boundary area, which can be decribed as a reduction in grain boundary energy. The distribution of grain boundary and its energy determined by texture and microstructure of primary recrystallization are very important data to understand the optimization point. The hot band annealing is known to be a prerequisite for the high quality product in grain-oriented silicon steel. It can affect the evolution of microstructure and texture in primary recrystallization. In this study, the influence of the hot band annealing temperature on the texture and grain size distribution in primary recrystallization was investigated in detail.

1. Introduction

The grain size and texture after primary recrystallization annealing are very important for controlling the good magnetic properties in grain-oriented silicon steel. The uniform grain size distribution of primary recrystallization is considered to be helpful to improve the quality of final product in silicon steel [1]. The average grain size and distribution of grain size in grain-oriented si steel is determined by recrystallization and grain growth during primary recrystallization annealing process. The driving force for grain growth is the reduction of grain boundary area, and thus the total energy of the system is lowered. The driving force for the grain growth comes from the difference in grain sizes between neighboring grains. Thus, as growth occurs with rising temperatures the nuber of grains decreases as larger grains consume smaller grains. As smaller grains begin to shrink, their area diminishes and toteal energy is lowered. As grains are continually growing and shrinking over time, the mean grain size increases with forming a certain grain size distribution.

Hot band annealing is an important process to improve the magnetic properties of the product by modifying the microstructure, texture and precipitate distribution of the hot rolled sheet. It is well known that hot band annealing plays an important role to improve the inhomogeniety of the hot rolled sheet. Because the grain growth can be inhibited in the presence of fine precipitates which pin grain boundaries in place, the fine precipitations and their distribution should be considered for controlling the distribution of grain size during grain growth. The uniformity of grain size distribution is also an important factor to control the Goss texture in silicon steel[1,2]. In grain growth, the grain size distribution of the sample was significantly influenced by the temperature of hot band annealing.

In this study, the effects of the hot band annealing temperature on the texture and microstructure changes during primary recrystallization annealing were investigated.

2. Experimental procedure

The starting materials in this study were the 2.3mm-thick hot-rolled sample of Fe-3%Si, which employed AlN as an inhibitor to produce the grain-oriented (GO) electrical steel. The specimens were heated to high temperature to dissolve the fine precipitations which has non-uniform distribution, which is called hot band annealing process. The hot rolled sample were heated at 1000 ~ 1100°C, cooling to 850 ~ 950°C by furnace cooling, holding for 2 min, and air-cooling below 800°C and quenching in boiling water. Following the hot band annealing, the samples were cold rolled corresponding to approximately 89% thickness reduction using a laboratory mill. The cold-rolled samples underwent the primary recrystallization at 810 ~ 830°C for 2 min in a wet hydrogen atmosphere for decaburization and recrystallization with grain growth.

For a sample analysis, the TD (transverse direction) cross sections of sample sheets were prepared by micro-polishing. Microstructure and orientations of the primary recrystllized samples were investigated by electron backscattered diffraction (EBSD), attached to a field emission scanning electron microscopy (FE-SEM, JEOL JSM - 6500F). The data related to microstructure and texture were obtained and analyzed using the orientation imaging microscopy (OIM 8) software packages. 220 × 3000μm^2 area scans with 1㎛ step size were used, enough to measure the microstructure with an average grain size of the samples in this paper. In order to calculate grain size from EBSD scan data, the grain boundaries were defined within the 5 degree tolerance angle and edge grains were excluded for the grain size calculation. About 300 ~ 500 points per a grain were used, if calculated from the average grain size in the analysis area.

3. Results and discussion

3.1 Microsturucture and texture of primary recrystallized samples

Figure 1 illustrates EBSD data including an orientation map with image quality map. The EBSD data shows the microstructure after primary recrystallization annealing at 810°C, 820°C, and 830°C. The microsturucture is showing that grain growth occurs and the decaburization and recrystallization were already finished due to 89% cold rolling process and high soaking temperature to recrystallization. The average grain size of samples is increased from 16 to 22㎛ with increasing soaking temperature from 810°C to 830°C. As can be seen in the microstrucure of figure 2, the frequency of fine size grain is decreased with increasing soaking temperature.

(a)
(b)
(c)

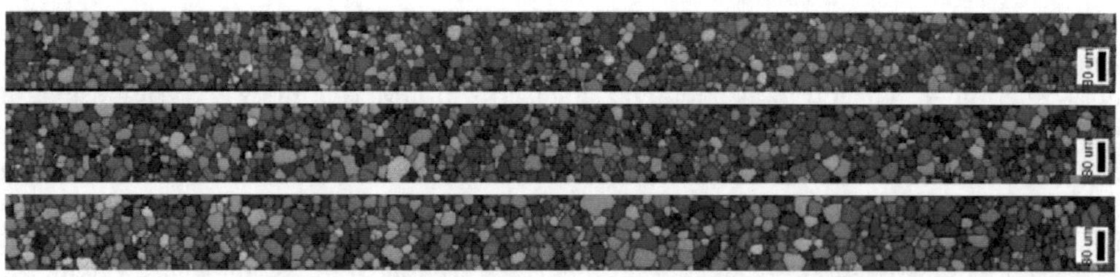

Figure 1. IPF maps(TD) of samples annealed at (a) 810℃ (b) 820℃ (c) 830℃ for 2min.

In order to compare the area fraction of grains by size, the grain size distribution for samples after primary recrystalization annealing at 810°C, 820°C, and 830°C is shown in figure 2. As can be seen in figure 2, the grain size distribution shifted to right with increasing the annealing temperature. As the the soaking temperature increases, the area fraction of smaller grains than 25㎛ was decreased and the one of larger grains than 30㎛ was increased. As the the soaking temperature increases from 810℃ to 830℃, the area fraction of smaller grains than 25㎛ is reduced by 48.7%, 40.2% and 29.4%. On the other hand, the area fraction of larger grains than 40㎛ is increased by 10.9%, 17.9%, 25.7%. This graph indicates that the large grains grow while consuming smaller-sized grains at the temperature of 820, 830℃

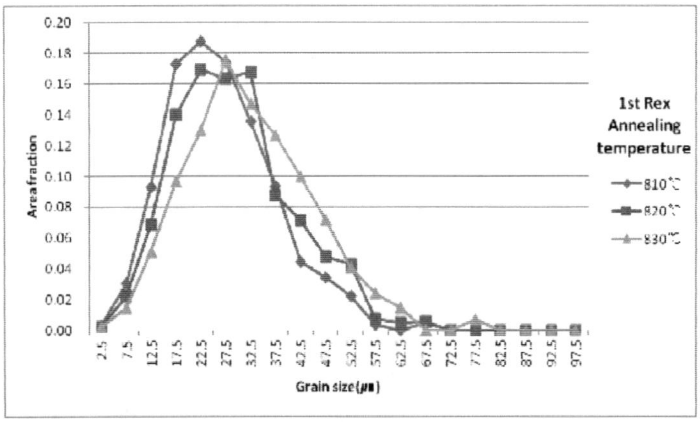

Figure 2. Grain size distribution of samples after primary recrystallization annealing at (a) 810°C, (b) 820°C, and (c) 830°C

Figure 3(a)~(c) shows the texture of samples after primary recrystallization annealing at 810°C, 820°C, and 830°C, respectively. The primary recrystallization texture was developed similar after annealing due to 89% cold rolling process. All of the ODFs in figure 3(a) ~ (c) have the strong {111}<112> or {554}<225> components, and {411}<148> and {100}<012> components. The maxium intensity at these components in each sample was between 5 to 7. The Goss component, {110}<001> was not shown in ODFs because the volume fraction of Goss components in primary recrystallized samples was too minor. Calculated from EBSD scan data, the area fraction of Goss components in each sample was between 0 to 1.5%. As the maximum intensity of main texture components increased, the Goss components decreased. It should be noted that the EBSD scan area for each sample includes 1,300 to 2,600 grains. The number of grains in EBSD scan area depends on the size of the average grain. It was still small number of grains to acquire quantitative texture data [3]. The texture data from TD section showed some deviation from the measured position and the Goss component was too small to acquire reliable data.

Figure 3. ODFs represented in Phi=45degree sections (intensity levels: 1, 1.5, 2, 2.5…) of samples after primary recrystallization annealing at (a) 810°C, (b) 820°C, and (c) 830°C

3.2 Microsturucture and texture of primary recrystallized samples with different hot band annealing temperature

In order to compare the effects of the hot band annealing temperature on the grain size after primary recrystallization, the average grain size of primary recrystallized samples is shown in figure 4. As shown in figure 4, the average grain size of two samples were similar at 810℃. Meanwhile, as temperature rose up to 830℃, the grains of sample with hot band annealing at 1000℃ grew faster than the sample with hot band annealing at 1100℃. In other words, there was the retardation of grain growth in the sample with hot band annealing at 1100°C. This retardation seemed to have resulted from the inhibition of grain growth by fine AlN precipitate formed during hot band annealing process.

In order to compare the area fraction of grains by size, the grain size distribution for samples with hot band annealing at 1000 and 1100℃ were shown in figure 2. As can be seen in figure 5(b), in the case of hot band annealing at 1100, the grain size distribution of samples annealed at 810 and 820℃ remained unchanged. Even when the annealing temperature was raised from 810°C to 820°C, small size of grains less than 20㎛ remained unconsumed by larger grains. So, the large grains did not grow further and the grain size of samples remained same as around 17㎛ after annealing at 820°C. As the annealing temperature rose up to 830°C, the grains of sample with hot band annealing at 1100°C began to grow late. In contrast, as shown in figure 5(a), the distribution of grain size in samples was shifted sequentially to the right at 810, 820, and 830°C for hot band annealing at 1000°C. As the annealing temperature rose up to 830°C, in particular, the grain size distribution has broadened. And also, by consuming small grains less than 25㎛, the area fraction of grains over 50㎛ was increased considerably in figure 5(a). As the annealing temperature rose up from 810°C to 830°C, the area percentage of small size grains (less than 25㎛) was reduced from 49% to 32% in figure 5(a). At the same time, the area percentage of large size grains (over 40㎛) was increased from 11 % to 24%. In the sample with hot band annealing at 1000°C, the area percentage of small size grains (less than 25㎛) was reduced from 49% to 40%. At the same time, the area percentage of large size grains (over 40㎛) was increased from 11 % to 20%.

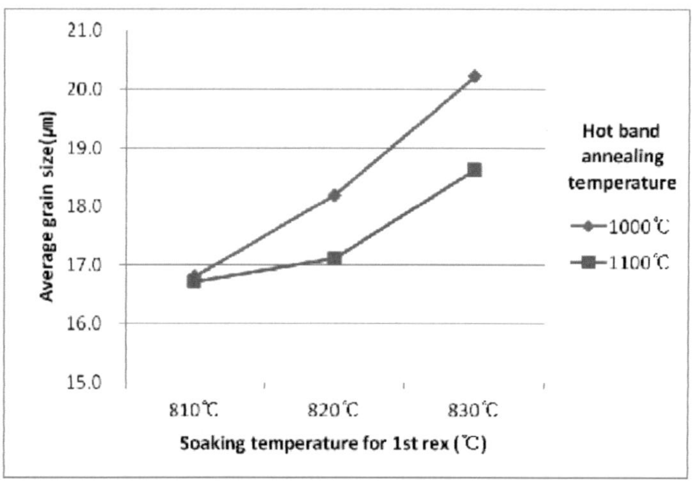

Figure 4. Grain size of primary recrystallized samples with hot band annealing at 1000℃ and 1100℃

In the grain growth, the main difference between the two samples is whether or not the shrinkage of small size grains is inhibited at same temperature annealing. The differences in the inhibition of grain growth could be explained by differences in the precipitates size or volume fraction. [4] There was no difference in the volume fraction of precipitates between two samples, so the sample with hot band annealing at 1100°C might have finer precipitates. Generally, fine precipitates can dissolve at lower temperature than coarse precipitates due to Gibbs-thompson effect. therefoer, the grain coarsening temperature in steel depends on the size of AlN precipiates [5]. The finer precipitates may have a pinning effect and they could dissolve and lose their pinning force at lower temeprature than the other coarser precipitates. As can be seen in figure 4 and 5(b), the fine precipitates in the sample with hot band annealing at 1100°C seems to dissolve and lose their pinning force to preserve the small grains during annealing at 830°C. On the other hands, the sample with hot band annealing at 1000°C seems not to have the pinning force to preserve the small grains due to the lack of fine precipitates during primary recrystallization annealing. The difference in size of precipiates between two samples can be explained as follows. Higher temperature during hot band annealing the fine precipitates in the hot-rolled sample can dissolve more at higher temperature and they formed fine precipitates again during further annealing process. The higher temeperaure of hot band annealing can form higher density of fine precipitates. Therfore the sample with hot band annealing at 1100°C could have finer precipitates than the sample with hot band annealing at 1000°C.

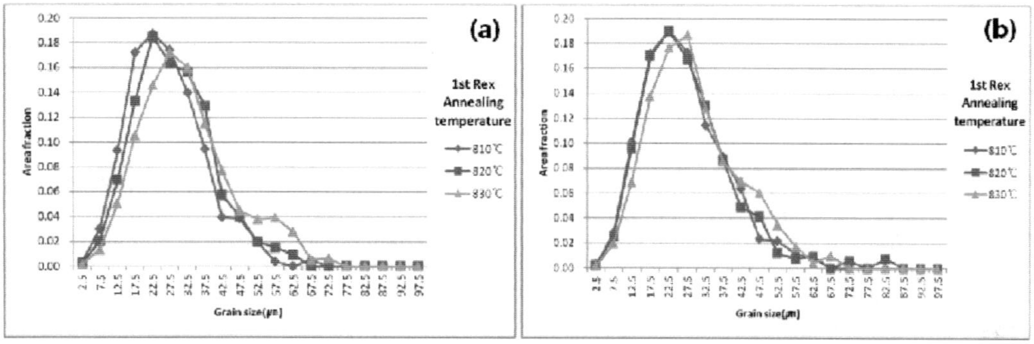

Figure 5. Grain size distribution of samples with hot band annealing at (a) 1000℃ and (b) 1100℃

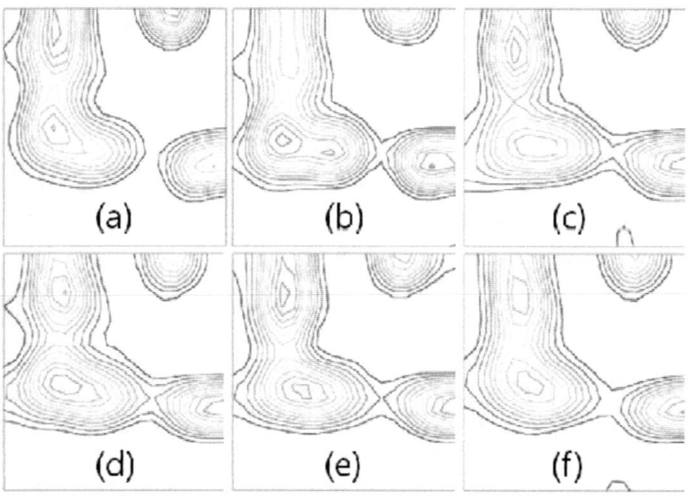

Figure 6. ODFs represented in Phi=45degree (intensity levels: 1, 1.5, 2, 2.5…) primary recrystallized samples at (a, d) 810°C, (b, e) 820°C, (c, f) 830°C with hot band annealing at (a,b,c) 1000°C and (d,e,f) 1100°C.

The primary recrystallization texture should be considered to control the Goss texture in grain-oriented silicon steel product [6-8]. Figure 6 shows the texture of samples after primary recrystallization annealing at 810°C, 820°C, and 830°C, respectively. All of the ODFs in figure 6 have the strong {111}<112> or {554}<225> components, and {411}<148> and {100}<012> components. The Goss component, {110}<001> was not shown in ODFs because the volume fraction of Goss components in primary recrystallized samples was too minor. Calculated from EBSD scan data, the area fraction of Goss components in each sample was between 0 to 0.8%. the primary recrystallization texture between these samples were not changed due to same cold rolling process (89% thickness reduction ratio).

The temperature of hot band annealing did not have much effect on the primary recrystallization texture but it had a significant effect on the grain growth during the primary recrystallization annealing. In grain growth, the grain size distribution of the sample was significantly influenced by the fine precipitates which were formed after the high temperature of hot band annealing.

4. Summary

The influence of the hot band annealing temperature on the texture and grain size distribution in primary recrystallized samples was investigated in detail. The temperature of hot band annealing did not have much effect on the primary recrystallization texture but it had a significant effect on the grain growth during the primary recrystallization annealing. The grains of sample with hot band annealing at 1000℃ grew faster than the sample with hot band annealing at 1100℃. In grain growth, the grain size distribution of the sample was significantly influenced by the fine precipitates which were formed after the high temperature of hot band annealing.

References

[1] Woo J S, Han C H and Hong B D 1998 *Acta Mater.* **46** 4905
[2] Hayakawa Y 2017 *Sci. Technol. Adv. Mater.* **18** 480
[3] Wright S I and Adams B L 1990 *Metall. Mat. Trans. A*, **38** 1845
[4] Park C S, Na T W, Kang J K, Lee B J, Han C H and Hwang N M 2013 *Phil. Mag.* **93** 4198
[5] Wilson F G and Gladman T 1998 *Int. Mater. Rev.* **33** 221
[6] Lin P, Palumbo G, Harase J and Aust K T 1996 *Acta Mater.* **44** 4677
[7] Rajmohan N, Szpunar J A and Hayakawa Y 1999 *Acta Mater.* **47** 2999
[8] Ko K J, Rollett A D and Hwang N M 2010 *Acta Mater.* **58** 4414

IOP Conf. Series: Journal of Physics: Conf. Series **1270** (2019) 012007 doi:10.1088/1742-6596/1270/1/012007

Evolution of recrystallization texture in non-oriented electrical steels during final annealing – influence of shear stress after cold rolling

B Bacroix[1], J Schneider[2], A Franke[3]

[1] LSPM – CNRS, UPR3407, Université Paris 13, 99, Av. J.B. Clément, 93430 Villetaneuse, France.
[2] Institute of Metal Forming, Technische Universität Bergakademie Freiberg, Bernhard-von Cotta-Str.4, D-09596 Freiberg, Germany.
[3] Stahlzentrum Freiberg e.V., Leipziger Straße 34, 09599 Freiberg, Germany.

Abstract. The fabrication process of non-oriented electrical steels comprises hot rolling, annealing of the hot band (optional), cold rolling and final annealing. Starting from different hot band, quite distinct deformation substructures have been observed after cold rolling of ferritic FeSi steels. The recrystallization texture after annealing of the cold rolled material depends on the deformation structure obtained after cold rolling. In the present paper, a simple method to evaluate the energy stored during cold deformation of Fe-Si steels is presented. An estimate of the stored energy has then been made for various deformation modes (in-plane compression superimposed with shear stress). It is shown that, by adding a shear component, a more appropriate explanation of the experimental observed recrystallization textures may be obtained. The existence of shear components produces an inversion of the soft and hard orientations with respect to the case of a pure in-plane strain compression.

1. Introduction

It is now well established that the magnetic properties of the electrical steels strongly depend on their crystallographic texture and microstructure. The existence of a strong texture determines their magnetization behaviour (and especially at high values of applied external magnetic field) as well as the remanent induction value. Large values of grain size are also necessary to reach low values of the specific magnetic losses. These two features (strong texture and large grain size) are obtained in FeSi steels through a fabrication process, which comprises several steps: reheating of the slab, pre-rolling, final hot rolling, annealing of the hot band (optional), cold rolling and final annealing. The microstructure after cold rolling reflects mostly a complex deformation structure across the whole thickness, which is quite different from those of conventional steels. The ongoing structural changes during the final annealing step then strongly depend on the deformation structure obtained after heavy cold rolling. In this case, one usually observes in the very same sample, zones composed of deformation bands (which generally correspond to highly homogeneously deformed grains during cold rolling in plane strain compression) and zones between the deformation bands containing shear bands. Sometimes these shear bands are regions, where slip may be concentrated within one grain [1]. However, for large amounts of cold rolling, like for the regarded ferritic FeSi steels, the bands can cross several grains [1]. When hot rolling is supplemented by an additional heat treatment, either immediately after hot rolling or as a separate processing step before cold rolling, the amount of zones with shear bands after cold rolling is higher than for a material subjected to hot and cold rolling without intermediate annealing stage [2]. As a high amount of zones with shear bands has also been observed to lead to the final development of the good magnetic texture and to enhanced grain growth [2], it is important to understand the influence of the shear bands on the nucleation and growth process at recrystallization during final annealing.

It is well recognized that the difference in the energy stored during plastic strain in various crystallographic orientations presenting different microstructures is one of the major factors affecting the behaviour of each orientation during recrystallization [3,4]. This stored energy (SE) strongly depends on the local strain mode and exhibit an orientation dependence [5]. It is generally assumed that cold rolling can be regarded as a plane strain compression deformation mode. However, if the deformation is considered to be homogeneous plane strain compression through the sample, one cannot explain the complete resulting figure for the texture for the regarded feritic FeSi steels [2,3]. Therefore, it becomes important to study the effect of shear bands in addition to in plane strain compression on the stored energy and its orientation dependence. In the present paper, we will describe the features of the microstructure after cold rolling by means of two typical examples. We will then calculate, with the simple Taylor model, the effect of adding one shear component to uniform plane stress for specific texture components, and, using the Taylor factor as estimate of stored energy, we will propose a possible scenario for active recrystallization mechanisms during final annealing and their effect of the final textures.

2. Microstructure after cold rolling

Figure 1 shows some typical orientations maps measured by EBSD on the section perpendicular to TD after cold rolling, on two FeSi 2.4 samples. Details of the fabrication of the hot rolled and cold rolled materials are given in [2,6]. Sample 1 was rapidly quenched immediately after hot rolling, whereas sample 2 experienced a thermal treatment (TT) immediately after the last hot rolling pass. These two hot bands with a thickness of 2mm were then cold rolled to the same final thickness of 0.50 mm.

It is noticed that in these steels (unlike conventional steels), a rather inhomogeneous microstructure containing deformation and shear bands is generally obtained after rolling. Regions with low IQ, associated with sheared grains, alternate with regions of high IQ, comprising the deformation bands. But, although the two samples have experienced the same amount of hot and cold rolling, the zones with shear bands are broader in sample 2, which may be due to grain growth during the additional TT before cold rolling. The EBSD map for sample 1 shows well-defined deformation bands with very small zones containing shear bands in between, whereas in sample 2, we see quite broad zones containing shear bands. Some deformation bands are also present, also thicker than in sample 1.

From the inspection of these EBSD figures and of the associated data file, we can conclude that for both samples, the orientations associated with deformation bands generally belong to the θ fiber (see Figure 2 for definition) with a maximum of rotated cube oriented grains (in red in Figure 1). The orientations associated with shear bands belong mainly to the $\alpha*$ fiber (with a spread of colour between red and blue), and secondarily to the γ fiber (in blue), essentially close to the {111}<112> orientation. A high orientation spread is also present in these shear-banded zones, especially in sample 2, adding minor components to the description of the texture in these zones. We may now consider that the actual deformation of the deformation bands is closer to plane strain compression, while the sheared zones were additionally subjected to a shear component.

3. Estimation of the Taylor Factor for varying boundary conditions

In the case of different microstructures in differently oriented grains, we should choose, in order to evaluate the stored energy as a function of orientation, a model capable of predicting, from given macroscopically imposed boundary conditions, different states of stress and strain in each grain of a polycrystalline sample. For this purpose, we can use e.g. Crystal Plasticity Finite Element modelling, which can take into account the polycrystalline nature of the material and the initial texture, but which does not treat satisfactorily the question of different grain sizes or of the fragmentation of grains due to the formation of dislocation structures at very large strains. This limitation reduces significantly their predictive capacity and renders the associated computing efforts useless.

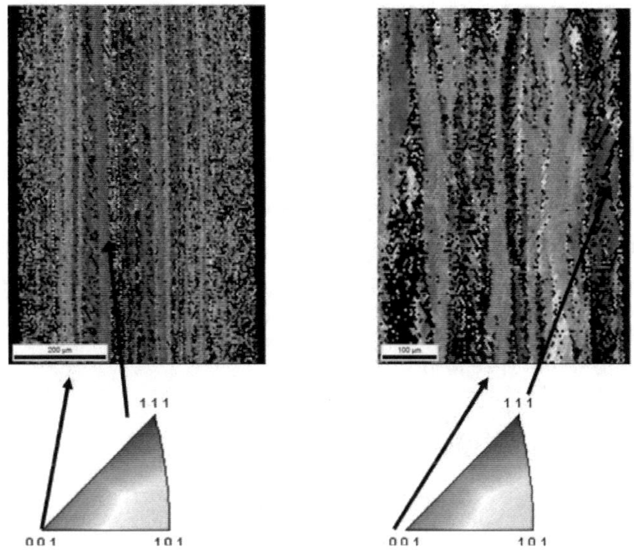

Figure 1: EBSD orientations maps obtained for the two samples (left sample 1 and right sample 2) after cold rolling (CR). The color indicates the orientation of the normal direction of the sample, according to the color code indicated by the triangles.

We thus adopt here a more direct approach, based on the use of the simple Taylor model and the recognition that the influence of the orientation on the stored energy can be indirectly assessed from the calculation of the Taylor factor from the final orientations after cold rolling. It has indeed been shown that, for a given model, these two quantities (stored energy and Taylor factor) evolve in the same way with crystal orientation [5]. The Taylor model is based on the main hypothesis that each grain within a polycrystalline sample will be subjected to the very same strain rate tensor as the one macroscopically imposed. As a consequence, the strain rate tensor becomes totally imposed at the level of the grain, and thus, each grain can then be treated separately. It becomes then easy to study the influence of varying boundary conditions imposed to each grain on the response of differently oriented grains, and to reproduce in a simple way the fact that within the same sample, different grains experience different strain tensors (with more or less shear) according to their direct environment and orientation – dependent resistance.

If we assume that cold rolling does correspond to pure plane strain compression (PSC), the strain rate tensor $\dot{\boldsymbol{\varepsilon}}_g$ imposed to each grain labelled g can then be written in the macroscopic reference frame linked to the rolling process (RD, TD, ND)[1] as:

$$\dot{\boldsymbol{\varepsilon}}_g = \dot{\varepsilon}_0 \begin{pmatrix} 1 & 0 & 0 \\ 0 & 0 & 0 \\ 0 & 0 & -1 \end{pmatrix} \quad (1)$$

In this equation, $\dot{\varepsilon}_0$ is the macroscopic strain rate along RD. But, if an additional shear component is added to account for the observations of in-grain shear zones, this strain rate tensor becomes

$$\dot{\boldsymbol{\varepsilon}}_g = \dot{\varepsilon}_0 \begin{pmatrix} 1 & 0 & K \\ 0 & 0 & 0 \\ K & 0 & -1 \end{pmatrix} \quad (2)$$

[1] RD = Rolling Direction, TD = Transverse Direction and ND = Normal Direction (normal to the rolling plane).

with $K = \dot{\varepsilon}_{13}/\dot{\varepsilon}_{11}$, the ratio characterizing the relative amount of additional shear (imposed along the rolling direction in the rolling plane, as observed). Plastic strain within each grain then takes place by slip on several systems, each labelled by index s. If we adopt a rate-dependent formalism to describe plastic slip, a widely accepted phenomenological viscoplastic power law relates within each grain g the shear rate $\dot{\gamma}_g^s$ on each slip system to the associated resolved shear stress τ_g^s as

$$\dot{\gamma}_g^s = \dot{\gamma}_0^s \left(\frac{\tau_g^s}{\tau_{0_g}^s}\right)^n \quad (3)$$

In this expression, n characterizes the material rate sensitivity[2]. The parameter $\dot{\gamma}_0^s$, usually called a reference shear rate, will be considered the same for all systems in all grains (i.e. $\dot{\gamma}_0^s = \dot{\gamma}_0$). The reference shear stress of slip system s in grain g, $\tau_{0_g}^{\ s}$ normally evolves with strain [7]. Also, $\tau_g^s = \sigma_{g_{ij}}.R_{g_{ij}}^s = S_{g_{ij}}.R_{g_{ij}}^s = \boldsymbol{S}_g.\boldsymbol{R}_g^s$ is the associated resolved shear stress on system s, $\boldsymbol{S}_g$ being the symmetrical deviatoric stress tensor derived from the stress tensor $\boldsymbol{\sigma}_g$ and $\boldsymbol{R}_g^s$ the orientation tensor of system s in grain g. The components of this last tensor read

$$R_{g_{ij}}^s = \frac{1}{2}\left(n_{g_i}^s b_{g_j}^s + n_{g_j}^s b_{g_i}^s\right) \quad (4)$$

with $\vec{n}_g^s$ and $\vec{b}_g^s$ characterizing respectively the slip plane normal and the slip direction for system s in grain g. The components of the strain rate tensor $\dot{\boldsymbol{\varepsilon}}_g$ in grain $\boldsymbol{g}$ are then expressed as:

$$\dot{\varepsilon}_{g_{ij}} = \sum_s \dot{\gamma}_0 \left(\frac{\tau_g^s}{\tau_{0_g}^s}\right)^n R_{g_{ij}}^s = \sum_s \dot{\gamma}_0 \left(\frac{\boldsymbol{S}_g.\boldsymbol{R}_g^s}{\tau_{0_g}^s}\right)^n R_{g_{ij}}^s \quad (5)$$

This equation represents a unique relationship between strain rate and deviatoric stress tensors at the level of grain $\boldsymbol{g}$, which allows to determine both tensors $\dot{\boldsymbol{\varepsilon}}_g$ and $\boldsymbol{S}_g$ from the boundary conditions imposed to the considered grain. In this framework, the Taylor factor is then defined as

$$M_g = \frac{\dot{\boldsymbol{\varepsilon}}_g.\boldsymbol{S}_g}{|\dot{\varepsilon}_g||\tau_0^s|} = \frac{\dot{W}_g}{|\dot{\varepsilon}_g||\tau_0^s|} = \frac{\sum_s \dot{\gamma}_0 \left(\frac{\boldsymbol{S}_g.\boldsymbol{R}_g^s}{\tau_{0_g}^s}\right)^n \boldsymbol{S}_g.\boldsymbol{R}_g^s}{|\dot{\varepsilon}_g||\tau_0^s|} \quad (6)$$

In other words, this factor is equal to the plastic work rate $\dot{W}_g$ divided by a normalization strain rate $|\dot{\varepsilon}_g|$ and by a normalization reference stress $|\tau_0^s|$ (taken here as the average on all systems of the references shear stresses τ_0^s reached after severe cold rolling). This normalization renders M_g dimensionless, which allows the influence of orientation on the capacity of deforming easily or not to be assessed, for a given model and given boundary conditions. Thus, when it is calculated at the end of the cold rolling process, it does not depend much on the strain path, on the actual strain rate or strain tensor or on the hardening state of the orientation [5].

In order to calculate it, two families of slip systems are selected to account for the usually mentioned slip capacities of carbon steels, the 12 {110}<111> and 12 {112}<111> systems and value of the initial (annealed state) reference stress τ_0^s is taken equal for the 24 systems in all grains, and as its evolution with strain is neglected, this stress simply becomes

$$|\tau_0^s| = \tau_0 \quad (7)$$

[2] By using only odd values of the exponent n, we avoid in the formulation the use of absolute values. Thus, the more classically chosen value of 20 will be replaced here by 21.

IOP Conf. Series: Journal of Physics: Conf. Series **1270** (2019) 012007 doi:10.1088/1742-6596/1270/1/012007

The reference strain rate $|\dot{\varepsilon}_g|$ is taken here equal to the von Mises equivalent strain rate

$$|\dot{\varepsilon}_g| = \dot{\varepsilon}_{VM} = \sqrt{\frac{2}{3}\sum \dot{\varepsilon}_{ij}\dot{\varepsilon}_{ij}} = \dot{\varepsilon}_0\sqrt{\frac{4}{3}(1+K^2)} \tag{8}$$

Then, by taking $\dot{\varepsilon}_{VM} = \tau_0 = 1$ and $n = 21$, the Taylor factor has been calculated for a discrete set of orientations describing the $\varphi_2 = 45°$ section of the Euler space, which contains all the main orientations composing the rolling and annealing textures. Some typical values are presented in Figure 2, in the $\varphi_2 = 45°$ section, which contains most of the information related to the rolling and annealing textures.

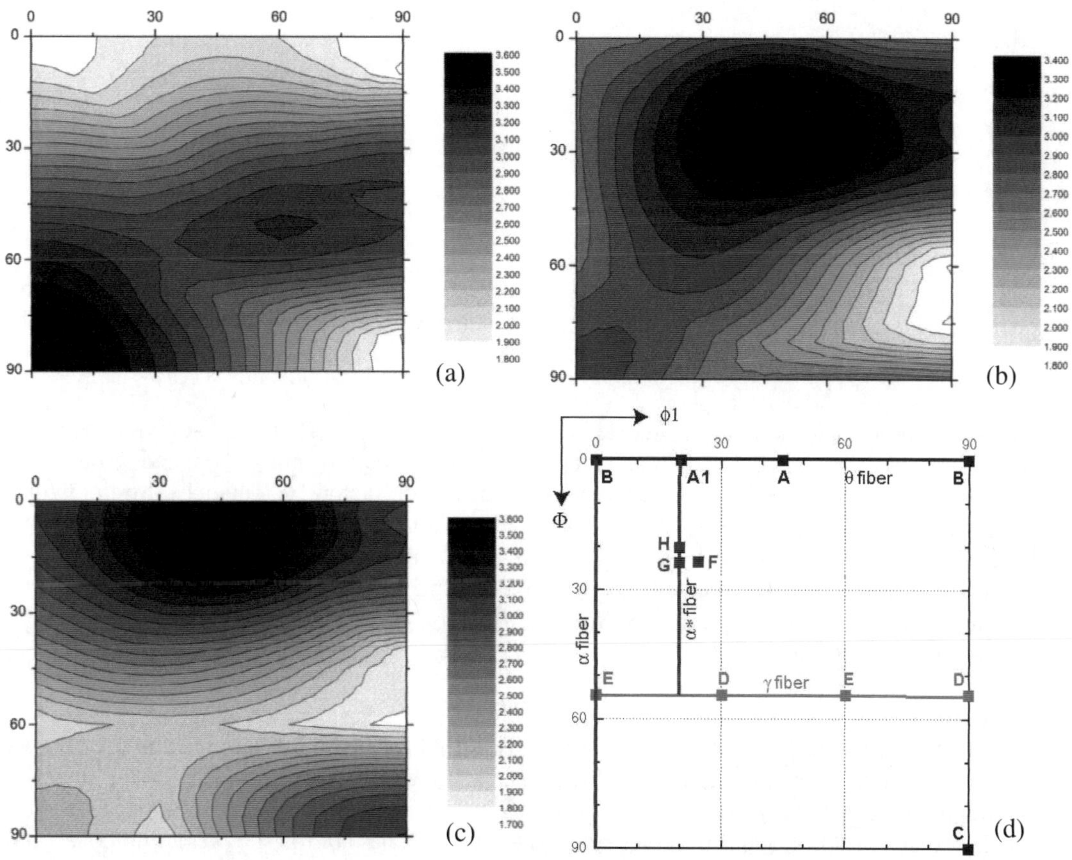

Figure 2. Predicted M_g values in the $\varphi_2 = 45°$ section for (a) $K = 0$ (plane strain compression), (b) $K = 1$, (c) $K = 5$ and (d) principal orientation fibers of interest: α = {hkl}<110> , α^* = {h,1,1}<1/h,1,2>, γ = {111}<uvw>, θ = {100}<uvw>.

In the case of non-oriented electrical steels, a low intensity of the α = {hkl}<110> and γ = {111}<uvw> fibers and a high intensity of the θ = {100}<uvw> fibre texture (and especially the Cube or Rotated Cube components) are desirable. Additionally, a high intensity of the α^* = {h,1,1}<1/h,1,2> fiber is also beneficial for a better magnetization behaviour [8].

When adding more and more shear to plane strain compression, there is a shift of the position of the maximal and minimal values in the Euler Space. For plane strain compression ($K = 0$), the orientations associated with a maximal value of the Taylor factor, i.e. the "hard" orientations, are the {110}<110> orientation (Euler angles (0,90,45)), which is generally absent from the rolling or annealing textures of steels, and also orientations located along the γ fiber. The softest components are the rotated Cube (B) and Goss (C) which are not always present in the deformation textures. When $K \geq 5$, there is a complete inversion, since the rotated Cube (B) and Goss (C) orientations become the hardest, whereas the {110}<110> orientation and the γ fiber become the softest components.

Between these two extreme cases, if we consider only the components of interest for magnetic properties, it is clear from Figure 3, which represents the evolution of M_g with K along the fibers of interest, that there is a shift of hard and soft orientations for K between 0.5 and 1. It is interesting to note, that although the maximal value of M_g is reached for the Cube orientations for $K > 1$, the components of the α* fiber are indeed very close.

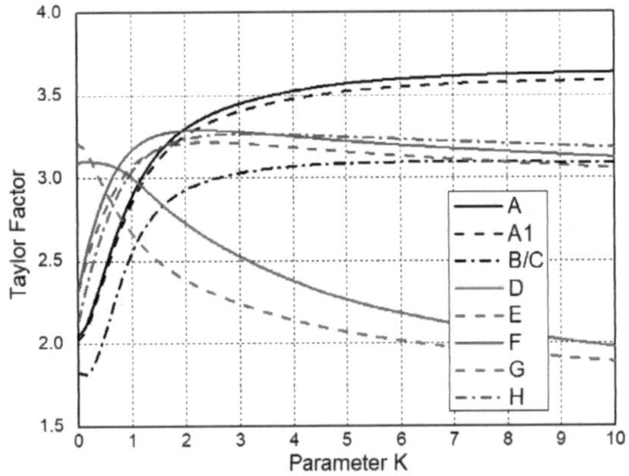

Figure 3. Taylor Factor of specific texture components as a function of the K parameter.

Now, if we accept that there is a link between Taylor factor and energy stored during cold deformation, it may be hypothesized that recrystallization occurs by nucleation and growth after deformation in the orientations associated with a high Taylor factor [4,5] and by Strain Induced Boundary Migration (SIBM) in the orientations associated with a low Taylor factor [4]. Of course, this simple reasoning does not allow to predict which one will be more active when two possible mechanisms are simultaneously expected. The relevance of this hypothesis can now be investigated by taking a closer look at the textures obtained after deformation and annealing and more particularly at the components with high shear.

4. Discussion

We thus examine some experimental data obtained after cold rolling and final annealing for the two FeSi2.4 samples, and especially the recrystallization texture at the very first steps of the annealing process, i.e. before extensive grain growth appears. The ODFs calculated from EBSD measurements performed on maps of 500 * 500 μm (representing the total thickness of the samples) after cold rolling and annealing at two different temperatures during 20s are presented in Figure 4 for the two samples.

Before checking the validity of our hypothesis on these data, it is worth recalling that the CR textures (although associated with different microstructures) measured before annealing were quite similar for both samples and mainly composed of two strong fibers α = {hkl}<110> and γ = {111}<uvw> quite classical for these steels [2]. The α* components found within the shear bands can be considered as part of the spread around these fibers. It has also been shown that the evolution of the microstructures during recrystallization was quite different within the deformation bands (in which recovery was seen first) and within the sheared zones (in which nucleation was much more active than in deformation bands) [2]. This has been shown clearly for sample 2, in which the shear zones are large and the process of nucleation more active. On the contrary, due to the small size of these shear bands in sample 1, nuclei were not visible after annealing at 700°C during 20s. For annealing at 760°C during 20s, new grains were visible in both samples, even in deformation bands in which these new grains can be produced by recovery and growth of subgrains. These already published observations, as well as the textures presented in Figure 4 are indeed consistent with the expected mechanisms listed in Table 1.

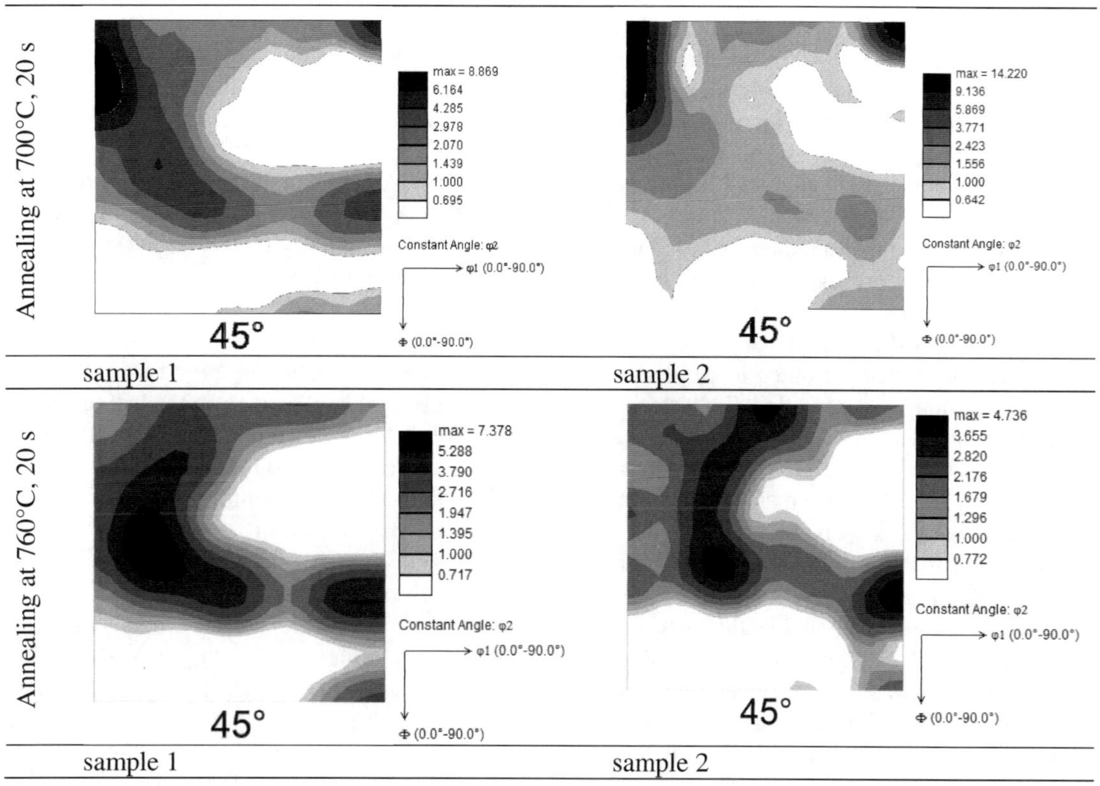

Figure 4. ODF φ_2=45° section for samples 1 and 2 after annealing at 700°C for 20s (top) and at 760°C for 20s (bottom); (total cross section).

Table 1. Expected favoured recrystallization mechanisms in different zones of the CR samples 1 and 2, as deduced from the Taylor factor calculations.

Sample	1	2
Deformation bands ($K = 0$)	Rot. Cube (Low M_g) = > SIBM	Rot. Cube (Low M_g) = > SIBM
Sheared zones ($K \cong 1$)	γ fiber (Low M_g) = > SIBM	(1) α* fiber (High M_g) = > Nucleation (2) γ fiber (Low M_g) = > SIBM

The observed evolution of texture for samples 1 and 2 after annealing at 700°C for 20s and at 760°C for 20s (see figure 4) can be summarized as follows:

- for annealing at 700°C, we observe a limited texture evolution in sample 1 with an increase of the Rotated Cube component. This can be explained by SIBM around this component during annealing for 20s. SIBM, also expected in the small {111} shear zones, will necessarily imply smaller changes of the ODF value due to the spread of orientations along the fibre. Nucleation, hardly visible in sample 1, could be completely absent at this stage, as deduced from our analysis. For sample 2, we observe an increase of the rotated Cube component which could again be due to SIBM in deformation bands, whereas nucleation of the α^* components within the shear bands could be responsible for the increased spread of orientations around the γ and α^* fibers.

- for annealing at 760°C, nucleation of minor α^* components within the sheared zones occurs in sample 1. The texture observed in sample 2 confirms the growth of the α^* nuclei.
- the Goss component becomes also visible within the texture of both samples, especially after annealing at 760°C. This could also be due to SIBM around this minor component, as already observed in some other non-oriented Fe-Si steels [4].

It is worth noting that the annealing conditions (annealing temperature and time) may also have a significant effect on the occurrence of both mechanisms (nucleation, SIBM). This is not considered in this case. The present analysis has also neglected the effect of temperature and strain gradients during processing, which also play an important role on the developing microstructures. Especially, additional shear is generally more present within the surface region of the sample. Fluctuations of Taylor factor across the thickness of the samples should also be considered to interpret more precisely the recrystallization processes. This will be addressed in a forthcoming paper.

5. Conclusions and perspectives

The present paper has presented a simple method to evaluate the distribution of stored energy due to cold deformation in cold rolled FeSi steel. Starting from different hot bands, quite distinct rather inhomogeneous deformation substructures have been observed after cold rolling. An estimate of the stored energy has then been made for various deformation modes (in-plane compression superimposed with shear stress). It has been shown that adding a shear component produces an inversion of the soft and hard orientations with respect to plane strain compression. By considering that the deformations bands were deformed by in-plane strain compression while the sheared zones were subjected to an additional shear component, it has been possible to explain the texture evolution at low annealing temperatures, where recrystallization is dominant and grain growth has not started. Recrystallization seems to occur by different processes: SIBM associated with low values of M within the deformation bands (subjected to PSC) and nucleation associated with high values of M, within the shear bands (subjected to PSC + Shear). Further research work is then desirable to clarify the mechanisms and driving forces during normal grain growth in these ferritic FeSi Steels and especially the interplay between recrystallization and grain growth, which appears across the thickness of the sheets.

References

[1] W B Hutchinson, *Deformation Substructures and Recrystallisation.* Materials Science Forum, 2007. **558-559**: p. 13-22.

[2] A Franke, J Schneider and R Kawalla. *Evolution of Texture at Final Annealing by Recrystallization and Grain Growth in Non-Oriented Ferritic FeSi.* in *WMM18, 8th International Conference on agentism and Metallurgy*. 2018. Dresden, Germany.

[3] J J Sidor, K Verbeken, E Gomes, J Schneider, P R Calvillo and L A I Kestens, *Through process texture evolution and magnetic properties of high Si non-oriented electrical steels.* Materials Characterization, 2012. **71**: p. 49-57.

[4] F Gregori, K Murakami and B Bacroix, *The influence of microstructural features of individual grains on texture formation by strain-induced boundary migration in non-oriented electrical steels.* Journal of Materials Science, 2014. **49**(4): p. 1764-1775.

[5] B Bacroix, A Miroux and O Castelnau, *Simulation of the orientation dependence of stored energy during rolling deformation of low carbon steels.* Modelling and Simulation in Materials Science and Engineering, 1999. **7**(5): p. 851-864.

[6] J Schneider, A Stöcker, R Kawalla and A Franke, *Evolution of Optimum Grain Size for Low Loss Ferritic FeSi Steels.* Steel Research International, 2017. **87**.

[7] B Bacroix and R Brenner, *A phenomenological anisotropic description for dislocation storage and recovery processes in fcc crystals.* Computational Materials Science, 2012. **54**: p. 97-100.

[8] J Schneider, A Franke, A Stöcker and R Kawalla. *Quantification of the Effect of Texture on the Magnetization Behavior. in Conference Magnetic Measurements.* 2017. Prague.

IOP Conf. Series: Journal of Physics: Conf. Series **1270** (2019) 012008 doi:10.1088/1742-6596/1270/1/012008

Influence of grain size before cold rolling on development of crystallographically equivalent main texture component in recrystallization texture of 3% Si-Fe

T Imamura and Y Hayakawa

Steel Research Laboratory, JFE Steel Corporation, Kawasakidori 1-chome, Mizushima, Kurashiki, Okayama 712-8511, Japan

E-mail: t-imamura@jfe-steel.co.jp

Abstract. 3%Si-Fe sheet produced by high cold rolling reduction over 80% has strong {111}<112> recrystallization texture. This orientation has two crystallographically equivalent orientations such as (111)[1 -2 1] and (111)[-1 -1 2]. However, there are a few previous works in which these orientations are investigated separately. In the present study, the recrystallization behavior of the crystallographically equivalent orientations of {111}<112> was investigated using two samples with different grain sizes before cold rolling. Both recrystallized samples had strong {111}<112> component texture. The sample with a small grain size before cold rolling had almost the same fraction of each equivalent orientations, while the sample with a large grain size had substantially different fractions of each equivalent orientations. Spatial distribution of two crystallographically equivalent orientations in the sample with a small grain size before cold rolling was almost homogeneous, however that in the sample with a large grain size before cold rolling was inhomogeneous. It is considered that these differences are explained by the assumption that a certain large grain before cold rolling will generate only single crystallographically equivalent orientation after cold rolling and annealing.

1. Introduction

Texture control is very important for improving mechanical or magnetic properties of several kinds of steel products. Therefore, rolling and annealing conditions are usually controlled precisely. For example, <111> // ND (normal direction of a sheet plane) component in the texture of final product provides good deep drawability [1, 2]. Therefore, this component is suitable for steel sheets for automobile. In the case of electrical steels, since <100> direction can be magnetized easily, this direction is aligned with an excitation direction on the steels. Texture control is applied not only to final process but also to intermediate processes. In grain-oriented electrical steel sheets, {111}<112> component is needed on a primary recrystallization texture in order to obtain a strong {110}<001> (Goss) texture after secondary recrystallization. There are many studies which attempt to explain a reason why {111}<112> component is important for the strong Goss texture development [3-13]. For example, Hayakawa et al. proposed a mechanism that grain boundaries having misorientation between 20° and 45° in primary recrystallized sheet play an important role in a selection of secondary recrystallization textures [3-6]. Since {111}<112> orientation has 35° of misorientation toward Goss orientation, the {111}<112> component in the primary recrystallization texture is highly effective for Goss texture development in a final product. Since <111>//ND component, including {111}<112> orientation, is important as described above, there are many studies which are trying to clarify a formation mechanism of this component. Abe et al.

mentioned that recrystallized grains with {111} were preferentially formed in initial grain boundary regions [14]. Senuma et al. [15] and Inagaki [16] obtained almost same results independently that {111} recrystallized grains were formed from grain boundary regions of deformed grains. Furthermore, other results [17,18] demonstrating that small initial grain size gave rise to a strong <111>//ND recrystallization texture implied that the nucleation would occur on initial grain boundaries. However, Vanderschueren et al. showed that <111>//ND nuclei did not generate only at grain boundaries, they also formed in the interior of deformed γ-fiber grains [19]. Also, Barnett confirmed that nuclei having near {554}<225> orientation, which is considered to be one of <111>//ND component, generated by in-grain shear banding [20]. In view of rolling reduction on recrystallization texture, a high cold rolling reduction seems to result in a {111}<112> texture development during recrystallization and grain growth [21, 22], whereas {111}<110> orientation become major in the case of a relatively low rolling reduction [14, 23].

{111}<112> orientation has two crystallographically equivalent orientations such as (111)[1 -2 1] and (111)[-1 -1 2] in rolling process. However, there are a few previous works in which these orientations are investigated separately. In the present study, the generation behavior of the crystallographically equivalent orientations of {111}<112> was investigated using two samples with different grain sizes before cold rolling.

2. Experiment

3.1 mass%Si-0.1mass%Mn-Fe alloy ingot was hot-rolled to 2.3 mm thickness after heating to 1473 K (1200 °C). Then the two hot-rolled samples were heated to 1173 K (900 °C) and 1373K (1100 °C) respectively for 30 seconds under nitrogen atmosphere in order to obtain different recrystallized grain size. Both heated samples were cold-rolled to 0.23 mm thickness with the cold rolling reductions rate of 90%. Then an annealing was performed on the cold rolled samples at 943K (670 °C) without soaking time and at 1073K (800 °C) for 10 seconds under nitrogen atmosphere. The samples subjected to the annealing at 943 K (670 °C) were hardly recrystallized, whereas the samples subjected to the annealing at 1073 K (800 °C) were fully recrystallized. This procedure is shown in Table 1.

An optical microscope was used to obtain the microstructure of the samples before cold rolling in order to evaluate their grain size. The orientation distribution function (ODF) of the fully recrystallized samples was calculated from 110, 200 and 211 pole figures measured by X-ray diffraction (XRD) method using Rigaku RINT-2000. XRD measurement was performed on the center layer of the samples. The ODF calculation was performed using the discrete method [24] with orthorhombic lattice system. In this paper, ODF representation of $\phi_2 = 45°$ cross section in Bunge notation [25] is used.

Table 1. Experimental procedures of this study.

Ingot with 3.1 mass%Si-0.1mass%Mn-Fe

↓

Heating at 1473K (1200°C) for 1hour
then hot rolling to 2.3mm

↓

Soaking for 30 seconds

at 1173K (900°C) <sample A>		at 1373K (1100°C) <sample B>	
↓		↓	
Cold rolling to 0.23mm <sample Ac>		Cold rolling to 0.23mm <sample Bc>	
↓ Annealing		↓ Annealing	
at 943K (670°C) without soaking time <sample A-1>	at 1073K (800°C) for 10 seconds <sample A-2>	at 943K (670°C) without soaking time <sample B-1>	at 1073K (800°C) for 10 seconds <sample B-2>

Microstructures and grain orientations of the annealed samples were investigated by electron backscatter diffraction (EBSD) method using scanning electron microscope (SEM, JEOL JSM-7001F) and EBSD data analysis software (TSL OIM Analysis 7). EBSD measurement was carried out on the surface (perpendicular to normal direction) of the fully recrystallized samples and the cross section (perpendicular to transverse direction) of the samples subjected to the annealing at 943 K (670 °C).

3. Results and discussion

The microstructures of the heated samples before cold rolling are shown in figure 1. Both samples are fully recrystallized and have relatively equiaxed grains. Sample A subjected to heat treatment at 1173 K (900 °C) has small grain size, sample B subjected to heat treatment at 1373 K (1100 °C) has huge grain size.

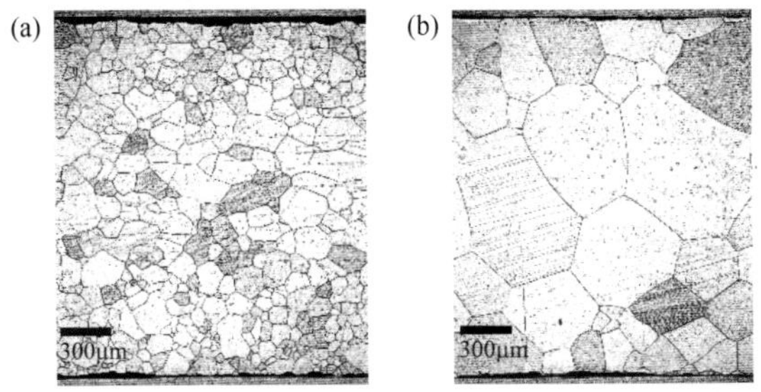

Figure 1. Optical micrographs of sample A (a) and B (b) (TD cross section).

Figure 2 shows textures of cold-rolled (sample Ac, Bc) samples. The cold-rolling textures have strong α-fiber and {111}<110> components, which are well known as major components in a deformation texture in the rolling process [26, 27], and are very similar each other. It is generally accepted that the initial grain size has little influence on cold-rolling texture [14, 26]. Same tendency was also observed in this study. In the recrystallization texture in sample A-2 and B-2, both textures had main component of near {111}<112> orientation, which is frequently observed in recrystallization textures of low carbon steels and silicon steels.

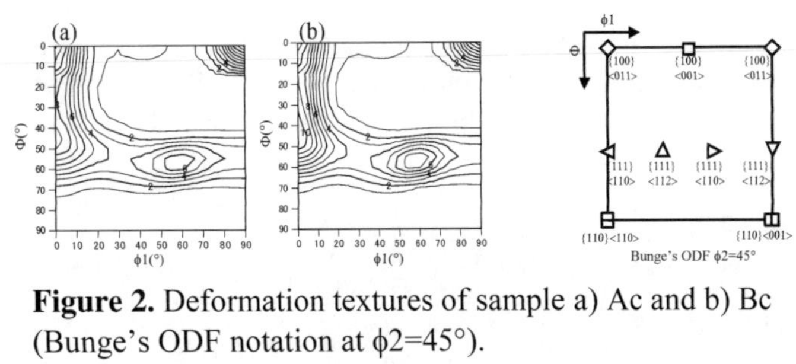

Figure 2. Deformation textures of sample a) Ac and b) Bc (Bunge's ODF notation at ϕ2=45°).

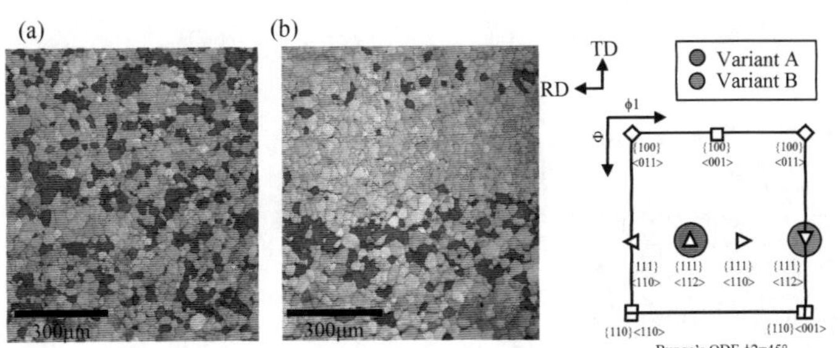

Figure 3. Orientation Image Map of Variant A (colored in red) and B (colored in green) with 15° tolerance. a) sample A-2, b) sample B-2.

IOP Conf. Series: Journal of Physics: Conf. Series **1270** (2019) 012008 doi:10.1088/1742-6596/1270/1/012008

{111}<112> orientation has two crystallographically equivalent orientations (111)[1 -2 1] and (111)[-1 -1 2] in rolling process. In order to investigate spatial distributions of each crystallographically equivalent orientations separately, crystal orientation analysis was performed on the fully recrystallized samples by EBSD. The result is shown in Figure 3. In this analysis, two crystallographically equivalent orientations were named as Variant A and B. Variant A and B are defined as the orientations represented as Euler angles (ϕ1, Φ, ϕ2) = (30, 55, 45) and (90, 55, 45) in ODF space, respectively. Figure 3(a) shows Image Quality (IQ) map of sample A-2. Recrystallized grains having Variant A and B were colored in red and green respectively with 15° tolerance and are placed uniformly. In the case of sample B-2, however, spatial distribution of Variant A and B is inhomogeneous as shown in Figure 3(b).

In order to evaluate spatial distribution precisely, nearest neighbor analysis was applied [28]. In this analysis, the degree of spatial dispersion in the distribution is evaluated on minimum distances between two points with a numerical value called as the nearest neighbor index (NNI), which is calculated using a following equation,

$$\mathrm{NNI} = \frac{Ad}{Ed} \tag{1}$$

where Ad is the average of nearest neighbor distance defined as,

$$Ad = \frac{1}{n}\sum_{i=1}^{n} di \tag{2}$$

where n is the number of points and di means the nearest neighbor distance of i-th point. Ed is the expected value of the average nearest distance in random distribution. Ed is defined as,

$$Ed = \frac{1}{2\sqrt{n/S}} \tag{3}$$

where S is the area that the points exist.

NNIs of the grains with Variant A on both sample A-2 and B-2 were calculated with above equations using the measured EBSD data including grain positions and measurement area. In this analysis, the number of intended grains of Variant A is approximately 280 in both samples. The calculated NNIs were 0.943 and 0.868, respectively. According to the nearest neighbor analysis, a distribution is

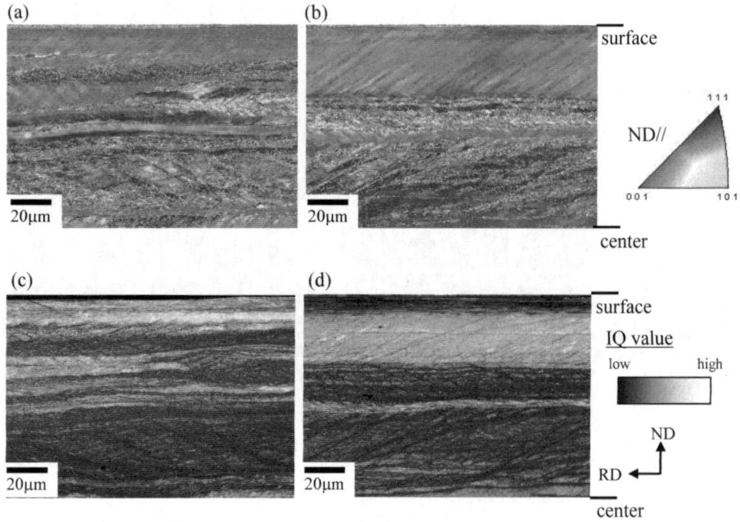

Figure 4. IPF maps and IQ maps of sample A-1 (a) and c)) and B-1 (b) and d)). (TD cross section)

classified three types such as random, clustered and dispersed patterns [29]. When NNI is near 1, the distribution of points is almost random pattern. When NNI is lower than 1, the distribution is clustered. Consequently, it is statistically determined that the grain distribution of Variant A on sample A-2 had almost random pattern and that on sample B-2 was clustered.

In order to explain a reason why the spatial distributions of the recrystallized grains were different with the grain size before cold rolling, microstructures and their crystal orientation distributions in deformed state, which was prior to recrystallization, were investigated using sample A-1 and B-1 by EBSD method. Figure 4 shows invers pole figure (IPF) maps and IQ maps obtained with the EBSD measurement on the cross section of the samples. Each IPF is parallel to the normal direction (ND). There are many elongated deformed band structures toward the rolling direction (RD) in sample A-1 as shown in Figure 4 (a) and (c). On the other hand, sample B-1 has a few elongated deformed band structures as shown in Figure 4 (b) and (d). Thickness of them is very high, which is considered to reflect the grain size before cold rolling. IPF maps (in Figure 4 (a) and (b)) display orientations of these deformed band structures for both samples. Orientation of these band structures mainly consist of <111>//ND (colored in blue) and <100>//ND (colored in red), which are well known as popular components in a heavily cold-rolled specimen [26, 27]. Many previous works have revealed that recrystallized grains that have {111}<112> orientation mainly generate in deformed grains with <111>//ND orientation [30-33]. Hence, deformed grain with <111>//ND orientation in the sample which had large grain size before cold rolling in Figure 4 (b) were examined in detail. Figure 5 shows the particular region of Figure 4(b) with <111>//ND. There are many small band-like areas with low IQ value (shown with arrows in figure 5 (a)), which implies strain accumulation. Many of these areas have Variant A as shown in figure 5 (b). Moreover, the deformed band structure has Variant B. In order to investigate the orientation relationship between the small band-like area and the deformed band structure, these orientations in selected areas shown with a rectangle in Figure 5 are represented by pole figures in Figure 6. These orientations have a common <211> axis as shown near the center of the 211 pole figure. This fact implies that crystal orientation rotation around the <211> common axis would occur. Similar orientation rotation relationship has been reported by Muraki et. al [31].

Since the many small band-like areas had the same orientation of Variant A, this orientation rotation probably occurred in the entire region of the deformed band structure shown in figure 5. Assuming the small band-like areas are main candidates of recrystallization nuclei, recrystallized grains with Variant A will dominate after the annealing. Hence, the special distribution differences of crystallographically

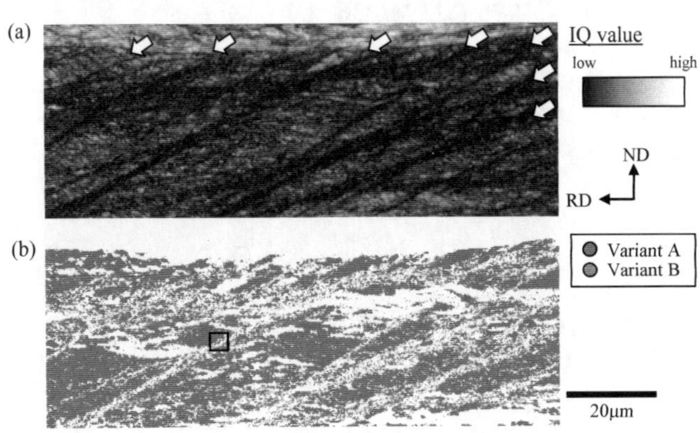

Figure 5. IQ map (a) and orientation image map (b) of sample B-1. In the orientation image map, Variant A and B were colored in red and green with 15° tolerance, respectively. Arrows show several small band-like areas. Black rectangle shows the area that is used for the pole figures in Figure 6.

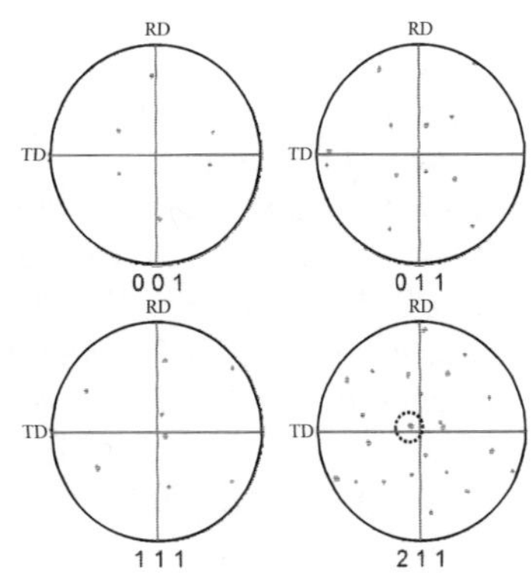

Figure 6. Pole figures of orientations in local area shown in Figure 5. Each color is correspond to the kind of the variants.

equivalent orientations of <111>{112} can be explained by the assumption that a certain large grain before cold rolling will generate only single crystallographically equivalent orientation after cold rolling and annealing.

Conclusion

The recrystallization behavior of the crystallographically equivalent orientations of {111}<112> was investigated using two samples with different grain sizes before cold rolling. Both recrystallized samples had strong {111}<112> component texture. But spatial distributions of two crystallographically equivalent orientations of {111}<112> in the recrystallized samples were different. It is considered that these differences are explained by the assumption that a certain large grain before cold rolling will generate only single crystallographically equivalent orientation after cold rolling and annealing.

References

[1] Katoh H, Takechi H and Nagashima S 1966 *J. Jpn. Soc. Technol. Plast.* **7** 13
[2] Barnett M and Jonas J 1997 *ISIJ Int.* **37** 706
[3] Hayakawa Y and Szpunar J 1997 *Acta Mater.* **45** 1285
[4] Hayakawa Y and Szpunar J 1997 *Acta Mater.* **45** 4713
[5] Hayakawa Y, Muraki M, Szpunar J 1998 *Acta Mater.* **46** 1063
[6] Hayakawa Y and Kurosawa M 2002 *Acta Mater.* **50** 4527
[7] Yoshitomi Y, Ushigami Y, Harase J, Nakayama T, Masui H and Takahashi N 1994 *Acta Metall. Mater.* **42** 2593
[8] Harase J, Shimizu R and Takahashi N 1990 *Acta Metall. Mater.* **38** 1849
[9] Shimizu R, Harase J and Dingley D 1990 *Acta Metall. Mater.* **38** 973
[10] Harase J, Shimizu R and Dingley D 1991 *Acta Metall. Mater.* **39** 763
[11] Homma H and Hutchinson B 2003 *Acta Mater.* **51** 3795
[12] Lee S, Hwang N, Han C and Yoon D 1998 *Scr. Mater.* **39** 825
[13] Park H, Kim D, Hwang N, Joo Y, Han C and Kim J 2004 *J. Appl. Phys.* **95** 5515
[14] Abe M, Kokabu Y, Hayashi Y and Hayami S 1980 *J. Jpn. Inst. Met. Mater.* 44 84
[15] Senuma T, Yada H, Shimazu R and Harase J 1990 *Acta Mater.* **38** 2673
[16] Inagaki H 1994 *ISIJ Int.* **34** 313
[17] Abe M, Kokabu Y, Hayashi Y and Hayami S 1982 *Trans. Jpn. Inst. Met.* **23** 718
[18] Hutchinson W 1989 *Acta Metall.* **37** 1047
[19] Vanderschueren D, Yoshinaga N and Koyama K 1996 *ISIJ Int.* **36** 1046
[20] Barnett M 1998 *ISIJ Int.* **38** 78
[21] Hashimoto N, Yoshinaga N and Senuma T 1998 *ISIJ Int.* **38** 617
[22] Verbeken K, Kestens L and Jonas J2003 *Scr. Mater.* **48** 1457
[23] Barnett M and Kestens L 1999 *ISIJ Int.* **39** 923
[24] Pawlik K 1986 *Phys. Stat. Sol. (b)* **134** 447
[25] Bunge H 1965 *Z. Metallk.* **56** 872
[26] Hutchinson W 1984 *Int. Met. Rev.* **29** 25
[27] Hutchinson W 1999 *Phil. Trans. R. Soc. Lond. A* **357** 1471
[28] Bailey T and Gatrell A 1995 *Interactive Spatial Data Analysis* (Harlow Essex, England : Longman Scientific & Technical ; New York, NY : J. Wiley)
[29] P Clark and Evans F 1954 *Ecology* **35** 445
[30] Inagaki H 1984 *Trans. ISIJ.* **24** 266
[31] Muraki M and Toge T 1999 *Tetsu-to-Hagane* **85** 751
[32] Okada T, Sakaguchi M, Ashida O, Tagami M and Inoko F 2004 *Materials Transactions* **45** 415
[33] Duggan B, Tse Y and Liu G 2010 *Materials Science and Technology* **26** 404

IOP Conf. Series: Journal of Physics: Conf. Series **1270** (2019) 012009 doi:10.1088/1742-6596/1270/1/012009

Retaining {100} fibre texture in electrical steel via strain-induced boundary migration recrystallisation.

M Ji, C Davis and C Slater

Warwick Manufacturing Group, University of Warwick, Coventry, CV4 7AL, UK

Correspondence: m.ji.1@warwick.ac.uk

Abstract: In this paper, the feasibility of retaining the preferred {100} fibre texture for non-grain oriented (NGO) electrical steel from the as-cast columnar {100} grains by encouraging recrystallisation via the strain induced boundary migration (SIBM) mechanism was investigated. Rolling with intermediate annealing stages before the final anneal was used to reduce the stored energy through recovery before the final recrystallisation step. A strong {100} fibre recrystallisation texture, i.e. 32% area fraction, was seen in the sample which was warm deformed to 1.6 strain with 5 intermediate annealing stages then annealed at 950 °C for 5 minutes (to observe the early stages of recrystallisation) or 1 hour. Whereas only 12 - 14% {100} fibre component was observed after 1.2 strain with a single or with no intermediate annealing stage. It was found that the {100} fibre recrystallisation texture was formed in regions adjacent to parental deformed grains with {100} fibre texture due to the SIBM recrystallisation mechanism. EBSD imaging analysis was used to follow the recrystallisation process in the same microstructural region using interrupted heat treatments and it was seen that the {100} texture grains carried very low stored energy after deformation, which meant the subgrains in these grains bulged into the surrounding grains with higher stored energy. Due to its slow recrystallisation nucleation rate, a low overall stored energy is preferred to encourage {100} texture recrystallised grains to form via the SIBM recrystallisation mechanism.

1. Introduction

Texture control is crucial for electrical steel production. For non-grain oriented (NGO) electrical steels, the {100} fibre texture is desirable to achieve optimum magnetic properties. It is well known that the {100} fibre is typically a minor component after recrystallisation when the subgrain growth (SGG) mechanism occurs due to its low nucleation rate and growth rate [1]. Strain-induced boundary migration (SIBM) is a recrystallisation mechanism where a subgrain with a low dislocation density will bulge into one with higher dislocation density and the dislocations will be swept up during bulging [2]. Hutchinson observed SIBM in bicrystal iron with {100}<011> subgrains growing into the neighbouring grain after 72 % cold reduction [3]. Takashima reported a similar observation in partially recrystallised NGO electrical steel where unrecrystallised grains with {001}<210> texture, in the central region of the sheet, grew into the surrounding matrix after light rolling (12% reduction) [4]. Stojakovic also reported that the {100}//ND texture was strengthened by SIBM in electrical steel via light rolling (12 % reduction) and intermediate annealing at 760 °C [5]. Although numerous studies have been carried out examining texture evolution during recrystallisation in electrical steels, with a general consensus that the {100} fibre texture can be enhanced from low strain and / or intermediate annealing prior to recrystallisation, no direct observation of SIBM during recrystallisation has been reported. In this paper, the recrystallisation behaviour in a high Si content (Fe-6 wt.% Si) has been considered, with direct observation of the recrystallisation mechanism: as-cast thin slab with columnar grains with a strong {100} orientation was produced, warm deformed with intermediate annealing, then final annealed.

2. Experiment

Fe-6 wt. % Si was produced in a vacuum induction melting furnace and then cast to 15 mm thick slab, which resulted in material with columnar grains of approximately 1.2 mm width with a strong {100} orientation. No equiaxed region was observed. This microstructure is representative of that present in belt cast or twin roll cast material [6]. Cylindrical compression samples, measuring 8 mm in diameter and 12 mm in height, were machined from the columnar grains by electrical discharge machining. The uniaxial compression axis was parallel to the columnar grain growth direction, i.e. the {100} direction. Compression tests were carried out in a Gleeble HDS-V40 simulation machine. The samples were heated to the deformation temperature of 450 °C at a heating rate of 20 °C/s, then warm deformed to 1.2 or 1.6 strain, with a strain rate of 1 s^{-1}, in one, two or five passes, with intermediate annealing at 750 °C for 30 seconds between each pass for the multiple pass deformations. Due to the compression load limitation, 1.2 strain could only be achieved for the one / two deformation pass samples, yet 1.6 strain was achieved for the sample with five passes and intermediate annealing, which is more representative of the commercial condition. Final annealing was carried out at 950 °C for either 5 or 60 minutes in an argon atmosphere. A section perpendicular to the compression axis was prepared for microstructure and texture determination. The samples were ground and polished with a Buehler AutoMet polisher. Electron backscattered diffraction (EBSD) analysis was performed in a JEOL 7800 FEG-SEM with an OXFORD instrument EBSD system to examine the micro texture evolution. The step size was 10 µm, and a 15-degree tolerance was allowed to calculate the ideal orientations. The same areas in each sample were examined after the sequential deformation and annealing steps to enable texture evolution measurement.

3. Results and discussion

The microstructure of the samples, which were warmed deformed to strain 1.2 with one or two passes, are shown in figures 1 and 2 respectively. Areas of misindexing are seen in the sample with one pass, figure 1 (a), which is attributed to its high stored energy. The microstructures after 5 minutes of annealing at 950 °C for both samples (same location as deformed image) have also been examined, figure 1 (b) and 2 (b) and show full recrystallisation; the recrystallised grain sizes are inhomogeneous in both samples, which is due to the varying local strain concentration in the deformed material in the initially different oriented grains (based on their Taylor factor), which agrees with literature observations [7].

The microstructure and local misorientation maps of the as-warm deformed (strain 1.6), partially recrystallised samples after holding at 950 °C for 5 and 60 minutes are shown in figure 3. Local misorientation has been used as an indication of the stored energy level and the deformed microstructure, figure 3 (a) and (b) shows strain inhomogeneity with the grain boundary regions being more deformed than the grain interior. The coarse grain at the top left side of the micrographs shows lower local misorientation, i.e. much less stored energy, compared to the finer grains in the centre, which is consistent with previous observations on the effect of grain size and orientation on stored energy after deformation: where small grains are known to have higher stored energy than larger grains; and {111}//ND grains (blue grains) have a higher stored energy than the {100}// ND grains (red grains) [7, 8]. Therefore, it is expected that the grain boundary regions and the fine grain regions with a {111}//ND orientation will recrystallise quicker than the other grains.

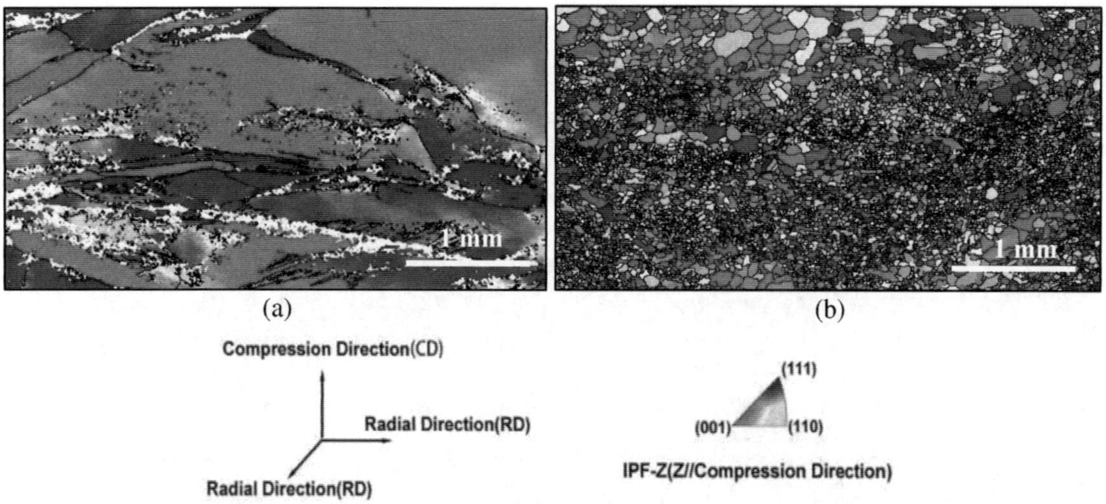

Figure 1. EBSD IPF Z images of samples deformed to 1.2 strain in one pass (a) and recrystallised (b)

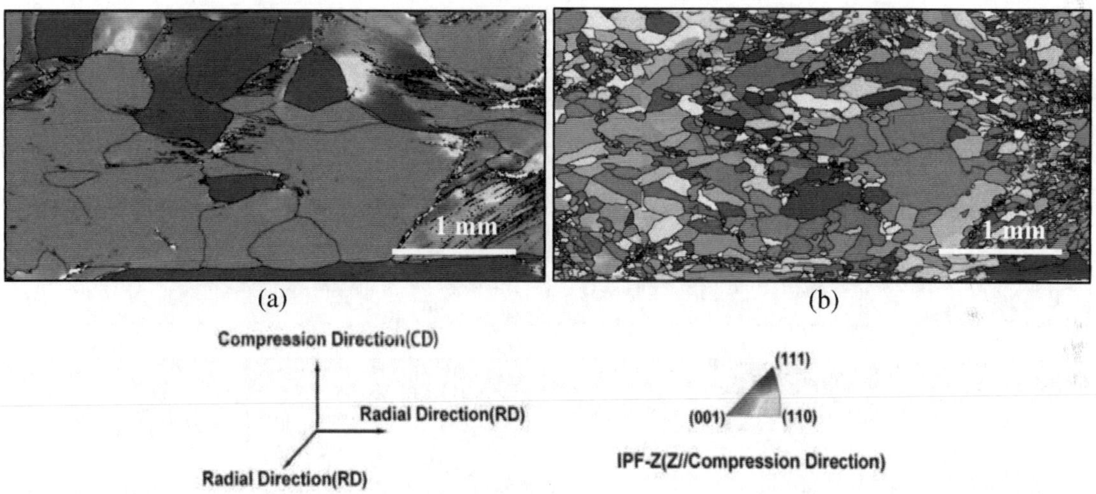

Figure 2. EBSD IPF Z images samples deformed to 1.2 strain in two passes (a) and recrystallised (b)

At the early stage of recrystallisation (5 minutes), it can be seen from figures 3 (c) and (d) that the more heavily deformed regions start to be replaced by recrystallising grains. In region 1, the recrystallising grains are close to {111}//ND and {011}//ND (i.e. coloured blue and green respectively). The area along the grain boundaries in region 2 has also recrystallised, but with a smaller recrystallised fraction and the recrystallised grains are close to a {100}//ND orientation (mostly red colour). It is generally considered that {100} texture grains are the least favourable to form on recrystallisation in bcc material due to their low nucleation rate. However, it can be seen that a high fraction of {100} texture grains have nucleated along the grain boundaries in figures 3 (c) and (d), where eleven recrystallising grains with an orientation of {100}//ND are present compared to ten recrystallising grains with a {111}//ND orientation, which is generally reported to have a high nucleation rate during recrystallisation [8].

It is therefore suggested that the recrystallisation nucleation rate of {100}//ND grains is comparable to that of grains with a {111}//ND texture in the 6 wt.% Si steel on annealing in this case. This observation of similar number of recrystallisation nuclei for the different textures was seen in other regions that were examined as well as having been reported in the literature for electrical steel after similar deformation conditions [9]. After annealing for 60 minutes, over 70% of the sample has recrystallised, although the coarse grain on the top left side of the micrograph remains unrecrystallised due to its low stored energy, figure 3(e) and (f). It can be seen in region 3 that the grain with a {100}//ND orientation is bulging into the neighbouring grain, this is indicative of SIBM; this can also be seen in other regions, (indicated by arrows in figure 3(e)).

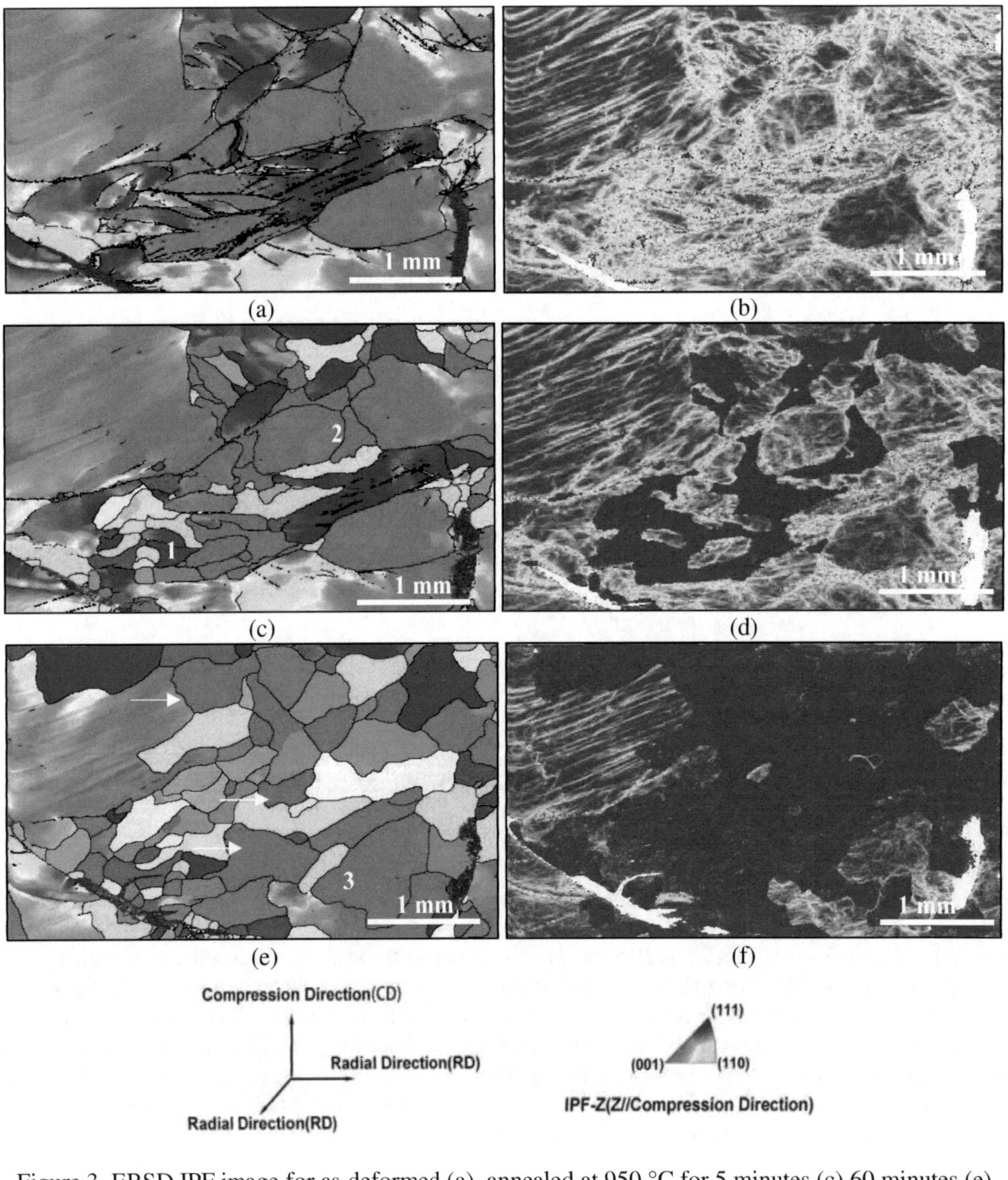

Figure 3. EBSD IPF image for as-deformed (a), annealed at 950 °C for 5 minutes (c) 60 minutes (e) Local misorientation mapping for as-deformed (b), annealed at 950 °C for 5 minutes (d) 60 minutes (f)

The recrystallised grain size is significantly finer (mode: 100 - 200 μm) in the sample after one / two passes than the sample with five passes (mode: 800 μm) despite the larger total strain imposed, which is attributed to their stored energy difference at the point of recrystallisation (reducing the stored energy by recovery via intermediate annealing). Only 12 - 14% area fraction of {100}//ND texture is present in the recrystallised samples without / with one intermediate annealing step, compared to 32% of {100}//ND texture shown in the sample with five intermediate annealing steps, summarised in table 1.

Table 1. Summary of texture fraction evolution in the samples with various number of passes.

Number of passes	Deformed		Recrystallised	
	{100}	{111}	{100}	{111}
1	40%	31%	14%	11%
2	41%	25%	12%	11%
5	37%	11%	32%	12%

The decreasing amount of {111}//ND texture present in the one / two pass compared to the five passes samples is attributed to the extent of recovery during deformation [10]. The low percentage of both {100}//ND and {111}//ND in the one / two passes samples suggested that recrystallisation did not retain or favour either texture in these cases, i.e. random recrystallisation texture is produced. It is noted that to achieve the desired final grain size for NGO electrical steel magnetic performance (typically 150 μm) a refined starting grain size would be required if low stored energy SIBM recrystallisation is desired to achieve high {100} texture. In summary, it is suggested that the low stored energy sample, i.e. five passes, promoted the formation of {100}//ND texture during recrystallisation via SIBM, with the high stored energy sample (deformed with one or two passes) showing a random recrystallised texture, which is consistent with sub-grain growth recrystallisation.

Hutchinson pointed out that the recrystallisation texture is dependent on both the deformation texture and the grain boundary mobility of the recrystallising nuclei [3]. Therefore, the recrystallisation texture is a compromise between the frequency of available nuclei and their respective growth velocities. In a material starting with {100} fibre texture, low strain, i.e. low stored energy, promotes the retention of the {100} texture in two ways: firstly, the low stored energy decreases the recrystallisation nucleation rate for all nuclei, which allows the formation of nuclei with less favourable texture, such as {100}//ND. The misorientation angles among subgrains decreases with decreasing stored energy, thus both the mobility of the subgrain boundaries and the recrystallisation driving force difference among nuclei decreases. As a result, recrystallisation nuclei with a lower misorientation to the deformed microstructure can grow during recrystallisation. Secondly, to promote the SIBM mechanism during recrystallisation, a low stored energy gradient is required. The gradient after recovery is sufficient to ensure the nuclei with {100}//ND texture bulge into the deformed neighbouring grains, but is low enough not to form new recrystallisation nuclei itself (i.e. does not favour the sub-grain growth recrystallisation mechanism). In this work intermediate annealing at 750 °C for 30 seconds between deformation passes was sufficient to allow recovery to reduce the stored energy and promote SIBM recrystallisation. Further work is required to define the processing conditions range to allow SIBM recrystallisation.

4. Conclusion

The {100}//ND texture was promoted via SIBM recrystallisation mechanism. The {100} texture grains, which carried very low stored energy after deformation, bulged into the surrounding grains during recrystallisation. Due to its slow recrystallisation nucleation rate, a low overall stored energy is preferred to encourage {100} texture recrystallised grains to form via SIBM recrystallisation mechanism. The SIBM recrystallizing grains have been directly observed from the deformed {100} grains using interrupted EBSD imaging.

Acknowledgments
The authors would like to thank WMG at the University of Warwick for the provision of research facilities, and EPSRC for the research funding support.

References

[1] Liu H, Wang Y, An L, Wang Z, Hou D, Chen J and Wang G 2016 *J. Magn. Magn. Mater.* **420** 192

[2] Beck P A and Sperry P R 1950 *J. Appl. Phys.* **21** 150

[3] Hutchinson W B 1989 *Acta Metall.* **37** 1047

[4] Takashima M, Komasubara M and Morito N 1997 *ISIJ Int.* **37** 1263

[5] Stojakovic D, Doherty R, Kalidindi S and Landgraf F 2008 *Metall. Mater. Trans. A* **39A** 1738

[6] Liu H, Li H, Wang H, Liu Y, Gao F, An L, Zhao S, Liu Z and Wang G 2016 *J. Magn. Magn. Mater.* **406** 149

[7] Holscher M, Raabe D and Lucke K 1991 *Steel Res. Int.* **62** 567

[8] Ji M, Davis C and Strangwood M 2019 *Metals* **9** 369

[9] Shan N, Liu J, Sha Y, Zhang F and Zuo L 2019 *Metall. Mater. Trans. A* **50A** 2486

[10] Barnett M and Kestens L 1999 *ISIJ Int.* **39** 923

IOP Conf. Series: Journal of Physics: Conf. Series **1270** (2019) 012010 doi:10.1088/1742-6596/1270/1/012010

In-situ observations of dislocation recovery and low angle boundary formation in deformed aluminium

H W Ånes[1], A T J van Helvoort[2] and K Marthinsen[1]

[1]Department of Materials Science and Engineering, Norwegian University of Science and Technology, N-7491 Trondheim, Norway

[2]Department of Physics, Norwegian University of Science and Technology, N-7491 Trondheim, Norway

E-mail: `hakon.w.anes@ntnu.no`

Abstract. An experimental study of the recovery of dislocations and low angle boundary formation in aluminium is presented. By combining *in-situ* annealing with orientation mapping in the transmission electron microscope, maps of geometrically necessary dislocation estimates derived from orientation measurements and subgrain formation can be quantitatively analysed. A thin foil of a commercially pure aluminium alloy cold-rolled to a true strain of $\varepsilon = 2.3$ and annealed *in-situ* in four steps of increasing temperatures from 170 °C to 560 °C was studied. An increase in the subgrain size and low angle boundary misorientation was accompanied by a halving of the dislocation density from $1.2 \times 10^{16}\,\mathrm{m}^{-2}$ to $0.6 \times 10^{16}\,\mathrm{m}^{-2}$. Limited boundary migration was observed and the increased subgrain size was attributed to the dissolution of dislocations within the low angle boundaries upon annealing.

1. Introduction

When an f.c.c. metal with a high stacking fault energy such as aluminium is deformed to moderate to large strains, energy is stored mostly in dislocation networks in cell walls and within the cells [1]. Upon annealing recovery may occur, leading to the rearrangement of the dislocation networks into sub-boundaries and eventually into a fully developed subgrain structure [2]. For a subgrain to become a viable recrystallization nucleus it needs a size advantage and a high local misorientation to its surroundings [3]. The main mechanisms of nucleation of recrystallization are thought to involve growth of subgrains by low angle boundary migration in an orientation gradient or strain induced boundary migration (SIBM) of existing boundaries. However, the dislocation recovery often required before boundary migration can occur is complex and not very well understood and a full experimental investigation requires a three-dimensional *in-situ* technique able to resolve dislocations [4].

With the developments of automated crystal orientation mapping in the transmission electron microscope (TEM) [5] and calculation of geometrically necessary dislocation (GND) densities from orientation differences in neighbouring map points [6], details of the deformation structure in highly deformed materials can be characterised [7]. In the present work we combine this characterisation with *in-situ* annealing. The objective is to provide measurements of the dislocation recovery and the mechanisms of low angle boundary (LAB) formation and migration in deformed Al, useful for development of recovery and recrystallization models.

2. Experimental

The investigated material was commercial purity Al with a composition of 0.5wt.%Fe and 0.15wt.%Si. The material was direct chill cast and a sample from the billet's centre region was homogenised at 450-600 °C for 120 h and water quenched. The sample had an initial grain size of 58 µm and was cold rolled to a true strain of $\varepsilon = 2.3$.

The hardness and average subgrain size of the deformed sample were determined, the former from a Vickers hardness instrument with a 1 kg load. The subgrain size was determined from electron backscatter diffraction (EBSD) maps from the sample's centre region viewed along the transverse direction (TD). Each map covered an area of 100×100 µm^2 with a step size of 0.1 µm, acquired on a Zeiss Ultra 55 scanning electron microscope operated at 20 kV and equipped with a NORDIF UF1100 detector. EBSD patterns were indexed with the EDAX TSL DC 7.3 software.

Thin foils from the centre region viewed along the normal direction, defined by the rolling direction (RD) and transverse direction, were prepared for TEM observations by mechanical grinding and electropolishing in a Struers Tenupol 3 twin jet polishing unit. The unit was operated at 20 V using an electrolyte of parts 1:2 nitric acid:methanol kept at −30 °C. A JEOL 2100F TEM operated at 200 kV with a Gatan HHST4004 heating holder was used. Scans of precession electron diffraction (PED) patterns were acquired using the NanoMEGAS ASTAR system to map the dislocation recovery and formation and migration of LABs during annealing. In the experiment presented here, a thin foil was annealed in steps according to the procedure in figure 1. Scans were obtained in between the steps at room temperature to ensure that the foil stayed at eusentric height during acquisition.

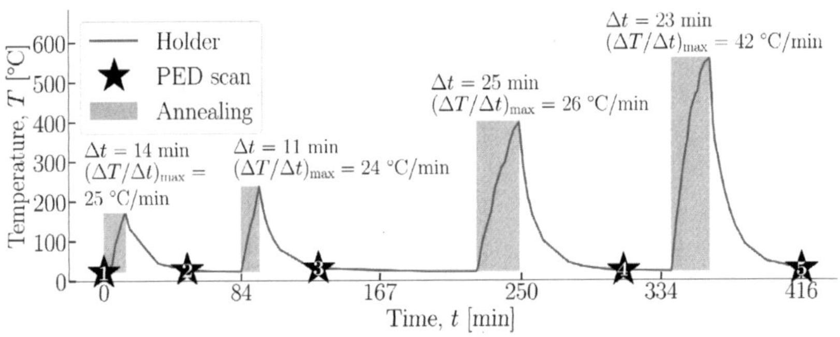

Figure 1. Annealing history for the *in-situ* TEM observation. The annealing interval Δt and max. heating rate $(\Delta T/\Delta t)_{\mathrm{max}}$ for each step and times for PED scans (#1-5) are indicated.

PED patterns were obtained with the microscope operated in nanobeam diffraction mode with a probe convergence semi-angle of 1 mrad and a precession angle of 0.5° and frequency of 100 Hz. A Stingray camera recorded the patterns off of the microscope's fluorescent screen. Five scans of an 8×8 µm^2 region of interest (ROI) with step sizes of 33-40 nm were obtained at times indicated in figure 1. Geometric distortions due to projection optics, shifts of the direct beam and background noise in the patterns were corrected for using the pyXem 0.7 software [8]. NanoMEGAS' template matching was used for pattern indexing.

Orientation data was processed and analysed with the MTEX 5.2 software [9]. Orientations were lightly averaged with the half quadratic filter [10] and grains were reconstructed with a misorientation angle threshold of 0.75° and 0.5° for data from EBSD and PED, respectively. The quality of the orientation averaging and grain reconstruction was assessed by inspection of maps with a quaternion based disorientation colour key [11]. The GND densities were estimated with MTEX based upon work by Pantleon [6]. The $\{111\}\langle110\rangle$ slip system was assumed active with 12 edge dislocations with line energies $u_{\mathrm{e}} = 1$ and 6 screw dislocations with $u_{\mathrm{s}} = 1-\nu = 1-0.347$, where ν is the Poisson ratio for Al.

3. Results

The deformed sample had a Vickers hardness of 51 and the average subgrain size determined from EBSD was 0.84 µm. Bright field (BF) TEM images of the ROI before annealing and after the final annealing step in figure 2 show that annealing led to the rearrangement of dislocation tangles into boundaries and a more developed subgrain structure. No recrystallization was observed. The numbered (1-3) coloured dots highlight the same areas. Some subgrains disappeared during recovery, like the dark grains of high diffraction contrast below (1) and (3) in (a). Other subgrains developed narrower and more defined boundaries to adjacent subgrains, like the subgrain in (2). Note the development of stable boundary sections with 120° angles near the subgrain highlighted by the arrow in (b).

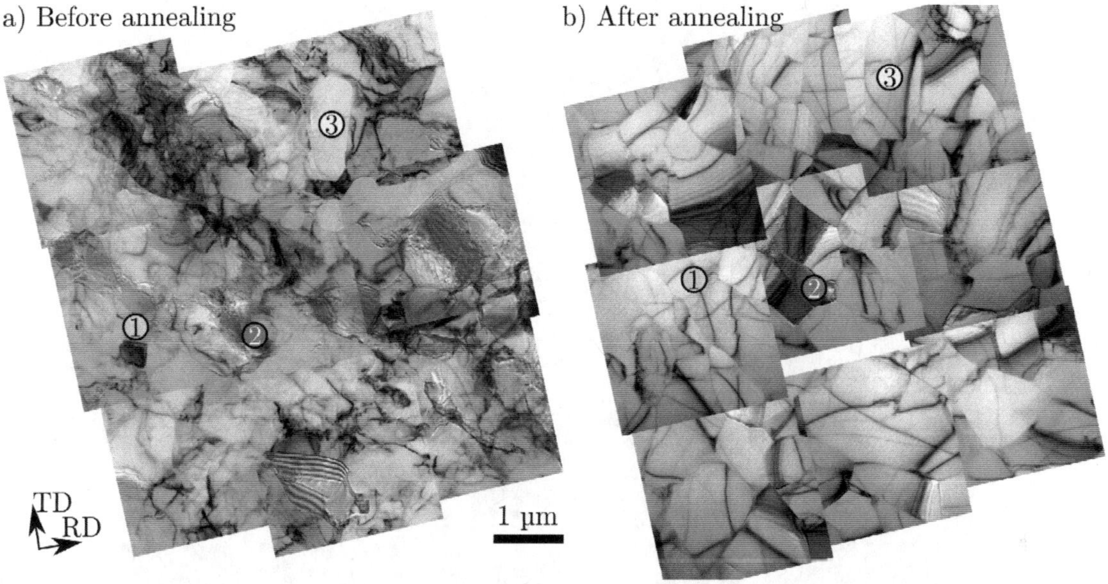

Figure 2. BF TEM images of the ROI (a) before annealing and (b) after the final annealing step. The numbered dots (1-3) indicate the same positions.

Results from the five PED scans are shown in figure 3, with the scale bar in (a) applying to all maps. The correlation index maps indicate the degree of correlation between experimental and simulated diffraction patterns from template matching, with higher correlation in brighter regions. The grain boundary (GB) and disorientation colouring maps show LABs and high angle boundaries (HABs) with misorientations $0.5° < \theta < 15°$ and $> 15°$, respectively. The average grain orientation is grey and deviations from this orientation follow the disorientation colour key as shown in the map in (a). This representation reveals multiple cells or subgrains with boundaries of misorientations below 0.5° that were not reconstructed, clearly demonstrating its usefulness when characterising deformation substructures. The dislocation density maps clearly show the dislocation recovery taking place upon annealing.

Three observations are highlighted in the shapes in figure 3, the first of the subgrain surrounded by the ellipse in (a) with an initial HAB to a large subgrain below and a LAB to its parent grain above. After the 1st annealing step both its boundaries straighten out and the high dislocation density inside appears to be annealed out. However, the dislocation density increases again after the 2nd annealing step and the LAB dissolves after the 3rd annealing step. Simultaneously, while the HAB misorientation stays at 60° the LAB misorientation decreases from 4° before annealing to 1° after the 2nd annealing step and below 1° after the 3rd step.

IOP Conf. Series: Journal of Physics: Conf. Series **1270** (2019) 012010 doi:10.1088/1742-6596/1270/1/012010

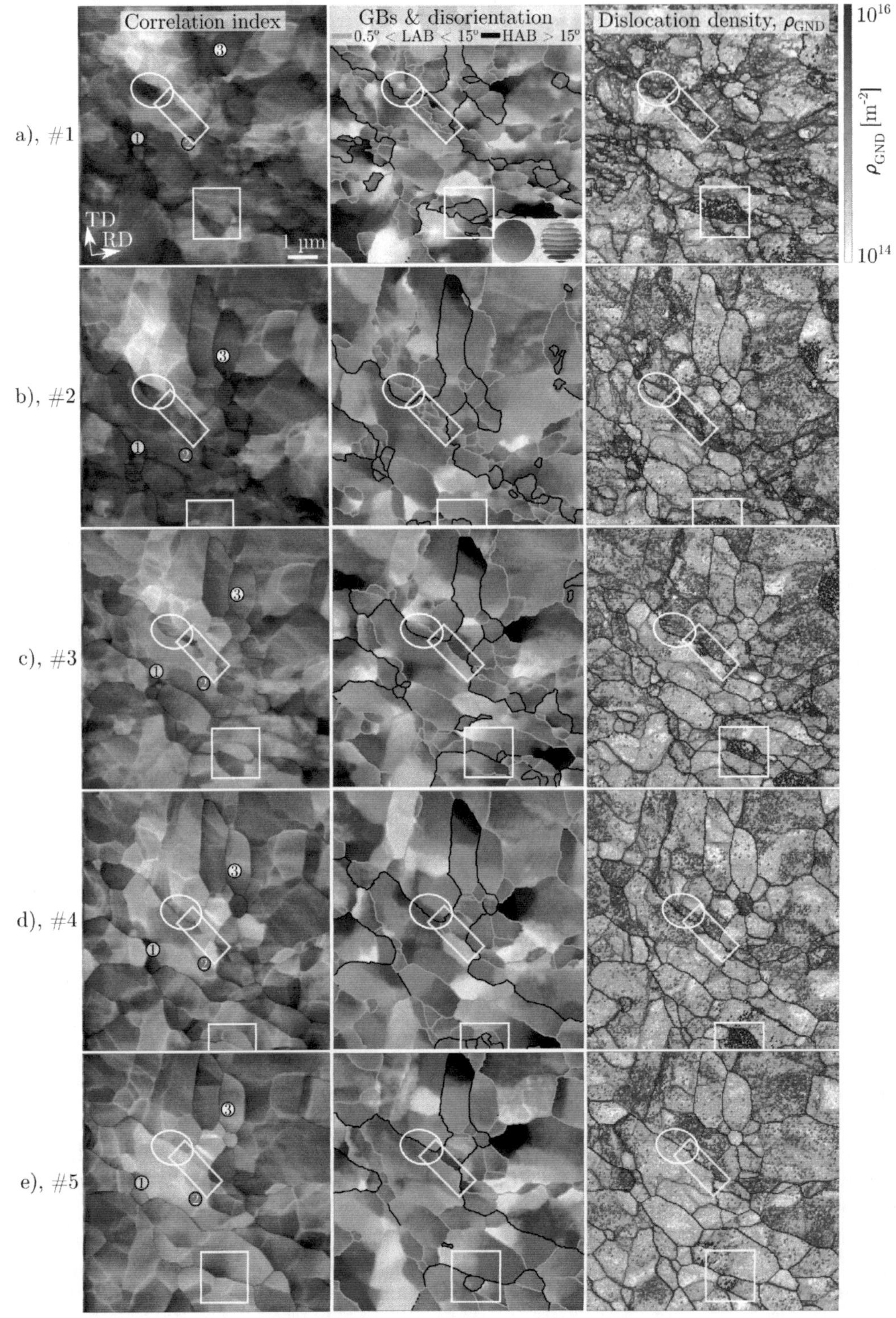

Figure 3. Substructure evolution from the PED scans (#1-5). (Left) Maps of correlation index, (middle) grain boundaries and disorientation colouring and (right) dislocation densities ρ_{GND} (a) before annealing, after (b) the 1st, (c) 2nd and (d) 3rd annealing step and (e) after the final annealing step. See figure 2 and the text for descriptions of dots and shapes, respectively.

After the final step the subgrain is gone and the HAB has moved upwards. Note also in (e) that the parent grain obtains an increased dislocation density during the final step. The second observation is of a group of subgrains surrounded by the rectangle in (a) that upon annealing disappear and leave behind a well defined LAB in (c). The misorientation of this boundary increases at the same time from 6° to 11° after the 2nd annealing step and decreases somewhat to 8° after the final step. The third observation is of a region surrounded by the rectangle in (a) of high dislocation density with a HAB to a subgrain of lower dislocation density above. Upon annealing this region shrinks in size and after the final annealing step it is entirely inside the above subgrain.

The evolution of the average subgrain size δ_{ECD}, given by the equivalent circular diameter (ECD), misorientation θ across LABs and dislocation density ρ_{GND} determined from the PED scans are shown in figure 4 (a-c), respectively. The change in misorientation with subgrain size is shown in (d), with a linear fit with a slope of 10. The subgrain size was observed to increase from 0.42 µm to 0.62 µm, the LAB misorientation increased from 5.3° to 6.2° and the dislocation density halved from $1.2 \times 10^{16}\,\mathrm{m}^{-2}$ to $0.6 \times 10^{16}\,\mathrm{m}^{-2}$.

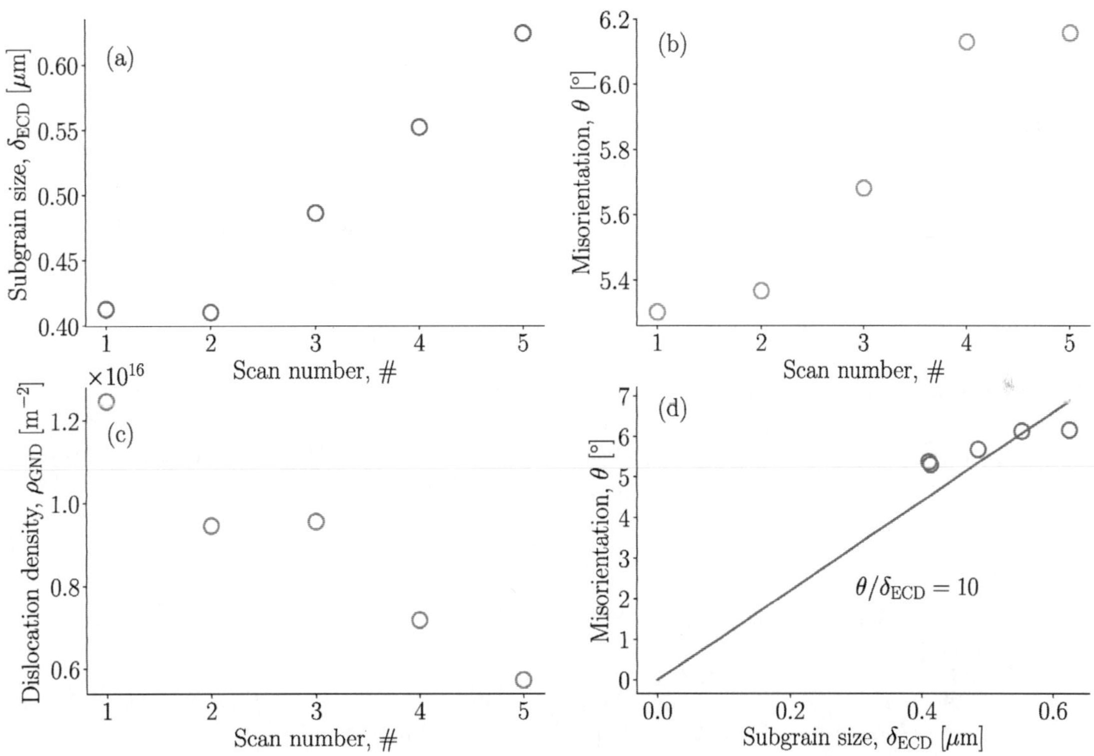

Figure 4. Evolution of substructure parameters from PED scans (#1-5) in figure 3. (a) The average subgrain size δ_{ECD}, (b) LAB misorientation θ, (c) dislocation density ρ_{GND} and (d) change in misorientation with subgrain size.

4. Discussion

Limited LAB migration is observed, and the increased subgrain size in figure 4 (a) is attributed to the dissolution of LABs upon annealing. The first two observations from the PED scans in the ellipse and rectangle in figure 3 indicate such dissolution, as reported in earlier works on a similar alloy by for example Sandström *et al.* [12]. In the first observation, the dislocations in the LAB appear to have been pulled out by the stress field in the adjacent larger subgrain interior, thus contributing to the increased dislocation density observed in this subgrain after the final annealing step. In the second observation, the dislocations in the LABs appear to have been absorbed by the nearby LAB that experienced an increased misorientation upon annealing. The migration of the HAB in the square in (a) is an example of SIBM, which also seems to be the operating mechanism in the HAB migration in the first observation.

The combination of scanning precession electron diffraction with *in-situ* annealing in the TEM provides details of the dislocation recovery and substructure of metals deformed to large strains not readily available from EBSD in the SEM, exemplified by the results presented here. However, care must be taken when comparing results from thin foils to results from bulk since microstructures experience slower growth upon annealing in thin foils. When performing recrystallization experiments the foil thickness must be chosen according to the amount of stored energy and the size of the substructure. Also, increased resolution leads in this case to reduced statistics. For example, the ROI in this experiment, chosen for its multiple HABs and thus an increased possibility for boundary migration to occur, most likely has a higher dislocation density than the sample overall. The observed dislocation densities $\propto 1 \times 10^{16}\,\mathrm{m}^{-2}$ are much higher than the densities $\propto 1 \times 10^{13}\,\mathrm{m}^{-2}$ reported previously for a similar alloy [13]. These aspects will be considered in future work expanding upon the presented methodology.

5. Conclusions

A commercial purity Al alloy cold rolled to a true strain of $\varepsilon = 2.3$ was *in-situ* annealed in steps in the TEM. Scans of precession electron diffraction patterns were collected to characterise the dislocation recovery and formation of LABs upon annealing. An increase in the subgrain size and LAB misorientation was accompanied by a halving of the dislocation density from $1.2 \times 10^{16}\,\mathrm{m}^{-2}$ to $0.6 \times 10^{16}\,\mathrm{m}^{-2}$. Limited LAB migration was observed and the increased subgrain size was attributed to the dissolution of LABs upon annealing.

Acknowledgments

The authors would like to thank Hydro Aluminium Sunndalsøra for providing the material. HWÅ acknowledges NTNU for financial support through the NTNU Aluminium Product Innovation Centre (NAPIC). The Research Council of Norway is acknowledged for the support of the Norwegian Centre for Transmission Electron Microscopy (NORTEM, Grant No. 197405).

References

[1] Humphreys F J, Rohrer G S and Rollett A D 2017 *Recrystallization and Related Annealing Phenomena* 3rd ed (Oxford: Elsevier)
[2] Nes E 1995 *Acta Metallurgica Et Materialia* **43** 2189–2207
[3] Doherty R D, Hughes D A, Humphreys F J, Jonas J J, Jensen D J, Kassner M E, King W E, McNelley T R, McQueen H J and Rollett A D 1997 *Materials Science and Engineering: A* **238** 219–274
[4] Humphreys F J 2004 *Materials Science Forum* **467** 107–116
[5] Rauch E F, Portillo J, Nicolopoulos S, Bultreys D, Rouvimov S and Moeck P 2010 *Zeitschrift fur Kristallographie* **225** 103–109
[6] Pantleon W 2008 *Scripta Materialia* **58** 994–997
[7] Ghamarian I, Liu Y, Samimi P and Collins P C 2014 *Acta Materialia* **79** 203–215
[8] pyXem developers 2019 pyXem v0.7 `https://github.com/pyxem/pyxem`
[9] Bachmann F, Hielscher R and Schaeben H 2010 *Solid State Phenomena* **160** 63–68
[10] Hielscher R, Silbermann C and Schmidl E 2018 *J. Appl. Crystallogr.-In Rev.*
[11] Thomsen K, Mehnert K, Trimby P W and Gholinia A 2017 *Ultramicroscopy* **182** 62–67
[12] Sandström R, Lehtinen B, Hedman E, Groza I and Karlsson S 1978 *Journal of Materials Science* **13** 1229–1242
[13] Nord-Varhaug K, Forbord B, Benestad J, Pettersen T, Rønning B, Bardal A, Benum S, Marthinsen K and Nes E 2000 *Materials Science Forum* **331** 1387–1392

IOP Conf. Series: Journal of Physics: Conf. Series **1270** (2019) 012011 doi:10.1088/1742-6596/1270/1/012011

Novel applications of centreless X-ray diffractometers for non-destructive pole figure measurements

M Sepsi, V Mertinger and M Benke

Institute of Physical Metallurgy, Metalforming & Nanotechnology, University of Miskolc, Miskolc-Egyetemváros H3515, Hungary

femsepsi@uni-miskolc.hu

Abstract. Centreless X ray diffractometers are specially designed for non-destructive residual stress measurements. The similarity between the X-ray diffraction-based stress and texture measurement methods led us to introduce a new method on non-destructive (sample cutting-free) stress measurement devices to studies on crystallographic texture. The method that determines texture characteristics from residual stress data has recently been introduced. Afterwards, a method with which pole figures (fully equivalent to conventionally measured pole figures) at high Bragg-angles can be constructed with centreless diffractometers were developed. The paper gives an illustration of the new methods' utilization on the field of archaeometry. In that area, only non-destructive testing modes are permitted. In the case of archaeological finds, knowing the crystallographic texture assists to reconstruct the ancient process technology (hammering, rolling, annealing) that was applied during the manufacturing. The new methods are suitable to be used on valuable and unique objects both in the automotive industry, archaeometry and in the space applications.

1. Introduction

For archeological investigations, only non-destructive testing methods are permitted [1-2]. Neutron diffraction is a commonly used method for examining archeological objects, although this characterization method is not easy to access, and the sample is activated during the test. Neutron diffraction gives information about the whole, or a large volume of the object [3]. If local investigation is needed, the use of X-ray diffraction is a good solution [4]. If the goal is to obtain information from a very high resolution (from small volume), electron back scattered diffraction (EBSD) is a commonly used test method [5]. In the latter cases, however, a sample must be cut. A new, non-invasive texture characterization technique based on X-ray diffraction using centreless diffractometers have been developed [6]. Such diffractometers are designed especially for residual stress measurements which causes displacement in crystalline lattice and results in a shift of Bragg-angles. Centreless diffractometers are able to detect the shifts, the peak intensity and full width half maximum precisely. For accuracy, the operational range for centreless diffractometer is typically at high Bragg-angles (2θ). Since tilting and rotation are required for both texture and stress measurements, it led to the idea of developing the texture test method for centreless diffractometers [7-8].

In metals, crystallographic textures develop after plastic deformation and heat treatments. One of the most common ways to characterize texture is to measure pole figures. Pole figures indicate the angular distribution of given crystallographic planes on a reference coordinate system.

In the industrial practice it is common to optimize the technological parameters based crystallographic texture for better material properties.. In archeometry, the interpretation of texture is suitable to determine and reconstruct the technological processes of question [2,5,9,10].

The Seuso treasure is one of the most famous silver collection from the Late Roman period. The body and lid of the Perfume Box were both manufactured from a single sheet of silver. No evidence of any joints has been detected. Ultrasonic measurements indicate that the Perfume Box is made of thin silver sheets with only 0.5 to 1.5 mm in thickness. Since the areas of high relief are not thinner than the background, an extensive working of from both outside and inside is supposed [11,12]. However, no other evidence about the deformation processes during shaping had been revealed so far. Because of the very high value of the objects, non-destructive investigation is the only acceptable method to reveal the microstructure. In an earlier work of the authors [13] the residual stress dataset was used to obtain information about the texture. There are many examples in the literature that not the full pole figure, but only specific components are necessarily needed for adequate information about the texture [14-16]. In this paper, data from specific directions are compared to pole figures from the Perfume Box of the Seuso treasure.

2. Experimental procedures

The standardized $\sin^2\psi$ residual stress measurement method was used to determine the interferential function in different tilting positions. The measuring parameters for the applied Stresstech G3R centreless diffractometer are as the following: exposure time is 5 s; accelerating voltage 28 kV; current 8 mA; collimator 4 mm Ø; detector arc 50 mm in modified ψ mode; Cr source, tilting range -45 / + 45 °; number of tilting: 5/5. The {222} and {311} reflections of silver were measured for qualitative texture analysis. To interpret the intensity data of residual stress measurements in terms of texture, it must be transformed to be comparable with the standard representation method. The effect of tilting on the intensity value was corrected with silver powder. Due to the standard residual stress data acquisition, only a few intensity data points are available, therefore, the χ-sections cannot be reconstructed.

The method of the calculation and visualization of the specific section of the pole figure is introduced through an example of a cold rolled aluminum sheet (figure. 1). If the measurement points of an object that were determined during stress measurement were measured in the same measuring direction related to the coordinate system of the object, the following analysis can be performed [13]. Refer the selected measuring direction to a particular direction of the pole figure, for example the rolling direction. Calculate and transform the measured point into the Cartesian coordinate system. Perform the intensity correction. Visualize and compare the projection curves for each measurement point. The signal of the two detectors were identical, therefore only the projection curve of detector A is shown.

By extending the stress measurement method, it is also possible to determine pole figures. This is possible in a variety of ways. One is to determine the measurement directions of the centreless diffractometer in accordance with the points of the conventional pole figure, which is the so-called modified X (CHI) method [6]. The other option is when the pole figure coordinates of the applied stress measurement positions (tilting and rotations) are determined, called the reverse modified X (CHI) method. This mode is described in detail in another manuscript of this journal [17]. Since the beam path differs from centreless and conventional diffractometers, the coordinates (tilting and rotation angles) used in the tests are not the same of the two measurement methods, so the equivalent positions for both modified and reversed modified cases had to be determined.

3. Results and discussion

The Perfume Box consists of a cylindrical body and a conical cover. The edges of the body and cover are overlapped on the gridded photos (figure. 2) where the residual stress measurement locations and the direction are marked according to the tilting positions. Measurement points 206-212, 214-223 and 224-228 are from the edge, side and the bottom, respectively. Axial measured direction was applied on the body (edge and side) while tangential direction was used on the bottom of the body. Since the measured directions were identical at every measurement point on the same part, the calculated data are comparable.

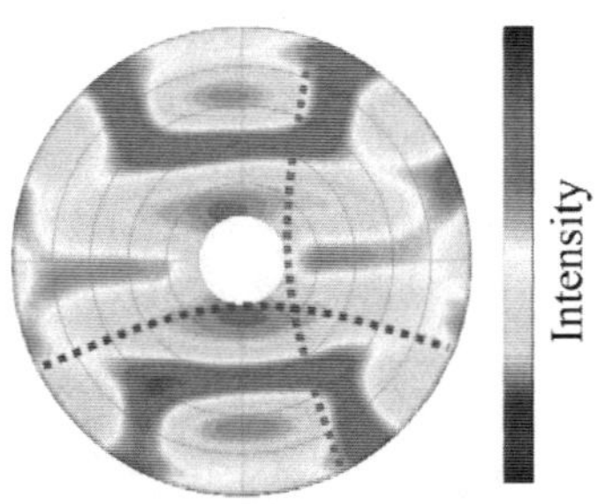

(a) {311} pole figure of rolled aluminum sheet,

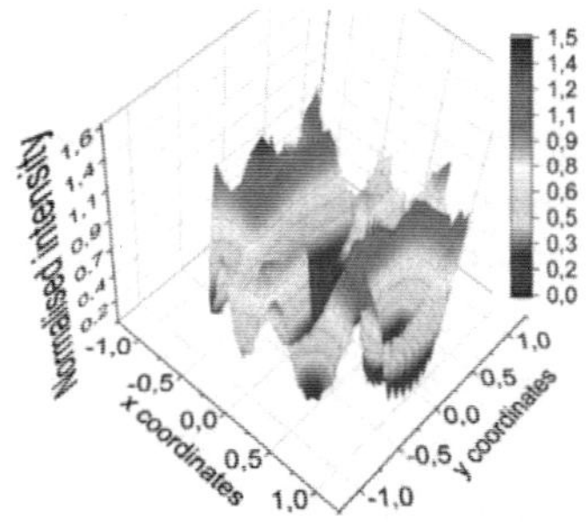

(b) The same pole figure in Cartesian coordinate system

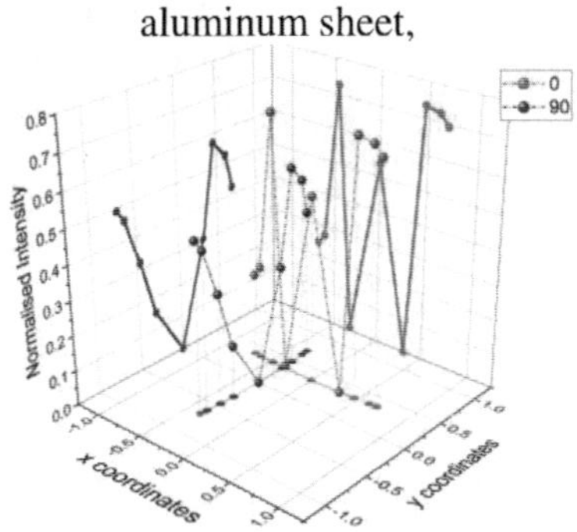

(c) Solid lines are the projection curves of the dashed lines in the Cartesian system

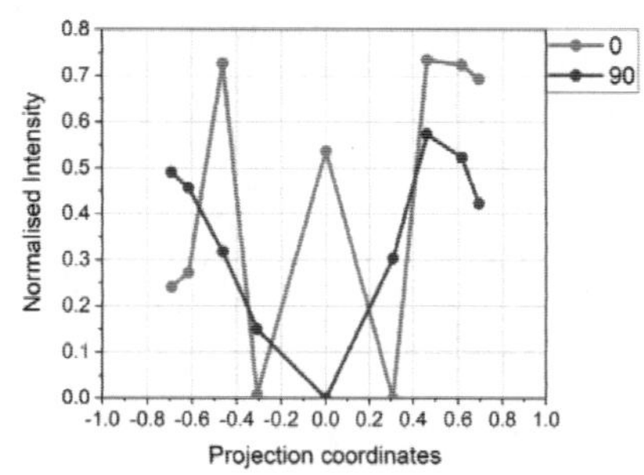

(d) Comparison of the projection curves

Figure 1. Procedure for qualitative texture description [13]. Dashed lines represent data obtained from stress measurements.

(a)

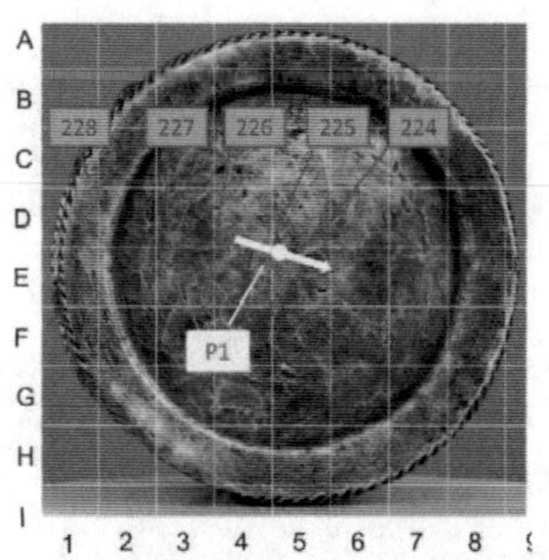

(b)

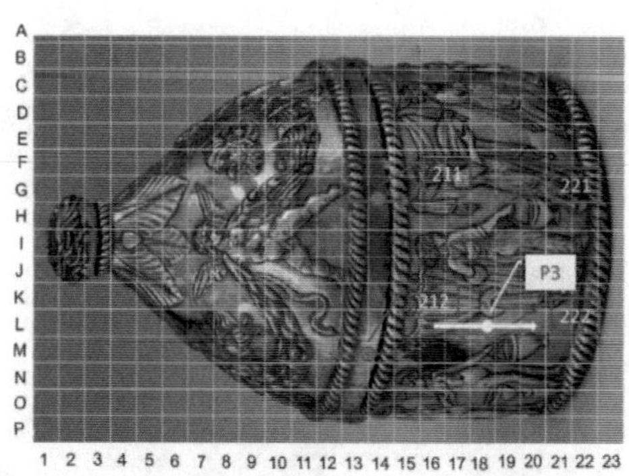

(c)

Figure 2. Investigation of the Perfume Box. (a) Onsite texture measurement with centreless diffractometer. Measurement points and tilting directions (b) on the bottom (224-228 stress, P1 pole figure) (c) on the side (221,222 stress, P3 pole figure) and on the edge (211,212 stress) of the body.

Pole figure measurements have been also carried out on the bottom (point P1) on the edge (P5) and on the side (P3) of the body. The starting point of the pole figure (conventionally called rolling direction) is also indicated on the gridded photos. The P3 and the 222 measurement point give data from the absolutely same location of the side of the body. While P1 and 225 are identical on the bottom.

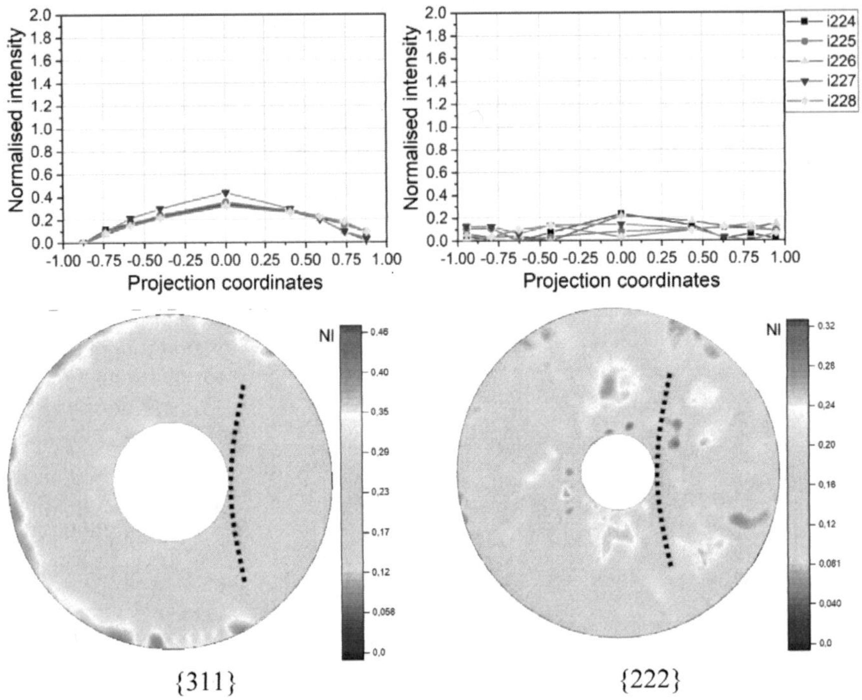

Figure 3. Projections curves and pole figures of the bottom part of Parfume Box

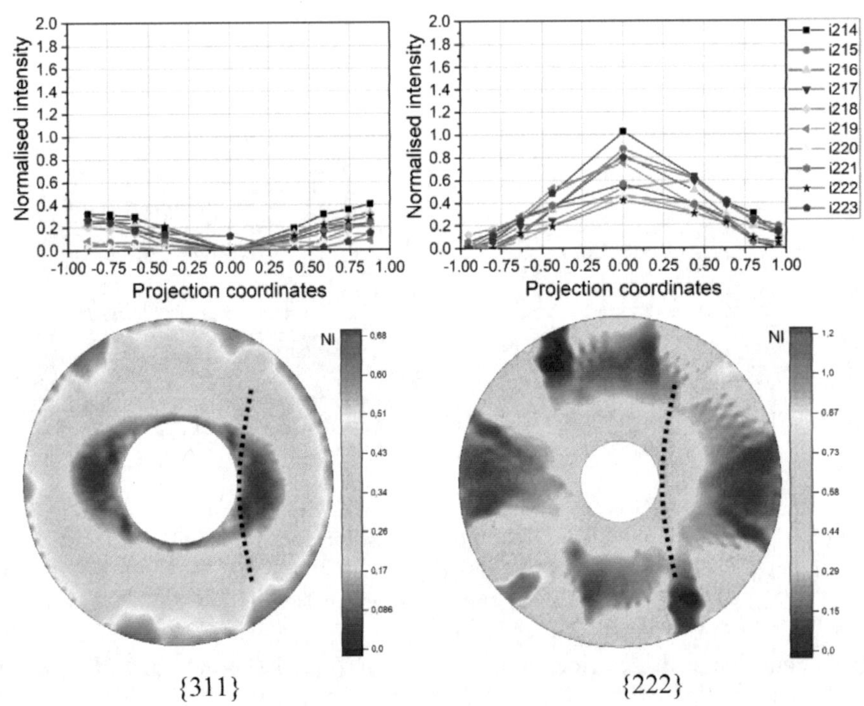

Figure 4. Projections curves and pole figures of the side part of Parfume Box

figure 3-5 show projection curves and pole figures from different parts of the object. The measurement points of the projection curves are marked on the pole figures by the dotted curves. The bottom plate data are distinctly different from the results from other parts. Both plane series show a quasi-isotropic structure (figure. 3). The results of pole figures show similar results. The non-textured nature is in line

with the assumed final hammering (patenting) of the bottom part. There are two reasons for this. One is that the deformation did not have a distinct orientation, which is a logical consequence if thinning was done with hammering. The other possibility is that heat treatment was applied after the deformation.
Data from the side, however, show a clear deformation feature (figure. 4). Compared to the bottom data, the pole figures are much more characteristic. The direction of generatrix is the direction of the poles for 12 hours. The pole figure does not resemble the known forming operations. It can be clearly stated that the deformation of the side differs from the shaping operation of the bottom and that the side is certainly not shaped by hammering.
The strongest forming characteristics are found at the edge of the body (figure. 5). Comparing to the pole figures on the side, the polarity of the figures is the same for the same index but a 90° rotation can be observed. It is evident from the fact that the starting position during the measurement was also rotated by 90°.
The shaping technology of these objects is radically different from today's industrial technologies, so the practice in those industrial problems does not help in understanding the resulting texture. However, it is clear from the results that the bottom is in isotropic, thus mildly shaped or partially softened state. While in contrast, a texture-like pattern of unidirectional formation was visible on the side, which increased even more strongly along the edges, that is likely to be the result of a metal spinning operation.

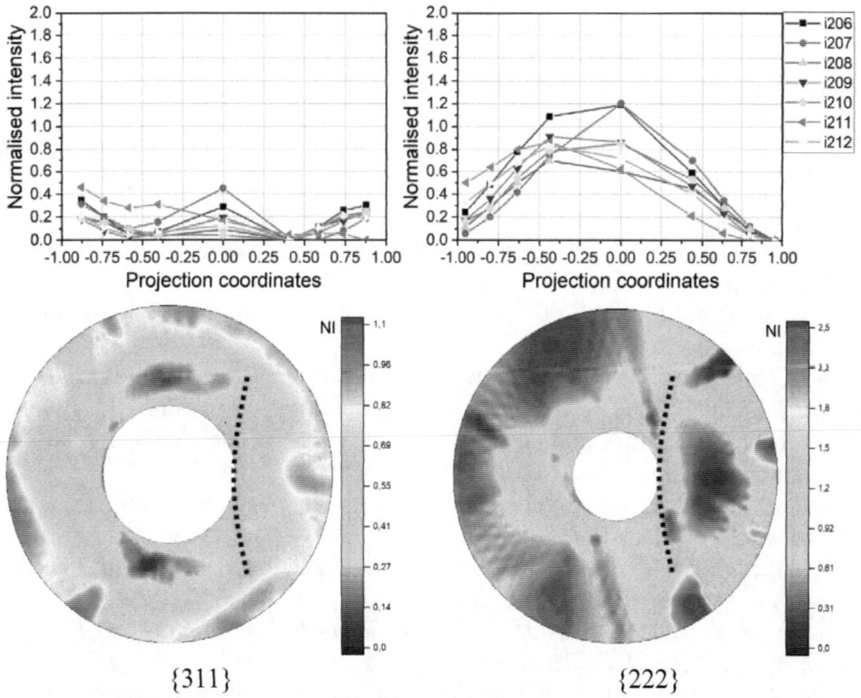

Figure 5. Projections curves and pole figures of the edge part of Parfume Box

4. Summary

A fast texture characterisation method by projection curves was developed based on the data of non-invasive residual stress measurement performed with a centreless diffractometer. The method was used to describe the texture feature of the Perfume Box, the one part of the Seuso treasure. It can be concluded that the method provided consistent and relevant texture information on the different part of the Perfume Box. It is possible to distinguish clearly the metallic state of the bottom, the side and the edges of the box. The bottom plate data are distinctly different from the results obtained at the other measurement sites, both plane series show a non-characteristic, a quasi-isotropic structure. The side of the body and the cover of the box have a clear deformation feature for each of the plane series. The strongest forming characteristics are shown by both edge of body and the cover, earing was also visible. The results of these evaluations give guidance to the production technology.

Acknowledgements

The paper was supported through the National Research, Development and Innovation Office – NKFIH K119566 project and was carried out as part of the EFOP-3.6.1-16-00011 "Younger and Renewing University – Innovative Knowledge City – institutional development of the University of Miskolc aiming at intelligent specialisation" project implemented in the framework of the Szechenyi 2020 program. The realization of this project is supported by the European Union, co-financed by the European Social Fund. This paper was supported by the Janos Bolyai Research Scholarship of the Hungarian Academy of Sciences. Authors thank the Hungarian National Museum, Dr. Zsolt Mráv and Viktoria Mozgai for the object and the photographs. Archaeological findings were investigated within the framework of contractual cooperation with the Institute of Astronomy and Earth Sciences, Geological and Geochemistry of the Hungarian Academy of Sciences, as part of the Seuso Research Project.

References

[1] Bayley J, Crossley D and Ponting M 2008 Metals and Metalworking, A research framework for archaeometallurgy (The Historical Metallurgy Society Occasional Publication) No 6

[2] Giacomo C, Philippe S and Arlen H 2016 Non-conventional applications of a noninvasive portable X-ray diffraction/fluorescence instrument *Appl. Phys. A* **122:990** 2-17

[3] Yanxia Xie, Lutterotti L, Wenk H R and Kovacs F 2004 Texture analysis of ancient coins with TOF neutron diffraction *J. Mater.Sci.* **39** 3329 – 3337

[4] Duran A, Herrera L, Jimenez de Haro M, Justo A and Perez-Rodriguez J 2008 Nondestructive analysis of cultural heritage artefacts from Andalusia, Spain, by X-ray diffraction with Göbel mirrors, *Talanta* **76**, Issue 1, pp 183-188

[5] Wanhill R J H 2005 Embrittlement of Ancient Silver *J. Failure Anal. Prev.* **1** 41-54

[6] Sepsi M, Mertinger V, Benke M 2019 Sample cutting-free pole figure measuring method for centreless diffractometers in modified X mode *Mat. Char.* **151** 351–357

[7] ed Totten G, Howes M, and Inoue T 2008 *Handbook of Residual Stress and Deformation of Steel* (USA ASM International)

[8] Kocks F, Tomé C and Wenk H 1998 *Texture and Anisotropy- Preferred Orientations in Polycrystals and their Effect on Materials Properties* (Cambridge: Cambridge University)

[9] Artioli G 2007 Crystallographic texture analysis of archaeological metals: interpretation of manufacturing techniques *Appl. Phys. A* **89**, 899–908

[10] Artioli G, Dugnani M 2004 Crystallographic texture analysis: applications in mineralogy and archaeometry *Per. Mineral.* **73**, 5-16

[11] Mango M and Bennett A 1994, *The Sevso Treasure. Part One* (USA Rhode Island: Journal of Roman Archaeology Supplementary) Series 12.1.

[12] ed Visy Zs 2012 *The Sevso treasure and Panonia, Scientific contributions to the Sevso treasure from Hungary*, Vil.1, (Pécs:GeniaNet)

[13] Sepsi M, Mertinger V, Benke M 2018 Utilization of spatial resolved qualitative texture assessment method on an object of the Seuso Treasure *IOP Conf. Ser.: Mater. Sci. Eng* **375** 012036

[14] Mao W, Chen L, Yang Ping, Feng H 2004 Rapid texture measurement of annealed aluminum sheet based on X-ray diffraction *Chinese Sci. Bull.* **49** No. 19 2112—2114

[15] Ma Q, Mao W, Feng Ha, Yu Y Rapid texture measurement of cold-rolled aluminum sheet by X-ray diffraction *Scripta Mater.* **54** 1901–1905

[16] Sepsi M, Hlavacs A, Mertinger V, Benke M 2018 Industrial application of a quick, non-destructive anisotropy characterisation method *IOP Conf. Ser.: Mater. Sci. Eng.* **426** 012042

[17] Mertinger, Sepsi, Benke Nondestructive texture measurement methods for centreless X-ray diffractometers in reverse modified X (CHI) mode *IOP Conference series,* (in press)

Non-destructive texture measurement methods for centreless X-ray diffractometers in reverse modified X (CHI) mode

V Mertinger, M Sepsi, M Benke

Institute of Physical Metallurgy Metalforming and Nanotechnology, University of Miskolc, Miskolc, Hungary

femvali@uni-miskolc.hu

Abstract. Conventional X-ray diffraction-based pole figure measurements have been carried on with dedicated instrument which are accompanied with certain limitations such as sample size, geometry, and no possibility for on-site measurements. These restrictions limit the availability of texture measurements to situations in which cutting a small sample from large parts is difficult or even not allowed at all. The present paper introduces new texture measurement method developed on mobile centreless X-ray diffractometers which are originally applied for residual stress measurements. The essence of this new method, named reverse modified X mode is that it uses the data obtained by residual stress measurement to describe anisotropic characteristics, even to the determination of pole figure. Using this method, pole figures can be obtained with all the benefits of centreless diffractometers: no need for sample cutting, flexibility in case of large components with complex shapes, short measuring time and portability. The obtained pole figures are equivalent to the pole figures determined by conventional diffractometers. The presentation describes the measurement method and includes the validation with conventional pole figure measurements and provides instances of applications of the new technique.

1. Introduction

The definition of crystalline anisotropy, also known as texture, is the key in many technological and solid-state processes since many properties such as mechanical properties, deform-ability, magnetic properties, or several metallic processes such as plastic deformation, recrystallization are texture dependent. To control or to modify the texture is also crucial question during all thermomechanical treatments. The conventional techniques to measure crystallographic texture are based on neutron, X-ray or electron diffraction, which involve the determination of several pole figures and ODF (Orientation Distribution Function) for investigated materials. The general differences between these methods are the size of radiated volume and the spatial resolution [1-4]. Except for neutron diffraction, which is limited by source and the activation process, other methods are destructive. Sample cutting is usually necessary for conventional texture measurements by X-ray diffraction (XRD) and a very dedicated sample preparation technique is required for TEM (transmission electron microscopy) and electron backscatter diffraction (EBSD) investigations. The full 3D representation for crystallographic texture is given by the ODF which can be achieved by evaluation of a set of non-equivalent pole figures. Non-destructive determination of the pole figures by XRD method in many cases, such as when the test object is unique or valuable or when the sampling is forbidden by many other reasons can be beneficial [5-7]. The present paper introduces new texture measurement methods developed for mobile centreless X-ray diffractometers which are originally developed for residual stress measurements [8]. The pole figure measurements can be realized with centreless diffractometers in modified X (CHI), Ψ (PSI) and Ω (OMEGA) modes. Using these methods, pole figures can be obtained with all the benefits of centreless diffractometers: no need for sample cutting, flexibility in case of large components with complex shapes, short measuring time and portabilityű

2. Description of the new method

2.1. *Scheme of beam paths of X-ray goniometers*

In general, three different modes can be distinguished based on goniometer tilting relative to sample: Ω (OMEGA), X (CHI) or Ψ (PSI) and modified X or modified Ψ. The term "Ψ mode" is widespread in Europe, while "X mode" is more commonly used in North America. The names refer to the axis around which the goniometer circle (holding the source and detectors) is tilted. Here, a universal nomenclature system (Figure.1.) is applied in which the three different modes can be univocally distinguished [9]. Axes are represented with Greek uppercase letters while tilting/rotation angles with Greek lowercase letters. The X axis is the intersection of the plane of sample surface and the plane of the goniometer circle. Tilting around X axis is indicated by the angle χ. It is the angle between the normal of the sample surface and the normal of the diffracting plane series (i.e, the measuring direction). The Ω axis lies within the plane of sample surface and it is perpendicular to the goniometer circle. The Ω axis is thus perpendicular to the plane of primary (incident) and diffracted beam. The ω angle is interpreted as the angle between the primary beam and the sample surface (or the X axis). While rotating/tilting around Ω axis,ψ angle forms between the measuring direction and the normal of the sample's surface. The Φ (PHI) axis is normal to the sample surface. Rotation around Φ axis is marked with φ, which increases clockwise.

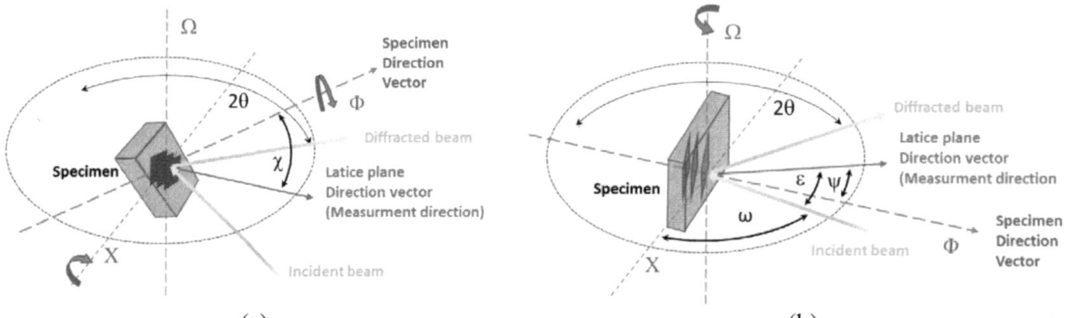

(a) (b)

Figure 1. General scheme of beam path, tilting axes and relevant angles of X-ray goniometers (a)$\chi \neq 0°$, (b) $\chi = 0°$. [9]

2.2. *Comparison of centreless and conventional diffractometers*

Different setup of centreless diffractometer is available on the market. Such devices do not require sample cutting and virtually have no sample size limit. Those are designed to determine residual stress in a non-invasive manner even in surface mapping conditions. Mechanical stress can be represented by a stress tensor and, when is presence in materials resulting in variations of interatomic plane distances. As the distances are modified along different directions in the crystal, the angle of diffraction satisfying the Bragg-condition changes. This situation results in a shift of reflections measured by the detector. In practice, the stress on the sample could be represented by a tensor, i.e., each orientation in the crystal lattice of the sample has its own independent value for strain, and shear stress could also be present. By collecting the diffracted X-ray beam with the sample rotated at different orientations along a particular angular direction referred to as the "tilt angle" Ψ, and by comparing the deviation or shift in diffraction angle with respect to the expected diffraction angle of an unstressed sample, the stress tensor can be characterized. Tilting and rotating is generally performed by modified X or Ω modes. In contrast, the diffraction cones are not distorted in the textured materials, but the intensity distribution on the cone surface will be different. In practice, this texture is often represented using a pole figure, in which a specified crystallographic axis (or pole) from each of a representative number of crystallites is plotted through stereographic projection, along with directions relevant to the processing history of materials. During the determination of stereographic projection tilting and rotating of the sample is performed, conventional pole figure measurements are carried out in X mode.

During operating the centreless diffractometer in contrast to conventional one the measured object is stationary while the diffractometer is tilted around the object. However, the main difference in stress and conventional texture measurement is not the tilting of the sample but the beam path. Therefore, the transition between the two methods (stress and texture test) is not obvious. While describing the solution of transition, axes, angles, directions regarding the centreless diffractometer system are distinguished by apostrophe (') from the conventional ones. There are basically two ways to find the solution. First case is when the coordinate (φ rotating and χ tilting angles) of the conventional pole figure is fixed and used to calculate the appropriate φ'rotating and χ' tilting angles to be applied on the centreless diffractometer. This modified X method is described in detail in a previous article [9]. The other case is a much more convenient solution for the user, when the pole figure coordinates of the applied rotating and tilting angles of the centreless diffractometer (φ', χ') are calculated. This is the reverse modified X method and is described in this manuscript. The advantage of this method is that in this case stress values can also be determined from the measurement data, which in many cases can also provide useful information. To make the difference between the modified X mode and the reverse modified X mode easier to understand, we will use the same graphics and nomenclature. Figure 2 shows the beam path and the different set ups. The main parts of the conventional and the centreless diffractometer setup are indicated: C-collimator, A, B detectors, S_0, S_0' incident beam, S_A, S_B, S_A', S_B' diffracted beams, m_A, m_B, m_A', m_B' -measurement directions or diffraction vector or normal of the plains in Bragg position.

2.3. *Reverse modified X mode*

It is possible to determine the pole figure coordinates (φ, χ) from the data obtained during residual stress measurements, if the position of the sample (β- angle between the Ω and Ω'or the physical rotation of the centreless diffractometer around the Φ' axis), the value of the tilting (χ') and the Bragg-angle (2θ) are known. The transition from the centreless system to the conventional one is not obvious, because of the different beam passes of the two systems. The pass of the measuring direction vector is represented by concentric cones series during operation in both systems. For simplicity, let us consider the length of the cone generatrix unity. The two rotating axes of the two cones series are perpendicular to each other (Figure 3.). Determination of the coordinates of the intersections (P at χ'=0 and Q at χ'≠0°,) is the crucial point when measuring pole figures with centreless diffractometers. The questions are the φ and χ value at these positions. Equations 1-11 give the solution. Notice that the R_i value is difference between the modified X and the reverse modified X mode.

Since 2ψ' is the complementary angle of 2θ' (Figure 2.), thus, ψ' can be calculated from the Bragg-angle (2θ') by (1) [9].

$$\psi' = (180 - 2\theta)/2 \quad (1)$$

M' is the perpendicular projection of m_B' on the tilting axis (X) and can be calculated according to (2) and Figure 3 [9].

$$M' = \sin\psi' \quad (2)$$

U' is the distance of P or Q from the X and can be calculated according to (3) [9].

$$U' = \cos\psi' \quad (3)$$

L_i is the distance of the vertical projection of P or Q from X and can be calculated by (4) [9].

$$L_i = \sin\chi'_i * \cos\psi \quad (4)$$

R_i, the perpendicular projection of m_B' on the sample surface can be calculated according to (5).

$$R_i = \sqrt{M'^2 + L_i^2} = \sqrt{(\sin\psi)^2 + (\sin\chi'_i * \cos\psi)^2} \quad (5)$$

α is the angle between R and Ω axis and can be calculated according to (6).

$$\alpha_i = \arcsin\left(\frac{M'}{R_i}\right) = \arcsin\left(\frac{\sin\psi'}{\sqrt{(\sin\psi')^2 + (\sin\chi'_i * \cos\psi')^2}}\right) \quad (6)$$

χ_i can be calculated using equation (7).

$$\chi_i = \arcsin(R_i) = \arcsin\left(\sqrt{(\sin\psi')^2 + (\sin\chi'_i * \cos\psi')^2}\right) \quad (7)$$

Then, φ_j can be calculated according to equations (8), (9), (10) and (11) within the I., II., III. and IV. quarter of the current coordinate system, respectively (Figure 3.), where β is the angle between Ω and Ω' or the physical rotation of the centreless diffractometer around Φ' axis [9].

$$\varphi_j = \alpha_i + \beta_j \quad (8)$$

$$\varphi_j = (180 - \alpha_i) + \beta_j \quad (9)$$

$$\varphi_j = (180 + \alpha_i) + \beta_j \quad (10)$$

$$\varphi_j = (360 - \alpha_i) + \beta_j \quad (11)$$

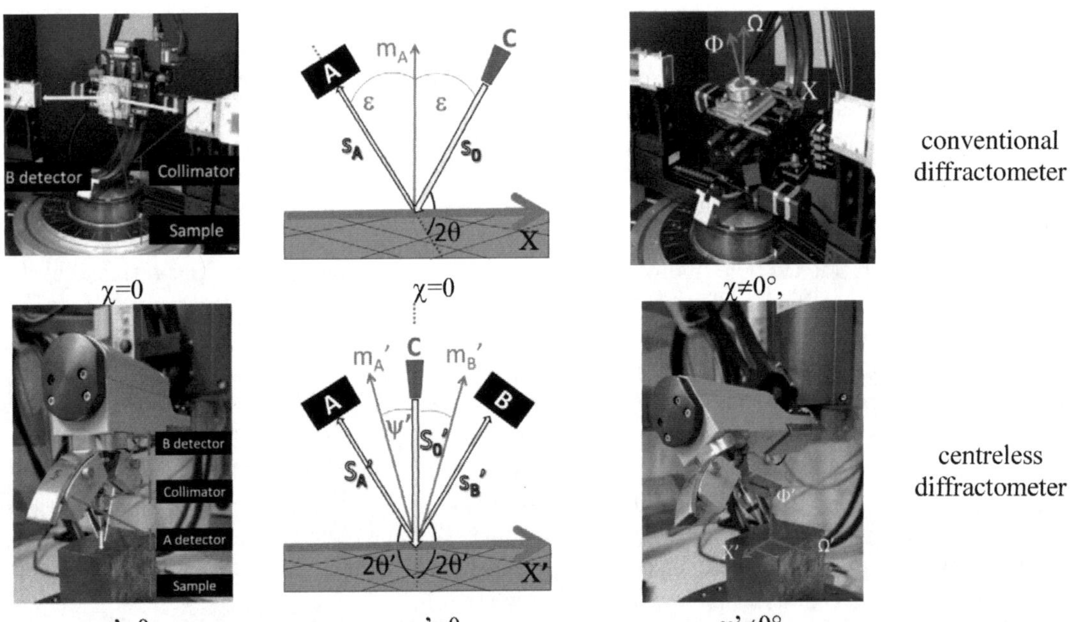

Figure 2. Beam path of different diffractometer configurations at χ≠0°, and χ=0°. [9]

After absorption (defocusing) correction, the measured intensity values are available as a function of χ and φ, from which the pole figure can be constructed.

3. Results and the validation of the new method

A 3104 type cold rolled aluminium sheets were chosen for the validation. Figure 4 shows the {222} pole figures determined by a conventional and the centreless diffractometers. The cold rollings were performed by a Von Roll experimental rolling mill, having 220 mm roll diameter, using 50 m/min rolling speed and mineral oil-based liquid lubricant. The sheets were cold rolled from 4.8 mm thickness to 1 mm (H20K) and from 4.8 to 0.2 mm (H10K), respectively. Conventional pole figure measurements were carried out by Bruker D8 Advance diffractometer equipped Eulerian cradle and Co Kα radiation source operating with 40kV tube voltage and 40 mA tube current. Pole figure were measured up to 75% tilting with 5° increments and visualised by the TexEval software. The new method was introduced using a Stresstech XStress 3000 G3R type centreless diffractometer with CrKα source operating with 30 kV tube voltage and 8 mA tube current. The used χ' values were -59°+59°, with 5° increments, β was chosen as -90…+90°, with 5° increments. The absorption corrected pole figures were visualised by Origin software. For both equipment, absorption correction was performed on Al powder. It is evident that the relevant information of the pole figures of the two methods is the same.

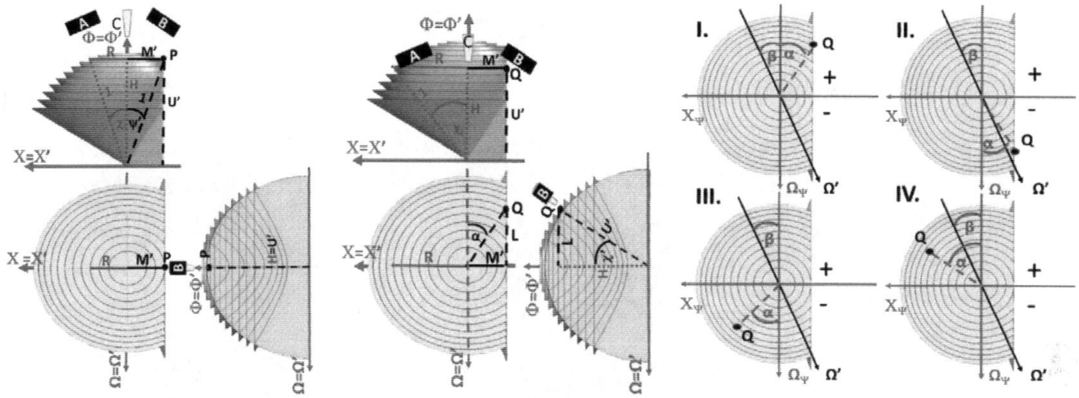

at χ'=0° tilting at χ'≠0° tilting.
(front) view from the direction perpendicular to the tilting axis (X'), (top) view from the primary beam, (side) view from the tilting axis (X').

at χ'≠0° tilting.
Intersection is in the I., II., III., the IV. quarter.

Figure 3. Theoretical view of measurement directions of different configuration X and X' modes. Q and P are the intersections. Interpretation of parameters. [9]

4. Summary

The reverse modified X mode as a new non-invasive pole figure measurement method was introduced based on the non-destructive residual stress measurement performed with centreless diffractometer. The centreless diffractometers generally operate by modified X (CHI) and Ω (OMEGA) modes. The full method to use modified X mode for pole figure determination had already shown [9]. In this paper the mathematical background demonstrated for the reverse modified X mode. It means that the centreless diffractometer operates as during residual stress measurements and the mathematical correlations describing the connection of the data set from the residual stress measurement with the coordinates of the pole figure have been shown. Each of the two measurement methods can be implemented in any case where the sample cut is not allowed, the surface is not even flat and the object to be examined is too large. The limit of the methods is that the centreless diffractometers operate at high value of 2θ diffraction angles based on the principle of residual stress measurements so that these pole figures can be detected. The lack of the method can be eliminated by a diffractometer system attached to the robot arm (X mode).

5. Acknowledgement

The present paper was supported through the National Research, Development and Innovation Office – NKFIH K119566 project and by the ÚNKP-18-2-I. New National Excellence Program of the Ministry of Human Capacities. Marton Benke was supported by the Janos Bolyai Research Scholarship of the Hungarian Academy of Sciences.

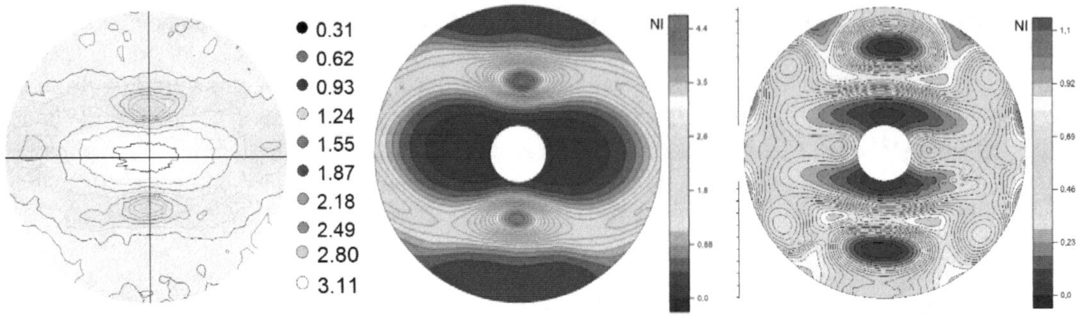

(a) {111} pole figures of sample H20K (b){222} pole figures of sample H20K (c) {311} pole figures of sample H20K

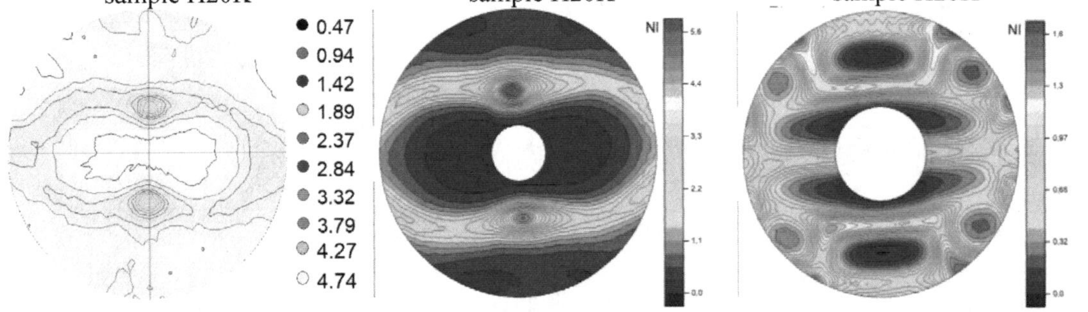

(d) {111} pole figures of sample H10K (e){222} pole figures of sample H10K (f) {311} pole figures of sample H10K

Figure 4. {hhh} pole figures and pole density levels of cold rolled 3104 type aluminium sheets after different degrees of cold rolling. (a, d) pole figures measured with a conventional diffractometer (max χ=75°); (b,c,e,f) pole figures measured with a centreless diffractometer (max χ'=60°) using reverse modified X mode

References

[1] Engler O, Randle V *Introduction to Texture Analysis – Macrotexture, Microtexture and Orientation Mapping*, CRC Press. 2010
[2] Suwas S, Kumar Ray R. *Crystallographic Texture of Materials*, Springer-Verlag, London, 2014.
[3] Rollet A, Humphreys F,.Hatherly M, *Recrystallization and Related Annealing Phenomena 3rd ed.* , Elsevier, 2017
[4] Kocks U. F, Tomé C. N, Wenk H.-R, *Texture and Anisotropy*, Cambridge University Press 2005.
[5] Duran A, Herrera L, Jimenez de Haro M, Justo A and Perez-Rodriguez J 2008 Nondestructive analysis of cultural heritage artefacts from Andalusia, Spain, by X-ray diffraction with Göbel mirrors, *Talanta* **76,** Issue 1, pp 183-188
[6] Adriaens A 2005 Nondestructive analysis and testing of museum objects: An overview of 5 years of research *Spectrochimica Acta Part B* **60** pp 1503 – 1516
[7] Giacomo C, Philippe S, Arlen H, (2016) Non-conventional applications of a noninvasive portable X-ray diffraction/fluorescence instrument *Appl. Phys. A* **122:990** 2-17
[8] Fitzpatrick M, Fry A, Holdway P, Kandil F, Shackleton J, and Suominen L. 2005. A National Measurement Good Practice Guide, Determination of Residual Stresses by X-ray Diffraction - Issue 2 No. 52 (Teddington: Crown)
[9] Sepsi M, Mertinger V, Benke M 2019 Sample cutting-free pole figure measuring method for centreless diffractometers in modified X mode *Mat. Char.* **151** 351–357

IOP Conf. Series: Journal of Physics: Conf. Series **1270** (2019) 012013 doi:10.1088/1742-6596/1270/1/012013

Determination of static recrystallization and recovery parameters for steel by fitting model to stress relaxation data

A Pohjonen, O Seppälä, A Jokiranta, A Kaijalainen, M Somani, D Porter, J Larkiola and J Kömi

Materials and Mechanical Engineering, Faculty of Technology, University of Oulu, PL4200, 90014, Oulu, Finland

E-mail: Aarne.Pohjonen@Oulu.fi

Abstract. A model for static recrystallization by Zurob et al. [1] has been fitted to experimental stress relaxation [2] data obtained on a low-alloyed steel using a Gleeble thermomechanical simulator. The model has been implemented as an algorithm that calculates the stress relaxation as a function of time, including physical descriptions of the recovery and recrystallization processes. The activation energy and volume were used as fitting parameters for recovery, and the activation energy of diffusion and nucleation site density were used as the fitting parameters for recrystallization. The four fitting parameters were determined from the experimental data by applying the Nelder-Mead algorithm within Matlab software. It can be concluded from the preliminary results that Zurob's model can be successfully fitted to the stress relaxation data in order to illustrate the static restoration characteristics and kinetics in carbon steels using these fitting parameters.

1. Introduction

In order to control the microstructural evolution during thermomechanical processing, computational models are needed to accurately capture the microstructural phenomena that occur during processing. In the hot rolling of steel, during the interpass time between the rolling passes, both static recovery and static recrystallization take place. Since the resulting austenite microstructure affects the phase transformation occurring during the subsequent cooling and the resultant mechanical properties, computational models provide a way to simulate and optimize the processing conditions that lead to the desired outcome.

Although recovery dominates stress relaxation immediately after deformation and recrystallization mainly dictates stress relaxation at longer times, there is a noticeable overlap of the two phenomena in the intermediate time scales. One of the methods of determining the time dependence of recrystallization is stress relaxation testing, where the time region of recrystallization is interpreted from the shape of a stress relaxation curve. The fraction of recrystallization is determined by separating the recovery and recrystallization and fitting the corresponding part of the curve with an Avrami type function [2]. On the other hand Zurob et al. presented a coupled model to describe simultaneously occurrence of recovery, recrystallization and precipitation [1]. In the current study, we focus our attention on an implementation of the Zurob's model of recrystallization and recovery by fitting the model parameters to stress relaxation data.

IOP Conf. Series: Journal of Physics: Conf. Series **1270** (2019) 012013 doi:10.1088/1742-6596/1270/1/012013

2. Experiments

A Gleeble 3800 thermomechanical simulator coupled with a Hydrawedge module was used to carry out the stress relaxation experiments on a low-alloy steel of the following composition (in wt.%): 0.16 C, 0.97 Mn, 0.33 Si, 0.22 Mo, 0.009 V, 0.001 Nb. Cylindrical test specimens of the dimensions ϕ 10 x 12 mm were first reheated at 10 °C/s to 1200 °C, held for 5 minutes, followed by cooling at 2 °C/s to the temperature at which the samples were deformed and relaxed for 200 s before subsequently quenching to room temperature. The deformation temperatures were 950, 1000 and 1050 °C and the strain rates were 1 and 10 /s and strains were 0.2, 0.3 and 0.4.

3. Modelling and simulation

In order to concisely describe the model and the constructed algorithm, the nomenclature, constants and function definitions of the model are given in tables 1 and 2, while the initial conditions and the algorithm are presented with the flow charts shown in figures 1 and 2.

Table 1: Nomenclature and constant values used in the model [1,3–6]

$\sigma_{rec}(t)$ The recovery stress, i.e. the stress as function of time due to recovery only
$\sigma_{gg}(t)$ The stress as function of time due to grain growth only
$\sigma_{rel}(t)$ The relaxation stress, i.e. the stress as function of time due to recovery, recrystallization and grain growth processes
v_d Debye frequency
$M_T = 3.1$ Taylor factor
$\alpha_r = 0.15$ Constant
U_a Activation enthalpy of recovery (fitting parameter)
V_a Activation volume of recovery (fitting parameter)
ρ Dislocation density
σ_y Yield stress of fully recrystallized material
$b = 2.5$ Å Length of the Burgers' vector
G Driving force of recrystallization
A_c Area of single nucleus on the grain boundary
γ_{gb} Grain boundary energy
N_{rex} Number of recrystallization nuclei per unit volume
k Geometrical factor (fitting parameter)
S_v Surface to volume ratio (fitting parameter)
M_{pure} Grain boundary mobility for pure iron
$\delta = 0.5$ nm grain boundary width
$D_{gb_0} = 6.35\times10^{-15}$ Self diffusion constant for iron grain boundary [3]
D_{gb} Self diffusion coefficient for austenite grain boundary
$Q_{Fe} = 55.7$ kJ/mol Activation energy of self diffusion [3]
$V_{m_{Fe}} = 55.845\times10^{-3} / \rho_\gamma$ Molar volume of iron, ρ_γ is the temperature dependent density of austenite calculated according to [4]
E Modulus of elasticity (temperature dependent, calculated according to [6])
$\mu = \frac{3}{8}E$ Shear modulus
$R_g = 8.3145$ J/mol Universal gas constant
$k_B = 1.3807\times10^{-23}$ Boltzmann constant
M_{gb} Grain boundary mobility
Q_d Activation energy for the diffusion of alloying elements (fitting parameter)
$I(t) = \int_0^t M_{gb}(t')G(t')dt'$, Integral
X_{rex} Fraction of recrystallized volume

Table 2: The equations applied in the calculation of recovery, recrystallization and relaxation [1,3,5]

$$\frac{d\sigma_{rec}}{dt}(t,T,\sigma_{rel}) = \frac{64\,(\sigma_{rel}-\sigma_y)^2\, v_d}{9M_T^3\alpha_r^2 E}\exp\left(-\frac{U_a}{RT}\right)\sinh\left(\frac{(\sigma_{rel}-\sigma_y)V_a}{k_B T}\right)$$

$$\rho(t,\sigma_{rec}) = \frac{(\sigma_{rec}-\sigma_y)^2}{(M_T\alpha_r\mu b)^2}$$

$$G(t,\rho) = \frac{1}{2}\rho\mu b^2$$

$$A_c(G) = 4\pi\left(\frac{2\gamma_{gb}}{G(t_0)}\right)^2$$

$$\gamma_{gb} = 1.3115 - 0.005T$$

$$N_{rex}(k,S_v,A_c) = \frac{kS_v}{A_c}$$

$M_{pure}(t,T) = \frac{1\cdot\delta D_{gb} V_{m_{Fe}}}{10b^2 R_g T}$ [1], where $D_{gb} = D_{gb0}\exp\left(-\frac{Q_{Fe}}{R_g T}\right)$ [3]

$$M_{gb}(t,T,M_{pure}) = M_{pure}\exp\left(-\frac{Q_d}{R_g T}\right)$$

$$I(t,M_{gb},G) = \int_0^t M_{gb}(t')G(t')dt'$$

$$X_{rex}(t,I) = 1-\exp\left(-N_{rex}I^3\right)$$

$$\sigma_{elp}(t,X_{rex},\sigma_{rec},\sigma_{gg}) = (1-X_{rex}(t))\sigma_{rec}(t) + X_{rex}(t)\sigma_{gg}(t)$$

$\sigma_{gg}(t) = \sigma_y + b\,log_{10}(t)$ (function fitted to relaxation (t > ~10 s) data, σ_y and b are fitting parameters)

Slightly differently to Zurob's model, the temperature dependent Debye frequency $v_d = \left(\frac{6\pi v N}{V}\right)^{\frac{1}{3}}$ [7] was calculated using values for unit cell volume V [8] (number of atoms / unit cell N = 4) and the speed of sound $v = \sqrt{E/\rho_\gamma}$, which is in turn calculated using the elastic modulus E [6] and austenite density ρ_γ [4], providing temperature dependent quantities instead of constant values. Also, the grain boundary mobility is calculated using $M_{gb} = M_{pure}\exp\left(-\frac{Q_d}{R_g T}\right)$, since Nb precipitation is not relevant for the current steel composition.

The initial definitions and the algorithm used for calculating the stress, recovery and relaxation as a function of time are presented as flow charts in figures 1 and 2, respectively.

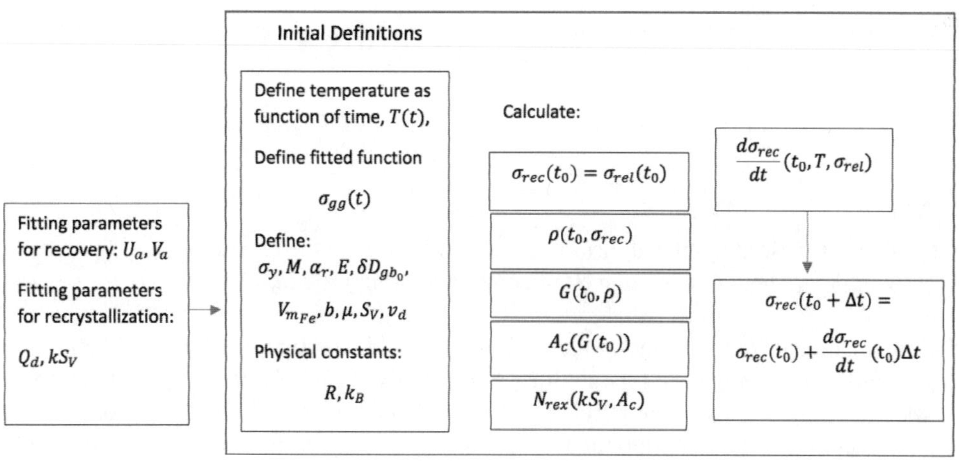

Figure 1: Initial definitions of the algorithm, which are calculated before the time loop described in figure 2.

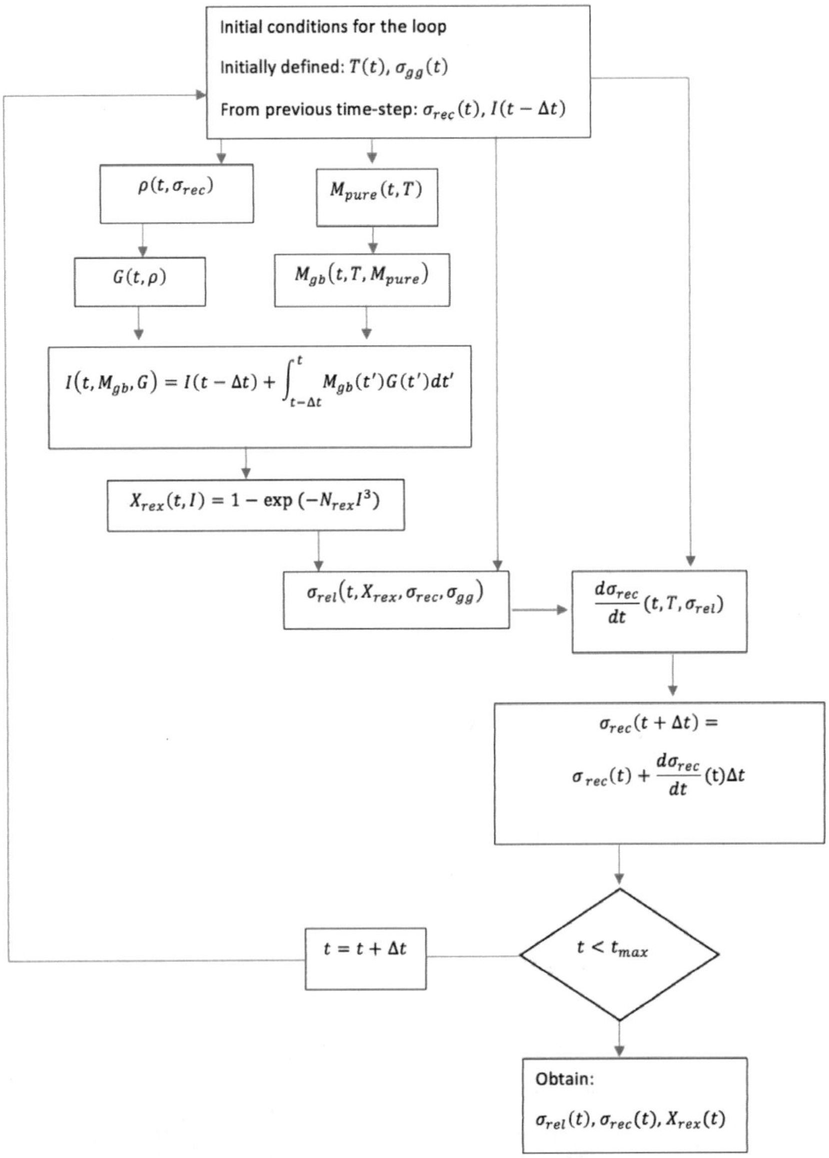

Figure 2: Flowchart describing the algorithm used for calculating the relaxation stress as a function of time corresponding to the recovery, recrystallization and grain growth.

The fitting parameters used in the model were the activation enthalpy U_a and activation volume V_a of recovery, the activation energy for the diffusion of the alloying elements Q_d and the product of k and S_V, which is proportional to the number of recrystallization nucleation sites. In order to obtain an evenly distributed amount of fitting points in the stress vs. ln t diagram, additional fitting points were created for $t < 1$ s by linearly interpolating the data between the measurement points. This ensured that for the fitting, the data in the stress vs. ln t curve, when $t < 1$ s were equally emphasized as the data when $t > 1$s. The lower bound of 5×10^{-5} was set for parameter k in the fitting, as proposed in [1].

4. Results and discussion

We implemented the model described in section 3 to simulate the recovery and recrystallization processes after deformation. The model was numerically fitted to the experimental data by applying the Nelder-Mead algorithm [9], using the Matlab software *fminsearch* function [10]. The initial fitting parameters were U_a and V_a for the recovery and Q_d and kS_V for the recrystallization. The representative comparison of calculated vs. experimental data for some of the tests is shown in figure 3 a). The comparison of all the simulation results against the experimental data is shown in figure 3 b).

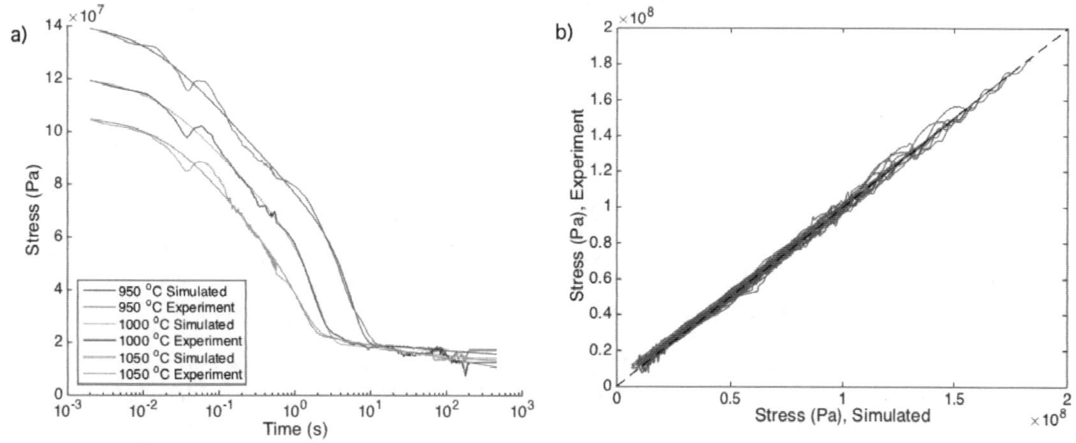

Figure 3: a) Calculated results compared to experimental results for 0.3 strain, 1/s strain rate for different temperatures. b) Calculated values (with parameters shown in table 3) compared to simulated values.

The fitted parameter values are given in table 3 for all the cases where fitting was successful. The results show that the implemented Zurob's model describing the recovery and recrystallization can be adequately fitted to the experimental data using the chosen fitting parameters. It can be seen from the obtained parameter values that the parameter kS_V is significantly affected by the experimental conditions, while the variation of the other parameters is relatively small. Though small, a noticeable increase of activation volume can be seen with the decrease of strain.

The obtained values for the activation enthalpy U_a and activation energy V_a are close to the values used in [1]; $U_a = 286\times10^3$ J/mol and $V_a = 45b^3 = 7.03\times10^{-28}\text{m}^3$. In [11], the effect of solute alloying elements on the apparent activation energy of recrystallization is described by $Q_{rex} = Q_0 + Q_{alloy}$, where $Q_0 = 148636.8$ J/mol would be the activation energy in the absence of alloying and $Q_{alloy} = -71981.3[\%C] + 21180[\%Mn] + 56537.6[\%Si] + 121243.3[\%Mo] + 64469[\%V] + 109731.9[\%Nb]^{0.15}$. The calculated value according to [11] is then $Q_{alloy} = 93760$ J/mol, which is very close to the values obtained for Q_d from our fitting shown in table 3. In [1] it is estimated that, the parameter k can vary between 5×10^{-5} and 5×10^{-2}, where the former was used as a lower bound in the fitting in the current study. Assuming a reasonable value of $S_V = 80\times10^3$ 1/m implies that the product kS_V would vary between 4 and 4000, which is consistent with the fitted parameter values shown in table 3.

Table 3: Model parameter values obtained from fitting.

Strain	U_a (J/mol)	V_a (m^3)	Q_d (J/mol)	kS_V (1/m)
10/s, Temp 950 °C				
0.2	3.05×10^5	9.17×10^{-28}	1.18×10^5	4179.41
0.3	3.09×10^5	8.11×10^{-28}	0.960×10^5	4.04
0.4	3.12×10^5	7.91×10^{-28}	1.20×10^5	4769.55
1/s, Temp 950 °C				
0.2	3.09×10^5	8.99×10^{-28}	1.17×10^5	3185.09
0.3	3.13×10^5	7.99×10^{-28}	1.12×10^5	355.69
0.4	3.10×10^5	7.76×10^{-28}	1.15×10^5	2093.75
10/s, Temp 1000 °C				
0.2	3.07×10^5	9.14×10^{-28}	0.893×10^5	4.04
0.3	3.01×10^5	7.01×10^{-28}	0.903×10^5	4.04
0.4	2.95×10^5	5.72×10^{-28}	1.06×10^5	565.88
1/s, Temp 1000 °C				
0.2	3.09×10^5	9.44×10^{-28}	0.878×10^5	4.04
0.3	3.11×10^5	7.57×10^{-28}	1.16×10^5	4882.72
0.4	2.95×10^5	4.73×10^{-28}	1.07×10^5	1317.69
10/s, Temp 1050 °C				
0.2	2.97×10^5	6.74×10^{-28}	1.13×10^5	3088.08
1/s, Temp 1050 °C				
0.2	3.15×10^5	9.30×10^{-28}	1.18×10^5	5998.31
0.3	3.01×10^5	5.01×10^{-28}	1.13×10^5	3371.02

5. Conclusions

A model based on Zurob's model was successfully fitted to data obtained from stress relaxation tests for different temperatures, strains and strain rates. The obtained model parameter values were in reasonable agreement with those found in literature. In future, the model will be fitted to larger set of experimental data in order to see the parameter dependence on different conditions, and we are planning to couple the model with finite element modelling in order to predict the recovery and recrystallization in different locations of deformed material.

6. Acknowledgements

Financial support from the Academy of Finland through Genome of Steel project is gratefully acknowledged.

References

[1] Zurob H S, Hutchinson C R, Brechet Y and Purdy G 2002 *Acta Mater.* Vol. 50, 3075–3092
[2] Karjalainen L P and Perttula J 1996 *ISIJ Int.* **36** 729–36
[3] Inoue A, Nitta H and Iijima Y 2007 *Acta Mater.* Vol. 55, No 17, pp. 5910-5916
[4] Miettinen J 1997 *Metall. Mater. Trans. B* Vol. 28B, 281-297
[5] Seppälä O 2016 *Rekristallisaatio kuumataotussa kappaleessa* (University of Oulu)
[6] *ANSI/AISC specification for structural steel buildings 2005*
[7] Kittel C *Introduction to Solid State Physics, 4th Edition* (John-Wiley & Sons)
[8] Seki I and Nagata K 2005 *ISIJ International* Vol. 45, No. 12, pp. 1789-1794
[9] Lagarias J C, Lagarias J C, Reeds J A, Wright M H and Wright P E 1998 *SIAM J. Optim.* Vol. 8, No 1, pp. 112-147
[10] MATLAB documentation https://se.mathworks.com/help/matlab/, retrieved 11.4.2019
[11] Gómez M and Medina S F 2011 *Int. J. Mater. Res.* Vol. 102, No 10, pp. 1197-1207

IOP Conf. Series: Journal of Physics: Conf. Series **1270** (2019) 012014 doi:10.1088/1742-6596/1270/1/012014

Analysing mobility from austenite grain growth data

Winfried Kranendonk

Tata Steel, P.O.Box 10.000, 1970 CA IJmuiden, the Netherlands

Winfried.Kranendonk@tatasteeleurope.com

Abstract. To obtain information on mobility of grain boundaries in austenite grain growth data has been collected from the open literature, to fit a physics-based grain growth model, which relates the mean grain size to the annealing time. The suitability of this model for predicting the grain size as a function of time is discussed and some preliminary results on the reduced mobility are presented. For the compositions in the database it has been found that the activation energy ranges from 100 to 600 kJ/mol. Such high activation energies cannot be explained well with many solute drag models.

1. Introduction

Metallurgical models have an increasingly important role in product development and process control in modern steel production. Both performance and reduction of CO_2 emissions from cars nowadays drives the development of steels with higher strength, combined with a good formability. To meet these requirements, new steel grades, like Advanced High Strength Steels and modern HSLA steels lead to an increasing variety in chemical compositions and process conditions. Because of this large variety the preference is more for physics-based models rather than empirical models, because properly constructed physics-based models are in general more valid over a wide range of compositions and process parameters (temperatures, rolling reductions etc.) and are more reliable in extrapolation outside the validation range.

In the production of these new steel grades recrystallisation and grain growth are two metallurgical phenomena, which also determine the final microstructure and, hence, the mechanical properties of these steels. For modelling the grain growth and recrystallisation in metals, the mobility of grain boundaries is essential, because it relates the migration rate of the grain boundaries directly to the driving force. Thus, knowledge of the mobility is of great value for developing models, especially the relation between the mobility and the temperature and the chemical composition of alloys.

2. Aim of the project

Despite the importance of modelling the mobility, its research on even single-phase grain boundary migration for steel is quite limited. Most models on grain growth and recrystallisation are empirical [1-3]. Although several models for the role of solutes on mobility have been formulated [4-8] these models have hardly been tested for steel. The only known study is by Yogo et al [9, 10], who presented an analysis of Hillert's model but tested it on a very small database.

This project aims to test and validate the solute drag theories, and mobility models in general, using data from the open literature. Here, a brief introduction will be given on the database, the methodology, and some preliminary results. A more detailed analysis of the database and discussion of the results will appear in a later publication.

3. The database

Data on grain growth in austenite have been collected from the open literature, i.e. scientific journals and monographs. For the analysis of the mobility and testing the models only data are used where grain sizes were measured for several holding times under isothermal conditions. By far the most experimental data are only available in the form of plots. These figures were digitized and the time-grain size data points were extracted from the images.

Data have been taken from 23 sources. An impression of the database, which was used for the preliminary analysis as presented in these proceedings, is given in table 1. Because grain growth kinetics can be influenced by Zener pinning from precipitates, the information of the database has been split up with respect to the presence of precipitate forming elements.

Table 1. The statistics of the database, used for the analysis. Row 2 gives the overall information and rows 3 to 8 give more detailed information with respect to the presence of precipitate-forming elements.

Composition	Number of compositions	Number of isotherms	Number of data points
All compositions	72	309	1815
Without Al, Nb, V, Ti	23	83	469
With only Al	30	141	832
With Al + Nb	4	15	109
With Al + V	10	48	294
With Al + Ti	2	10	50
With Nb, V and/orTi, without Al	3	12	61

4. The model

The model as proposed by Grey and Higgins was chosen to model the grain growth data [11]. The Grey-Higgins model is a mean value model, where the rate of the mean grain radius $\overline{R_g}$ is related to a driving force, based on the mean radius and the mean grain boundary energy γ. The mobility M is the proportional factor between the migration rate and the driving force. The time t is the isothermal annealing time. The model assumes that for very long times the mean grain size goes to a limit value R_{limit}. Grey and Higgins formulated their model based on grain growth experiments in zone refined metals and diluted alloys, where all the experimental time-grain size curves tended to have a limit to a finite grain size.

$$\frac{d\bar{R}_g}{dt} = 2 \cdot \gamma \cdot M \left(\frac{1}{\bar{R}_g} - \frac{1}{R_{\text{limit}}} \right) \quad (1)$$

This model is one of the simplest models, which still has a physical basis and the reduction of the evolution of the grain size distribution to one of a mean value guarantees a limited computing time. The reduced mobility (γM), the limit grain size (R_{limit}) and the initial grain size were fitted for each isotherm (with a minimum number of 4 data points) using a least squares fit program, in which equation (1) was approximated numerically using a 4th order Runge-Kutta algorithm.

5. Preliminary results

Here, two preliminary results of the analysis are briefly discussed: the suitability of the Grey-Higgins model to describe the (literature) data and some examples of the reduced mobility as a function of the temperature.

5.1. Suitability of the model

The goodness-of-fit of the model is illustrated in figures 1 and 2 as representative examples. In most of the isotherms the Grey-Higgins model gives a good representation of the experimental data, especially for short times. For long times, however, the Grey-Higgins model predicts an asymptote for the grain size evolution. Data for long holding time are very rare: the only available data have been measured by Miller [13]; up to 5000 hours for some isotherms. One example of these data (steel 1) is replotted in figure 3 with the original double-logarithmic axes. Even after soaking for 300 hours no sign of an asymptotic behaviour is observed, so the Grey-Higgins model represents these experimental data less accurately, as is shown in figure 4, where the same data for steel 1 were used as in figure 3. The absence of an asymptotic value limits the use of the Grey-Higgins model to short times.

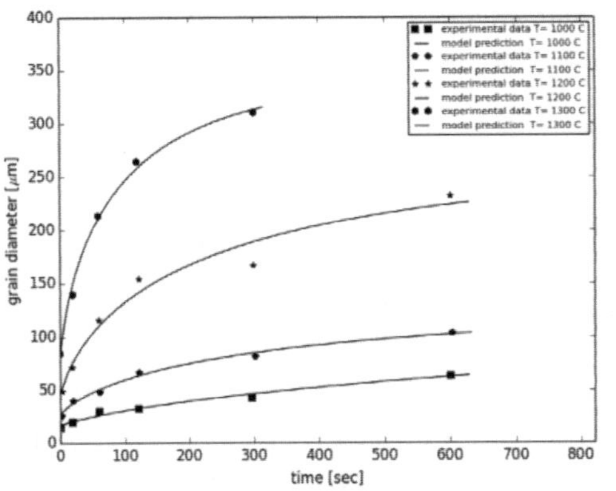

Figure 1. Isothermal grain growth data and model fit for a Mn-Ni-Cr-Mo steel (steel G) [3].

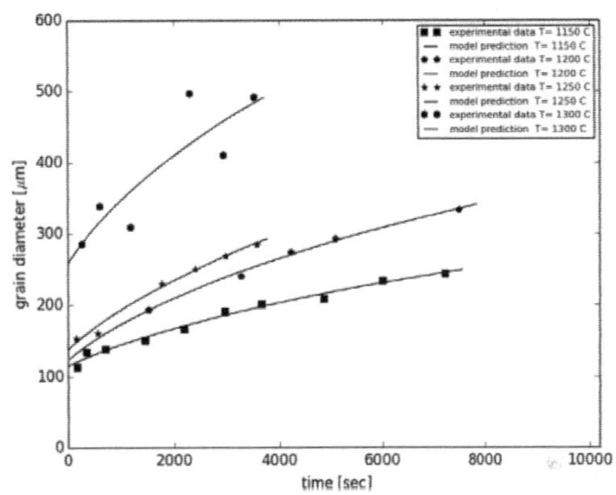

Figure 2. Isothermal grain growth data and model fit for a C-Mn steel (steel 1) [12].

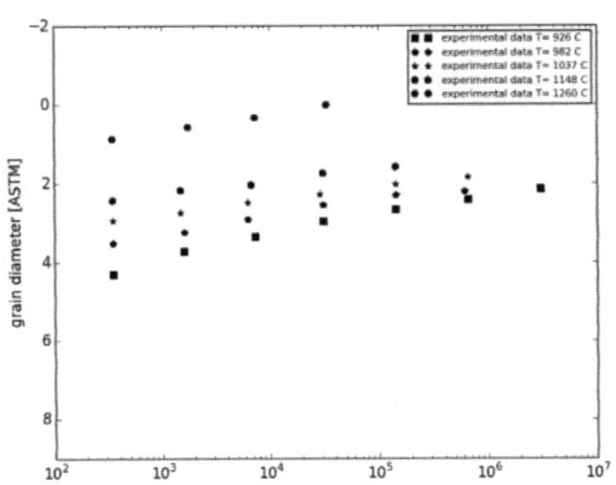

Figure 3. The isothermal grain growth data for steel 1with the double logarithmic axes as presented in [13].

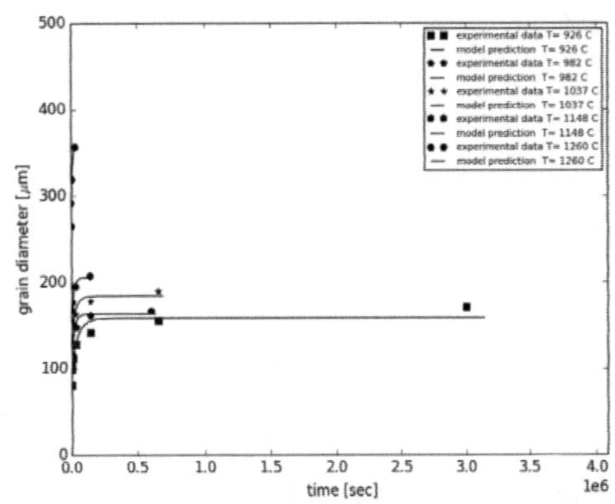

Figure 4. The isothermal grain growth data and model fit for steel 1 from [13].

IOP Conf. Series: Journal of Physics: Conf. Series **1270** (2019) 012014 doi:10.1088/1742-6596/1270/1/012014

Short times are quantified here using the characteristic time, as defined by Grey and Higgins:

$$t_{\mathrm{char}} = \frac{R_{\mathrm{limit}}^2}{2\gamma M} \tag{2}$$

When t_{max} is the longest soaking time in an isotherm, the ratio $t_{\mathrm{max}}/t_{\mathrm{char}}$ is a reasonable indicator for the applicability of the Grey-Higgins model: for values smaller than 10 a satisfactory goodness-of-fit for the Grey-Higgins model is found. Deviations between the experimental data and the model were found for ratios larger than 10, which only occur for the data of Miller [13].

5.2. The reduced mobility

The reduced mobility (equation (1)) has been estimated for each isotherm with more than 3 data points using a least squares fit. The reduced mobilities have been plotted as a function of the reciprocal of the temperature 1000/T for each composition; three examples are shown in figures 5 to 7. The compositions for these figures correspond to those in figures 1 to 4.

The reduced mobility versus 1/T plots can be well represented by an Arrhenius function for most compositions. Taking into account the estimated errors in the reduced mobility the logarithm of the reduced mobility is a linear function of 1/T:

$$\gamma M = \gamma M_0 \exp\left(\frac{-Q}{RT}\right) \tag{3}$$

However, some compositions deviate from a linear ln(M) versus 1/T relation, and a further analysis of these cases has yet to be done.

The Arrhenius function (3) has been applied to all compositions with two or more isotherms and the pre-factor γM_0 and the activation energy have been estimated from a (second) least squares fit. An overview of the γM_0 versus the activation energy of all compositions is given in figure 8.

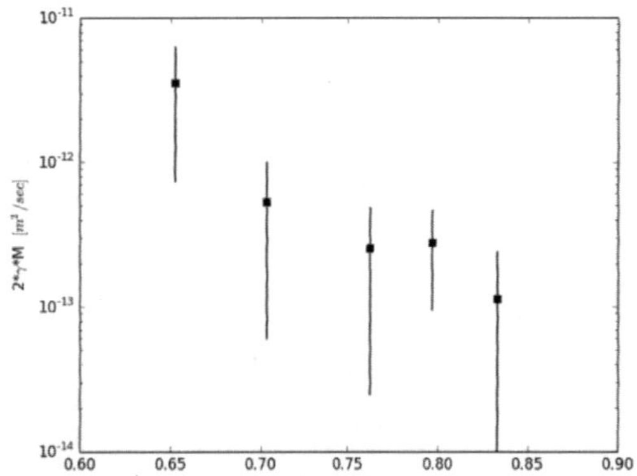

Figure 5. The reduced mobility versus 1/T for steel 1 from ref. [13].

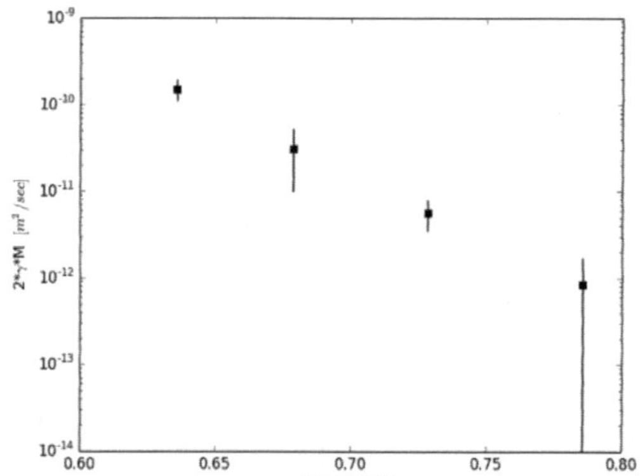

Figure 6. The reduced mobility versus 1/T for a Mn-Ni-Cr-Mo steel (steel G) [3].

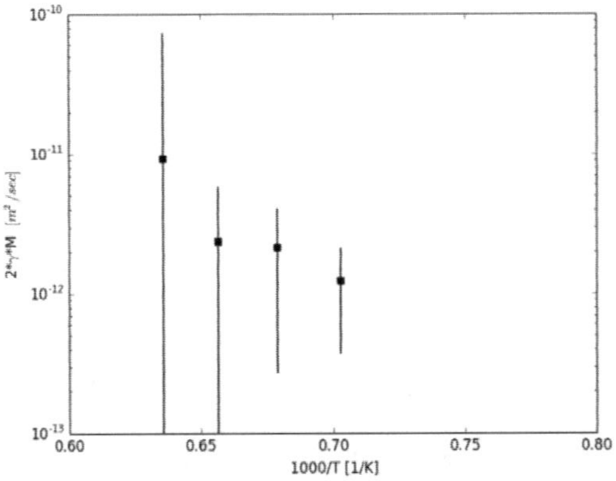

Figure 7. The reduced mobility versus 1/T for a C-Mn steel (steel 1) [12].

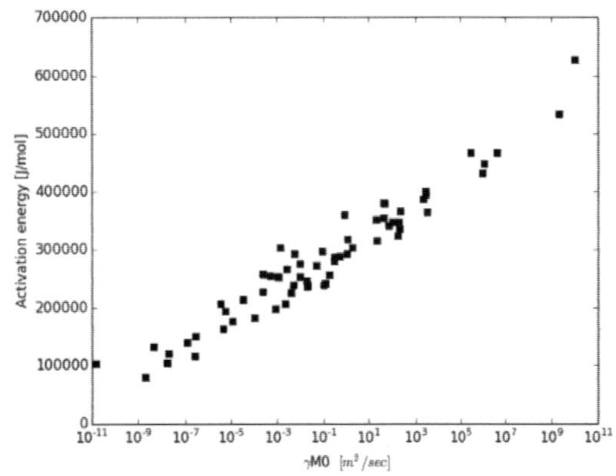

Figure 8. The activation energy versus the pre-factor of the reduced mobility for the compositions in the database.

6. Discussion and conclusions

This paper presents the preliminary results of an analysis on a collection of grain growth data in austenite. Most of the isotherms can be accurately represented by a model, which was proposed by Grey and Higgins, although data from one source indicates that this model fails for very long soaking times. A more severe test of the Grey-Higgins model for short times, namely a study of the distributions of the residuals as a function of time, has to be done yet.

For most of the compositions the reduced mobility can be described as an Arrhenius function of the temperature. However, the most remarkable finding is the large variety in the values for the activation energy of the mobility; they range from 100 kJ/mol to more than 600 kJ/mol (figure 8). Especially the very high values should be compared with the activation energies of the diffusivities of solutes (in the order of 250 kJ/mol), which enter the classical solute drag models of Cahn and Hillert [4, 5]. These models cannot, therefore, explain the highest activation energies for the reduced mobility, unless rather unrealistic values are used for the physical properties in these solute drag models. Alternative explanations are not available yet.

7. Acknowledgements

The author wants to thank dr.ir. P. van der Wolk for critically reading the manuscript and for writing the program to digitize the figures. Fruitful discussions with Prof.dr.ir. J. Sietsma of Delft University of Technology are much appreciated.

References

[1] Lee S-J and Lee Y-K 2008 Prediction of austenite grain growth during austenitization of low alloy steels *Materials and Design* **29** 1840-4
[2] Lee S-J 2013 Predictive model for austenite grain growth during reheating of alloy steels *ISIJ International* **53 (10)** 1902-4
[3] Uhm S, Moon J, Lee C, Yoon J and Lee B 2004 Prediction model for the austenite grain size in the coarse grained heat affected zone of Fe-C-Mn steels: considering the effect of initial grain size on isothermal growth behavior *ISIJ International* **44** 1230-7
[4] Cahn J W 1962 The impurity-drag effect in grain boundary motion *Acta Metallurgica* **10** 789-98
[5] Hillert M and Sundman B 1976 A treatment of the solute drag on moving grain boundaries and phase interfaces in binary alloys *Acta Metallurgica* **24** 731-43

[6] Machlin E S 1962 Theory of solute atom limited grain boundary migration *Transactions of the Metallurgical Society of AIME* **224** 1153-67
[7] Hersent E, Marthinsen K and Nes E 2013 The effect of solute atoms on grain boundary migration: A solute pinning approach *Metallurgical and Materials Transactions A* **44** 3364-75
[8] Westengen H and Ryum N 1978 On the effect of solute atoms on grain-boundary migration *Philosophical Magazine A* **38(3)** 279-95
[9] Yogo Y, Tanaka K, Ikehata H, Iwata N, Nakanishi K and Ishikawa T 2011 Calculation for grain growth rate of carbon steels by solute drag model considering segregation effect of each substitutional element *Materials Science and Technology* **27** 1593-8
[10] Yogo Y, Tanaka K, Ikehata H, Iwata N, Nakanishi K and Ishikawa T 2012 The solute drag model to calculate grain growth rate at high temperatures in carbon steels *Materials Science Forum* **706-709** 1574-9
[11] Grey E A and Higgins G T 1973 Solute limited grain boundary migration: A rationalisation of grain growth *Acta Metallurgica* **21** 309-21
[12] Kobayashi H 1977 Effect of vanadium and niobium on austenite grain growth kinetics in low carbon high strength steel *Tetsu-to-Hagane* **63** 73-9
[13] Miller O O 1951 Influence of austenitizing time and temperature on austenite grain size of steel *Transactions of the American Society for Metals* **43** 260-89

IOP Conf. Series: Journal of Physics: Conf. Series **1270** (2019) 012015 doi:10.1088/1742-6596/1270/1/012015

Effect of pearlite morphology on recrystallization behaviour in cold rolled high strength automotive steel sheets

S Janakiram[1], J Gautam[1] and L A I Kestens[2]
[1] School of Engineering Sciences and Technology, University of Hyderabad, Gachibowli, Hyderabad, Telangana 500046, India
[2] Department Electromechnaical Systems and Materials, Ghent University, Ghent, Belgium

Email: sabavath.janakiram@gmail.com

Abstract The present study is mainly focused on recrystallization behaviour and associated microstructure and texture evolution with the effect of second phase as pearlite. Two cold rolled sheets with 80% cold reduction consisted of coarse and fine pearlite bands along with ferrite. Cold rolling was followed by interrupted annealing at different soaking time to study the kinetics and recrystallization texture evolution. Results are analysed with the help of micro-Vickers hardness, microstructure and texture analysis through SEM and Electron Backscattered Diffraction (EBSD). Results clearly show the effect of pearlite band size on recrystallization kinetics and texture evolution. Faster recrystallization is observed for fine pearlite band compared to coarse band. Heterogeneous nucleation of grains took place at ferrite pearlite interface, grain boundaries of RD-ND orientations and within the deformed ferrite. Fragmentation and spherodization of pearlite lamellae takes place inside the colony of pearlite with increase in soaking time.

1. Introduction

High strength steels (HSS) generally consist of second phase mainly to enhance the strength. HSS have high strength to weight ratio and low formability, which is the major drawback for these steels in automotive applications. Formability of steels is determined by their microstructure, texture, grain size and second phase distribution [1-2]. Formability can be enhanced by formation of {111}//ND fibre orientation on recrystallization in single phase or extra low carbon steels [2]. As HSS consists of two phases, continuous efforts are being made to increase the formability of these steels by varying initial microstructures ferrite with pearlite/ bainite/martensite [3-4]. The microstructure evolution, recovery, recrystallization texture of initial ferrite pearlite microstructure to final ferrite martensite microstructure of dual phase steels has been reported [5]. After that study it was concluded that the effect of cold rolled ferrite –pearlite banded microstructure on recrystallization is topic of interest, as the pearlite banded structure was completely transformed to martensite on intercritical annealing [5]. The banded structure consists of carbon rich pearlite and low carbon ferrite and on intercrictal annealing the carbon rich region of pearlite was transformed to martensite. The reasons for banding and their effects are reported in [6]. This banded structure was not well understood in terms of recrystallization. However, Recrystallization kinetics, grain size and texture are influenced by the presence of second phase [7]. Recrystallization behaviour is affected in terms of driving force for recrystallization, sites of nucleation, size and distribution of second phase [7]. Whereas least importance was given to the softening and recrystallization behaviour with 80% cold reduction of initial ferrite-pearlite banded microstructure.

The aim of the present work is to study the recrystallization behaviour and texture evolution of 80% cold rolled C-Mn high strength steels with the effect of second phase as pearlite having variation in its morphology.

2. Experimental procedure

Material used in this investigation is high strength steel with composition C-0.09, Mn-1.65, Si <0.028, Nb-0.001 and Cr-<0.057. The alloy was casted and pre rolled up to 24 mm thick plate at Tata Steel Europe. Further hot rolling was carried out, first one at high temperature having recrystallization of austenite to occur and second one at lower temperature where austenite recrystallization did not occur. Both hot rolling scheme were followed by coiling at pearlite start temperature to create ferrite pearlite microstructure. Coarse grained ferrite-pearlite microstructure (here after referred as CP) was obtained by hot rolling above the austenite recrystallization temperature, whereas fine banded ferrite-pearlite microstructure (here after referred as FP) was obtained by hot rolling below the austenite recrystallization temperature. Subsequently hot rolled sheets are cold rolled to 1.2 mm thickness with 80% reduction. Interrupted annealing was carried out in salt bath furnace at 700°C with varying soaking time. Hardness was measured by micro Vickers indenter with 500 gm load and 20 seconds dwell time. Microstructural details and texture data were acquired from field emission gun scanning electron microscope combined with EBSD of TSL OIM. Microstructural observations were performed on the Rolling and Normal direction (RD-ND) plane of the sheet. Samples were polished as per the standard procedure for microstructural observation and etched with 2% nital. TSL OIM version 7.2 was used to analyse the data. Grain orientation spread criteria (GOS) was used in the range of 0.2-1.4 and grain size greater than 1 micron was used in partitioning of the deformed and recrystallized ferrite. Partitioning of the recrystallized grains from deformed grains by GOS criteria reported by [9].

3. Results

3.1 Effect of pearlite morphology on recrystallization annealing

Cold rolled microstructures consists of ferrite and pearlite. Pearlite band thickness was varied as shown in figure 1(a) coarse pearlite (CP), 1 (b) fine pearlite (FP). The variation of hardness in ferrite pearlite and recrystallization fraction of only ferrite in CP and FP microstructures after interrupted annealing at 700°C for different soaking times are shown in figure 1(c). Hardness values are reported from the bulk of the samples consisting of ferrite-pearlite. Whereas recrystallization fraction of only ferrite calculated from EBSD maps using GOS criteria is reported in the figure 1(c). Hardness curve shows two characteristics differences, increase in hardness from initial condition for first 10 s and starts to decrease thereafter with increasing soaking time, partitioned by dashed line as in figure 1. Increase in hardness form cold rolled condition for first 10 s is designated as part A, indicates hardening of ferrite-pearlite microstructure. Hardness decrease after 10 s is designated as part B, shows initiation of softening in ferrite pearlite microstructure. Softening consists of both recovery and recrystallization. This hardening behaviour is similar in CP and FP microstructures for first 10 s. Softening (recovery and recrystallization) behaviour varies with changing the behaviour of pearlite morphology with CP and FP microstructure with increase in soaking time.

In case of coarse pearlite (CP) microstructure it was observed that hardness starts decreasing below the cold rolled condition after 150 s of soaking time. In case of fine pearlite (FP) hardness decreases below cold rolled condition slightly before 60 s of soaking. Recrystallization of ferrite fraction shows that recovery dominates up to 60 s of soaking time and recrystallization starts thereafter in case of CP, whereas in case of FP, recovery starts at 10 and finishes by 20 s of soaking time, recrystallization starts thereafter from 30 s. Two horizontal dotted lines were drawn to track changes of recrystallization in initial 10 %, intermediate 38% and final stage 55 % recrystallization fraction. In initial stage, time taken to reach 10% recrystallization fraction of ferrite with two different pearlite morphology was different. In case of CP time taken was 180 s, where as in FP it reaches in 60s of soaking time as indicated by the dotted line in figure 1. In intermediate stage, the time taken to reach the 38% recrystallization was 300 s for CP and 150 s for FP microstructures. In final stage, the time to reach 55% recrystallization fraction was 420 s in CP and 180 s in FP. This indicates that rate of recrystallization is faster in the FP microstructure, whereas recrystallization in CP microstructure is slower. In summary, hardening takes for first 10 s of annealing time in CP and FP microstructures. Changing the morphology of pearlite showed significant changes in the softening behavior. Recovery time decreases and recrystallization fraction increases in time with the FP microstructure, whereas recovery time increases, followed by slower recrystallization in the CP microstructure.

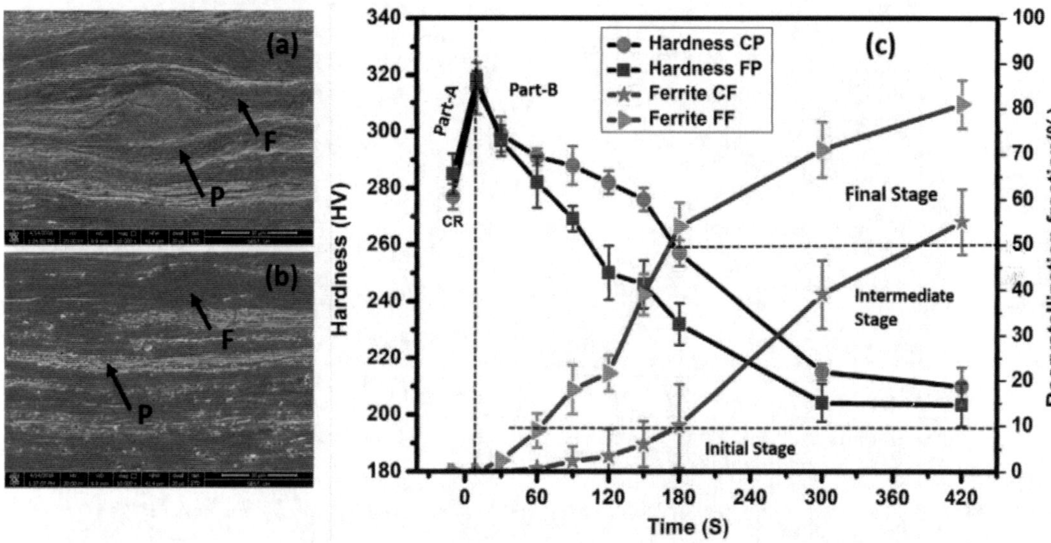

Figure 1. Shows (a) coarse pearlite (CP) (b) fine pearlite (FP) cold rolled microstructure, (c) variation of hardness in ferrite-pearlite and recrystallization fraction of ferrite after annealed at 700°C for different soaking time in coarse pearlite. (P-Pearlite) and (F-Ferrite)

3.2 Evolution of Microstructure during recrystallization:

Microstructures of CP and FP after interrupted annealing of specimens at 700°C for different soaking time are shown in figure 2. High resolution inserts of the pearlite phase are placed on the top right corner of each image, allowing to compare the changes of pearlite in hardening and softening stages. Changes in ferrite and pearlite morphologies were tracked after hardening, at initial 10% recrystallized, intermediate 38% recrystallized and at final stage 55% of recrystallized volume fraction. The microstructure consisted of ferrite in grey colour, a pearlite colony of white colour, recrystallized ferrite and fragmented pearlite.

The microstructures of figure 2 show coarse and fine severely pancaked ferrite-pearlite microstructure after hardening stage; the ferrite pearlite microstructure remains unchanged after 10 s of soaking time compared to the cold rolled condition. The closely spaced lamellar morphology of the CP microstructures is retained as shown in figure 2(a). In the case of FP microstructures, a fragmented cementite morphology appears, cf. figure 2(a). In the initial stage of recrystallization, ferrite recrystallizes at the ferrite-pearlite interface, as indicated by dotted circles figure 2 (b). In the same figure it is observed that fragmentation of the lamellar structure near the ferrite-pearlite interface, inside the colony of the lamellar structure is still maintained as observed in the high magnified image.

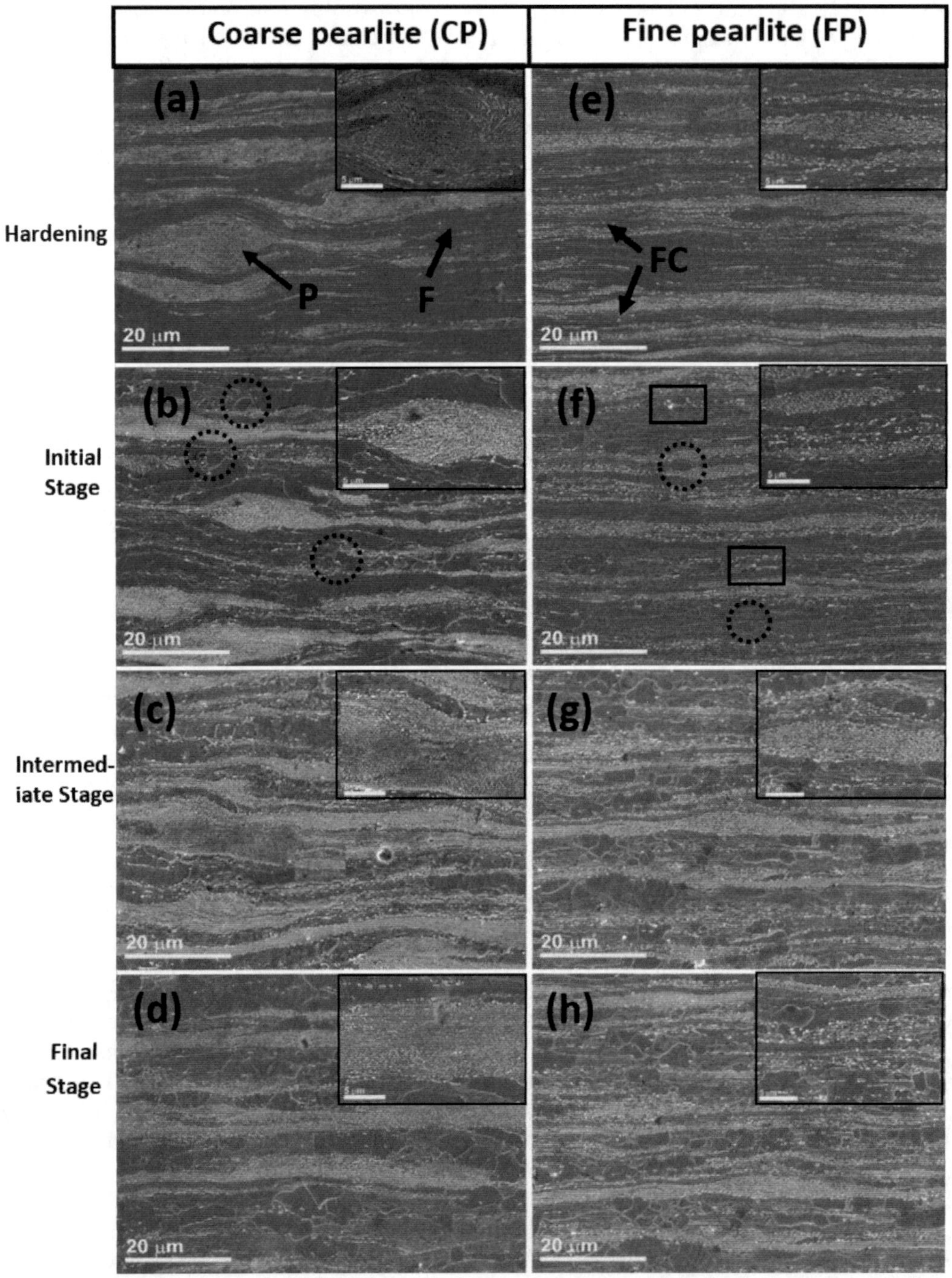

Figure 2. Microstructure of annealed (a) 10 s (b) 180 s (c) 300 s (d) 420 s time in coarse pearlite CP and (e) 10 s (f) 60 s (g) 150 s (h) 180 s of soaking time in fine pearlite FP
DF- deformed ferrite P- pearlite FC- Fragmented Cementite

In the FP microstructure a fragmented cementite morphology was present in the pearlite colony with coarse cementite particles at the ferrite-pearlite interface; recrystallization of ferrite takes place at the interface as indicated by the dotted circles in figure 2 (f). In the same figure it can be seen that ferrite grains recrystallize at the fragmented cementite particles indicated by the rectangular box. At the intermediate stage with increasing soaking time, the ferrite recrystallization fraction increases to 38% and the lamellar structure is getting fragmented in fine spherical cementite particles and pearlite spheroidizes, cf. figure 2 (c). In the FP sample, as the recrystallization fraction increases, ferrite grains recrystallize at the ferrite pearlite interface as well as in the vicinity of fragmented cementite particles, while pearlite colonies remain unchanged. In the final stage of recrystallization, recrystallization fraction increases and, as shown in figure 2(d), the lamellar pearlite structure is now completely spheroidized. The recovered ferrite is present, still mainly adjacent to pearlite and in between two pearlite colonies. In the FP microstructure, fragmentation of plates takes place and coarsening of cementite plates is observed mainly at the interface, cf. figure 2(h). In the overall microstructure there was no recrystallization observed in the pearlite lamellar structure neither with coarse nor fine pearlite bands.

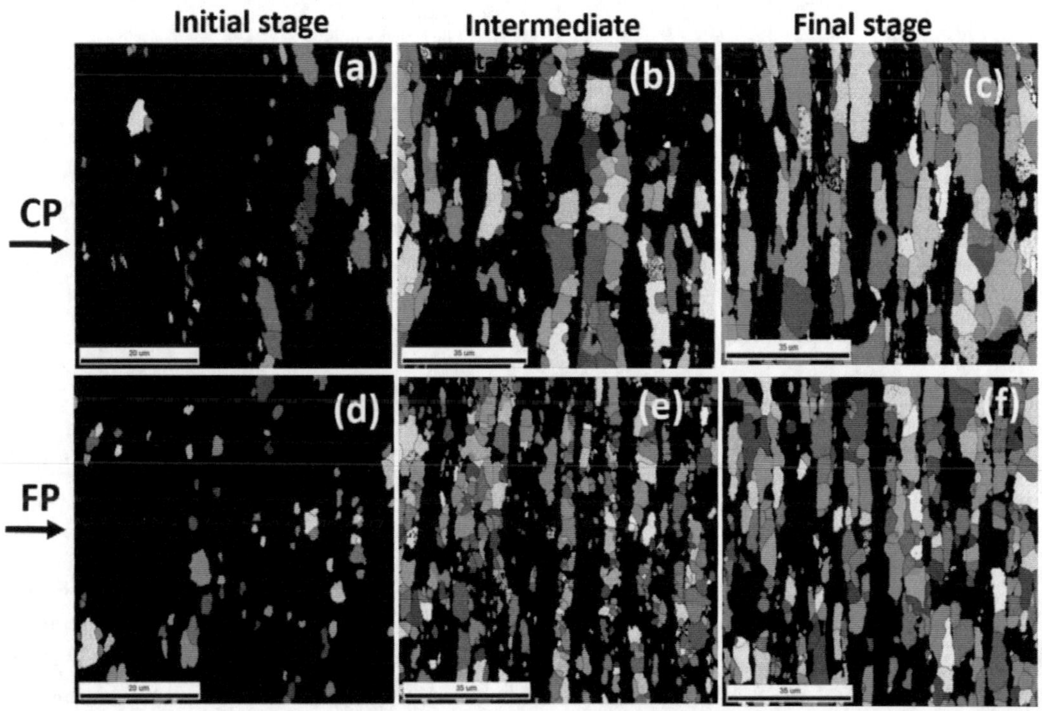

Figure 3. IPF of recrystallized ferrite grains (a) 180 (b) 300 (c) 420 s soaking time in CP and (d) 60 (e) 150 (f) 180 s of soaking time in FP microstructures

3.3 Evolution of Recrystallization texture:

The recrystallization textures of CP and FP specimen were investigated by EBSD. The ODFs were plotted for $\varphi_2 = 45^0$sections of both deformed and recrystallized grains, cf. figure 4.

*3.3.1 In Coarse pearlite microstructure***,** in the initial stage of recrystallization, where the recrystallized fraction was 10 %, fragmentation of pearlite lamellar structure appears at the ferrite-pearlite interface as in figure 2(b). A strong texture appeared at cube and near cube orientations with secondary maxima on the RD fibre near to {112} <1-10>. In the intermediate stage, with 38% of recrystallized volume fraction, where pearlite lamellar structure gets fragmented, cf. figure 2 (c), a weak texture appears near {111} <1-10> and {111} <0-11> on the ND fibre and on {112} <1-10> on the RD fibre, along with some minor components of {114} <5-91> and {115} <0-51> orientations, cf. figure 4 CP (d). In the final stage of recrystallization (55%) where complete sperodization of pearlite lamellar morphology takes place, as shown in figure 2 (d), the overall texture intensity decreases on the RD fibre with maxima at {114} <1-10>. The ND fibre weakens and the Goss component appears. With the increase in recrystallization fraction, we see that deformed ferrite and pearlite consists of a strong RD fibre and very weak ND fibre texture components as shown in figure 4(a-c-e).

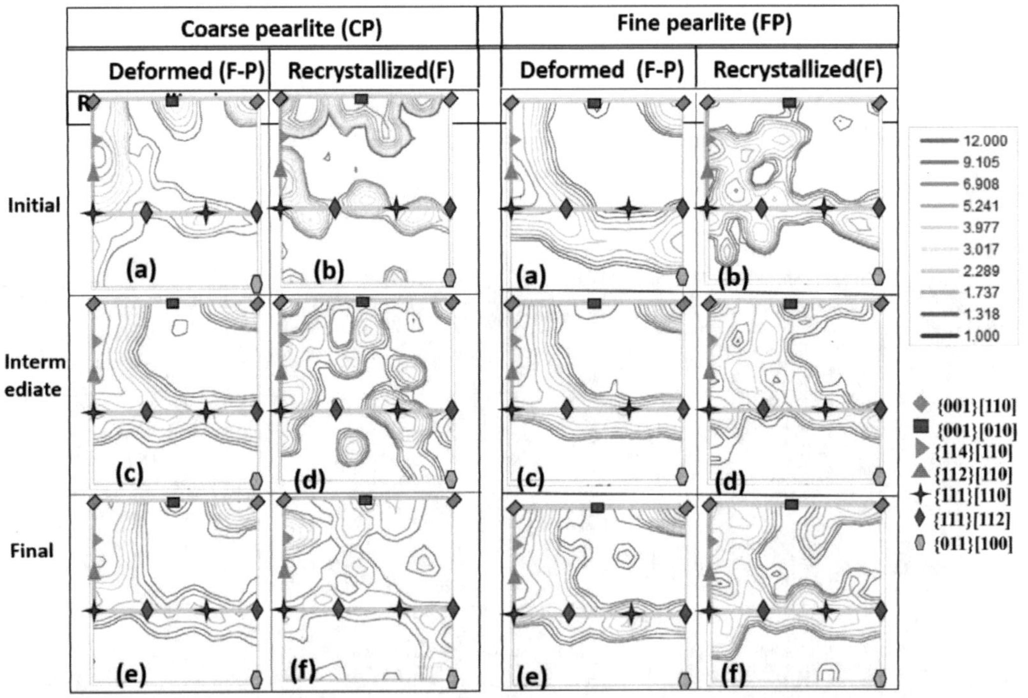

Figure 4. ODF of ϕ_2 =45 section of deformed (ferrite-pearlite) and recrystallized (ferrite) in coarse and fine pearlite microstructures
Ferrite-pearlite (F-P), Ferrite (F)

3.3.2 In the fine pearlite microstructure, in the initial stage of recrystallization with 10% of recrystallized volume fraction, the RD fibre texture dominates with texture maxima at {113} <1-10> with 9.2x mrd. Whereas weak ND fibre components are present with {111} <1-10> and near to {111} <-1-1 2> with 3.2x mrd. Along with it some minor components of {114} < 1-92> and {223} <7-10 2> orientations, cf. figure 4 FP (b). In the intermediate stage, with 38% recrystallized volume fraction, weakening of the texture takes place as shown in figure 4 (d), with texture maxima at {114} <1-10> and {114} <2-61> on the RD fibre. Whereas maxima intensity on the ND fibre appeared at {111} <1-10> and {111} <0-1 1> with 3x mrd. In the final stage (55% recrystallized volume fraction), coarsening of cementite at the ferrite-pearlite interface, the intensity increases to 4x mrd on same components as mentioned above for 55% recrystallized volume fraction.

4. Discussion:

Cold rolled microstructure with two distinct pearlite morphologies of coarse and fine pearlite bands showed characteristic differences in terms of recrystallization kinetics, microstructure and texture in initial, intermediate and final stage of recrystallization at 700°C. In the coarse pearlite (CP) microstructure, the pearlite exhibits a lamellar morphology, whereas in the fine pearlite (FP) microstructure a fragmented pearlite morphology is shown, with RD-ND fibre textures after 80% cold rolling, irrespective. The cold rolled microstructures and textures of CP and FP microstructures are well reported in [9]. The cold rolled pearlite band size and morphology had a clear effect on the annealing kinetics, cf. figure 1(c), whereby two distinct stages A and B are present. In part A, after 10 s of soaking time, hardening takes place within the ferrite pearlite microstructure. The reason for hardening can be attributed to pinning of dislocations by the carbon atoms, similar to bake hardening behaviour [10]. In part B, softening takes place both in CP and FP microstructures, which consists of recovery and recrystallization. The recrystallization was slower and sluggish till the hardness reaches to its cold rolled value. Once the hardness drops below the cold rolled condition, the recrystallization fraction increases with soaking time. From ferrite recrystallization curves as shown in figure 1, it was evident that slower and more sluggish recrystallization takes place in the CP samples as compared to FP samples. In the CP microstructure, cf. figure 2 (b) and figure 3 (a), it was confirmed that concurrent recovery and recrystallization taking place, whereby recovered grains exhibit bigger grain size when compared to the recrystallized grains. It was further confirmed with ODF maps of figure 4 (b) in the CP microstructure, that recovered grains were of Cube ({001}<010>) and near to cube orientation. As these orientations are having low stored energy, recrystallization of these grains is not possible. The grains recovered faster due to higher amount of cold deformation and low substructure present in them. The reason for concurrent recovery and recrystallization, might be due to spherodization of the pearlite lamellar structure at the ferrite-pearlite interface. Spherodization leads to carbon dissolution into the ferrite grains and pin the dislocation by reprecipitation and inhibit the progress of recrystallization. It is also the reason for the increase in initial hardness from cold rolled condition [10-11]. Initially, due to prolonged recovery in the CP microstructure the amount of stored energy decreases for nucleation and growth. Spheroidized particles also pin the migration of high angle boundaries with Zener solute drag effect and retard recrystallization mainly at the interface of ferrite and pearlite [11]. With the increase in recrystallization fraction the amount of carbon in ferrite is increasing due to spherodization and it weakens the ND fibre and strengthens the Goss component ({110}<001>), cf. figure 4 (f), which is in good agreement with results reported by [12-13]. In the final stage of recrystallization overall weakening of texture takes place where the pearlite lamellar structure is spheroidized.

5. Conclusions

In the present work two distinct types of pearlite bands consisting of different morphologies were studied during recrystallization at 700°C with varying soaking time

- At the very early stage of annealing, a sudden increase in hardness has been observed.
- The micro Vickers hardness results shows different softening behaviour for coarse and fine pearlite. Steel sheet with fragmented lamellae of fine pearlite bands soften faster in comparison to lamellar, coarse pearlite bands.
- The recrystallization fraction of ferrite is influenced by the pearlite morphology. In the case of coarse pearlite recrystallization, the fraction is lower in comparison to fine pearlite.
- Recovery and recrystallization is slower and more sluggish in case of coarse pearlite due to spherodization of cementite. Whereas recrystallization dominates in the case of fine pearlite band.
- Weakening of the texture takes place with increase in recrystallization fraction mainly due to solute carbon and C-Mn interaction.
- In the very early stage of annealing (part A) hardening occurs, which might be due to dislocation pinning by carbon diffusion to free mobile dislocations.

6. References

[1] Hutchinson W B 1984 *Int. Metal Rev* **29** 25
[2] Ray R K, Jonas J J and Hook R E 1994 *Intel. Metal Rev* **39** 129
[3] Ray R K 1985 *Journal of materials sciences and letters* 67-70

[4] Gautam J, Miroux A, Moerman J, Barbatti C and Kestens L A I 2012 *Materials Science Forum* 778-781

[5] Peranio N, Li Y J, Roters F and Raabe D 2010 *Materials Science and Engineering A* **527** 4161–4168

[6] Francisca G Caballero, Andrea Garci Junceda, Carlos Capdevila and Carlos Garcia de Andres 2006 *Material Transactions* 2269- 2276

[7] Humphrey F J and Hatherly M 2004 *Recrystallization and Related Annealing Phenomena* Elsevier Oxford GB 136

[8] Dong Kyu Kim, Won-woong Park, Ho won lee, Seong hoon kang and Yong Taek 2013 *Journal of Microscopy* **252** 204–216

[9] Janakiram S, Prakash Gautam J, Miroux A, Moerman J and Leo Kestens A I 2019 *Diffusion Foundations* **22** 84-93

[10] Timokhina I B, Hodgson P D, Ringer S P, Zheng R K and Pereloma E V 2009 *steel research Int.***80** 7

[11] Ghiabakloo H and Kazeminezhad M 2017 *Met. Mater Int* **23** 5 984-993

[12] Hutchinson W B and Ushioda K 1984 *Scand. J. Metal* **13** 269

[13] Ushioda K, Hutchinson W B, Agrenu J and Schlippenbach U V 1986 *Materials Science and Technology* **2** 807-815

Acknowledgements Author would like to thank DST FIST and DST PURSE for funding instrumentation facility FESEM-EBSD. UGC and Ministry of Tribal affairs for providing fellowship. Tata Steel Research, Development and Technology (RD&T), IJmuiden, The Netherlands for supplying the materials.

Effect of initial microstructures prior to cold-rolling and intercritical annealing on ferrite recrystallization and ferrite-to-austenite phase transformation in Nb bearing low-carbon steels

Toshio Ogawa[1,*], Hiroyuki Dannoshita[2], Yoshitaka Adachi[1] and Kohsaku Ushioda[3]
[1] Department of Materials Design Innovation Engineering, Graduate School of Engineering, Nagoya University, Aichi, Japan
[2] Department of Mechanical Engineering, Materials Science, and Ocean Engineering, Graduate School of Engineering Science, Yokohama National University, Kanagawa, Japan
[3] Nippon Steel Research Institute Corporation, Tokyo, Japan

*Corresponding author's e-mail address: ogawa.toshio@material.nagoya-u.ac.jp

Abstract. This study investigates the relationship between the initial microstructures of Nb bearing low-carbon steels and their microstructural evolution during subsequent intercritical annealing. The initial microstructures of specimens P, B, and M comprised ferrite/pearlite, bainite, and martensite, respectively. In the specimen lacking of Nb, the time for ferrite recrystallization during annealing decreased in the order of M > P > B. Recrystallized ferrite grains in specimen M are fine and equiaxed. The ferrite-to-austenite phase transformation during intercritical annealing occurred more rapidly in specimen M than in specimens P and B. The high austenite fraction in specimen M is presumably caused by the large number of austenite nucleation sites due to the ferrite grain refinement. In the Nb bearing specimens, the time for ferrite recrystallization decreased in the order of M > B > P. The Nb addition retarded the ferrite recrystallization and accelerated the ferrite-to-austenite phase transformation. The high austenite fraction in the Nb bearing specimen is presumably caused by the large number of austenite nucleation sites owing not only to the ferrite grain refinement, but also to the increased number of non-recrystallized ferrite grains.

1. Introduction

Dual phase (DP) steels have good ductility and have been used in various automobile parts [1]. DP steels generally consist of ferrite and martensite, and their mechanical properties mainly depend on the volume fraction, size, morphology, and distribution of martensite [2-5]. Therefore, to ensure high ductility of DP steels, a method to control the above-mentioned microstructural features of martensite is demanded.

DP steels are typically manufactured by continuous casting, hot-rolling, cold-rolling, and subsequent annealing. Previously, we found that the volume fraction and distribution of martensite in low-carbon steels strongly depend on the initial microstructures prior to cold-rolling [6,7]. For instance, recrystallized ferrite grains are fine and equiaxed when the initial microstructure consists of martensite, which leads to the acceleration and homogenization of austenite formation at the early stage of intercritical annealing [6]. The austenite at intercritical temperature transforms into martensite during the cooling process. This implies an important role for the morphology of recrystallized ferrite grains in martensite formation in low-carbon steel. During intercritical annealing, the microstructure is highly dependent on the interaction between ferrite recrystallization and the ferrite-to-austenite phase transformation.

We also investigated the interaction between ferrite recrystallization and the ferrite-to-austenite phase transformation during intercritical annealing in low-carbon steels [8,9]. For instance, we confirmed that ferrite recrystallization is remarkably retarded when it accompanies a ferrite-to-austenite phase transformation during intercritical annealing [8]. In Nb bearing low-carbon steels, the Nb addition retards the ferrite recrystallization and accelerates the ferrite-to-austenite phase transformation during intecritical annealing [9]. Clearly, the Nb addition results complicated interaction of the ferrite recrystallization and the ferrite-to-austenite phase transformation during intercritical annealing. However, the relationship between the initial microstructures of microalloyed low-carbon steels and their microstructural evolution during annealing remains unknown. Designers of low-carbon steel materials would desire to understand the roles of the initial microstructures and microalloying elements on the microstructural evolution during annealing.

The present study aims to relate the initial microstructures of Nb bearing low-carbon steels to their microstructural evolution during subsequent intercritical annealing. In particular, it is dedicated to metallurgical phenomena (ferrite recrystallization and the ferrite-to-austenite phase transformation) during cold-rolling and intercritical annealing.

2. Experimental procedures

The chemical compositions of two steels, one Nb-less specimen and one Nb bearing specimen, are listed in Table 1. The results of the Nb-less specimen were reported in our previous articles [6,7]. The vacuum-melted ingots were rough-rolled to a thickness of 30 mm. The rough-rolled steels were hot-rolled to 3.0 mm at a finishing temperature of 900 °C in the austenite region, and subsequently water-cooled to 650 °C (the ferrite/pearlite specimen P), 500 °C (the bainitic specimen B), and below 100 °C (the fully martensite specimen M), and then cooled to room temperature in air. The hot-rolled sheets were cold-rolled to a thickness of 1.0 mm (reduction: 67%).

After cold-rolling, the specimens were heated to 500, 660, and 700 °C at 0.5 °C/s, and then water-quenched to room temperature within 2 s of removal from the furnace. The rapid quenching avoided the austenite-to-ferrite phase transformation. The austenite phase in these steels, present during intercritical annealing, was transformed to martensite during cooling. The Ac_1 and Ac_3 temperatures of the tested steels were approximately 670 °C and 820 °C, respectively.

The microstructure was observed by a scanning electron microscope (SEM). The microtexture of the annealed specimens was analyzed at the quarter-thickness position on the rolling plane by an electron backscatter diffraction/field emission scanning electron microscopy (EBSD/FEGSEM) system employing OIM software. The step size of the EBSD measurements was 0.2 μm. In the EBSD data analysis, the fraction of recrystallized ferrite grains was analyzed by kernel average misorientation (KAM) analysis. Regions with misorientation results below 1° in the KAM analysis were assumed as recrystallized ferrite grains. In our previous report [8], the area fractions of the recrystallized ferrite grains evaluated by KAM analysis and the point-counting method showed good agreement. The martensite fraction was estimated by the point-count method (ASTM E 562), as mentioned in our previous studies [6,7]. The martensite fraction was estimated from 2475 counting points.

Table 1. Chemical compositions of the steels (in mass%).

	C	Si	Mn	P	S	Nb	Al	N	O
Nb-less specimen	0.10	< 0.003	2.03	0.010	0.0029	< 0.003	0.027	0.0030	< 0.001
Nb bearing specimen	0.10	< 0.003	2.02	0.010	0.0030	0.050	0.027	0.0031	< 0.001

3. Results and discussion

Figure 1 shows the image quality (IQ) and KAM maps of the Nb bearing specimens heated to 660 °C (immediately below the Ac_1 temperature). After heating, non-recrystallized ferrite grains remained in the specimens P (Figure 1(a)) and B (Figure 1(c)), whereas ferrite recrystallization was almost completed in specimen M (Figure 1(e)). Previously, we reported that ferrite recrystallization in the Nb-less specimen was completed at 660 °C in specimens of all compositions [6]. Therefore, the Nb addition clearly retarded the ferrite recrystallization. Furthermore, the recrystallized fraction was larger in specimen B than in specimen P (Figure 1(b) and (d)). These results suggest that the ferrite recrystallization time during annealing decreased in the order of M > B > P. In the Nb-less specimen, we previously demonstrated that the ferrite recrystallization time during annealing decreased in the order of M > P > B [6]. It appears that ferrite recrystallization was significantly retarded by Nb addition in specimen P.

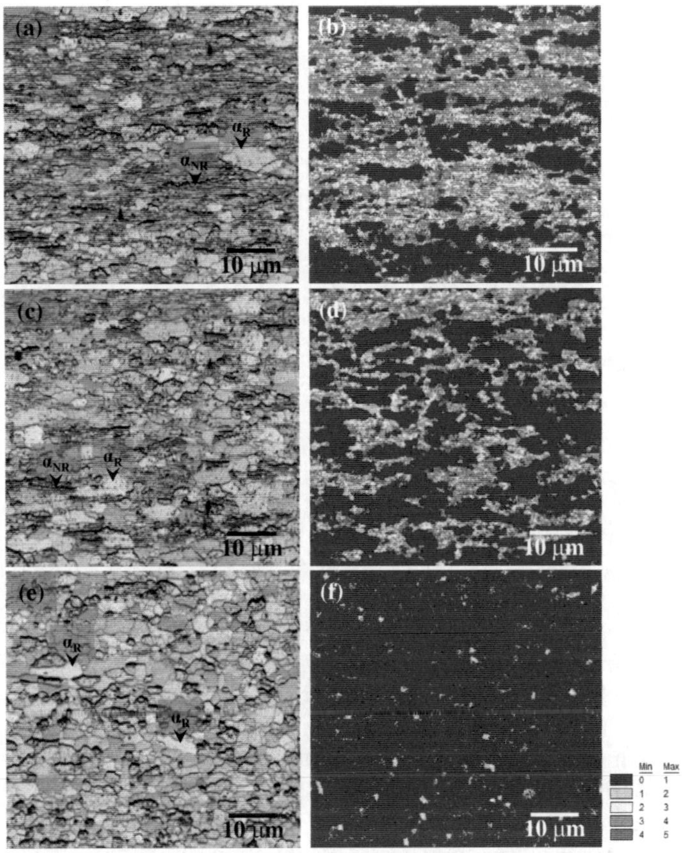

Figure 1. Image quality and kernel average misorientation maps of the Nb bearing specimens (a, b) P, (c, d) B and (e, f) M heated to 660 °C (α_R: recrystallized ferrite, α_{NR}: non-recrystallized ferrite).

Figure 2 shows the Vickers hardness changes in the Nb bearing specimens during isothermal holding at 500 °C. In specimen M, the hardness gradually decreased with holding time. Meanwhile, the hardness of specimen P slightly decreased after isothermal holding for 2000 s, while that of specimen B hardly changed. These results indicate that the recovery time of the specimens during annealing decreased in the order of M > P > B. We previously demonstrated that ferrite recovery and recrystallization occurs more rapidly in martensite than in bainite or pearlite, because the dislocation density and its heterogeneity after cold-rolling was larger in specimen M than in specimens P and B [6]. Furthermore, as more solute Nb increases prior to cold-rolling, the recovery progress becomes more retarded [10]. The amount of precipitates containing microalloying elements such as Nb and Ti increases with increasing cooling stop temperature after hot-rolling [11]. Therefore, the amount of solute Nb prior to cold-rolling was probably smaller in specimen P than in specimen B. This would explain the stronger retardation of recovery in specimen B than in specimen P.

Although the Nb bearing specimen B exhibited stronger retardation to recovery than specimen P, its ferrite recrystallized more quickly than in specimen P. The interaction between recovery and recrystallization in steels has been extensively studied [12-14]. Belyakov et al. [12] demonstrated that recrystallization is retarded when the recovery is rapid because the driving force of recrystallization decreases as the recovery accelerates. Therefore, the driving force of recrystallization is probably larger in specimen B than in specimen P. These results suggest that Nb addition can vary the effect of the initial microstructures on the recovery and recrystallization behaviors of ferrite.

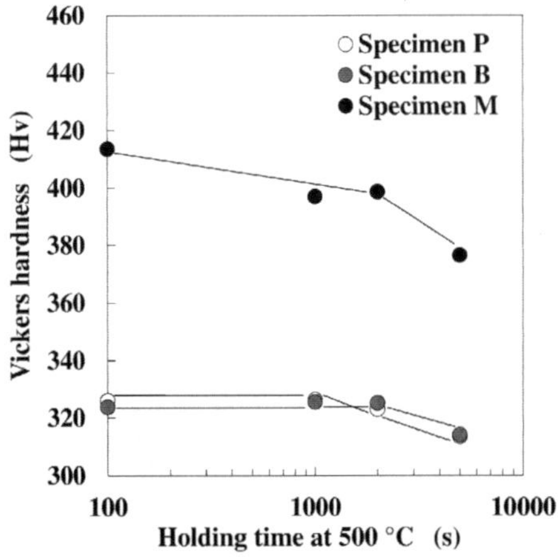

Figure 2. Changes in Vickers hardness of the Nb bearing specimens during isothermal holding at 500 °C.

Figure 3 shows the microstructures of the Nb bearing specimens heated to 700 °C (above the Ac_1 temperature). In specimens P and B, non-recrystallized ferrite grains and martensite connected along the rolling direction were observed (Figure 3(a) and (b)). In contrast, fine and equiaxed recrystallized ferrite grains were observed in specimen M, and the martensite was distributed rather homogeneously in this sample (Figure 3(c)). Figure 4 shows the fraction of martensite in the specimens heated to 700 °C. The martensite fractions in the Nb-less specimen were below 10%, irrespective to their initial microstructures. In contrast, martensite fractions in all Nb bearing specimens exceeded 10%. We previously demonstrated that the nucleation sites of austenite during intercritical annealing are mainly at the recrystallized ferrite grain boundaries and the subgrain boundaries in non-recrystallized ferrite grains [6]. In addition, we reported that increasing the number of austenite nucleation sites increases the fraction of austenite during intercritical annealing [7]. Accordingly, increasing the number of recrystallized ferrite grain boundaries and non-recrystallized ferrite grains also increases the fraction of austenite during intercritical annealing. In specimens P and B, the Nb addition increased the numbers of recrystallized ferrite grain boundaries and non-recrystallized ferrite grains (Figure 3(a) and (b)), but in specimen M, it increased only the recrystallized ferrite grain boundaries. The enhancement of recrystallized ferrite grain boundaries and non-recrystallized ferrite grains after Nb addition probably increased the fraction of austenite during intercritical annealing in all specimens. In particular, the austenite fraction was probably largest in specimen P, because the fraction of non-recrystallized ferrite grains (i.e., austenite nucleation sites) was largest in the specimen.

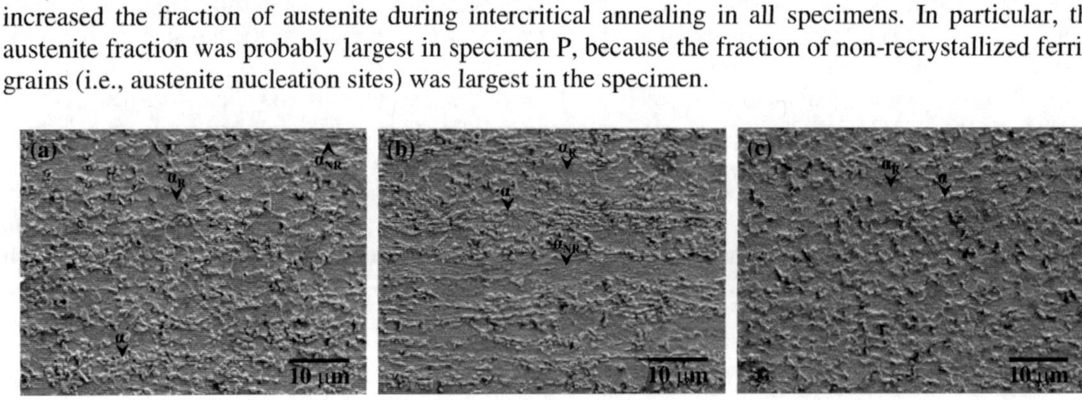

Figure 3. Microstructures of the Nb bearing specimens (a) P, (b) B and (c) M heated to 700 °C (α_R: recrystallized ferrite, α_{NR}: non-recrystallized ferrite, α': martensite).

IOP Conf. Series: Journal of Physics: Conf. Series **1270** (2019) 012016 doi:10.1088/1742-6596/1270/1/012016

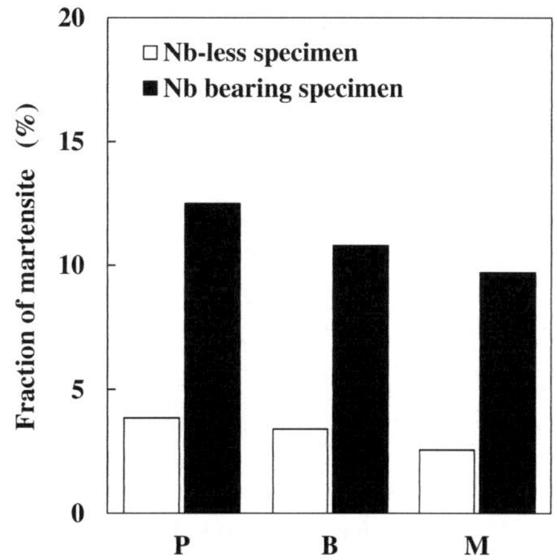

Figure 4. Fraction of martensite in the specimens heated to 700 °C (Data of Nb-less specimen were taken from a previous report [6]).

Figure 5 illustrates the effects of Nb addition and initial microstructures on the interaction between ferrite recrystallization and the ferrite-to-austenite phase transformation behaviors during heating of each specimen. The Nb and initial microstructures strongly influenced the ferrite recrystallization; in turn, the ferrite recrystallization behavior strongly affected the austenite formation behavior during intercritical annealing.

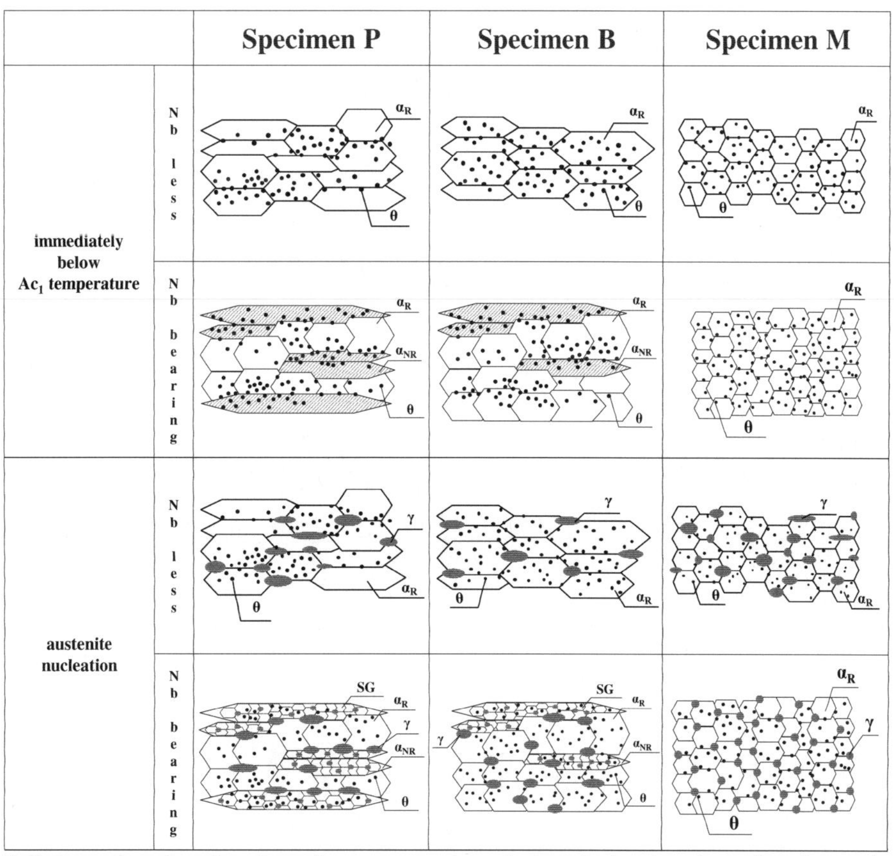

Figure 5. Schematics showing the microstructural evolution during heating of each specimen (α_{NR}: non-recrystallized ferrite, α_R: recrystallized ferrite, θ: spheroidal cementite, γ: austenite, SG: subgrain).

4. Conclusions

We investigated the effect of initial microstructures prior to cold-rolling and intercritical annealing on ferrite recrystallization and the ferrite-to-austenite phase transformation in Nb bearing low-carbon steels. The time of ferrite recrystallization during annealing decreased in the order of M > P > B in the Nb-less specimens, and in the order of M > B > P in the Nb bearing specimens. The Nb addition retarded the ferrite recrystallization during annealing and accelerated the ferrite-to-austenite phase transformation. This trend, which occurred in all specimens, was attributed to the increase in the number of recrystallized ferrite grain boundaries and non-recrystallized ferrite grains after the Nb addition, and the consequent increase in the austenite fraction during intercritical annealing.

References

[1] Takahashi M 2004 *Tetsu-to-Hagané* **100** 82
[2] Tomota Y, Tanimoto I and Kuroki K 1982 *Trans. Jpn. Soc. Mech. Eng.* A **48** 528
[3] Sugimoto K, Sakaki T, Fukusato T and Miyagawa O 1985 *Tetsu-to-Hagané* **71** 994
[4] Kondo D, Kunishige K and Ueji R 2006 *Tetsu-to-Hagané* **92** 457
[5] Jiang Z, Guan Z and Lian J 1995 *Mater. Sci. Eng.* A **190** 55
[6] Ogawa T, Dannoshita H, Maruoka K and Ushioda K 2017 *J. Mater. Eng. Perform.* **26** 3821
[7] Dannoshita H, Ogawa T, Maruoka K and Ushioda K 2019 *Mater. Trans.* **50** 165
[8] Ogawa T, Maruyama N, Sugiura N and Yoshinaga N 2010 *ISIJ Int.* **50** 469
[9] Ogawa T, Sato K, Dannoshita H, Maruoka K and Ushioda K 2016 *ISIJ Int.* **56** 2290
[10] Ogawa T, Sugiura N, Maruyama N and Yoshinaga N 2013 *Mater. Sci. Eng.* A **564** 42
[11] Sugiura N, Yoshinaga N, Kawasaki K, Yamaguchi Y, Takahashi J and Yamada T 2008 *Tetsu-to-Hagané* **94** 179
[12] Belyakov A, Wei F G, Tsuzaki K, Kimura Y and Mishima Y 2007 *Mater. Sci. Eng.* A **471** 50
[13] Stüwe H P, Padilha A F and Siciliano Jr F 2002 *Mater. Sci. Eng.* A **333** 361
[14] Zurob H S, Hutchinson C R, Brechet Y and Purdy G 2002 *Acta Mater.* **50** 3075

Acknowledgment

This work was supported by a research grant from the Hitachi Metals · Materials Science Foundation, Japan.

IOP Conf. Series: Journal of Physics: Conf. Series **1270** (2019) 012017 doi:10.1088/1742-6596/1270/1/012017

Microstructural and micro hardness variation during low temperature recovery annealing in C-Mn high strength Steels

S Janakiram[1], Jai Gautam[1], P Sudharshan Phani[2] and Leo A I Kestens[3]

[1] School of Engineering Sciences and Technology, University of Hyderabad, India
[2] International Advanced Research Centre for Powder Metallurgy and New Materials (ARCI) Hyderabad India
[3] Department of Electromechanical Systems and Materials, Ghent University, Ghent, Belgium

Email: sabavath.janakiram@gmail.com

Abstract Recovery was studied in four different grades of steel, ULC, HSLA and DP600 and DP800 with varying C-Mn content. All the steels samples were 80% cold rolled and annealed at 300 and 500°C for 30, 60,120, 300 minutes, mainly to study the hardness and microstructural changes. Specimens were further tested using micro Vickers hardness tester. Microstructural changes were observed using scanning electron microscope. The cold rolled microstructure consisted of ferrite in ULC and ferrite-pearlite in HSLA, DP600 and DP800 steels. All the grains were pancaked and aligned in the rolling direction. The Hardness values increases for annealed specimens at 300 and 500°C in HSLA, DP600 and DP800 compared to cold rolled condition, which can be attributed to hardening of the ferrite matrix due to pinning of dislocations by carbon atoms. The hardness values for annealed specimens in ULC and DP800 decreases at 300 and 500°C which is attributed to softening.

1. Introduction

High Strength steels like HSLA and DP containing carbon and manganese are commonly used in automotive application and may undergo different thermo-mechanical processing routes, including hot rolling and cold rolling followed by annealing (recovery, recrystallization and phase transformation). The microstructure of C-Mn steels consists of higher volume fraction of ferrite and pearlite colonies (second phase) as remaining constituent. Pearlite colonies are composed of lamellar layers of cementite and ferrite. Carbon is mainly added to increase the strength, which segregates in the form of cementite in the pearlite colonies. C-Mn steels are cold rolled and annealed in the intercritical temperature to transform the initial ferrite pearlite microstructure to a final ferrite-pearlite microstructure [1]. Competition between recovery and recrystallization was studied in terms of hardness. Hardness decreases due to recovery and recrystallization and increases upon phase transformation from the initial cold rolled condition [1]. E.g. increase in hardness was reported in severe plastic deformed low carbon steels, with 0.05% C, in the temperature range of 300-600°C for various soaking time [2]. Increase in yield strength was reported in DP and TRIP steels by pre straining from 2- 10% and paint baking at 140-220°C for various soaking time [3-4] However, in the pearlitic steel consisting of 0.7% carbon it was reported that pearlite lamellae spheroidized, leading to softening mainly due to cementite dislocation

interaction. The acceleration of spherodization depends on the amount of cold working and annealing temperature [5].
In all the above-mentioned investigations special processing practices were adopted compared to the conventional ones with the purpose to enhance the strength of the material at different temperatures and soaking time. C-Mn steels with different C-Mn content were cold rolled and annealed in the intercritical temperature range were observed to increase in hardness compared to the cold rolled condition [6].

The present study was designed to investigate the effect of lower annealing temperature and soaking time to understand the recovery behaviour with variation in carbon and manganese in particular grades of high strength steels like HSLA, DP600 and DP800.

2. Experimental procedure:

The chemical composition of the steels used in this study is reported in table 1. Hot rolling of all the steels was carried out above the austenite recrystallization temperature and coiled at pearlite start temperature, to obtain ferrite-pearlite hot bands, which are further cold rolled to 80% reduction. From the above steels specimen were cut to 10 mm width and 12 mm length in the rolling direction. All the samples were heat-treated in muffle furnace at 300 and 500°C for various soaking time 30, 60, 120, 300 minutes soaking time followed by furnace cooling. These specimens were mechanically polished and etched with nital to study the microstructural changes using FESEM. Hardness values were recorded on flat well-polished samples without etching, at least 15 indents were recorded for each specimen with 500gm load and 20 s dwell time. Nanoindentation tests were performed using a commercially available nanoindenter iMicro® with an InForce50 actuator from Nanomechanics Inc., Oak Ridge, USA. Multiple large indentation maps with 2500 indents per map on a 50 x 50 mm^2 area was collected using the NanoBliz 3D high speed mapping technique. An indentation spacing 1 μm was maintained and the load was changed such that the indentation depth was approximately 100 nm, which results in a depth to spacing ratio of 10 based on the recommendation of [7].

Table 1 chemical composition, area fraction of pearlite and hardness of ferrite-pearlite

GRADE	C	Mn	Si	Pearlite Area fraction	Cold Rolled Hardness
ULC	0.002	1.28	0.22		255±5.2
HSLA	0.06	1.32	<0.008	7.6± 0.5	268 ±5.8
DP600	0.09	1.65	0.28	29.6± 6.2	277± 4.5
DP800	0.13	2.00	0.25	33.5± 4.6	282± 5.9

3. Results

3.1. Cold rolled microstructure, morphology and hardness

Cold rolled microstructures consist of two phases, ferrite and pearlite, in all steel grades, except ULC steels which has only ferrite. With increasing amount of carbon, the amount of pearlite increases. Pearlite colonies consist of alternate lamellar layers of cementite and ferrite. The area fraction of pearlite in hot rolled condition is reported in table 1. The pearlite morphology in HSLA consists of lamellar, fragmented coarse and fine cementite particles at the interface after 80% cold rolling. With increasing amount of C-Mn in DP600 and DP800, pearlite appears in the form of bands and explained in [8]. The cementite lamellar morphology is retained in DP600 grade after 80% cold rolling, whereas the DP800 grade exhibits both lamellar and fragmented cementite morphology after cold rolling. As amount of carbon content increases the density of cementite increases, which leads to increases in the thickness of pearlite colony cf. figure 1, 30 min in HSLA, DP600 and DP800. With increase in carbon content the hardness values in the cold rolled condition increases from ULC to DP800 as shown in table 1.

3.2. Microstructure variation after low temperature recovery

3.2.1. Variation in microstructure at low temperature recovery 300°C

Microstructural variation after annealing at 300°C for different soaking time in four different grades of cold rolled steels are shown in figure 1. In three grades of steel the microstructure consists of ferrite (in grey) and a pearlite colony appears in white. In ULC, the cold rolled microstructure is retained and consisted of the ferrite phase only. In HSLA, for the first 30min annealing the cold rolled microstructure is retained, but with increasing soaking time, at 120 min the dissolution of cementite takes place and continues with increasing soaking time up to 300 min. The dissolution of cementite in the ferrite matrix can be confirmed by the thinning of the cementite lamellae as indicated in the rectangular box as in figure 1, (300 min). In DP600, the cold rolled microstructure is retained till 120 min of soaking time, at 300 min fragmentation of cementite slowly appears as indicated in the rectangular box in figure 1. In DP800, no microstructural changes in ferrite nor pearlite are observed till 120 min of soaking. However, with increasing soaking time to 300 min, partial spherodization of pearlite lamellae takes place in the pearlite colonies figure 1, DP800, 300 min.

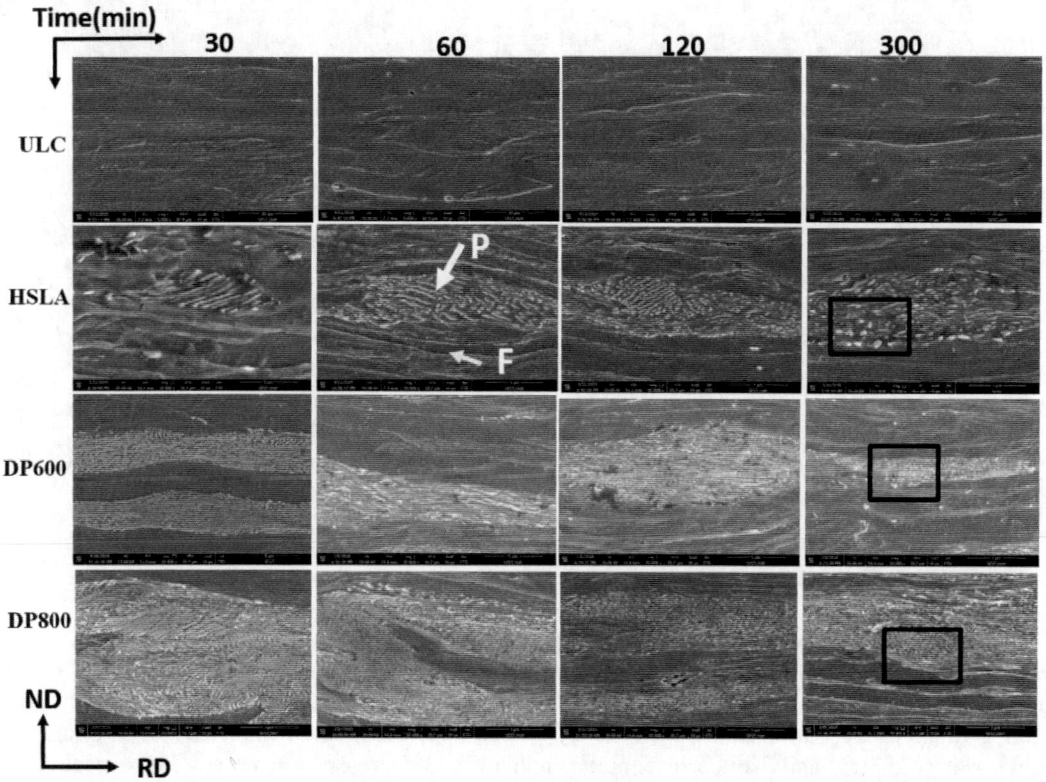

Figure 1. Shows variation of microstructure in all the four alloys with temperature at (a) 300° for 30, 60,120 and 300 min of soaking. F-ferrite, P- pearlite, Box indicating spherodization at ferrite-pearlite interphase

3.2.2. Variation in microstructure at low temperature recovery 500°C

The microstructural variation with annealing at 500°C for various soaking time in four different grades of cold rolled steels are shown in figure 2. In ULC, recrystallization of ferrite takes after 30 min of soaking time, as indicated by the circle in figure 2 and continues to increase with increase in soaking time. In the HSLA steel, the pearlite colony remains the same for the first 30 min of soaking time, but with increase in soaking time spherodization of pearlite lamellae takes place. In DP600, the ferrite pearlite microstructure is retained till 30 min of soaking time, but with the increase in soaking time, the complete lamellar structure is replaced by fine cementite particles (spheroidized) in the pearlite colony. In the DP800 grade, the lamellar cementite structure spheroidizes inside the pearlite colonies from first 30 min soaking time and continues with increase in time.

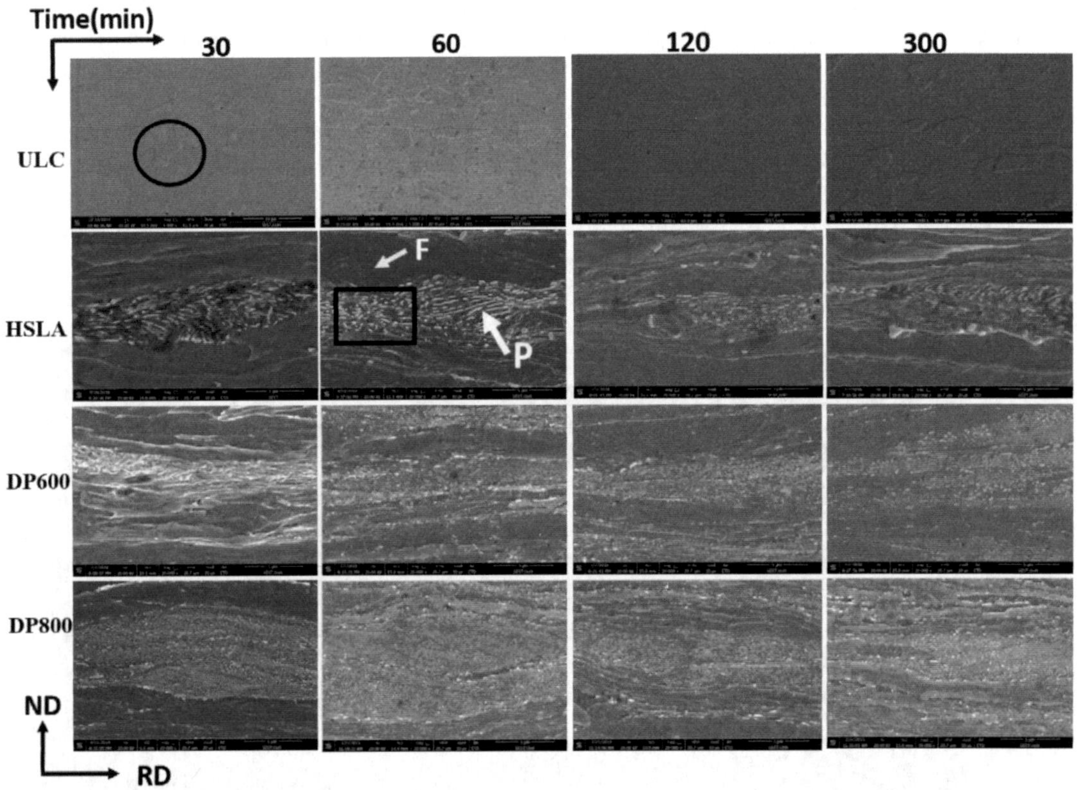

Figure 2. Variation in microstructure in all the four alloys at 500°C for 30, 60, 120 and 300 min of soaking time

3.3. Variation of hardness after low temperature recovery

3.3.1. Variation of hardness after low temperature recovery 300 °C

Cold rolled sheets of different C-Mn content are annealed at low temperature 300 °C for different soaking time. The variation of hardness as a function of C-Mn content for different soaking time is shown in figure 3(a) and 3(c) corresponding increase and decrease in hardness. At low temperature 300°C, cf. figure 3(a), ULC steel hardness increases after first 30 min of soaking time and then decreases with increase in time. This decrease in hardness can be attributed to recovery in ULC. In HSLA steel, the hardness first increases with 30min of soaking time, then slightly decreases at 60 min and saturated thereafter with increase in time. This increase in hardness is mainly due to increase in carbon content which leads to formation of pearlite phase, which was absent in the ULC. In DP600, hardness increases from cold rolled condition and is higher than HSLA steel as the area fraction of pearlite increases, cf. in table 1. The hardness increases with continuous increase in soaking time. In DP800, the hardness significantly increases till 120 min of soaking time in comparison with HSLA and DP600. With further increase in soaking time, the hardness decreases below the value of DP600 at 300 min, even though the area fraction of pearlite is higher than DP600. This decrease in hardness can be linked to morphological changes in DP800 as in figure 3(a), 300min. It was also observed that the hardness is still higher as compared to the cold rolled condition. In summary, the hardness decreases in ULC steel from cold rolled condition, whereas the hardness increases in HSLA, DP600 and DP800 with presence of ferrite and pearlite as second phase from initial cold rolled condition by recovery annealing at 300°C.

.

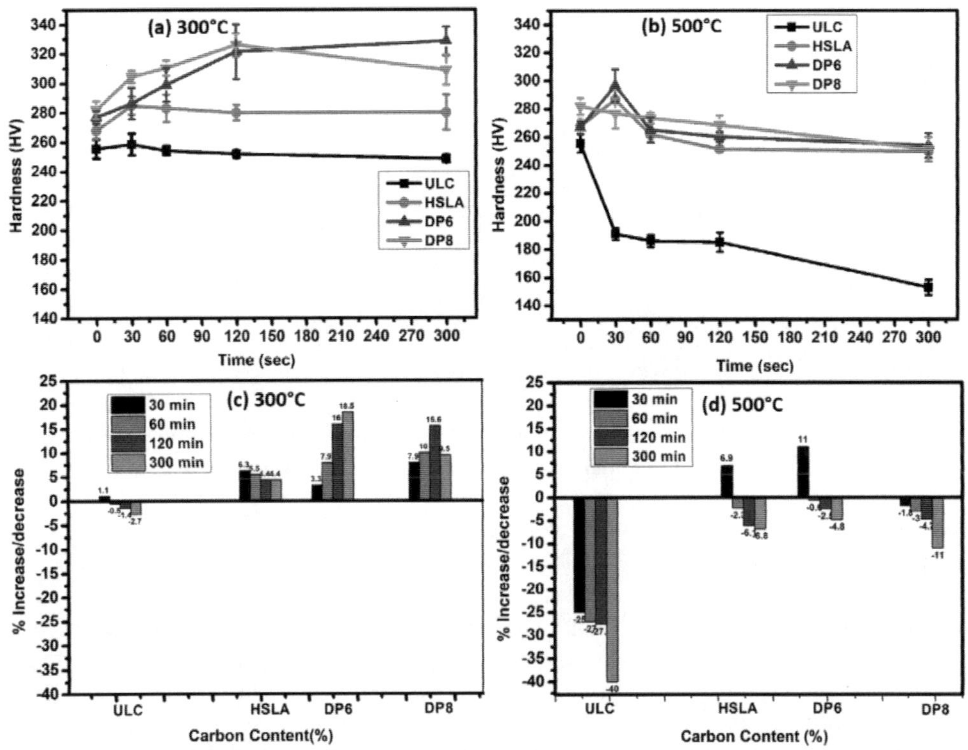

Figure 3. Shows variation of hardness as a function of C-Mn content with temperature at (a) 300°C and (b) 500°C, percentage increase and decrease of correspond Hardness variation (c-d)

3.3.2. Variation of hardness after high temperature recovery 500°C

At high temperature recovery 500°C, cf. figure 3(b) and (d), both increase and decrease in hardness are observed. In ULC steels, the hardness drop significantly and continues to decrease with increasing soaking time, which is attributed to recrystallization and grain growth. In HSLA steel, the hardness increases first for 30 min and then starts decreasing below the value corresponding to the cold rolled condition with increase in soaking time. The DP600 grade shows the similar kind of behaviour as HSLA for first 30 min. It drops back to the cold rolled hardness condition at 60 min and decreases further with increasing soaking time. In the DP800 grade, the hardness starts decreasing below the cold rolled condition after annealing for 30 min and continues decreasing thereafter with increasing soaking time. In summary, the hardness starts to decrease below the initial cold rolled condition in ULC and DP800 steel, right away after 30 min of soaking time. Alternatively, for grades HSLA and DP600 first the hardness increases for 30 min and the starts decreasing below the cold rolled condition.

3.4. Nano indentation mapping

As shown in figure 4, further investigation was carried out by nano indentation mapping showing variation in hardness from cold rolled condition to annealing at 300 and 500°C for 30 min in DP800. This results further confirms microstructure consisting of ferrite and pearlite in DP800, hardness increase from cold rolled condition to low temperature annealing at 300°C, the peak shifts to right side showing increase hardness as in figure 4(d) and reaches to its initial cold rolled condition after annealing at 500°C as in figure 4(d). It can be clearly seen that hardness increases in the pearlite colony and ferrite-pearlite interface.

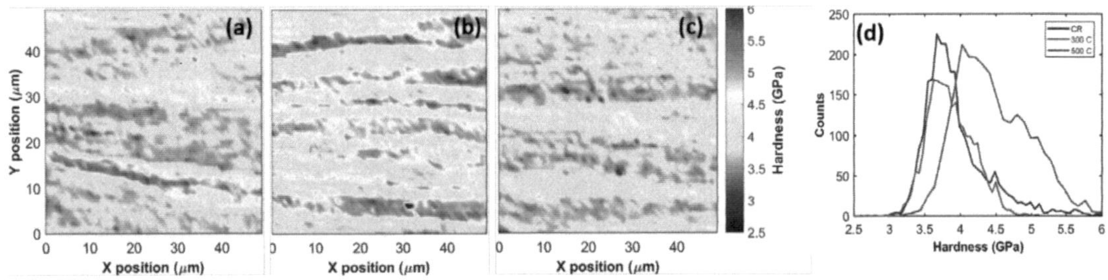

Figure 4. Variation in hardness form Nano indentation mapping in DP8 (a) cold rolled (b) 300°C (c) 500°C for 30min soaking time (d) pixel counts vs hardness map

4. Discussions

Recovery at low temperature by varying C-Mn content, resulted in microstructural and hardness variation occurring in four different grades of steels. Increasing the amount of C-content, the microstructure changes from ferrite to ferrite-pearlite. The changing pearlite area fraction resulted in different cementite morphology after 80% cold rolling, which resulted in microstructural and hardness variation after low temperatures recovery treatment. In ULC steels only the ferrite phase is present due to the minute amount of carbon and manganese. The hardness slightly decreases with increasing soaking time due to recovery. Additionally, the hardness significantly drops below the cold rolled hardness due to recrystallization at 500°C. From the microstructural analysis it was observed that the cold rolled microstructure is retained but hardness increases compared to the cold rolled condition on annealing at 300°C in HSLA, DP600 and DP800. This increase in hardness of C-Mn steels can be related to the strengthening mechanisms applicable to the bake hardening typical in low carbon steels after small cold rolling, which is essentially a strain aging phenomenon, whereby interstitial carbon atoms pin the mobile dislocations and form the Cottrell atmosphere associated with first stage of hardening in addition the precipitation of carbide would also lead to increase in yield strength as a second stage of bake hardening reported by [9]. As the carbon content in ULC steels is very low therefore least amount of carbon is available to pin the dislocation and henceforth only softening, recovery and recrystallization occur with increase in temperature from 300 to 500°C. Increase in amount of C-content where microstructure change from ferrite to ferrite-pearlite. Cold rolling to 80% leads to excess of dislocations mainly in pearlite colonies, at ferrite-pearlite interfaces and inside ferrite grains like happening during sever plastic deformation reported in [10]. The amount of dislocations is relatively higher inside the pearlite colony compared to inside the ferrite grain [10]. This increase in dislocation density by cold rolling and availability of more carbon atoms by increase C- content resulted in an increase of hardness by pinning of mobile dislocation. This was evident from the nano indentation mapping; in the cold rolled condition the hardness was higher in the pearlite colony than in the ferrite as in figure 4(a). With low temperature annealing, cf. figure 4(b) the hardness increases in the pearlite colony at the interface and inside the ferrite grains, which can be attributed to pinning of the dislocation by precipitation of carbon atoms. Increasing the amount of carbon in HSLA, DP600 and DP800 causes the area fraction of pearlite to increase and hardness keeps increasing inside the ferrite grain, in the pearlite colonies and at the ferrite-pearlite interface as well. The increase in hardness can also be due to pinning of the dislocations by the second phase particles like Fe_3C nitrides and carbonitrides [11]. This increase in hardness needs to be further investigated and confirmed by TEM studies. .

After recovery annealing at 500°C, it has been observed that retention of the cold rolled microstructure leads to an increase in hardness in HSLA and DP600 for the first 30 min of soaking time. With the increase in soaking time the lamellar structure is replaced by fine spherical particles (spherodization) and the hardness decreases below the cold rolled condition with increase in time. In DP800 steel, the hardness starts decreasing from the first 30 min of soaking time due to spherodization of cementite. It is well documented in the literature that spherodization of cementite leads to softening, which is in good agreement with the present result [5]. It was observed that hardness decreases in HSLA steel for 120 and 300 min of soaking time, in DP800 grade at 300 min of soaking time. This decrease in hardness can be attributed to softening taking place due to partial spheroidization. However, the hardness still remains higher than the cold rolled hardness, which can be linked to the microstructure in the pearlite colony, which consists of a combination of lamellar, fragmented and partial spherodized cementite lamellae. The lamellar and fragmented cementite contributes to the increase in hardness, whereas the spheroidization leads to a decrease in hardness. However, the question arises why does pearlite lamellae spheroidize at 300°C. It is has been reported that acceleration of spherodization is mainly due to a higher amount of plastic straining and annealing temperature [5]. In DP800, consisting of 34% of pearlite and due to 80% cold rolling, the interlamellar spacing of cementite decreases and becomes thinner and accelerates the spherodization by cementite dislocation interaction phenomena [12]. Due to finer and fragmented cementite morphology rapid sperodization takes place at 300°C after 30 min of soaking time. In a recent in-situ study under synchrotron in dual phase steels reported that dislocation of annihilation rate is almost negligible at 300C and softening starts at 500C [13].

5. Conclusions

The present results show the significant effect of carbon on variation in microstructure and hardness in C-Mn high strength steels containing pearlite and the following conclusions can be drawn

- In ULC steel, where the less amount of C-Mn is present, and hence the microstructures consists of a single ferrite phase, softening takes place and leads to normal recovery at 300°C and recrystallization at 500°C.
- In HSLA, DP600 and DP800 grades, the increasing pearlite content resulted in an increase in hardness after recovery annealing at 300°C. The cold rolled microstructure is retained and hardening takes place in the microstructure.
- Softening takes place with increase in annealing temperature to 500°C and increase in soaking time due to spheroidization of cementite.
- Nano indentation mapping confirms the hardness increase inside ferrite grain, pearlite colonies and at the ferrite-pearlite interface due to precipitation of carbon atoms and pinning of dislocation.

6. References

[1] Peranio N, Li Y J, Roters F and Raabe D 2010 *Materials Science and Engineering A* **527** 4161–4168
[2] Ghiabakloo H and Kazeminezhad M 2017 *Met. Mater. Int.*23 **5** 984-993
[3] Ramazani A, Bruehl S, Gerber T, Bleck W and Prahl U 2014 *Materials and Design* **57** 479–486
[4] Timokhina I B, Hodgson P D, Ringer S P, Zheng R K and Pereloma E V 2009 *steel research Int* **80** 7
[5] Chojnowski E A and MeG Tegart W J 1968 *Metal Science Journal* Vol **2**
[6] Janakiram S, Gautam J and Leo kestens A I 2019 manuscript submitted at ReX GG IOP conference series (under review)
[7] Oliver WC and Pharr GM. 1992 *J Mat Res.* **7** 1564-83.

[8] Janakiram S, Gautam J, Miroux A Moerman J and Leo Kestens A I 2019 *Diffusion Foundations* **22** 84-93
[9] De A K, Vandeputte S, Soenen B and De Cooman B C 2001 *Scripta Meterialia* **44** 695-697
[10] Shin D H, Kim Y S and Lavernia E J 2001 *Acta mater* **49** 2387–2393
[11] Ahmet Bulbula and Ramazan Kaçara 2017 *Materials Research* **20** 210-217
[12] Languillaume J, Kapelski G and Baudelet B 1997 *Acta mater* **45** 3 1201-1212
[13] Marc Moreno, Julien Teixeira, Guillaume Geandier, Jean Christophe Hell, Frederic Bonnet, Mathieu and Sebastien Allain Y P 2019 *Metals* **9** 1-13

The effect of microalloying elements on prior austenite grain growth of low-carbon slab material

J Hannula*, A Kaijalainen, D Porter, J Kömi

Materials and Mechanical Engineering, Centre for Advanced Steels Research, FI-90014 University of Oulu, Finland

*corresponding author: jaakko.hannula@oulu.fi

Abstract. The effect of microalloying elements on the prior austenite grain growth of slab materials during slab reheating has been investigated. The investigated materials were laboratory castings with two levels of carbon (0.05 and 0.09 wt.%) and different combinations of microalloying elements, such as V, Nb, Mo and Ti. Experimental results were compared to equilibrium Thermo-Calc® simulations predicting the solubility of different precipitates during slab reheating. Based on slab reheating experiments at temperatures between 1100 – 1250 °C, Ti is the only alloying element to hinder prior austenite grain growth effectively above 1200 °C. For the steel containing high carbon with Nb, V and Ti microalloying, the average prior austenite grain size was less than 50 μm after the slab reheating to 1250 °C, which can be considered very small. According to Thermo-Calc® calculations, stable Ti(CN) precipitates are formed from the liquid and are in solution until 1500 °C. Until 50-60 % of soluble NbC -precipitates, austenite grain size remained relatively small in Nb-alloyed compositions without Ti, but after that grain size increased drastically. For only V-alloyed steels, already at the soaking temperature of 1100 °C hardly any VC precipitates exists based on Thermo-Calc® simulations, and after that grain coarsening can be assumed to be controlled only by the coarsening rate of austenite grains.

1. Introduction

Controlling austenite grain growth during slab reheating can have a major effect on the final properties of thermomechanically rolled steels. It is well known that factors affecting the austenite grain growth are heating rate, reheating temperature and composition of the steels including different microalloying elements [1–6]. The use of microalloying, such as niobium, vanadium and titanium can effectively hinder the austenite grain growth at high temperatures [5,6]. The prior austenite grain size and structure can have a major effect on the formation and morphology of phases formed at lower temperatures, such as martensite, which can further be reflected in the mechanical properties of the steel [7,8]. By controlling the prior austenite grain size, mechanical properties, such as impact toughness of the steel can be significantly improved [7,9].

The solubility of different precipitates can be simulated using Thermo-Calc® software. By comparing the actual experimental data and simulated data from Thermo-Calc®, the accuracy of simulations can be evaluated and can be utilized in further studies.

2. Materials and experimental procedures

The materials used in this investigation are low-carbon steels containing nominally equal amounts of manganese (~1.3 wt.%), silicon (~0.17 wt.%) and chromium (~0.2 wt.%) but various combinations of the microalloying elements vanadium, niobium and titanium. Two different carbon levels are selected (0.05 and 0.09 wt.%) to evaluate the effect of carbon content on the hardness of the steels, see Table 1. All the steels contained V either 0.15 or 0.35 wt.%. Other variants had additions of Nb or Nb+Ti. The experimental steels were vacuum cast into approximately 70 kg ingots at the Tornio Research Centre of Outokumpu in Finland.

Table 1. Chemical compositions of investigated steels (in wt.%).

Steel	C	Si	Mn	P	S	Cr	Mo	Ti	Nb	V	N
loC-V	0.055	0.18	1.25	0.005	0.004	0.22	-	-	-	0.19	0.0054
loC-NbV	0.055	0.18	1.26	0.006	0.004	0.22	-	-	0.087	0.14	0.0059
loC-Nb-V-Mo	0.054	0.18	1.26	0.006	0.004	0.22	0.09	-	0.089	0.15	0.0054
loC-NbVTi	0.050	0.17	1.29	0.006	0.004	0.22	-	0.04	0.087	0.14	0.0053
hiC-V	0.089	0.17	1.30	0.007	0.004	0.22	-	-	-	0.38	0.0051
hiC-NbV	0.093	0.16	1.26	0.007	0.003	0.22	-	-	0.083	0.34	0.0060
hiC-Nb-V-Mo	0.087	0.16	1.25	0.006	0.003	0.22	0.09	-	0.082	0.33	0.0063
hiC-NbVTi	0.094	0.17	1.26	0.006	0.003	0.22	-	0.04	0.083	0.33	0.0060
hiC-NbV-LoTi	0.095	0.16	1.25	0.007	0.006	0.22	-	0.01	0.081	0.33	0.0056

The effect of microalloying elements on austenite grain growth behaviour was studied. Four different target soaking temperatures were used: 1100, 1150 1200 and 1250 °C. Austenite grain growth behaviour was studied using relatively slow heating rates to simulate the actual slab soaking treatment before laboratory hot rolling trials used in our laboratory. Samples (~15x15x30 mm) were heated along with the furnace reaching the peak temperature after ~2 hours prior to water quenching to room temperature. After heat treatments, samples for light optical microscopy (LOM) and laser scanning confocal microscopy (LSCM) were prepared using picral etching and average mean linear intercept grain sizes were measured using a minimum of 200 grains from each sample. Thermodynamic and kinetic simulations including equilibrium phase diagrams, were executed using Thermo-Calc® (software version of TC2017a) with the TCFE9 thermodynamic database for Fe-based alloys.

3. Results and discussion

3.1 Prior austenite grain structures

Figure 1 presents the measured average austenite grain sizes after the ~2 hours slab soaking treatments in the range 1100-1250 °C. At the soaking temperature of 1150 °C, clear differences in grain sizes among the investigated steels started to occur, 1100 °C producing relatively small grain sizes in all cases, as seen in Figure 1. At the soaking temperature of 1150 °C, the addition of Nb to V-alloyed steel produced a clearly smaller grain size, as seen in Figure 2. At the soaking temperatures higher than 1150°C, only Ti-alloying hindered austenite grain growth effectively, producing significantly small grain sizes even at 1250 °C. In the case of the hiC steels, surprisingly both titanium contents produced small grain sizes, even though Ti/N ratios are different, hiC-Nb-V-Ti steel having ratio of ~6.7 and hiC-Nb-V-loTi a ratio of ~1.8. It is known that generally the best effect of TiN particles is achieved with a Ti/N ratio close to the stoichiometric 3.42. Beyond the stoichiometric ratio the growth of TiN particles is greater leading to less effective austenite grain pinning [10,11]. Based on our results, both Ti/N ratios of ~1.8 and ~6.7 produced fine austenite grain structure even at the 1250 °C soaking temperature.

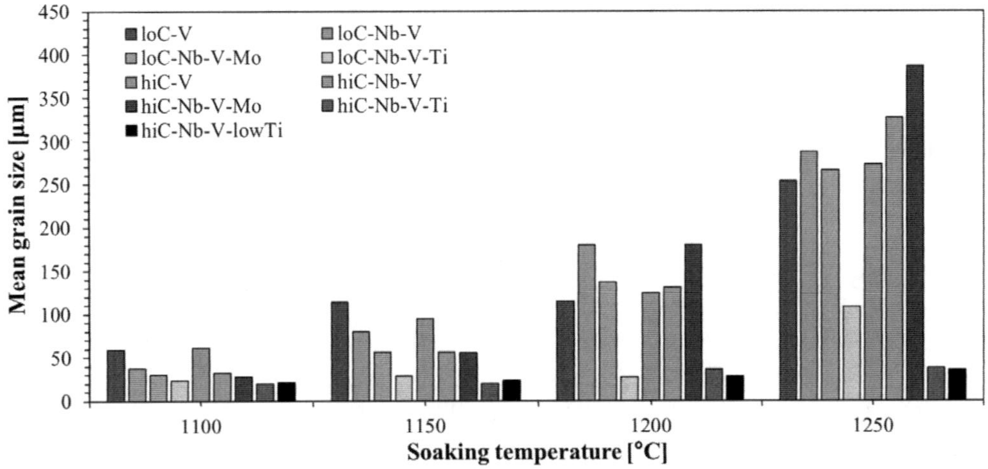

Figure 1. Mean austenite grain sizes based on more than 200 grains after heating at soaking temperatures 1100-1200 °C.

Figure 2 presents the prior austenite grain structures (LSCM) of the hiC steels with addition of vanadium, vanadium-niobium and vanadium-niobium-low titanium at soaking temperatures of 1150 °C and 1250 °C. The effect of niobium at 1150 °C can be clearly seen by comparing Figures 2a and 2b. However, at soaking temperature of 1250 °C, grains with the same size were discovered with average size of ~300 microns in case of hiC-V and hiC-Nb-V steels. Comparing Figures 2c and 2f to Figures 2a-2d, the hindering effect of titanium on grain growth is clearly evident even at the high temperature of 1250 °C.

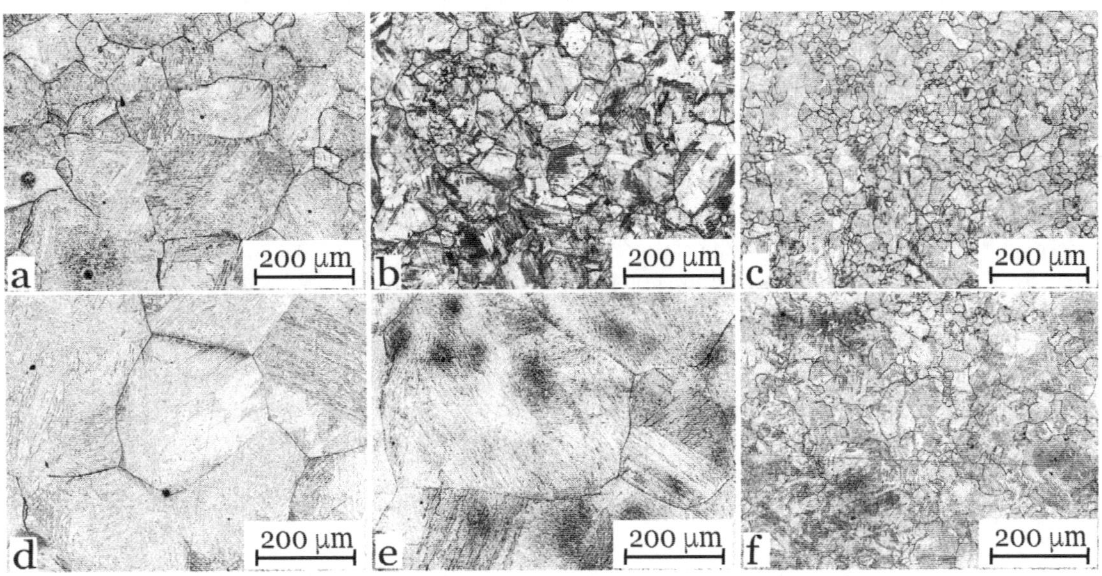

Figure 2. Prior austenite grain structure of a) & d) HiC-V, b) & e) HiC-Nb-V, c) & f) HiC-Nb-V-loTi after heating to 1150 °C (upper row) and 1250 °C (lower row). Picral etching.

It is well known that in certain temperatures abnormal grain growth can occur due to the dissolution of precipitates leading to a non-uniform grain structure [12,13]. For the Nb-alloyed steels without Ti, the occurrence of abnormal grain growth started to be evident at 1150 °C, although the average grain size was relatively small. Already at 1200 °C, the grain size of the Nb-alloyed steels without Ti was uniform and large indicating that the temperature range for strong abnormal grain growth was approximately 1150 °C ± 20 °C. Figure 3 presents the austenite grain structure of the hiC-Nb-V-Mo steel at the soaking temperatures of 1150 °C and 1200 °C. The non-uniform grain structure has started to form at 1150 °C, as seen from Figure 3a. Only for V-alloyed steels, the abnormal grain growth was not clearly evident in the temperature range of 1100 °C to 1250 °C implying that abnormal grain growth occurs at temperatures lower than 1100 °C.

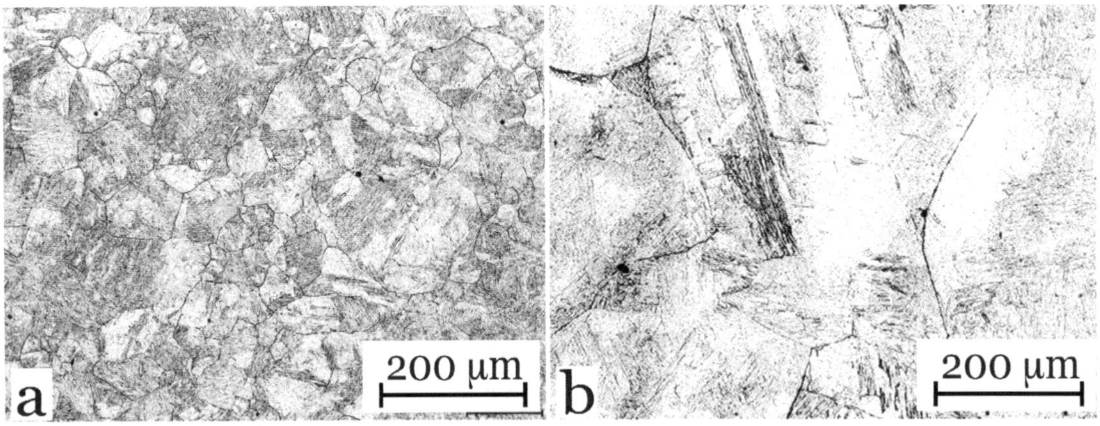

Figure 3. Prior austenite grain structure of HiC-Nb-V-Mo steel at soaking temperature of a) 1150 °C, b) 1200 °C. Picral etching.

3.2 Thermo-Calc simulations

Thermo-Calc simulations were executed for all investigated compositions to simulate the dissolution of different precipitations under equilibrium conditions. The results of the simulations were compared with the results of actual soaking treatments, which can be assumed to be close to equilibrium conditions. Figure 4a and 4b presents the dissolution of precipitates for loC-Nb-V and loC-Nb-V-Ti steels. It can be seen that vanadium carbides dissolve already at ~800-900 °C (lines 5 & 4) depending on the Ti content, and niobium carbides at ~1220 °C (lines 4 & 3). Figure 4b shows that Ti(CN) precipitates only dissolve above 1200 °C (line 5). Figure 4c and 4d presents the dissolution of precipitates for the hiC-Nb-V and hiC-Nb-V-Ti steels. Higher carbon content increased the dissolution temperatures of all precipitates. Alloying with 0.04 wt.% Ti reduced the dissolution temperature of VC precipitations by more than 100 °C, as seen from Figures 4c and 4d (lines 4 & 3). Table 2 summaries the solubility temperatures of the different precipitates. Higher carbon content produces more stable carbides leading to higher solvus temperatures. The effect of VC carbides hindering the grain growth can be considered to be practically limited to temperatures less than 1100 °C, as seen from the dissolution temperatures (Table 2). Above 1100 °C, the austenite grain growth in the case of the V-alloyed steels, is controlled by the austenite grain growth rate without any pinning precipitates. Based on Thermo-Calc simulations, the low titanium content produced almost three times less Ti(CN) compared to the higher titanium content (0.015 vs. 0.04 wt.%), however it was enough to hinder the austenite grain growth effectively, as seen from Table 2.

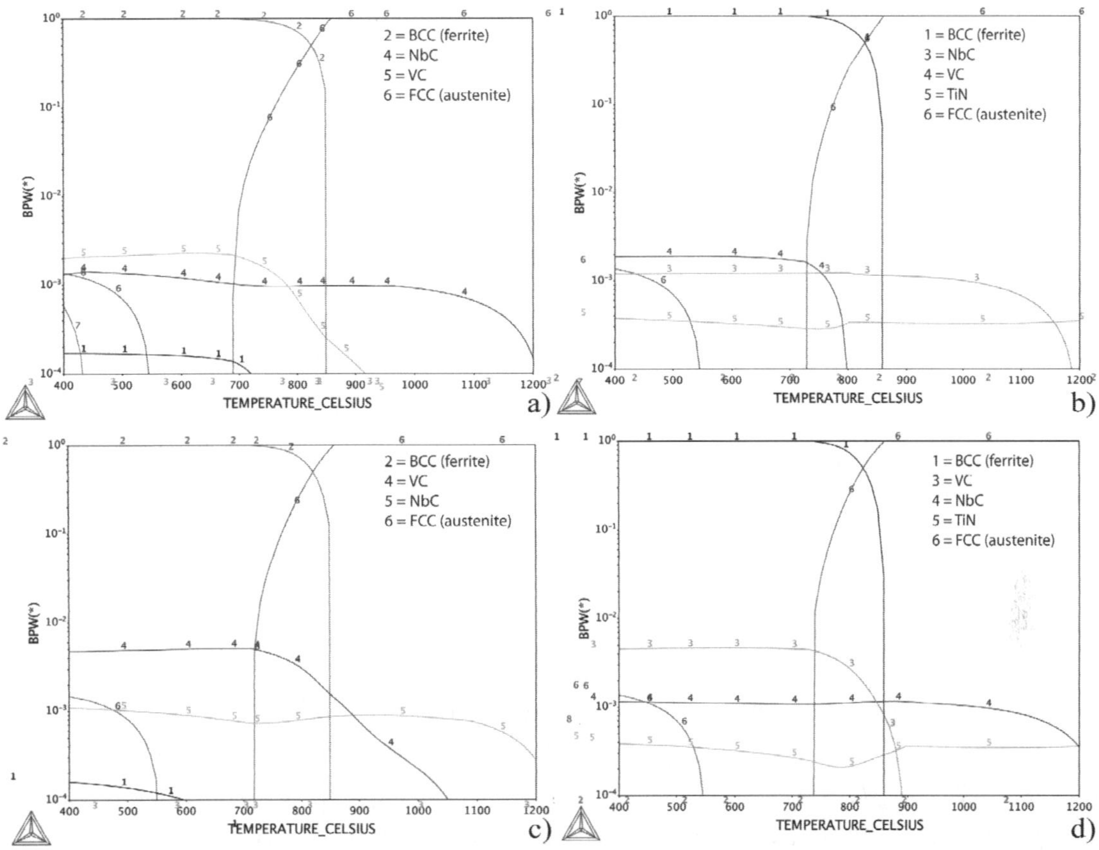

Figure 4. Thermo-Calc simulations for a) loC-Nb-V, b) loC-Nb-V-Ti, c) hiC-Nb-V, d) hiC-Nb-V-Ti steels.

Table 2. Solvus temperatures of different precipitates based on Thermo-Calc.

	Solvus temperature [°C]		
Steel	**VC**	**NbC**	**Ti(CN)**
loC-V	1085	-	-
loC-Nb-V	950	1210	-
loC-Nb-V-Mo	950	1220	-
loC-Nb-V-Ti	800	1200	1500
hiC-V	1135	-	-
hiC-Nb-V	1060	1235	-
hiC-Nb-V-Mo	1090	1235	-
hiC-Nb-V-Ti	895	1220	1500
hiC-Nb-V-low Ti	1010	1245	1500

Figure 5 shows the correlation between measured prior austenite grain sizes and calculated amount of NbC in solution. NbC precipitates hinder grain growth effectively until 50-60 % of NbC precipitates are in solution, as seen in Figure 5. The abnormal grain growth was clearly discovered at soaking temperature of 1150 °C in case of Nb-alloyed steels without titanium, and abnormal grain growth was assumed to be strongest in the temperature range of 1130-1170 °C. Based on the Thermo-Calc simulations, in this temperature range, the amount of NbC in solution is approximately 40-70 %. When the amount of NbC in solution is less or higher than this range, the grain growth phenomenon changes from abnormal to normal grain growth.

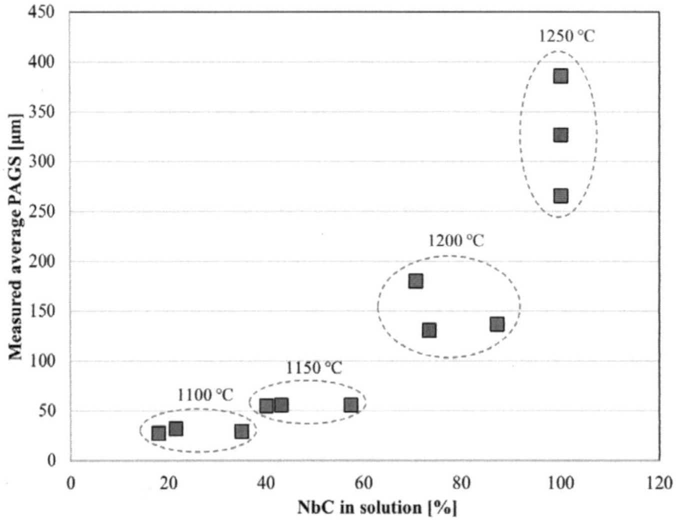

Figure 5. Measured grain sizes vs. calculated NbC in solution.

4. Summary

The effect of microalloying elements on the prior austenite grain growth of slab materials during slab reheating have been investigated. The investigated materials were laboratory castings with two levels of carbon (0.05 and 0.09 wt.%) and different combinations of the microalloying elements V, Nb, Mo and Ti. Based on the results, the following conclusions can be made:

- With both carbon contents, niobium effectively hinders austenite grain growth up to the soaking temperature of 1150 °C. Above 1150 °C, more than 60 % of niobium is in solution based on Thermo-Calc simulations, and no longer able to prevent grain growth.
- Titanium inhibited the austenite grain growth even at the soaking temperature of 1250 °C. For hiC steel, both low (~1.8) and high (~6.7) Ti/N ratios produced small grain sizes.
- Although Thermo-Calc simulations showed that NbC solvus temperatures were higher than 1200 °C, their grain boundary pinning ability disappears at lower temperatures, where more than ~60 % of the NbC is in solution.
- Abnormal grain growth is associated with the dissolution of NbC precipitates. For Nb-alloyed steels without Ti, the temperature range for abnormal grain growth seemed to be at approximately 1130-1170 °C. Thermo-Calc showed that in this temperature region, 40-70 % of NbC precipitates are in solution.

Acknowledgements

The financial support of CBMM (Companhia Brasileira de Metalurgia e Mineracao) and SSAB Europe Oy is gratefully acknowledged.

References

[1] AlOgab K A, Matlock D K, Speer J G and Kleebe H J 2007 The Effects of Heating Rate on Austenite Grain Growth in a Ti-modified SAE 8620 Steel with Controlled Niobium Additions *ISIJ Int.* **47** 1034–41

[2] Napitupulu R A M 2017 Influence of heating rate and temperature on austenite grain size during reheating steel *IOP Conf. Ser. Mater. Sci. Eng.* **237** 012038

[3] Zhao X and Chen L 2018 Effect of Reheating Temperature on Austenite Grain Size and Solid Solution of Second Phase Particles of the Pipeline Steel Slab *IOP Conf. Ser. Mater. Sci. Eng.* **392** 022019

[4] Zhao Y, Shi J, Cao W, Wang M and Xie G 2011 Kinetics of austenite grain growth in medium-carbon niobium-bearing steel *J. Zhejiang Univ. A* **12** 171–6

[5] Karmakar A, Kundu S, Roy S, Neogy S, Srivastava D and Chakrabarti D 2014 Effect of microalloying elements on austenite grain growth in Nb–Ti and Nb–V steels *Mater. Sci. Technol.* **30** 653–64

[6] Roy S, Chakrabarti D and Dey G K 2013 Austenite grain structures in Ti- and Nb-containing high-strength low-alloy steel during slab reheating *Metall. Mater. Trans. A Phys. Metall. Mater. Sci.*

[7] Białobrzeska B, Konat Ł and Jasiński R 2017 The Influence of Austenite Grain Size on the Mechanical Properties of Low-Alloy Steel with Boron *Metals (Basel).* **7** 26

[8] Prawoto Y, Jasmawati N and Sumeru K 2012 Effect of Prior Austenite Grain Size on the Morphology and Mechanical Properties of Martensite in Medium Carbon Steel *J. Mater. Sci. Technol.* **28** 461–6

[9] Wang J, Enloe C, Singh J and Horvath C 2016 Effect of Prior Austenite Grain Size on Impact Toughness of Press Hardened Steel *SAE Int. J. Mater. Manuf.* **9** 2016-01–0359

[10] Saastamoinen A, Kaijalainen A, Porter D and Suikkanen P 2016 Effect of titanium and nitrogen content on the microstructure, hardenability and mechanical properties of direct quenched and tempered ultra-high strength structural steel *TMP 2016 - 5th Int. Conf. ThermoMechanical Process. Adv. Program.*

[11] Wang S C 1989 The effect of titanium and nitrogen contents on the austenite grain coarsening temperature *J. Mater. Sci.* **24** 105–9

[12] Flores O and Martinez L 1997 Abnormal grain growth of austenite in a V-Nb microalloyed steel *J. Mater. Sci.* **32** 5985–91

[13] Cuddy L J and Raley J C 1983 Austenite grain coarsening in microalloyed steels *Metall. Trans. A* **14** 1989–95

IOP Conf. Series: Journal of Physics: Conf. Series **1270** (2019) 012019 doi:10.1088/1742-6596/1270/1/012019

Microstructure, texture, and mechanical behaviour of warm-rolled and annealed standard duplex stainless steel

F L Dias[1], T B Tavares[1] and D B Santos[1]*

[a] Metallurgical and Materials Engineering Department - Universidade Federal de Minas Gerais, 6627Antônio Carlos Avenue, Belo Horizonte, MG, 31270901, Brazil.

* Corresponding author: dsantos@demet.ufmg.br

Abstract. The microstructural evolution and mechanical behaviour of 2205-type duplex stainless steel were studied after plate warm rolling at 600 °C with 60% and 80% thickness reduction and subsequently annealed at 1050 °C for 300 s. With increasing thickness reduction level, ferrite-phase α-fibre became weaker and γ-fibre was developed. For the austenite phase, a weakening of Brass component and increases in Goss and Copper components was observed. A transition from bamboo type into a pearl structure from the 60% sample to the 80% warm-rolled annealed sample was observed. An average tensile strength of 794 MPa was assessed through tensile tests. Post deformation tests demonstrated an increase in microhardness toward the fractured areas and a deformation-induced martensitic transformation was observed in transmission electron microscope results in those regions for all conditions.

1. Introduction

The presence of approximately equal volume fractions of austenite (γ) and ferrite (α) phases results in good properties of duplex stainless steels (DSS). They are widely used in the pulp and paper, chemical, oil and gas, petrochemical, and energy industries [1,2].

Its conventional industrial processing consists of initial hot operation followed by solubilisation annealing, cold rolling, and a final annealing. Owing to the expensive thermal energy demanded from hot rolling, operating at a lower temperature (e.g. warm rolling) has been used as an alternative option. Thus, considering a large thickness reduction to the final application, cold rolling may be reduced or even eliminated and, in some cases, additional heat treatment may be avoided [3].

A significant investigation of the effects of warm rolling and annealing processes on the property evolution of an UNS S32205 duplex stainless steel is presented herein. Three conditions were analysed: as-received and 60% and 80% thickness reduction with annealing. They were characterised as microstructural, textural, and mechanically performing results such that the effect of the thermomechanical process on the phase morphology, its relationship with the properties, and the phase-transformation-induced plasticity in DSS can be understood.

2. Material and Methods

The as-received DSS used in the present study was an industrially produced hot-rolled and annealed 5.16 mm-thick sheet. The chemical composition of this alloy given by the manufacturer is shown in Table 1, which characterises it as UNS S32205 DSS.

Table 1. Chemical composition (wt.%) of UNS S32205 duplex stainless steel

Element	C	Mn	Si	Cr	Ni	Mo	N	P	S	Fe
Weight percent	0.026	1.83	0.25	22.43	5.44	3.04	0.15	0.03	0.0002	Balance

Using the Thermo-Calc™ software and the composition listed in Table, it is found that degraded DSS properties can occur owing to deleterious phases, as sigma (σ) and chi (χ) [2]. Therefore, the temperature of 600 °C was carefully selected to avoid the precipitation of these secondary phases in thermomechanical processing [4].

The as-received DSS strips were warm rolled in along the previous rolling direction in a Fröhling rolling mill. A conventional muffle furnace was used to preheat the samples at 600 °C for 1800 s to achieve homogeneity. After each warm rolling passed a holding time of 900 s, the same temperature was applied to the slabs. The samples reached strains (ε) of ε = 0.60 (thickness reduction from 5.16 to 2.16 mm in nine passes) and ε = 0.80 (thickness reduction from 5.16 to 1.22 mm in 16 passes). After the last warm-rolling pass, the samples were air quenched until the room temperature.

Subsequently, the warm-rolled samples were isothermally annealed in a box-type furnace at 1050 °C for 300 s, and subsequently air quenched. The final annealing was applied to control the proportions of ferrite and austenite, achieving a nearly equal volume fraction of the two phases, and to dissolve any secondary phase.

The schematic diagram for the orientations corresponding to the rolled product is shown in figure 1 where ND, RD, TD are the normal, rolling, and transverse directions, respectively.

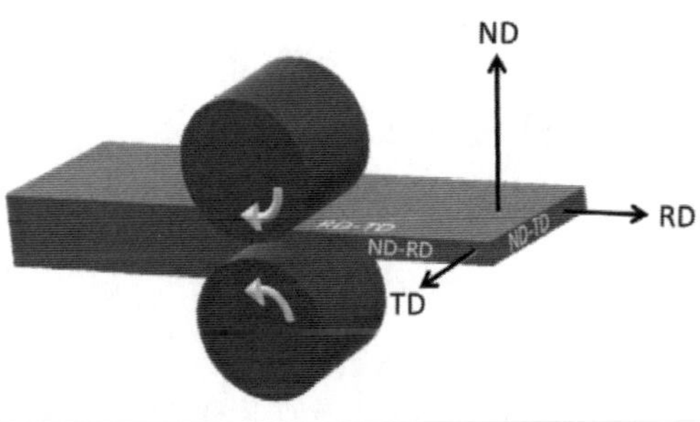

Figure 1. Schematic illustration of orientations corresponding to the warm-rolled product.

Microstructural characterisation was performed in the as-received and warm-rolled and annealed states. For the microstructural analysis, samples of ND-RD:RD plane were mechanically ground and polished with diamond paste according to the conventional metallographic procedure. Subsequently, an additional polishing with colloidal silica particles allowed for the investigation of the microstructure and microtexture in the electron backscatter diffraction system. The phase maps, orientation distribution function (ODF), and pole figure (PF) for each condition were determined using the data in the TSL-OIM™ software.

Thin foils from the RD-TD plane were prepared using the conventional metallographic procedure and a twin-jet polisher (solution of 5% perchloric acid and 95% methanol in –30 °C and –20 °C, tension of 20 V) for transmission electron microscopy (TEM) observation.

To evaluate the mechanical behaviour of the steel, tensile tests were performed under a strain rate of 10^{-3} s-1 at room temperature, using a 5582-model Instron machine. The specimens were obtained from the RD-TD plane in the RD direction according to the dimensions of the ASTM A-370-15 subsize [5]. In addition, Vickers microhardness was measured using the Future Tech FM700 equipment. Each hardness result was determined from an average of 20 measurements with a 2.98-N indentation load and dwell time of 10 s, along the central regions of the ND-RD plane.

3. Results and Discussion

IOP Conf. Series: Journal of Physics: Conf. Series **1270** (2019) 012019 doi:10.1088/1742-6596/1270/1/012019

3.1. Texture Evolution

Figure 2 illustrates the microstructure using the phase map. The microstructure of the as-received sample is formed by the alternated lamellae of austenite and ferrite, with elongated morphology, parallel to the RD. The 60% warm-rolled and annealed (60% WRA) sample exhibited a still lamellar microstructure, similar to the as-received sample, but with a finer thickness. This morphology consisting of elongated grains is referred to as a bamboo-like structure. Meanwhile, the 80% warm-rolled and annealed (80% WRA) sample exhibited the microstructure constituted by the bamboo-like and pearl structures (more globular). The transition from bamboo-like to pearl structure can be observed in some regions, as indicated by the black arrows in figure 2.

The transition from bamboo structure to pearl structure occurs owing to the migration of interphase boundaries between both phases. The increase in deformation level reduced the lamellae thickness and enhanced the interface between both phases. Thus, it can be inferred that the driving force for nucleation of the 80% WRA sample was superior to that of the 60% WRA sample, thus resulting in the difference in the microstructure morphology [6].

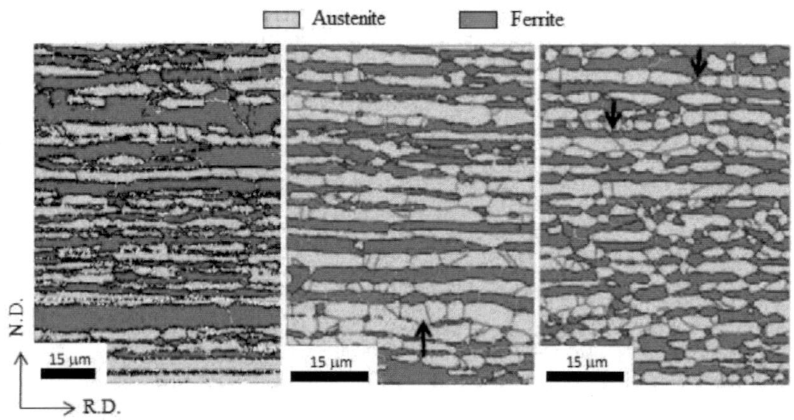

(a) (b) (c)

Figure 2. Phase maps of the samples (a) as-received condition, (b) 60% WRA, and (c) 80% WRA.

The orientation distribution function (ODF) with a $\varphi_2 = 45°$ section for the ferrite is shown in figure 3. Its microtexture is primarily characterised by the α-fibre (<011>//RD) in the as-received sample, and contributed to its stability after warm rolling, as shown in figure 4 (b,c). The thermomechanical process promoted a weakening of this fibre type and a moderate development of the γ-fibre (<111>//ND); these are slightly more intense in 80% WRA.

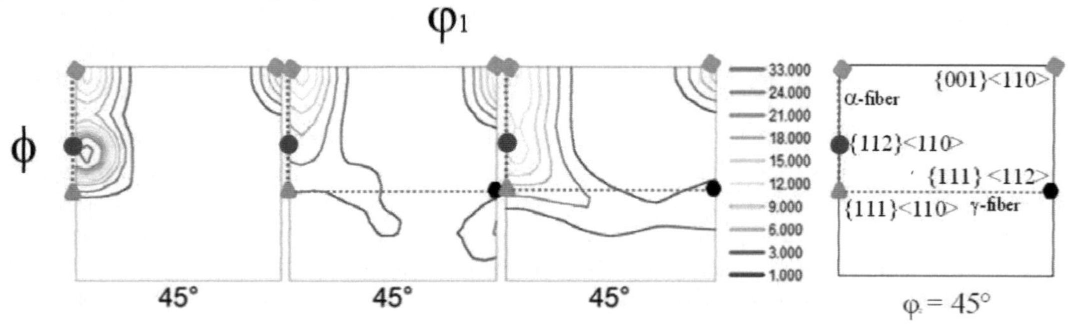

(a) (b) (c)

Figure 3. ODF sections of ferrite phase for (a) as-received, (b) 60% WRA, and (c) 80% WRA conditions.

IOP Conf. Series: Journal of Physics: Conf. Series **1270** (2019) 012019 doi:10.1088/1742-6596/1270/1/012019

Figure 4 shows the PF for the austenite phase. The as-received sample is constituted mainly by {110}<112> Brass and {123}<634> S components. After the warm rolling, the {110}<112> Brass component was weakened, while the {110}<001> Goss components became more intense. Furthermore, {112}<111> Copper was found after warm rolling.

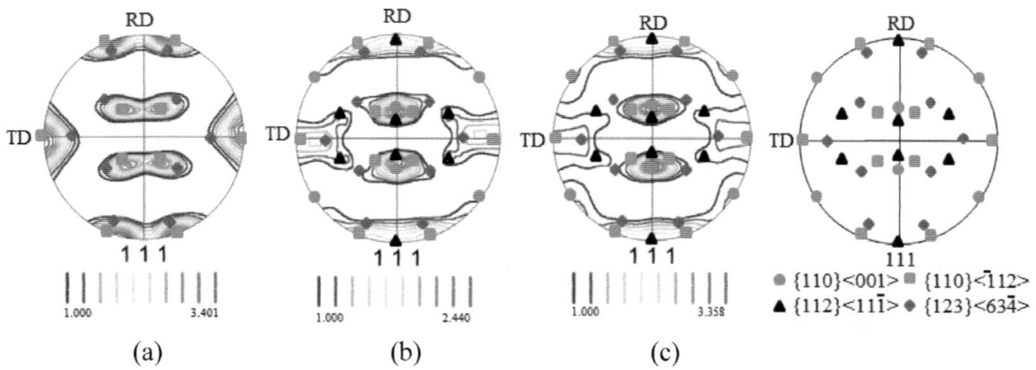

Figure 4. PF sections of austenite phase for (a) as-received, (b) 60% WRA, and (c) 80% WRA conditions.

It is highly difficult to recrystallise grains with α-fibre orientation because of the low energy stored; furthermore, they tend to be consumed during recrystallisation [7,8]. For the DSS, the as-received sample exhibited a high intensity of the α-fibre. This condition and the warm-rolling process resulted in a low stored energy that favoured the recovery process and the remaining α-fibre in the ferrite phase. Meanwhile, the austenite developed a {110}<001> Goss component after warm rolling. The presence of {110}<001> Goss and {112}<111> Copper after warm rolling and annealing, and a similar texture in the warm and cold rolled samples, were reported by the Odnobokova et al. [9] for 304L stainless steel processed at 300 °C.

3.2. Tensile and Hardness Tests

To evaluate the effects of warm rolling on the mechanical properties, tensile tests were conducted. The mechanical properties extracted from the stress–strain data are shown in Table 2, along with the stress–strain curve, indicating that the results of mechanical strength and ductility are similar for all the analysed conditions.

Table 2. Summary of yield strength and tensile strength, along with uniform and post-uniform elongations of the analysed conditions

Condition	Yield Strength (0.2%) (MPa)	Tensile Strength (MPa)	Uniform Elongation (mm/mm)	Tensile strain at Break (mm/mm)
As-received	540	781	0.319	0.483
60% WRA	559	794	0.251	0.340
80% WRA	519	806	0.292	0.412

To evaluate the effect of plastic deformation on the mechanical properties, a microhardness profile was performed on the specimens after the tensile tests, as represented in figure 5. A gradual increase in microhardness occurred near the fractured region in all samples, yielding a value of approximately 400 HV. These results indicate, beyond strain hardening, the formation of martensite induced by plastic deformation (TRIP effect), with a greater volume fraction for warm-rolled specimens because of their higher microhardness values [10].

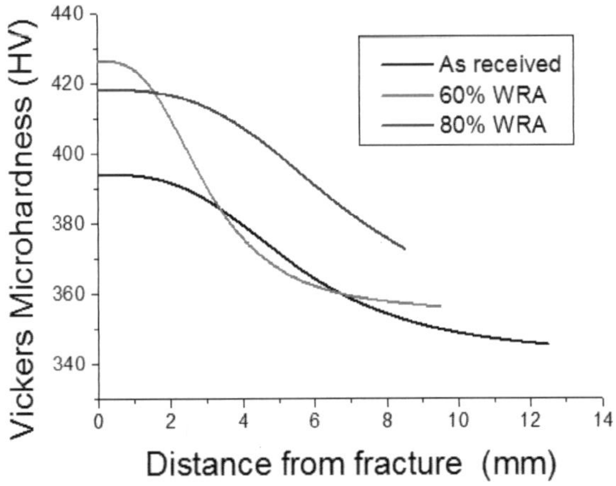

Figure 5. Evolution of microhardness profile as a function of distance from fracture after tensile tests.

3.3. Microstructural Analysis

Figure 6 shows the microstructures obtained from TEM near the fracture from the tensile tests. In all conditions, martensite is observed, with laths varying from 11 to 223 nm that confirm the findings obtained from the Vickers microhardness tests shown in table 3.

The transformation of austenite into martensite during deformation is more likely to occur when the SFE is lower than 20 mJ/m^2. From this value until 50 mJ/m^2, mechanical twinning is the predominant mechanism and dislocation slip for values above 50 mJ/m^2 [11]. Using the chemical composition of the austenite phase obtained through Thermo-Calc^{TM} and EDS, the values obtained for SFE were approximately 32 and 41 mJ/m^2 using Scharmm and Reed equations, respectively [12]. Although mechanical twinning had been reported as the predominant mechanism for this range of SFE, van Tol et al. [13] found a high SFE (50 ± 10 mJ/m^2) for a Mn-based TWIP steel with strain-induced α'-martensite formation.

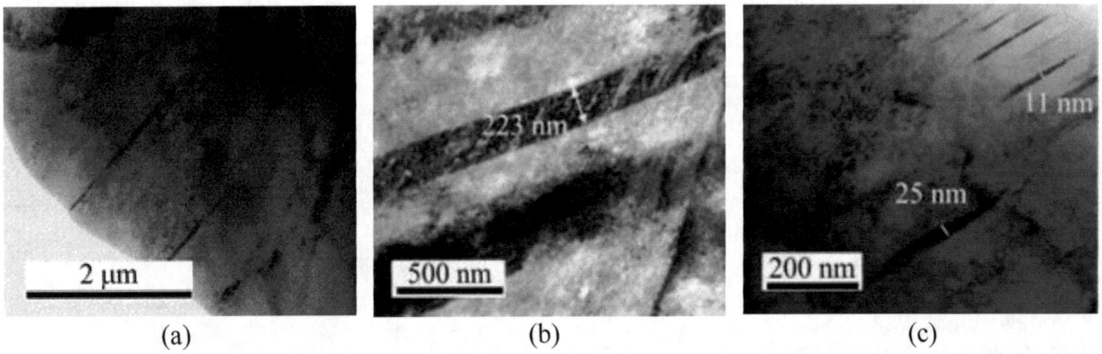

Figure 6. Bright-field images near the fracture for (a) as-received, (b) 60% WRA, and (c) 80% WRA conditions.

4. Conclusions

(i) The microstructure of as-received 2205 DSS demonstrated ferrite and austenite grains with elongated morphology parallel to the rolling direction. A more narrowed lamellar microstructure than

that in the as-received condition was observed in the 60% WRA sample. Meanwhile, bamboo-like and pearl structures were observed in 80% WRA.
(ii) The microtexture of the ferrite phase was primarily constituted by α-fibre (<011>//RD) for all samples, with a weakening of this fibre and a slight development of γ-fibre (<111>//ND) as the deformation level increased. The austenite phase, in the as-received condition, was mainly constituted by Brass ({110}<112>) and S ({123}<634>) components. For the warm-rolled samples, the Goss ({110}<001>) component became more intense as the Brass ({110}<112>) component weakened.
(iii) The stress–strain curve demonstrated a similar behaviour for all conditions. A gradual increase in microhardness was observed close to the fractured region after the tensile, thus proving the deformation-induced martensitic transformation (TRIP effect).
(iv) TEM analysis confirmed the formation of martensite laths near the fracture in all specimens, with the thickness varying from 11 to 223 nm.

Acknowledgements
The authors are grateful to CAPES, FAPEMIG, CNPq, UFMG Microscopy Center, and Aperam South America S.A. for their support in this study.

References

[1] Aguiar I V, Escobar D M P, Modenesi P J and Santos D B 2015 Microstructure characterization of a duplex stainless steel weld by electron backscattering diffraction and orientation imaging microscopy techniques *Revista Matéria* **20** pp 212–226
[2] Nilsson J -O 1992 Super duplex stainless steels *Mater. Sci. Technol.* **8** pp 685–700
[3] Akbari G H, Sellars C M and Whiteman J A 1997 Microstructural development during warm rolling of an IF steel *Acta Mater.* **45** pp 5047–5058
[4] Bhattacharjee P, Zaid M, Sathiaraj G D and Bhadak B 2014 Evolution of microstructure and texture during warm rolling of a duplex steel *Metall. Mater. Trans.* **45A** pp 2180–2191
[5] American Society For Testing And Materials 2011 Standard Test Methods and Definitions for Mechanical Testing of Steel Products. Philadelphia. **A370–10** p 47
[6] Keichel J, Foct J and Gottstein G 2003 Deformation and annealing behavior of nitrogen alloyed duplex stainless steels. Part II: annealing *ISIJ Int.* **43** pp 1788–1794
[7] Hutchinson B. 1999 Deformation microstructures and textures in steels *Phil. Trans. R. Soc.* **357** pp 1471–1485
[8] Humphreys F J and Hatherly M. 2004 *Recrystallization and Related Annealing Phenomena*. 2nd ed. Kidlington, Oxford, UK: Elsevier Ltd. p 605
[9] Chowdhury S G, Sudarsan D, Kumara B R, Deb P K and Ghosha R N 2007 Randomization of texture during recrystallization of austenite in a cold rolled metastable austenitic stainless steel *Mater. Sci. Eng*. A. **443** pp 114–119
[10] Nezakat M. Akhiani H, Sabet S M and Szpunar J 2017 Electron backscatter and X-ray diffraction studies on the deformation and annealing textures of austenitic stainless steel 310S *Mater. Charact.* **123** pp 115–127
[11] Marina O, Belyakov A, Enikeev N Molodov D A and Kaibyshev R 2017 Annealing behavior of a 304L stainless steel processed by large strain cold and warm rolling *Mater. Sci. Eng*. A **689** pp 370–383
[12] Schramm R E and Reed R P 1975 Stacking fault energies of seven commercial austenitic stainless steels *Metall. Trans. A*, **6A** pp 1345–1351.
[13] Van Tol R T, Kim J K, Zhao L, Sietsma J and De Cooman B C 2012 α′-martensite formation in deep-drawn Mn-based TWIP steel *J. Mater. Sci.* **47** pp 4845–4850

IOP Conf. Series: Journal of Physics: Conf. Series **1270** (2019) 012020 doi:10.1088/1742-6596/1270/1/012020

Microstructural evolution during continuous annealing of a 980 MPa cold rolled steel grade

M V P Arruda[1], T M F Melo[1], F S Costa[1], D B Santos[2]

[1] Research & Development Center of Usiminas, Ipatinga, Brazil
[2] Department of Metallurgical and Materials Engineering, University of Minas Gerais, Belo Horizonte, Brazil

Email: marcus.arruda@usiminas.com

Abstract. The development of advanced high strength steels (AHSS) is motivated by automotive industry demands for higher strength steels with better formability and specific stretch-flange formability to avoid fractures during parts forming. In response to this, the new steel grades must have special microstructures in terms of constitution, morphology, distribution and hardness. Thus, a better understanding of formation mechanisms and the characterization of these microstructures has shown to be of fundamental importance for the development of new steels. In this study the influence of heating rate and intercritical temperature on microstructural evolution and hole expansion ratio of a cold rolled 980 MPa AHSS grade was evaluated, using a Dilatometer and a Gleeble machine. The stretch-flange formability was evaluated by hole expansion tests and the microstructures were characterized using scanning electron microscopy, electron backscatter diffraction (EBSD) and nanohardness measurements. The increase of heating rate resulted in an overlap between ferrite recrystallization and austenite formation, changing the morphology of austenite from random to fibrous distribution. On the other hand, higher intercritical annealing temperature resulted in more homogeneous microstructures and higher hole expansion ratios. This microestructural homogeneity, evaluated by nanohardness, suggested that the difference in hardness between ferrite and second-phase affects micro-void formation and crack propagation during hole expansion test.

1. Introduction

The development of advanced high strength steels (AHSS) is motivated by automotive industry demands for higher strength steels with better formability and specific stretch-flange formability to avoid fractures during parts forming. In response to this, the new steel grades must have special microstructures in terms of constitution, morphology, distribution and hardness. In this way, the best choice for the chemical composition and the schedule processing of the cold rolled steel sheet, especially in the continuous annealing stage, becomes important to achievement of desired microstructure.

The continuous annealing process includes heating to an intercritical temperature, followed by an isothermal holding and a subsequent slow cooling to a quench temperature, from which the material is rapidly cooled to an isothermal overaging temperature. Few studies have been reported on the microstructure evolution of cold rolled dual phase (DP) steels during continuous annealing under conditions similar to those used in industrial lines [1-2]. They suggest that ferrite recrystallization during heating has a small effect on the subsequent austenite formation and decomposition. However, Huang et al. [3] have recently shown that the interaction between ferrite recrystallization and austenite formation can significantly affect the microstructure produced by the annealing. According to the authors, the heating conditions applied, specially the heating rate and the intercritical temperature can change the austenite formation kinetics and the resulting microstructure.

Thus, a better understanding of the austenite formation mechanisms and the characterization of these microstructures have shown to be of fundamental importance for the development of new steels. In this study the influence of the heating rate and the intercritical annealing temperature on the microstructural evolution and the final hole expansion ratio of a cold rolled 980 MPa AHSS grade was evaluated using a Dilatometer and a Gleeble machine.

2. Procedure

DP 980 (0.11C-2.3Mn wt. pct.) steel sheets received as industrially hot rolled material (HR) was used for this study. The HR steel was cold rolled from 3.3 mm to 1.6 mm (50 pct reduction) using a pilot scale rolling mill. For the austenite formation studies, dilatometry tests were conducted using test pieces of 4 x 10 x 1.6 mm cut out from the cold rolled sheets (CR) with the longitudinal direction of the test pieces aligned with the rolling direction. Continuous heating tests were performed with heating rates of 1°C/s and 100°C/s until 950°C. To analyze the microstructure during heating, samples were taken from interrupted heating at 760°C and 800°C and quenched with He gas. The cooling rate was approximately 230°C/s to ensure the complete austenite to martensite transformation during quenching.

In order to evaluate the influence of annealing temperature on microstructure and mechanical properties (tension, hole expansion and nanohardness), samples with 50 x 150 x 1.6 mm dimensions were taken from the CR sheets and subjected to continuous annealing cycles on a Gleeble thermomechanical simulator, with intercritical temperatures (IT) of 760°C, 800°C and 820°C. Figure 1 show a schematic diagram of the thermal cycle applied.

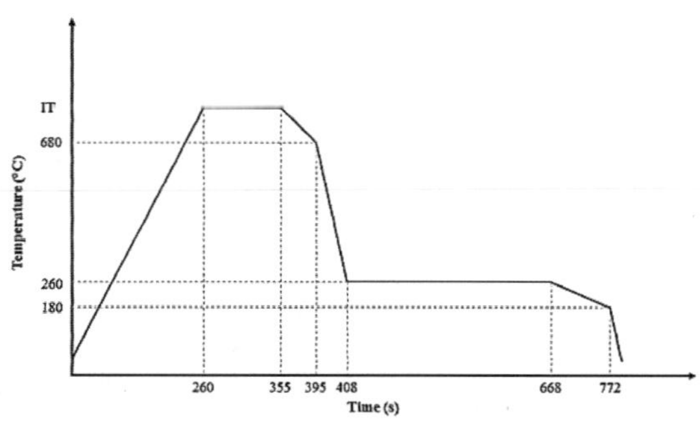

Figure 1. Schematic diagram of the thermal cycle simulated on Gleeble. IT: intercritical temperature

Standard metallographic techniques were used to prepare samples for the characterization of the microstructure using a field emission scanning electron microscope (FESEM) and an optical microscope (OM). The samples were etched with LePera and 2 pct. Nital for OM and FESEM, respectively. Electron backscatter diffraction (EBSD) measurements were used to generate band slope maps using the Oxford HKL acquisition and analysis software. The FESEM for the EBSD measurements was operated at 20 kV and a step size of 0.15 μm.

In order to quantify the strength of the individual phases in the microstructure, nanoindentation tests were performed using a Hysitron TriboScope TS75 coupled to an atomic force microscope (SPM9600). A 5 x 9 array of indents (in total 45) spaced 3 μm apart was performed using a piezo-automated method. Indents were performed with a Berkovich tip under load-control to 7500μN using a 10 sec. linear loading, 3 sec. holding and 10 sec. linear unloading. The hardness values were generated by the machine software using the unloading portion of the load vs. displacement curves.

IOP Conf. Series: Journal of Physics: Conf. Series **1270** (2019) 012020 doi:10.1088/1742-6596/1270/1/012020

Each indent must be categorized on the etched microstructure as corresponding to ferrite or second phase.

Tensile tests were performed according to ASTM A370 (sub-size specimens) [4]. The mechanical properties were determined as an average of three specimens tested along the rolling direction. Stretch flangeability was measured by conical hole expansion tests (HET) carried out in an Erichsen formability testing machine (model 145-60, 1.000 kN tensile force) according to ISO 16630:2003 [5]. The HET consists of forcing a conical punch at 15 mm/min into a pre-punched hole until any crack extends through the test piece thickness. The blank holder force used for both the hole punching and expansion was 200 kN. The expanded hole diameter (Df) was measured in two perpendicular directions and the limiting hole expansion ratio (HER) is calculated in relation to the original hole diameter (Di) as described in the follow equation:

$$\text{HER}\ (\%) = \frac{\text{Df} - \text{Di}}{\text{Di}} \times 100 \tag{1}$$

3. Results

3.1. Dilatometry tests - Influence of heating Rate on Austenite formation

Figure 2 shows the first derivate of dilation curves for the continuous heating at 1°C/s and 100°C/s in black and gray lines, respectively. The stages prior to austenite formation, such as the lattice expansion in ferrite-perlite structures and recovery of deformed structures, take place at heating with similar behavior for both heating rate. A first clear sharp drop for both heating rates which refers to pearlite to austenite transformation. The austenite formation start temperatures (Ac1) were similar for both heating rates. However, the heating rate increase shifts up the austenite formation finish temperature (Ac3) by approximately 14°C.

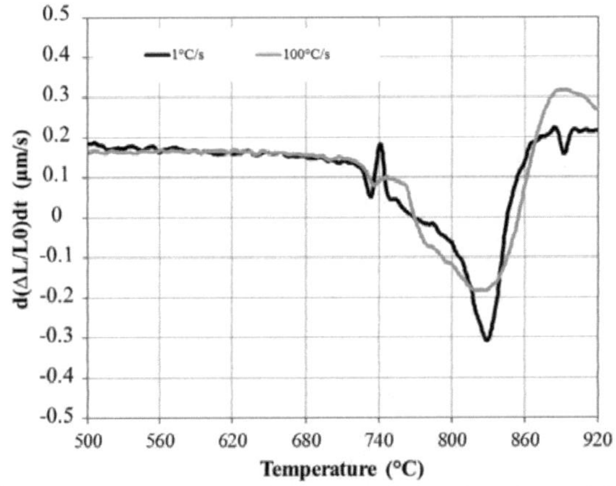

Figure 2. First derivate of dilation curves during continuous heating with different heating rates.

Figure 3 shows the microstructure observed at the specimens quenched at 760°C and 800°C for the heating rates evaluated. The increase in the heating rate delayed the recrystallization process, changing ferrite from fully to partially recrystallized. As can be seen in the Figure 3c, for the 100°C/s heating rate the ferrite grains are still recrystallizing while austenite starts to nucleate and grow within the pearlite colonies. At the slow heating rate, a few austenite particles are observed at the ferrite grain boundaries, which grow in competition with the austenite grains nucleated in the pearlite. As a result, the austenite morphology changes from a random to a fibrous distribution as the heating rate increase. The total austenite fraction (martensite after quenching) increases from 36 to 52 pct. as the temperature increases from 760°C to 800°C, respectively. At 760°C the austenite fraction is smaller for the high heating rate (100°C/s). At 800°C the difference in the total austenite fraction obtained at the two heating rates becomes insignificant [6].

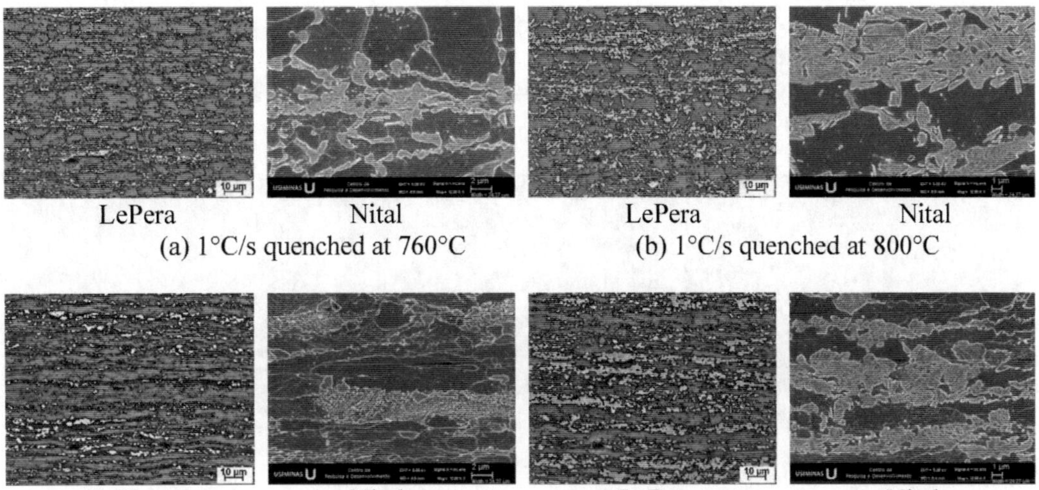

LePera Nital LePera Nital
(a) 1°C/s quenched at 760°C (b) 1°C/s quenched at 800°C

LePera Nital LePera Nital
(c) 100°C/s quenched at 760°C (d) 100°C/s quenched at 800°C

Figure 3. OM and FESEM micrographs with LePera OM and 2 pct Nital, respectively, with different heating rates.

3.2. *Influence of annealing temperature on mechanical properties and microstructure*

Figure 4 shows the measured yield strength (YS), ultimate tensile strength (UTS), total elongation (TE) and hole expansion ratio (HER) for the conditions evaluated. It is evident that raising IT will increase the strength (YS, UTS) and decrease the ductility (TE) of the steel, while the HER increases.

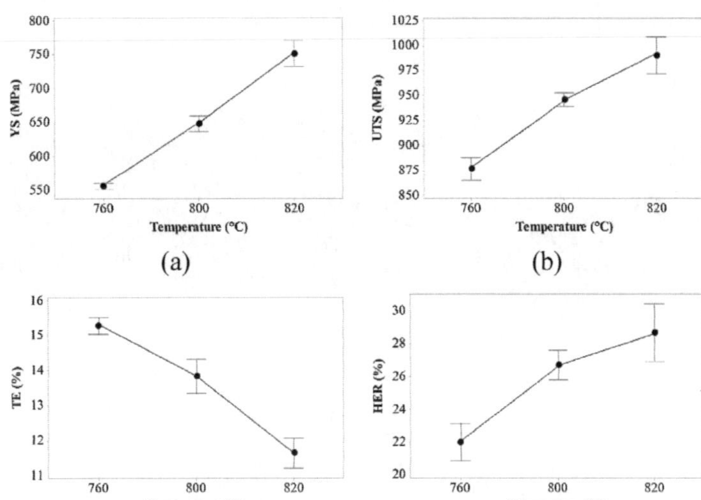

(a) (b)

(c) (d)

Figure 4. Mechanical properties (a) YS, (b) UTS, (c) TE e (d) HER with different intercritical temperature (IT).

As shown in Figure 5 at 760°C the microstructure consists of ferrite (F) and only martensite (M) as the second phase. For higher temperatures (800°C and 820°C) the microstructure shows a greater amount of second phase that includes M and bainite (B). These microstructures are in agreement with YS, UTS and TE results shown in Figure 4. On the other hand, the increase of HER for the condition of higher strength and lower ductility can be explained by the homogeneity of the microstructures. According to previous papers [7-8], a more homogeneous microstructure with a lower hardness difference between the constituents is appropriate to prevent the premature occurrence of voids and microcracks in the interfaces of the soft and hard phases.

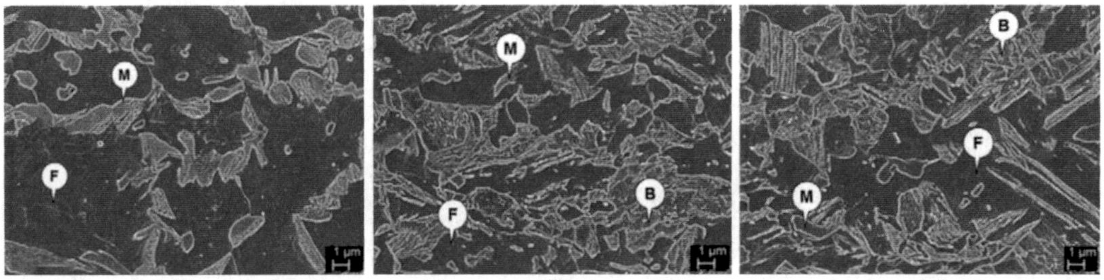

(a) (b) (c)

Figure 5. Microstructural at different intercritical temperature (IT). (a) 760°C, (b) 800°C and (c) 820°C.

In order to better evaluate the microstructure morphology, band slope maps were obtained by EBSD scans for the 760°C and 820°C intercritical temperatures (Figure 6). The band slope value is proportional to the sharpness of the Kikuchi Pattern, which is related to the presence of the crystalline defects. A highly dislocated lattice is expected to have a low band slope value and a high nanohardness [9]. The band slope values are graded on a gray scale, with 0 being the darkest and 255 being the lightest. Thus, the degree of misorientation can be associated with the present phases, with ferrite being the constituent with the highest band slope value, and martensite the lowest value. For the samples at 820°C, the highest volume of second phase formed for higher intercritical temperatures was easily perceived by band slope maps (fig. 6b), as well as their distribution shifted to lower BS values (fig. 6c).

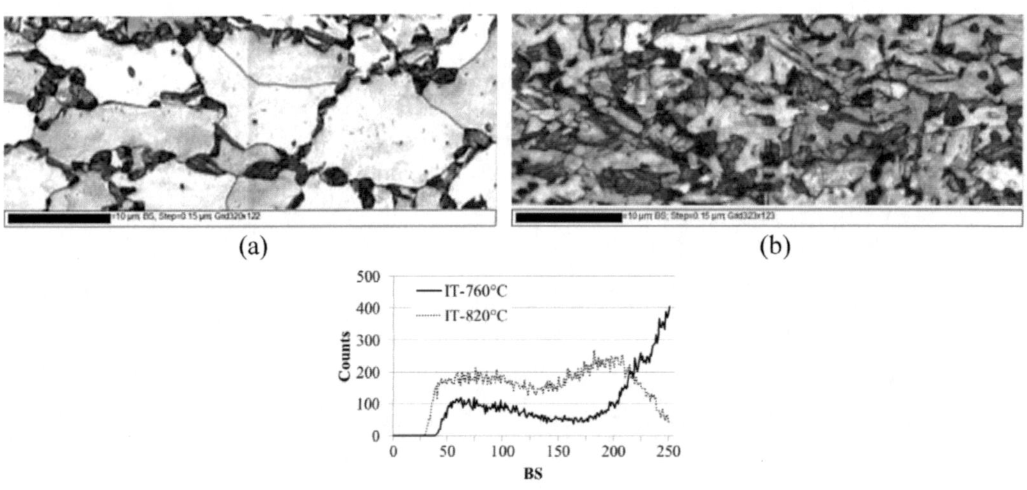

(c)

Figure 6. Band slope maps for (a) IT = 760°C, (b) IT = 820°C and (c) distribution curves of BS values in grey scale.

Average nanohardness and the standard deviation bars for ferrite and second phase constituents at 760°C and 820°C IT are shown in Figure 7. It can be seen that the ferrite nanohardness does not change significantly with the temperature increase. However, the nanohardness average and spread for the second phase for 820°C is lower than for 760°C. The ratio between second phase hardness and ferrite hardness is believed to influence the degree of strain partitioning in the microstructure [8] which explains the higher HER for the samples treated at 820°C.

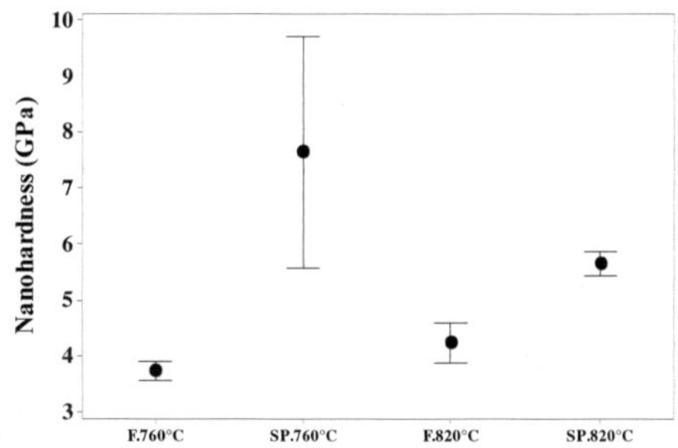

Figure 7. Average nanohardness values of the ferritic (F) and second phase (SP). Each error bar represents the standard deviation.

4. Conclusion

In this study the influence of heating rate and intercritical annealing temperature on the microstructure and the stretch-flangeability of a cold rolled 980 MPa DP grade was evaluated.

The increase of the heating rate shifts the finish austenite transformation (Ac3) to higher temperatures and resulted in an overlap between ferrite recrystallization and austenite formation, changing the morphology of austenite from a random to a banded distribution.

On the other hand, a higher intercritical annealing temperature resulted in a more homogeneous microstructure and higher hole expansion ratio. The microstructural homogeneity, evalueted by nanohardness and EBSD techniques, suggested that the difference in hardness between ferrite and second phase affects micro-void formation and crack propagation during hole expansion test.

References

[1] Girina O A *et al* 2003 *45th MWSP Conf.Proc* **41** pp 403-414
[2] Rocha R Melo T M F, Santos D B 2005 *Mater. Sci. Eng.* **313** pp 296-304
[3] Huang J Poole W J and Militzer M 2004 *Metall.Mater Trans.* **35** pp 3363-3375
[4] ASTM A370 2011 Standard Test Methods and Definitions for Mechanical Testing of Steel Products 47 p
[5] ISO 16630 2003 Metallic materials - Method of Hole Expanding Test 1st ed, 2003 12p
[6] Mohanty R R Girina O A Fonstein N M 2011 *Metall.Mater Trans.* **42** pp 3680-3690
[7] Hebesberger T Pichler A Pauli H 2008 *2nd Int. Conf. Steels in Cars and Trucks* pp 456-463
[8] Girina O A Fonstein N M Bhattacharya D 2008 *Int. Conf. Buenos Aires* pp 29-35
[9] Wu J *et al* 2005 *ISIJ International* **45** pp 254-262
[10] Taylor M D 2014 *Mater. Sci. Eng* **597** pp 431-439

IOP Conf. Series: Journal of Physics: Conf. Series **1270** (2019) 012021 doi:10.1088/1742-6596/1270/1/012021

Experimental and finite element analysis of Cu and Nb severely deformed by 3-axis forging

T Carnrike[1], N Bembridge[2], C Afangideh[3] and P N Kalu[1]

[1] Dept. of Mechanical Engineering, FAMU-FSU College of Engineering, Tallahassee, USA
[2] Formerly at FAMU-FSU College of Engineering, now at Medtronic Inc., MN, USA
[3] Dept. of Mechanical Engineering, Akwa-Ibom State University, Uyo, Nigeria

E-mail: tl12b@my.fsu.edu

Abstract. The evolution of microstructure during 3-Axis Forging (3AF) of OFHC Copper and RRR Niobium has been investigated using both experimental and numerical simulation techniques. The finite element analysis (FEA) aspect was carried out using DEFORM, a commercial software. The 3AF process was conducted at 25°C with a cross-head speed of 0.015 mm s^{-1} and a strain per pass of 0.2. Vickers hardness testing and orientation imaging microscopy (OIM) showed that the evolution of microstructure during 3AF can be classified into three regions: At low strain, region I, in which the high angle grain boundary percentage (HAGB%) declines due to the increase in low angle grain boundary (LAGB) resulting from dynamic recovery. At intermediate strain, region II, there is a progressive increase in HAGB%, due to the conversion of most of the LAGB to high angle grain boundaries (HAGB). This is followed by region III, in which the HAGB% remains relatively unchanged due to a stable microstructure formed by continuous recrystallization. Assessment of the grain size revealed that refinement occurred in regions I and II, resulting in a final grain size of about 2.0 μm and 1.4 μm in Cu and Nb, respectively. Flownet analysis of the first two cycles showed that four deformation zones similar to the hardness pattern were developed in both materials during the 3AF. These zones correlated well with the effective strain and hardness distributions. The highest effective strain and hardness were found in the core of the material (zone A), while the least hardness developed at the edges (zones C and D).

1. Introduction

Severe plastic deformation (SPD) techniques are a group of metal forming procedures, which may be used to impose large strains on materials resulting in significant grain refinement [1]. Some of the SPD techniques include equal channel angular pressing (ECAP), accumulative roll bonding (ARB), high pressure torsion (HPT), and 3-axis forging (3AF) or multi directional forging (MDF). The 3AF process is more attractive because it does not require any special devices, and can be scaled up to be used in industrial applications, in addition to producing nanocrystalline structures [1, 2]. Many studies on the 3AF process have resulted in ultrafine but inhomogeneous microstructures [1]. Wang et al. study on a 2A14 aluminum alloy using 3AF at 450°C utilized a finite element analysis (FEA) flownet approach and experimental techniques. The microstructure was inhomogeneous consisting of four zones [3]. The center (zone I) of the material exhibited the largest effective strain, then the corners (zone II), and finally the top/bottom (zone IV) and side (zone III) edges exhibited the least amount of effective strain [3]. Their study and many others have shown that an experimental approach is not enough to observe the microstructural evolution throughout the 3AF process thus fueling the need to use FEA. Our work established that the majority of grain refinement of the high purity copper (electronic applications) and niobium (super conducting radio frequency cavities) occurs during the first two cycles of the 3AF process. Therefore, the flownet analysis presented here using DEFORM (FEA software) was confined to the first two cycles and are compared with the results obtained by experimental investigations.

2. Finite Element Analysis

DEFORM-3D, a commercial FEA software was used to investigate 3AF of recrystallized C10100 OFHC copper (99.99%) and Residual Resistively Ratio (RRR) grade Nb (99.99%). The initial dimensions of the billet were 15 mm × 15 mm × 15 mm. The FEA model was constructed using the DEFORM-3D CAD features, in which the sizes of the top and bottom compression plates were as follows: 50.8 mm × 50.8 mm × 9.7 mm. These plates were considered rigid bodies while the billet of material was considered a rigid plastic. The mesh of the billet was generated based on a convergence test which indicated 4,352 8 noded brick elements was sufficient for each model. The process was performed at room temperature (25°C) with a friction coefficient of 0.25 and a strain per pass of 0.2. The analysis of effective strain distribution and flownet was carried out using the DEFORM-3D software for samples deformed to two cycles (cumulative strain of 1.2).

3. Experimental Procedure

The OFHC Cu and Nb samples, as described in the FEA, were subjected to 3AF. The die surfaces (made of hardened D2 tool steel) were finely ground using 3μm diamond solution and multiply coated with colloidal graphite lubricant before each forging pass. The process was carried out at room temperature (25°C) using an MTS hydraulic press at a speed of 0.015 mm s^{-1}. A strain per pass of 0.2 resulting in a 0.6 cumulative strain per cycle was used. The height of the billet of each axis was recorded in order to keep track of the cumulative strain. A total of 16 cycles were completed and this resulted to a cumulative strain of 9.6 and 10 for Cu and Nb, respectively. The samples were prepared, etched and sectioned perpendicular to the second forging axis for proper analysis [4, 5].

Vickers Microhardness Testing (indenter load was 300g), equipped with Computer Assisted Microhardness System (Newage Testing Instruments Inc) and Orientation Imaging Microscopy (OIM) were used for microstructural characterization. The OIM was used in conjunction with Electron Back Scatter Diffraction (EBSD) analysis software package designed by EDAX TSL to interpret the orientation data obtained [6, 7].

4. Results and Discussion

4.1. Orientation imaging microscopy

The starting microstructure of Cu and Nb based on OIM are shown in figures 1 and 2, respectively. The initial grain size (GS), Vickers hardness (HV), and high angle grain boundary percentage (HAGB%) for Cu were 26 μm, 50, and 61 %, respectively. The values for Nb were 134 μm, 50, and 53%, respectively. Upon the completion of the 16 cycles an inhomogeneous microstructure was developed in both materials as can be seen in figures 3 and 4. While the final values of the GS, HV and HAGB% for Cu were 2.0 μm, 116, and 40%, respectively, that of Nb were 1.4 μm, 118, and 59%, respectively. It is evident that the processing resulted in significant grain refinement coupled with hardness increase in both materials. The evolution of HAGB% over the course of processing can be seen in figures 5 and 6 for Cu and Nb, respectively. Three regions can be identified based on the development of HAGB%. Region I occurs at low cumulative strain and is characterized by decreasing HAGB%. Since the actual amount of high angle grain boundaries (HAGB) remains the same as the starting material in this region, it is reasonable to conclude that there is an increase in the total amount of low angle grain boundaries (LAGB), which leads to the HAGB% in the microstructure to decrease. The increase in the low angle grain boundary percentage (LAGB%) in this region is due to subgrain formation as a result of dynamic recovery [8]. At intermediate cumulative strain, region II, the HAGB% increases due to the conversion of the LAGB to HAGB. Finally, in region III, the structure has become stable, which suggests that discontinuous recrystallization cannot be occurring in this region and as such the microstructure regenerates through continuous recrystallization [8].

Figure 1. OIM image of the starting Cu material

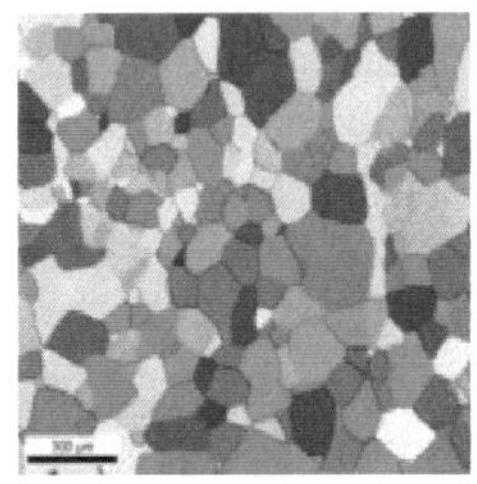

Figure 2. OIM image of the starting Nb material

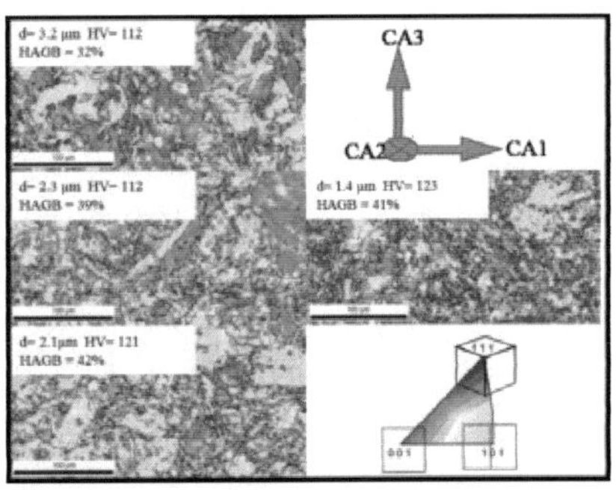

Figure 3. OIM images of the Cu sample after 16 cycles of 3AF at different locations in the sample (top,center,and bottom of left side and core)

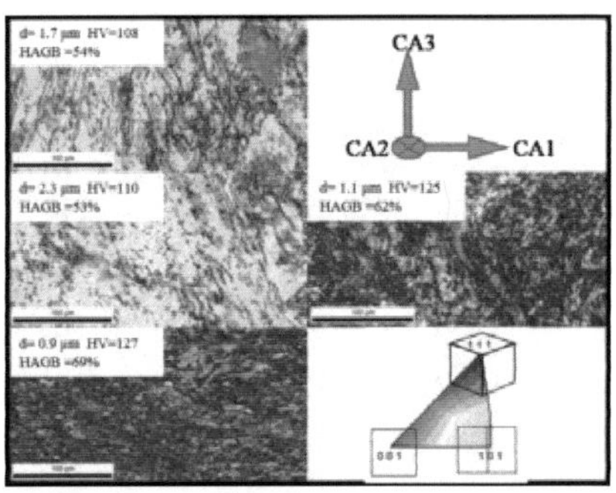

Figure 4. OIM images of the Nb sample after 16 cycles of 3AF at different locations in the sample (top,center,and bottom of left side and core)

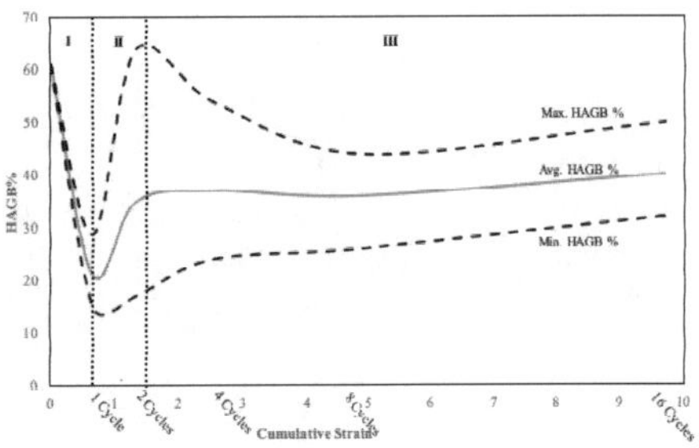

Figure 5. HAGB% of Cu during 3AF

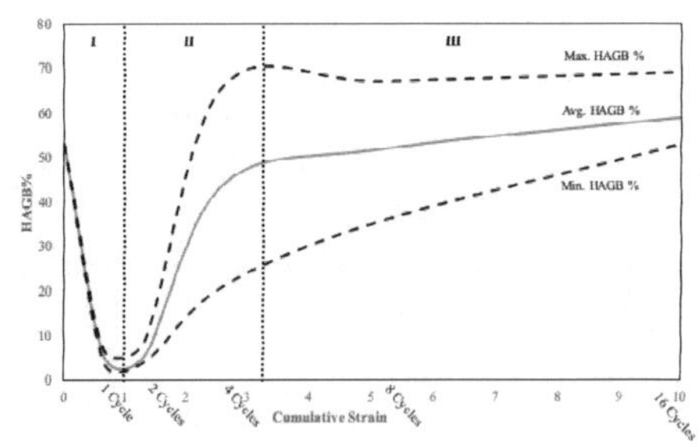

Figure 6. HAGB% of Nb during 3AF

4.2. Effects of 3AF on grain size

Figures 7 and 8 present the assessment of grain refinement in Cu and Nb, respectively. The minimum, maximum, and average GS curves are shown to reflect the inhomogeneity in GS. Superimposing the regions as defined by the HAGB%, it is evident that the grain refinement due to 3AF occurs in regions I and II. Region III is associated with the stabilization of the microstructure.

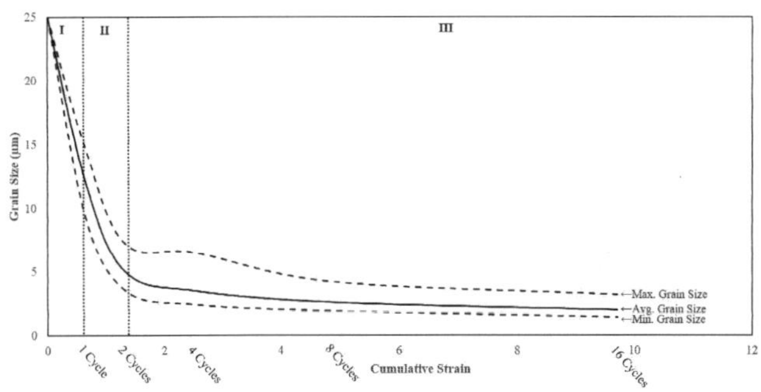

Figure 7. Average GS vs cumulative strain throughout 16 cycles of 3AF for Cu

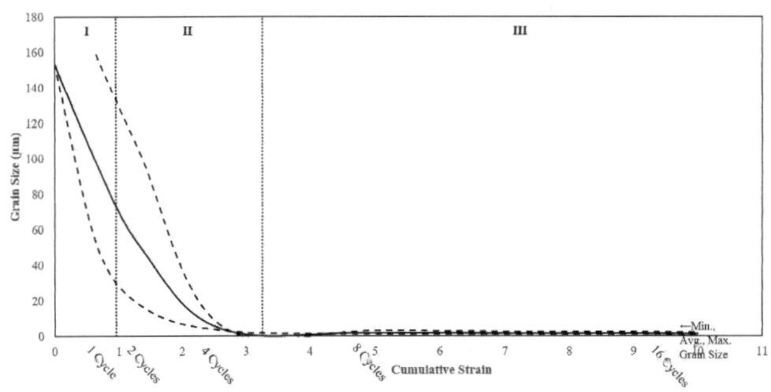

Figure 8. Average GS vs cumulative strain throughout 16 cycles of 3AF for Nb

4.3. Microhardness vs flownet analysis

Figures 9 and 10 show the hardness distributions after 1 cycle in Cu and Nb, respectively. In

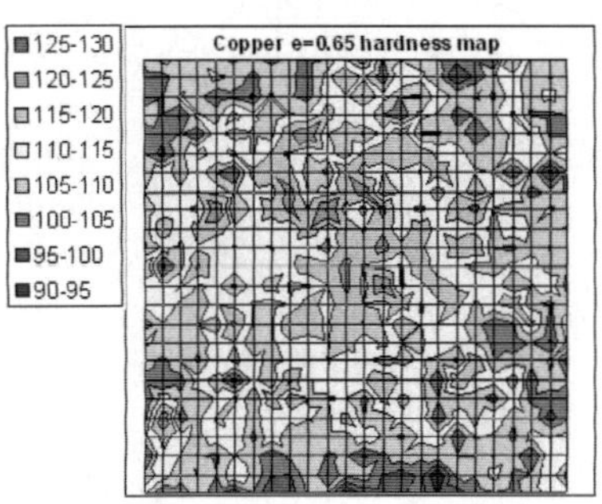

Figure 9. HV distribution after 1 cycle of 3AF for Cu

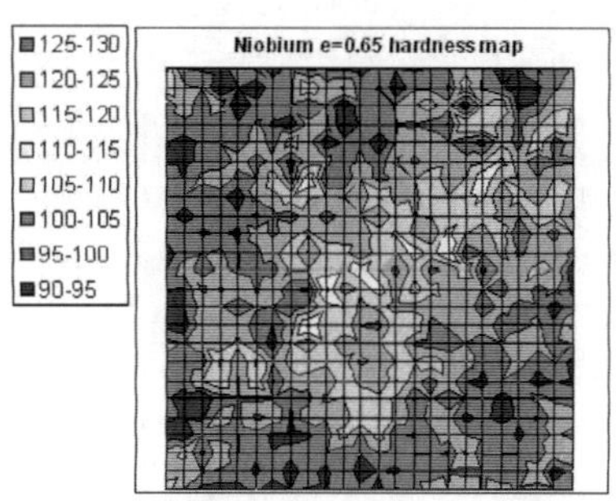

Figure 10. HV distribution after 1 cycle of 3AF for Nb

both cases, the hardness is inhomogeneous. While the highest hardness value was found in the core (center), the edges exhibited the lowest hardness. After one cycle of 3AF, it is evident that the Cu sample was harder than that of Nb. This means that the Cu sample strain hardened faster than Nb. Unlike the Cu, the niobium sample did not retain its original cubic shape, even after the first cycle. Therefore, there was a need to reshape the Nb after each cycle using a reshaping die. This explains why the hardness distribution shows a higher intensity in a diagonal pattern.

Figure 11 shows the FEA model created and processed using DEFORM software. A flownet analysis for both Cu and Nb has been constructed and that of the first cycle is presented in figures 12 and 13, respectively.

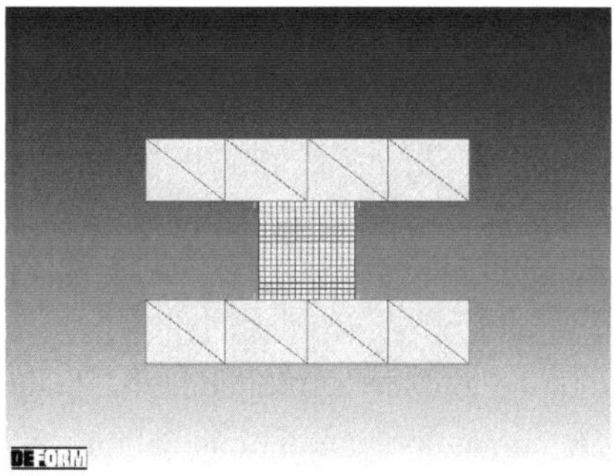

Figure 11. FEA DEFORM 3AF model

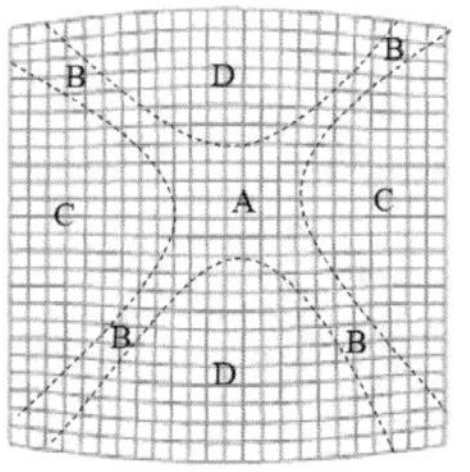

Figure 12. FEA flownet of Cu sample after 1 cycle of 3AF

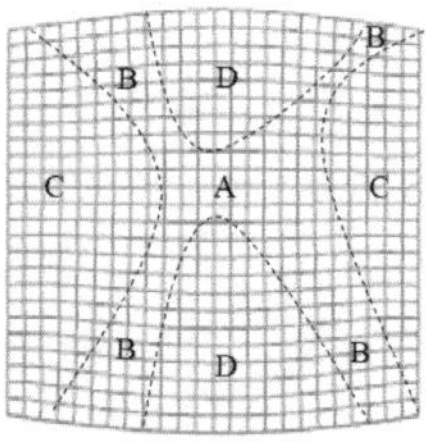

Figure 13. FEA flownet of Nb sample after 1 cycle of 3AF

In both cases four zones were established. These zones were constructed by plotting the flownet at each stage and seeing how it deviates from the starting flownet. The regions with similar flownet deviations were grouped together in a zone. The pattern of the zones in both materials are similar. In zone A, there is little or no deviation of the flownet from the starting flownet. In zone C the flownet protrudes inwards toward zone A while the flownet in zone D protrudes outwards away from zone A. Zone B is a combination of deviation from zones C and D. It is clear that the zones are not symmetrical and this is due to the inhomogeneity in the microstructure as well as the shifting of the billet, from the center of the plates, after each pass. The flownet analysis correlates well with the hardness distributions of figures 9 and 10.

4.4. Effects of cumulative strain on effective strain

Additional analysis of the FEA models was conducted to determine the effective strain distribution. Figures 14 and 15 show the effective strain distribution after the 1st cycle for

Cu and Nb, respectively. The data also prove the existence of the four zones similar to the flownet analysis. A comparison of the flownet analysis and the effective strain distributions clearly correlates the effective strain to hardness. Whereby, high effective strain equates to high hardness, and vice versa. Zone A exhibits the highest effective strain, as it is constantly being put under both tension and compression and is not in contact with the friction from the compression plates. Zone B has the second highest effective strain as it is generally not fully in contact with the compression plates thus the friction effect is not fully present. The smallest effective strain zones are C and D which is due to the presence of friction.

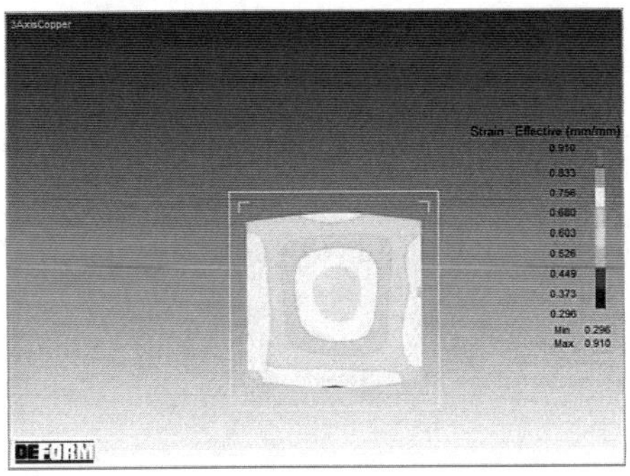

Figure 14. Effective strain distribution of Cu after 1 cycle of 3AF

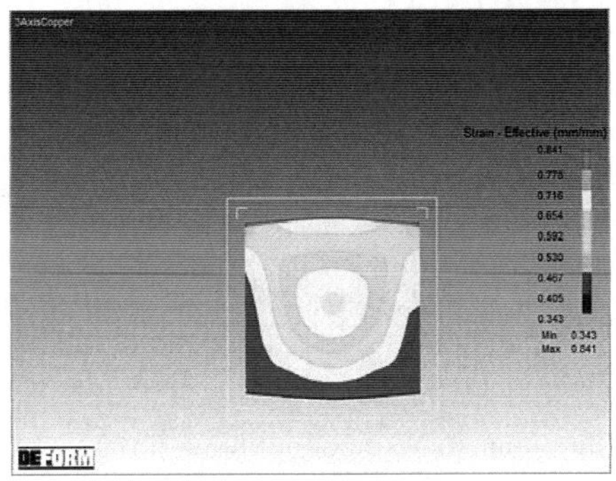

Figure 15. Effective strain distribution of Nb after 1 cycle of 3AF

5. Conclusion

This study used both experimental and FEA techniques to investigate the evolution of microstructure in high purity Cu and Nb subjected to 3AF. Experimental data showed that three regimes of microstructural evolution were present: region I was characterized by subgrain formation due to dynamic recovery. The subgrains were converted to HAGB in region II. The grain structure has been stabilized in region III, suggesting the lack of discontinuous recrystallization. Flownet analysis correlates well with the experimental results and showed that four hardness zones were observed when Cu and Nb were deformed by 3AF. The center (core), zone A, displayed the highest strain and hardness, while zones C and D (edges) had the lowest strain and hardness.

References

[1] Valiev R, Estrin Y, Horita Z, Langdon T, Zechetbauer M and Zhu Y 2006 *J. Miner. Met. Mater. Soc.* Producing Bulk Ultrafine-Grained Materials by Severe Plastic Deformation **58** 33-9
[2] Sitdikov O, Sakai T, Goloborodko A, Miura H and Kaibyshev R 2004 *Mater. Trans.* Effect of Pass Strain on Grain Refinement in 7475 Al Alloy during Hot Multidirectional Forging **45** 2232-8
[3] Wang M, Wang J, Liu W, Ma Y, Liu D, Guo L, Huang L and Huang B 2017 *Int. j. eng. res. eng. sci.* Experimental and Finite Element Analysis of Flow Behavior of 2A14 Aluminum Alloy during Multi-directional Forging **3** 23-32
[4] Bembridge N 2011 *Innovative Severe Plastic Deformation of Niobium with Application to Superconducting Radio Frequency Cavities* dissertation
[5] Vander Voort G 2010 *Metallography, principles and practice* (Materials Park - ASM International)
[6] Adams B, Wright S and Kunze K 1993 *Metall. Mater. Trans. A* Orientation Imaging: The Emergence of a New Microscopy **24A** 819-31
[7] Dingley D 2004 *J. Microsc.* Progressive steps in the development of electron backscatter diffraction and orientation imaging microscopy **213** 214-24
[8] Humphreys F J and Hatherly M 2004 *Recrysallization and Related Annealing Phenomena* edition 2 (Oxford: Elsevier) pp 187,453-5

Miniature tensile testing of SPD processed fine-grained aluminum

Harishchandra A Lanjewar[1,*], Soroosh Naghdy[1,2], Leo A I Kestens[1,3] and Patricia Verleysen[1]

[1]Department of Electrical Energy, Metals, Mechanical Construction and Systems (EEMMeCS), Ghent University, Ghent, Belgium
[2]Process models, ArcelorMittal, Ghent, Belgium
[3]Department of Materials Science and Engineering, Delft University of Technology, Delft, The Netherlands

*E-mail: harishchandra.lanjewar@ugent.be

Abstract. Manufacturing of ultrafine-grained (UFG) or nanocrystalline (NC) metals via a single step top-down approach imposing severe plastic deformation (SPD) is one of the most promising ways to achieve superior properties such as high strength and superplastic forming capability. Nonetheless, the lack of relevant data on their post-SPD performance in different test environments makes it difficult to fully understand their mechanical behavior. While characterizing the tensile behavior, almost all of the previous reports are limited to the discussion on the plastic performance of the material in terms of elongation to failure and corresponding strength, with only a few studies discussing the effect of grain fragmentation on work hardening response of the material. In the present work, a comprehensive analysis is presented in terms of the uniform and post-necking mechanical behavior of the ultrafine-grained material. Commercially pure aluminum is subjected to high pressure torsion (HPT) deformation with strains ranging from very low levels ($\gamma \approx 2.1$) to values well in the saturation regime ($\gamma \approx 25.1$). When tested in uniaxial tension, the strength increases monotonously. The uniform elongation improves with the imposed HPT strain, though remains lower than the value of the initial material. Based on the slopes of the stress-strain curve, three distinct zones are identified, i.e. uniform deformation, post-necking-1, and post-necking-2. With accumulating SPD deformation, the material shows enhanced pre-necking strength and ductility; while post-necking material fails early and at lower strength levels. The post-necking response is observed to be highly microstructure dependent: a lower grain size augments the resistance for micro-crack propagation and thus the ductility, however, once initiated, a crack propagates much faster in fine-grained than in coarse-grained HPT processed material.

1. Introduction

Fabrication of ultrafine-grained (UFG) or nanocrystalline (NC) metals via single step top-down approaches imposing severe plastic deformation (SPD) is one of the most promising ways to achieve superior properties such as high strength and superplastic forming capability [1]. Over the last two decades, many studies have been conducted on a variety of metals. Three methods, namely equal channel angular pressing (ECAP) [2], accumulative roll bonding (ARB) [3] and high pressure torsion (HPT) [4] have emerged as the most potent tools to generate refined structures. ECAP and ARB requires multiple passes to impose large plastic strains on the material, contrary to HPT where a significant amount of strain can be imposed on a disk-shaped sample in one single operation. Thus produced nanostructured (NS)/ultrafine-grained (UFG) metals, quite independent of the processing route, when tested in tension after pre-straining in simple shear mode, show extraordinary enhancements in strength but suffer from poor ductility [5]. However, with continued shear straining the uniform deformation improves to some extent and has been associated with the smaller size of the nucleating flaws during plastic instability [6]. In the case of ECAP Cu, increased rate sensitivity owing to refined grain size has been considered another contributing factor to improved ductility [7]. Methods to improve upon the combination of strength-ductility property have been proposed by either engineering a multi-component structure or post-SPD treatments [8]. In post-SPD annealing treatments, work hardening capacity of the material was seen to improve in conjunction with the reduction in overall dislocation density [9]. During such annealing treatments, non-equilibrium boundaries which are known to be present in large proportions in SPD material, convert into boundaries of equilibrium nature, i.e. with lowered geometrically necessary dislocation density and thereby enhanced ability to store the deformation [10].

In present work, the mechanical behavior of UFG metal is studied using uniaxial tensile testing of miniature dog-bone shaped specimen extracted from HPT deformed commercially pure aluminum (CP Al). The HPT deformation includes very small strains as well as strains in the saturation regime so that a continuous progression in the grain structure is obtained and later linked to their plastic flow properties in terms of elongation, strength and work hardening behavior. Based on the uniform deformation and post-necking mechanical behavior, preliminary conclusions can be drawn about the potential failure conditions specific to the UFG metals.

2. Material and methods

The cast block material of CP Al (Al - 99.6 wt%) was cold rolled to a thickness of 1.3 mm and annealed at 773 K for 30 s, producing an average grain size of 85 μm. HPT was performed, using 15 mm disks punched out of the annealed sheet, at a nominal pressure of 2.5 GPa and different rotations of ¼, ½, 2 and 3, at a rotational speed of 1 rpm. The equivalent strain applied during HPT is calculated using relation [11]:

$$\varepsilon_{eq} = \frac{r\varphi}{3^{1/2}h} \quad (1)$$

where r is the sample radius, φ is the rotation applied in rad and h is the sample initial thickness. As shown in figure 1 (a), two dog-bone shaped miniature tensile samples were extracted from each HPT processed disk using electrical discharge machining (EDM). The simple shear deformation geometry in terms of directions - radial (r), shear (θ) and shear plane normal (z) is also indicated.

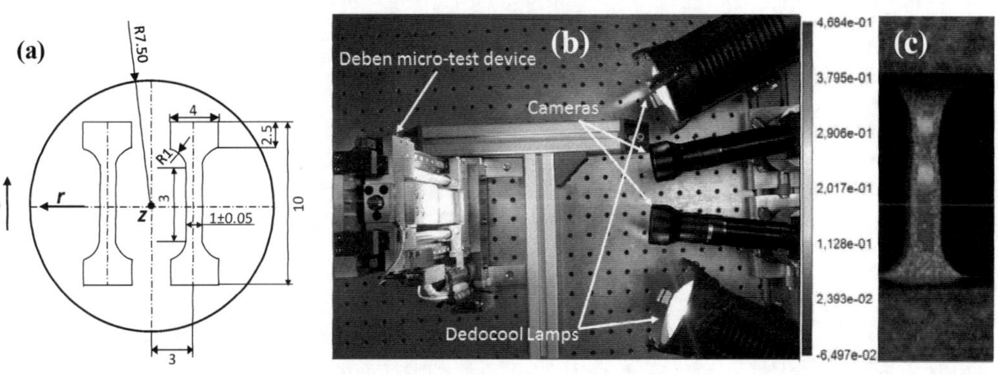

Figure 1. (a) Tensile sample extraction scheme with simple shear geometry, (b) tensile test and camera setup during an experiment and (c) full-field strain map using 3-dimensional digital image correlation (3D DIC).

The HPT ε_{eq} in the center of the tensile samples was 2.1, 4.2, 16.7, and 25.1. In addition, annealed material was also tested along different orientations with respect to the rolling direction, i.e. 0°, 45°, and 90°. A dual-screw Deben® micro-test device with 1 kN load cell was used and tests were performed at a strain rate of 0.0056 s^{-1}. For full-field strain mapping, 3D DIC a non-contact optical strain measurement technique was used. Here the tensile samples were spray painted with the speckle pattern (black dots over white background). Two high-resolution cameras captured the images of the deforming specimen during tension tests (figure 1 (b) and (c)). MacthID Grabber® was used to acquire the force and camera signals, while subsequent image correlations and calculations were performed using MatchID Stereo®, both are the modules of the commercial DIC software package MatchID®.

The microstructural characterization was carried out using electron backscatter diffraction (EBSD) in a scanning electron microscope (SEM) with a field emission gun (FEG). The observations were performed on the *r-θ* plane. During the analysis to determine the average grain size (*GS*), using OIM™ software, 1° and 15° were chosen as the minimum boundary misorientation and high angle boundary misorientation, respectively.

3. Experimental results and discussion

The microstructure experiences refinement in *GS* with the accumulating HPT strain and reduces to average values of 48, 18, 1.22, and 1.42 μm at ε_{eq} of 2.8, 4.2, 16.7, and 27.9, respectively. The engineering strain is obtained by calculating the extension between two horizontal lines marking the gage length of the specimen. The engineering stress-strain curves for annealed material along different orientations are shown in figure 2 (a). Generally, to describe the stress-strain behavior in metals Hollomon's power law can be used and in that case, the work hardening rate is given by:

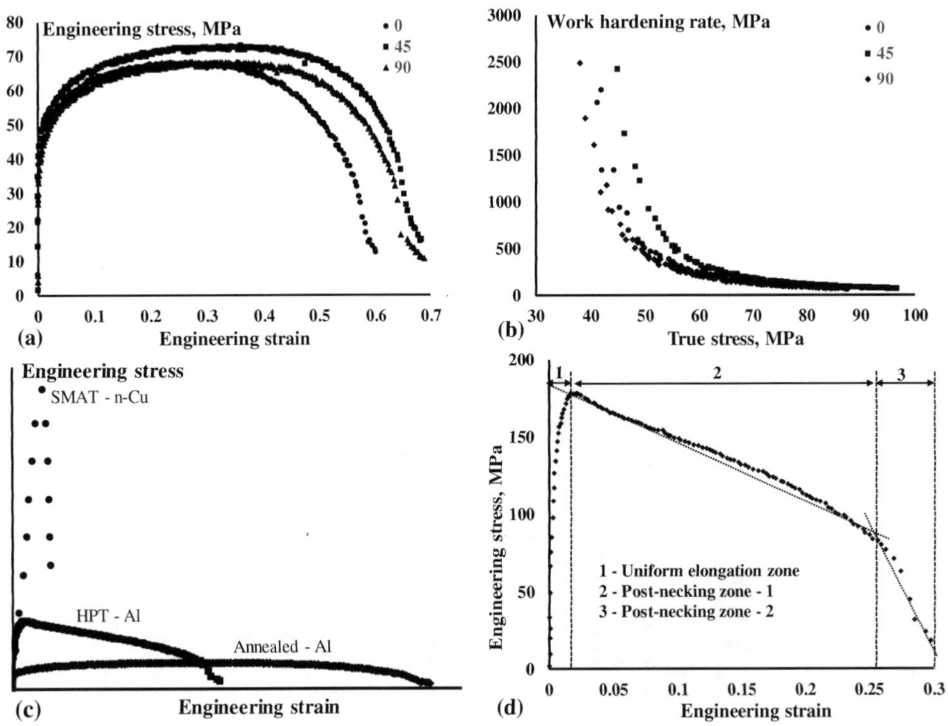

Figure 2. Engineering stress-strain curves (a) and work hardening rates along different orientations (b) for annealed material. (c) Engineering stress-strain curves for nano-Cu (*n*-Cu) from ref [12], annealed and HPT processed Al from present work and (d) partitioned engineering stress-strain curve for HPT Al.

$$\frac{d\sigma}{d\varepsilon} = n\frac{\sigma}{\varepsilon} \qquad (2)$$

where n is the strain hardening exponent, σ is the true stress and ε is the true strain. In the annealed material a slight anisotropy can be observed in both (figure 2 (b)), stress-strain and work hardening response. The material shows *stage III* and *IV* of work hardening. In *stage III* a rapid decline in work hardening is manifested due to the balance between dislocation multiplication and local dislocation annihilation process [13]. On the other hand, in *stage IV* where the work hardening rate is constant, the microstructure breaks into localized bands leading to deformation concentration in such bands.

Various studies conducted to evaluate the tensile properties of NC/UFG metals have reported two types of mechanical behavior with regard to the post-necking elongation, as shown by *n*-Cu from ref [12] and HPT-Al from present work, in figure 2 (c) [8, 12]. Though both show lower uniform elongation after reaching maximum load, *n*-Cu fails soon while HPT Al shows long tail-like post-necking elongation. As shown in figure 2 (d), the engineering stress-strain curve for HPT Al can be partitioned into three distinct zones: (i) uniform deformation zone (UEZ), (ii) post-necking zone-1 (PNZ-1), and (iii) post-necking zone-2 (PNZ-2). The PNZ-1 demarcates the diffuse necking deformation, whereas PNZ-2 covers the localized elongation in a very narrow band preceding failure.

The true stress-strain values for both the homogeneous and post-necking tensile deformation for HPT Al are shown in figure 3. As can be seen, the stress-strain values corresponding to the yielding and maximum load conditions show continuous improvement with ε_{eq}. The 0.2% offset yield strength which is ~42 MPa in annealed condition reaches a value of 146 MPa at ε_{eq} of 25.1. Contrary to the conventional forming processes, here in HPT, uniform deformation increase with accumulating shear strain in the material. Unlike pre-necking stress-strain values, the true stress values corresponding to the end of PNZ-1(σ_p) and PNZ-2(σ_f), either stay constant or decrease, respectively. To further compare

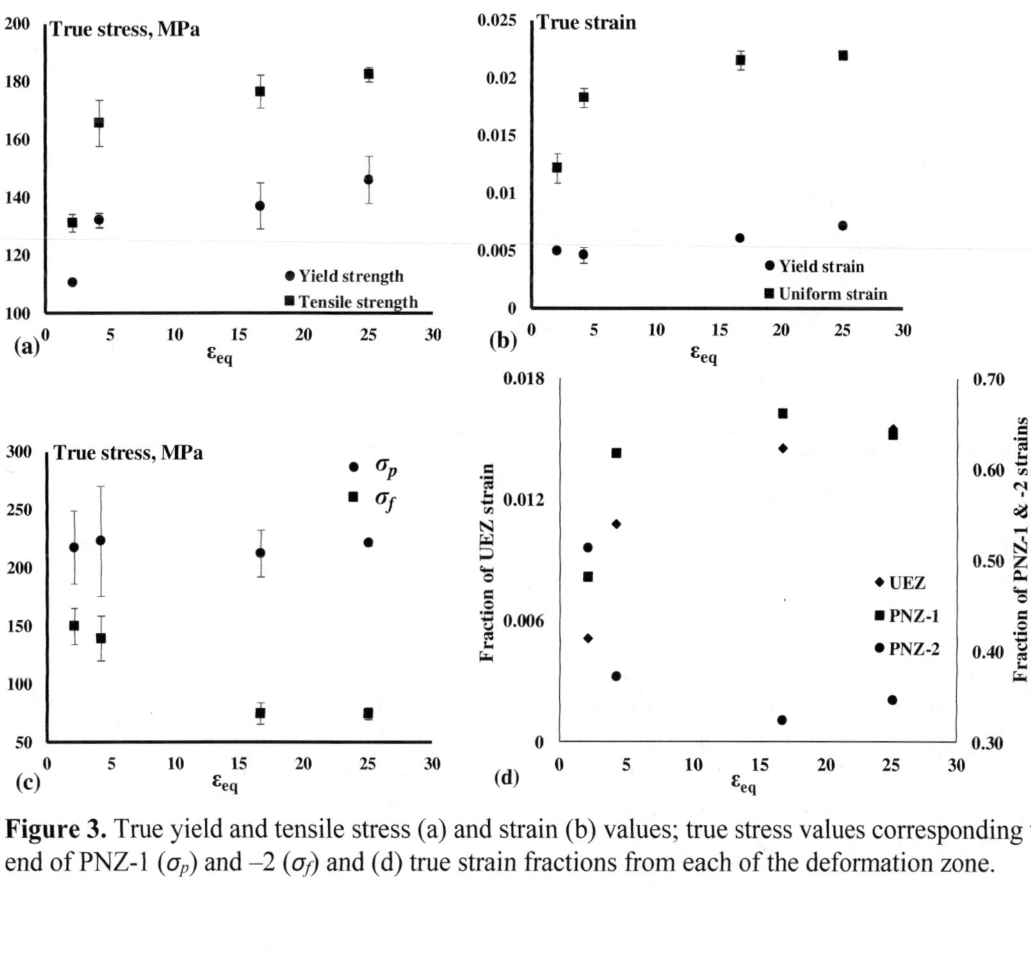

Figure 3. True yield and tensile stress (a) and strain (b) values; true stress values corresponding to the end of PNZ-1 (σ_p) and –2 (σ_f) and (d) true strain fractions from each of the deformation zone.

the pre- and post-necking tensile behavior, the average true strain values are utilized to determine the strain fractions from each deformation zone as shown in figure 3 (d). The true strain corresponding to the end of PNZ-1 is obtained from the volume constancy using the width strain value from a local strain field correlation near the fracture; while that for PNZ-2 is obtained from topographic measurements using SEM. From figure 3 (d), it can be observed that both UEZ and PNZ-1 strain fractions increase, while the strain fractions corresponding to PNZ-2 decrease with ε_{eq}. Thus it can be concluded that with ε_{eq} and refining *GS*, the pre-necking plastic performance improves, while post-necking plasticity deteriorates on account of the lower local ductility values.

Chan reported that in the case of fine-grained material, the stress required for micro-crack nucleation is lower than that required for propagation and therefore additional plastic work is required for crack propagation [14]. In present work, much better ductility observed in PNZ-1 indicates its role in providing the necessary plastic deformation and thereby strain hardening to locally produce the stress concentrations favorable for micro-crack propagation. This plastic work increases with decreasing grain size indicating an important role played by the refining *GS* in delaying the fracture. However, once initiated the finer-grained material fractures more swiftly than their coarse-grained counterparts.

The work hardening rates for HPT processed material are calculated using equation (2) and are plotted vs. true stress in figure 4 (a), while the linear slopes of the work hardening rates at different ε_{eq} are shown in figure 4 (b). It can be seen that unlike annealed material which exhibits both *stage III* and *IV* work hardening, HPT processed material shows only *stage III* hardening. The work hardening in severely deformed structures becomes difficult due to the presence of a large fraction of non-equilibrium boundaries which are known to possess enhanced diffusion properties [15]. In addition, dissociation of trapped dislocations at non-equilibrium boundaries is considered as one of the prevalent recovery mechanism in SPD processed materials [16-17]. However, from figure 4 (d) it can be seen that the work hardening ability is improving with ε_{eq}, which seems to be the source of enhancing uniform deformation in figure 3 (b).

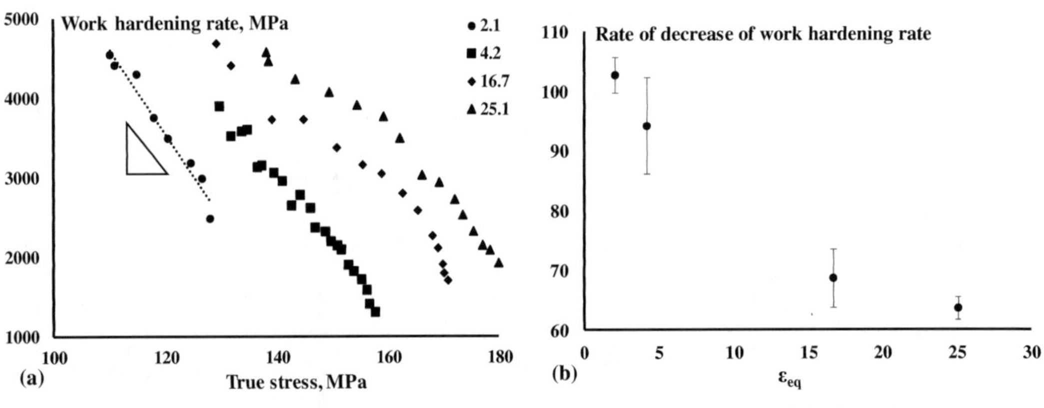

Figure 4. (a) Work hardening rates calculated for HPT processed material plotted vs. true stress and (b) the slope of *stage III* work hardening vs. ε_{eq}.

4. Conclusions

From the above experimental results and discussion following conclusions can be drawn:

1) The rolled and recrystallized material shows a slight anisotropy in its stress-strain and work hardening response. The annealed material exhibits *stage III* and *IV* hardening, while HPT processed material owing to efficient recovery mechanisms possess only *stage III* hardening.

2) With ε_{eq}, the pre-necking plastic properties show improvement, while the post-necking plastic performance deteriorates owing to reduced local ductility.

IOP Conf. Series: Journal of Physics: Conf. Series **1270** (2019) 012022 doi:10.1088/1742-6596/1270/1/012022

3) Refining *GS* causes the plastic work required for micro-crack propagation to increase and thereby delays the fracture. However, once initiated the finer-grained material fractures rapidly.

4) The work hardening ability of SPD processed material improves with the accumulating shear deformation.

Acknowledgements

The authors would like to acknowledge the Interuniversity Attraction Poles Program (IUAP) of the Federal Science Policy of Belgium and the partners of IUAP-VII-project P7/21 'Multiscale mechanics of interface dominated materials'. The funding sources did not play any role in design, collection, analysis and interpretation of the data, or in the writing of the report.

References

[1] Zhilyaev A P and Langdon T G 2008 *Prog. Mater. Sci.* **53** 893-979
[2] Valiev R Z and Langdon T G 2006 *Progr. Mater. Sci.* **51** 881-981
[3] Pirgazi H and Akbarzadeh A 2008 Int. J. Mod. Phys. B **22** 2840-47
[4] Naghdy S, Kestens L, Hertelé S and Verleysen P 2016 *Mater. Charact.* **120** 285-94
[5] Dalla Torre F H, Lapovok R, Sandlin J, Thomson P F, Davies C H J and Pereloma E V 2004 *Acta Mater.* **52** 4819-32
[6] Wang Y M, Ma E and Chen M W 2001 *Appl. Phys. Lett.* **80** 2395-97
[7] Valiev R Z 2003 *Adv. Eng. Mater.* **5** 296-300
[8] Ma E 2006 *J. Miner. Met. Mater. Soc.* **58** 49-53
[9] Stolyarov V V, Zhu Y T, Lowe T C and Valiev R Z 2001 *Mater. Sci. Eng.* A **303** 82-9
[10] Krasilnikov N, Lojkowski W, Pakiela Z and Valiev R 2005 *Mater. Sci. Eng.* A **397** 330-7
[11] Shrivastava S C, Jonas J J and Canova G 1982 *J. Mech. Phys. Solids* **30** 75-90
[12] Wang Y M, Wang K, Pan D, Lu K, Hemker K J and Ma E 2003 *Scr. Mater.* **48** 1581-6
[13] Verlinden B, Driver J, Samajdar I and Doherty R D 2007 *Thermo-mechanical processing of metallic materials* (Oxford, UK: Elsevier Pergamon Materials Series)
[14] Chan K S 1990 *Scr. Metall. Mater.* **24** 1725-30
[15] Valiev R Z, Kozlov E V, Ivanov Y F, Lian J, Nazarov A A and Baudelet B 1994 *Acta Metall. Mater.* **42** 2467-75
[16] Ko Y G, Shin D H, Park K -T and Lee C S 2006 *Scr. Mater.* **54** 1785-9
[17] Tucker G J and McDowell D L 2011 *Int. J. Plast.* **27** 841-57

IOP Conf. Series: Journal of Physics: Conf. Series **1270** (2019) 012023 doi:10.1088/1742-6596/1270/1/012023

A study of the static recrystallization behaviour of cast Alloy 825 after hot-compressions

M Al-Saadi[1,2], F Sandberg[1], C Hulme-Smith[2], A Karasev[2] and P G Jönsson[2]
[1] R&D, AB Sandvik Materials Technology, SE-811 81Sandviken, Sweden.
[2]KTH Royal Institute of Technology, Materials Science and Engineering, SE-100 44 Stockholm, Sweden.

E-mail : muniras@kth.se; munir.al-saadi@sandvik.com ; chrihs@kth.se

Abstract. The static recrystallization behaviour of a columnar and equiaxed Alloy 825 material was studied on a Gleeble-3800 thermo-simulator by single-hit compression experiments. Deformation temperatures of 1000-1200 °C, a strain of up to 0.8, a strain rate of $1s^{-1}$, and relaxation times of 30, 180, and 300 s were selected as the deformation conditions to investigate the effects of the deformation parameters on the SRX behaviour. Furthermore, the influences of the initial grain structures on the SRX behaviors were studied. The microstructural evolution was studied using optical microscopy and EBSD. The EBSD measurements showed a relaxation time of 95 % for fractional recrystallization grains, t_{95}, in both structures, was less than 30 seconds at the deformation temperatures 1100 °C and 1200 °C. However, fewer than 95% of recrystallized grains recrystallized when the deformation temperature was lowered to 1000 °C. From the grain-boundary misorientation distribution in statically recrystallized samples, the fraction of high-angle grain boundaries decreased with an increasing deformation temperature from 1000 °C to 1200 °C for a given relaxation time. This was attributed to grain coarsening.

1. Introduction

The Nickel-based alloy 825 is resistant to many corrosive environments [1-3]. Moreover, alloying it with small amounts of carbon provides strengthening. As shown in previous reports [4-5], dynamic recrystallization in either the columnar or the equiaxed compressed structure did not occur at any hot-working temperature. Industrial products, particularly billets or bars, manufactured from the alloy are subjected to high deformation temperatures and exposed deformation schedules which are varied to meet property requirements. The combined effects of the deformation temperature and relaxation time on the material lead to microstructural changes, which significantly affect the mechanical properties. In practice these microstructural changes can include recovery and static recrystallization. In literature, it has been shown that the deformation temperature and holding time have a large influence on the softening or recrystallization of the material structure [6-7]. The static-recrystallization kinetics in columnar and equiaxed grain structure have been mapped by using hot compression testing. The fraction recrystallized structure was evaluated by an EBSD study of rapidly quenched specimens. This method is based on the observation of microstructural evolution with relaxation time. This paper will focus on understanding the static recrystallization behaviour over a range of deformation temperatures (1000 °C, 1100 °C, and 1200 °C), relaxation time (30 sec, 180 sec, and 300 sec), at a constant strain rate of $1s^{-1}$ and a strain of 0.8, respectively.

2. Experimental material

The investigated alloy was prepared via melting in the Electric Arc Furnace, refining on the Argon Oxygen Decarburisation converter and then Continuous Casting of blooms. The chemical composition of this investigated heat is shown in table 1.

Table 1. Test certificate chemical analysis (wt. %) of Alloy 825.

C	Si	Mn	Cr	Fe	Mo	Ti	Cu	N	Ni
0.007	0.23	0.67	22.1	32.0	2.53	0.8	1.6	0.009	Bal.

The cylindrical compression samples of a 15 mm height and a 10 mm diameter were cut and machined with the compression axis parallel to the continuous cast axis from a constant radial position to the columnar and equiaxed grain zones. To study the softening behaviour of Alloy 825 at different temperatures, monotonic-hit compression was employed using a Gleeble 3500 Thermo-Mechanical Simulator. The deformation temperature tests were carried out at a deformation temperature of 1000 °C, 1100 °C and 1200 °C with a nominal strain of 0.8 and a constant strain rate of $1s^{-1}$. The deformed samples were held at the test temperature for relaxation times of 30, 180, and 300 seconds before undergoing the fast cooling. For more details about experimental procedure and similar flow stress curves the reader is referred to reference [5]. The percentage static recrystallized (SRX) grain and microstructural evolution at different testing conditions were studied by using electron backscatter diffraction and optical microscopy.

3. Results and discussion

3.1. Effect of deformation conditions on microstructure

Figure 1 shows an example of the microstructure of relaxed columnar and equiaxed structures of Alloy 825 samples, after the hot compression with different deformation conditions. A partial recrystallization occurred in both relaxed structures as shown in the micrographs, figure 1 (a and b), and recrystallized grains forms on grain boundaries and at large particles (as shown with PSN). This is shown by the presence of some grains that are much smaller than others, as shown by arrows. Additionally, static recrystallized grains observed in the equiaxed structure was larger than in the columnar structure and the fraction of static recrystallized grains in the equiaxed structure was larger than in the columnar structure.

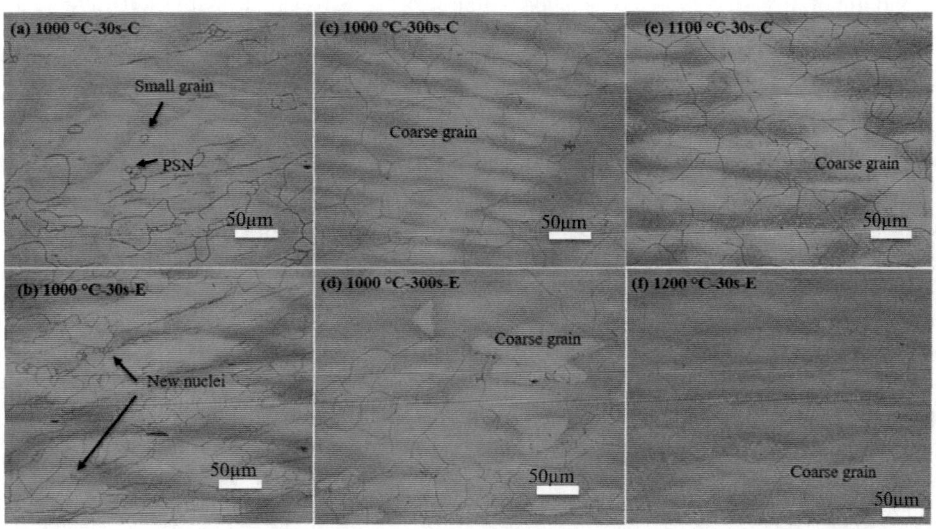

Figure 1. Optical microstructures at the center area of the columnar (C) and equiaxed (E) samples after completion of the SRX tests at the indicated deformation temperature and relaxation time.

After a relaxation time of 300s, the volume fraction of statically-recrystallized grains increased in both the relaxed columnar and equiaxed samples, figure 1(c and d). However, SRX grains became progressively much coarser in both the columnar and equiaxed specimens. A recovery and an almost full recrystallization occurred, as shown in the micrographs, figure 2e. Also, a small amount of static recrystallized grains was already coarsened. Additionally, statically-recrystallized grains observed in the columnar structure were coarser than in the equiaxed structure. A full recrystallization occurred as shown in the micrograph, figure 1f, and a large amount of static recrystallized grains were already coarsened. Additionally, statically-recrystallized grains observed in the columnar and equiaxed structure was similarly coarsened.

3.2. *Hardness of deformed columnar/equiaxed Alloy 825*

Table 2 shows variation of the Vickers microhardness with the relaxation time at given deformation temperatures of 1000 °C, 1100 °C, and 1200 °C in columnar and equiaxed relaxed samples, together with the standard deviations. The columnar and equiaxed recrystallized samples showed that the hardness level decreases with an increasing deformation temperature and relaxation time, reaching a lowest value of hardness indicative of softening via recovery.

Table 2. Hardness ($HV_{0.5}$) of recrystallized columnar/equiaxed Alloy 825, with ± standard deviation.

Structure	Deformation temperature, [°C]/Relaxation time, [sec]								
	1000			1100			1200		
	30	180	300	30	180	300	30	180	300
Columnar	160±5	157±7	154±4	162±4	141±7	143±6	138±2	131±7	131±11
Equiaxed	175±6	151±12	150±4	147±11	147±8	146±7	138±12	136±8	134±1

3.3. *Grain misorientation distribution of relaxed samples*

The recrystallized grain distribution of misorientation angles for columnar and equiaxed alloy 825 deformed microstructures, for a 5° grain definition, is shown in figure 2. Figure 3 supports the conclusions drawn from figure 2 (a and b) at a more quantitative level. In the columnar regions, figure 2a showed that the frequency of LAGB was somewhat higher than the frequency of HAGB, due to onset of the limited amount of dynamic subgrains recrystallization (DRX subgrains). At a relaxation time of 30 s, the frequency of HAGB increases with increasing deformation temperatures due to the SRX grains increases in size and a type of grain-growth occurs. However, the frequency of HAGB obtained at a deformation temperature of 1200 °C was approximately 75% lower than that obtained at 1100 °C due to grain coarsening. Beyond a relaxation time of 30 s, the amount of HAGB diminishes as the deformation temperature increases. This is indicative of increasing grain sizes as grain growth occurs and grains become coarser. This supports the conclusions drawn from Table 2 and figure 4b. In the equiaxed condition and at a given relaxation time, the decreasing fraction of HAGB with increasing deformation temperatures is due to grain coarsening. However, approximately 70% of boundaries detected in the equiaxed structure at a relaxation time of 180 s, figure 2c, were high angle boundaries (60°) and the remainder were low angle boundaries (<5°). This is due to low temperature, which leads to a mixture of both refined and coarse grains.

3.4. *Percentage static recrystallized grains and grain size, $d_{(SRX)}$*

The plot of measurements of the recrystallization percentage versus relaxation times in columnar and equiaxed compressed samples in the temperature range of 1000 °C and 1200 °C is shown in figure 3a. The fractional recrystallized grains of columnar and equiaxed specimens deformed to various deformation temperatures and relaxed at different relaxation times revealed that the fraction of SRX recrystallized grains is higher than 80%. As with the tests performed at 1000 °C, a clear sigmoidal shaped curve is shown for the columnar and equiaxed specimens with incomplete static recrystallization occurring until a relaxation time a relaxation time of 300 s, where more than 80% recrystallization was observed. It could be noted that the SRX recrystallization curve for the lowest deformation temperature

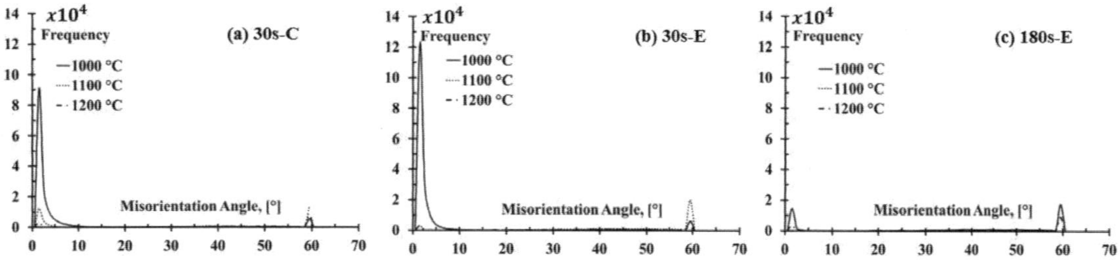

Figure 2. Grain-boundary misorientation distribution in the columnar (C) and equiaxed (E) samples after completion of the SRX tests at the indicated deformation temperature and relaxation.

1000 °C, showed sigmoidal behaviour with the rate of SRX recrystallization decreasing after a relaxation time of 180s. The fraction of SRX grains increases sharply at the intermediate deformation temperature 1100 °C up to a relaxation time of 30 s, where more than 80% and 90% recrystallization degrees in columnar and equiaxed, respectively. Thereafter, this becomes constant beyond 30 s which signifies to a fact that the relaxation times have no significant effect on SRX fraction. However, fractions of SRX grains are 88-90% for both columnar and equiaxed structures and more than 95% recrystallization was observed after 30 s. At deformation temperatures of 1100 °C and 1200 °C, full SRX recrystallization occurred within 180 s for both structures. Measurements of the mean recrystallized grain size versus deformation temperatures in columnar and equiaxed deformed samples at relaxation times of 30 s, 180 s, and 300 s compared to un-relaxed samples (the values obtained from our previous study[5]) are shown in figure 3b. As is shown, the average recrystallized grain size generally increases when increasing the deformation temperature at a given relaxation time. In both the columnar and equiaxed deformed samples, the average SRX grain size sharply increases with an increased relaxation times from 0 s to 30 s. It is obvious that the average SRX grain size in columnar deformed specimens for relaxation time 0 s was 15% lower than of equiaxed deformed samples. In columnar deformed specimens, the average recrystallized grain size is observed to be similar for relaxation times of 180 s and 300 s. It is apparent that the relationship between the average SRX grain size and the deformation temperature in both columnar and equiaxed specimens is almost linear.

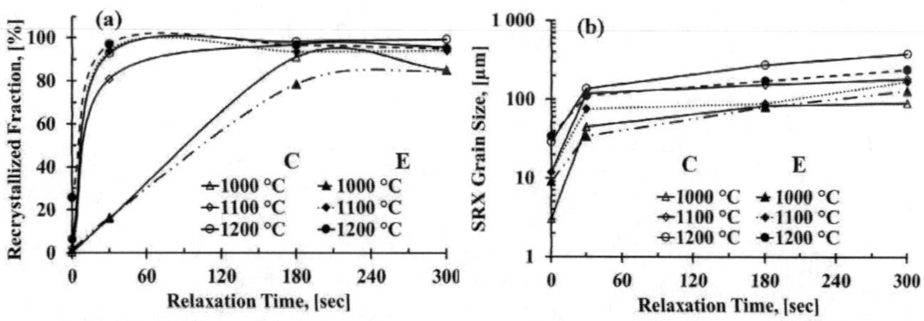

Figure 3. Plot of measurement of a) percentage of recrystallized SRX grains and b) recrystallized SRX grain size versus relaxation times in columnar and equiaxed compressed samples in the temperature range of 1000 °C, 1100 °C and 1200 °C at a strain rate of 1 s^{-1} and a strain up to 0.8. Open symbols and solid lines represent columnar samples. Closed symbols and dashed lines represent equiaxed samples.

3.5. *The inverse pole figure (IPF) maps*

Figure 4 shows a few of examples of the EBSD orientation mapping images of the relaxed specimens for relaxation times between 30 s and 300 s over a range of deformation temperatures of 1000 °C, 1100 °C, and 1200 °C and at a constant strain rate of 1s^{-1} and strains up to 0.8. The color of EBSD grains refers to the different orientation and the similar color shows that the grains have a similar orientation, as can be seen in the IPF coloring in figure 4g. The texture for both structures is a double fiber texture with <110> and <100> parallel to the compression direction. This is likely dominated by the large non-recrystallized grains, as shown in figure 4 (a and b). For both structures, subgrain structures form low angle grain boundaries are predominated during hot deformation at a temperature of 1000 °C with a relaxation time of 30 s (recrystallization by subgrain rotation may occur, [7]). At the 30 s relaxation time, large amounts of coarse and deformed grains (non-recrystallized) together with fine recrystallized grains are formed within both structures, (rotation mechanism, [7]), figure 4 (a and b). In the equiaxed structure, the SRX occurs through the so-called necklace mechanism, forming at some individual deformed grain boundaries, figure 4b. At a prolonged relaxation time of 300s, there were more statically recrystallized grains than after holding for 30s and the almost no deformed substructure was visible in the recrystallized microstructure, figure 4 (c-d). Furthermore, the SRX grains and size increased and a small amount of the non-recrystallized grains still could be found in the microstructure. At a temperature of 1100 °C and a relaxation time of 30 s, the large deformed grains almost disappear and the refinement of SRX grains and twins can predominately be seen in the microstructure, figure 4 (e). Large SRX grains (grain growth) can be observed and annealing twins can still be identified in the whole columnar microstructure, figure 5e. As shown in figure 4 (e and f), the number of statically-recrystallized grains decreases when increasing the deformation temperature at a given relaxation time because of grain growth. At a deformation temperature of 1100 °C, in a columnar relaxed specimen, the number of recrystallized grains significantly increases beyond a relaxation time 30 s. This is due to the nucleation of more SRX grains (figure 3a supported this) and then gradually decreases when reaching a relaxation time of 300s due to grain coarsening. The latter, the lowest deformation temperature combined with a long relaxation time, could lead to either the nucleation of more SRX grains or a growing of smaller SRX grains. Nevertheless, in equiaxed relaxed specimen, the amount of recrystallized grains slightly decreases from a relaxation time of 30s to 300s, which indicates that grain coarsening occurs. As previously discussed, the effect of increasing the relaxation time corresponds to an increase of the average recrystallized grain size and percentage of recrystallized fraction. It is obvious that the deformation temperature, below or at a relaxation time of 30 s, has a strong influence on the onset of SRX fraction and average recrystallized grain size in both columnar and equiaxed specimens. The relaxation time, after 30 s at the lowest deformation temperature of 1000 °C or before that time at the intermediate and highest temperature has only a slight effect on the SRX fraction at the later stages and/or on the average recrystallized grain sizes in both columnar and equiaxed specimens.

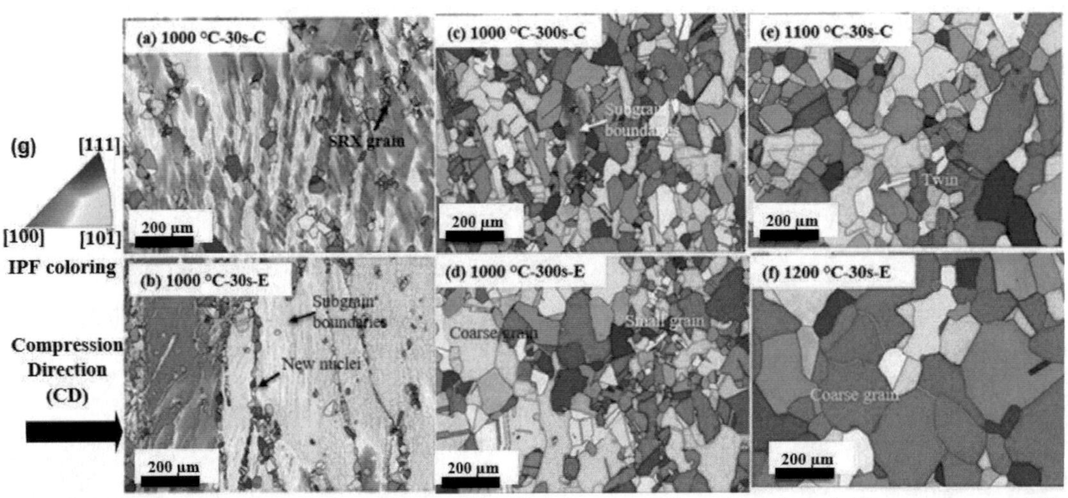

Figure 4. Microstructural evolution of the relaxed columnar (C) and equiaxed (E) specimens at the indicated deformation temperatures and relaxation times. (For interpretation of the references to colour in this figure legend, the reader is referred to the web version of this paper.)

4. Conclusions

The SRX behaviour in a columnar and equiaxed deformed structure of an Alloy 825 was studied in the temperatures range of 1000 to 1200 °C with different relaxation times: 30 s, 180 s, and 300 s. Tests were carried out using at a constant strain rate of 1 s^{-1} and a nominal strain of up to 0.8 through single-hit compression tests using a thermomechanical simulator. The microstructural evolution was studied using LOM and EBSD. The primary outcomes of the study are as follows:

1) The time of a 95% recrystallization, (t_{95}), in both samples commences the time for a complete recrystallization, was less than 30 s at the intermediate and highest deformation temperatures of 1100 °C and 1200 °C. However, a 95% recrystallization was not achieved at the lowest deformation temperature (1000 °C) for any relaxation time.
2) The average SRX grain size, in columnar/ equiaxed grain samples, increases 3-5% with increasing deformation temperatures. Beyond a 30 s relaxation time, the grain size increases almost linearly with an increased deformation temperature.
3) In general, the fraction of SRX and the recrystallized grain size increased with an increased deformation temperature combined with an increased relaxation time. At the lowest deformation temperature of 1000 °C, the relationship between the fraction of SRX and the relaxation time is a typical sigmoidal shape.
4) From grain-boundary misorientation distributions in both statically recrystallized structures shows that the decreasing amount of HAGB decreased with an increased deformation temperature at a given relaxation time, due to a grain coarsening.

Acknowledgement
The authors wish to thank Sandvik Materials Technology for the financial support, and the permission to publish this paper.

References

[1] ASTM International: www.astm.org, 2003, vol. 02, pp. 5–8.
[2] ASTM International: www.astm.org, 2015, vol. 03, pp. 3–7.
[3] ASTM International: ASTM B564 - Nickel Alloy Forgings, www.astm.org.
[4] Al-Saadi M, Sandberg F, Karasev A and Jönsson G P 2018 3rd Proc. Int. Conf. on Ingot Casting Rolling and Forging (Stockholm) p107-15
[5] Al-Saadi M, Sandberg F, Karasev A and Jönsson G P 2018 Procedia Manuf. 15 1626–34
[6] Verlinden B, Driver J, Samajdar I and Doherty R 2007 Thermo-mechanical processing of metallic materials vol 11, ed R W Cahn (Amsterdam:Elsevier) p 1-528
[7] Humphreys F J and Hatherly M 2004 Recrystallization and Related Annealing Phenomena (Amsterdam: Elsevier)

IOP Conf. Series: Journal of Physics: Conf. Series **1270** (2019) 012024 doi:10.1088/1742-6596/1270/1/012024

Recrystallization and grain growth at the interface of a bimetallic colaminated strip composed of two different Fe-Ni alloys

J Poncelet[1], T Baudin[1], M de Oliveira[1], T Waeckerlé[2], Y Ateba-Betanda[2], F Brisset[1] and AL Helbert[1]

[1]ICMMO, Univ. Paris-Sud, Université Paris-Saclay, UMR CNRS 8182, Orsay, France;

[2]Pierre Chevenard Research Center, Aperam Alloys Imphy, Imphy, France

Abstract. Roll bonding is a solid-state welding process widely used to manufacture layered metal composites. Particular properties may thus be obtained using the physical features of each material of the composite. Bimetal plates consisting of two different Fe-Ni alloys were made by roll bonding followed by heat treatment for 90 minutes at various annealing temperatures. The effects of post-rolling heat treatments on the bonding strength of a bimetal strip were investigated in relation to the interface microstructure evolution. Both recrystallization and grain growth took place at the interface during annealing. In particular, nucleation of new grains as well as growing grains crossing the interface may have contributed to the improvement of the bonding strength. Moreover, diffusion through the interface was found to drastically enhance the bonding strength from 850°C up to 1050°C. However, excessive grain growth associated to porosity occurrence probably caused the saturation of the bonding strength beyond 1050°C.

1. Introduction

Recent decades have been marked by the increasing production of clad metal materials. Such multi-materials aim to either save precious metal like electromechanical relays with gold-plated contacts or to achieve functional properties by combining the physical features of each material, like aluminum cookware cladded with stainless steel exhibiting high heat conductivity and food compatibility. One of the most widely used techniques to make the materials strongly connected is the cladding process, commonly obtained by co-rolling of two (or more) solid-state metals at high temperature (like hot rolling) or low temperature (preventing any surface oxidation). This paper deals with the latter technique, which can also be called "Roll bonding". One of the industrial applications is bimetal composed of two components with very different thermal expansion behaviour cladded together, that bends during overheating in case of current excess in electrical devices and thus interrupts the circuit [1].

It is generally supposed that bonding mechanisms rely on the film theory, first established by Vaidyanath and Milner [2], stating that the breaking of the surface layer is first required to bring "fresh" (which said out of any surface oxidation) metals into contact at the interface. This model called "fracture – extrusion – welding" was then refined by Bay [3] who proposed that surface expansion upon rolling breaks a contaminant film and a work-hardened surface layer produced by surface preparation. Due to roll pressure, fresh metal from each opposite material can extrude through the cracks and bond at the interface. In recent work, many parameters affecting welding quality have been reported [4]: surface preparation, thickness reduction rate, rolling temperature, deformation speed and normal pressure. Furthermore, post-heat treatments carried out after rolling usually provide an improvement in bonding strength, attributed to diffusion and short-range atomic movements [4]. However, heterogeneities induced by the rolling process may also affect bonding reinforcement during annealing. In this study, annealing treatments were performed on bimetallic strips of two iron-nickel alloys, to improve understanding of thermally activated consolidation mechanisms.

2. Experimental Method

Two different iron-nickel alloys Fe-47.5Ni and Fe-50Ni-8.6Cr (wt %), with an initial grain size of approximately 17 µm and 7 µm respectively, were used. These initial recrystallized laminations with a dimension of 75 x 10 x 1 mm^3 were first degreased with acetone. The surfaces of the sheets were then scratch-brushed in order both to remove contaminants and oxides and to form a work-hardened surface layer. After riveting the heads of the sheets, they were heated at 350°C for 15 minutes in air and rapidly roll bonded right out of the oven. Rolling was performed at a thickness reduction of 50% using a laboratory rolling mill with a capacity of 2.2 kW. The rolls diameter was 67 mm and the rolling speed was 48 mm/s. After rolling, the composite strips were annealed in air for 90 minutes at different temperatures ranging from 350°C to 1200°C. Bonding strengths were measured by the peeling test [5]. The peeling test was carried out using a Zwick/ Roell Z010 tensile testing machine with a crosshead speed of 100 mm/min and a 10 kN load cell. Zeiss Sigma HD scanning electron microscope (SEM) equipped with energy dispersive spectrometer (EDS) was used to analyze the interfacial bonding state under different annealing conditions with a step size of 0.5 µm. The microstructural interfacial evolution was monitored by the electron backscatter diffraction (EBSD) technique with a step size ranging from 50 nm to 0.5 µm. Microstructural investigations were performed in the transversal section which contained the transversal (TD) and normal (ND) directions. All EBSD orientation maps are ND-IPF.

3. Results and discussion

Figure 1 presents the experimental bonding strength as a function of annealing temperature. At low temperatures, the annealing treatment shows little impact on the bonding strength, which remains steady until 550°C. From 550°C, a significant growth of adhesion is achieved with the increase of temperature. Thus, the bonding strength goes up by a factor of 30 before reaching a plateau at 1050 °C. The influence of microstructural evolution at the interface on bonding strength was then investigated. The effects of annealing temperature on the recrystallized area fraction f_{Rex} and the recrystallized average grain size were also plotted in Figure 1. It appears that recrystallization occurs from 550°C and is complete at 650°C and 750°C for Fe-47.5Ni and Fe-50Ni-8.6Cr respectively (Figure 1a). Concerning the grain size, grain growth occurs as soon as recrystallization begins, at 550°C. A classical grain growth law is followed [6].

Figure 2 displays EBSD orientation maps carried out at the bonding interface of samples annealed at different temperatures. Before annealing, the microstructure of both alloys exhibits refined and elongated grains, as shown in Figure 2a. The coarse particles and non-indexed regions observed along the interface were found to be fragments of the work-hardened layer produced by scratch-brushing, broken and scattered during roll bonding. However, the welded interfacial areas were sufficient in number to ensure minimal bonding strength and allow the peeling test. For temperatures lower than 450°C, the grains show similar morphology to those of the roll-bonded sample. From 550°C, the area fraction of recrystallized grains in both components starts to rise rapidly with increasing temperature (see in Figure 1a). As recrystallization proceeds, substantial twinning occurs. The Fe-47.5Ni and Fe-50Ni-8.6Cr alloys achieve complete recrystallization at 650°C (Figure 2b) and 750°C (Figure 2c) respectively. The resulting microstructure comprises equiaxed grains, as it can be seen in Figure 2d. It is worth noting that the beginning of the increase in bonding strength coincides with the onset of grain

IOP Conf. Series: Journal of Physics: Conf. Series **1270** (2019) 012024 doi:10.1088/1742-6596/1270/1/012024

growth. From 650°C, both Fe-47.5Ni and Fe-50Ni-8.6Cr alloys undergo grain growth, reaching an average grain size of 28.5 and 14 µm respectively (see in Figure 1b, Figure 2d).

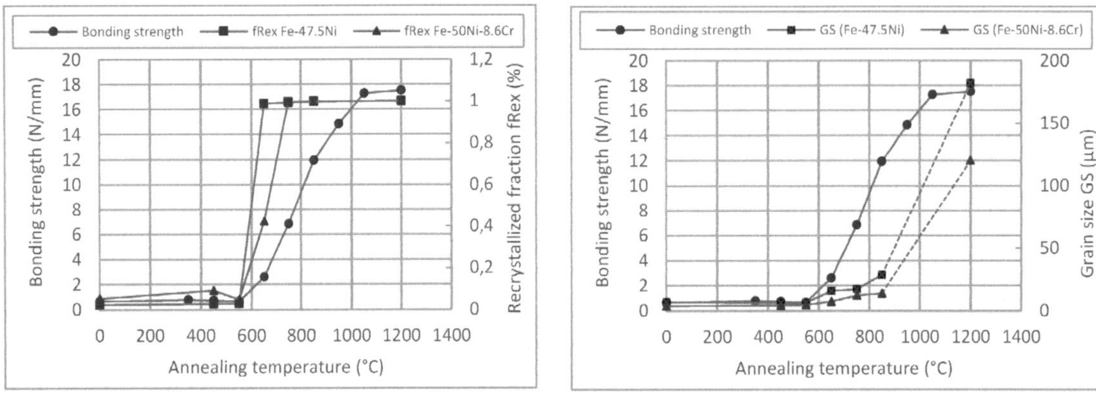

(a) (b)

Figure 1. The effect of the annealing temperature on the bonding strength and: (a) on the recrystallized area fraction; (b) on the average recrystallized grain size of each alloy. Both the area fraction and the grain size were measured in the base alloys at some distance of the interface. Due to the small number of grains analysed, the grain size in the samples annealed at 1200°C was estimated.

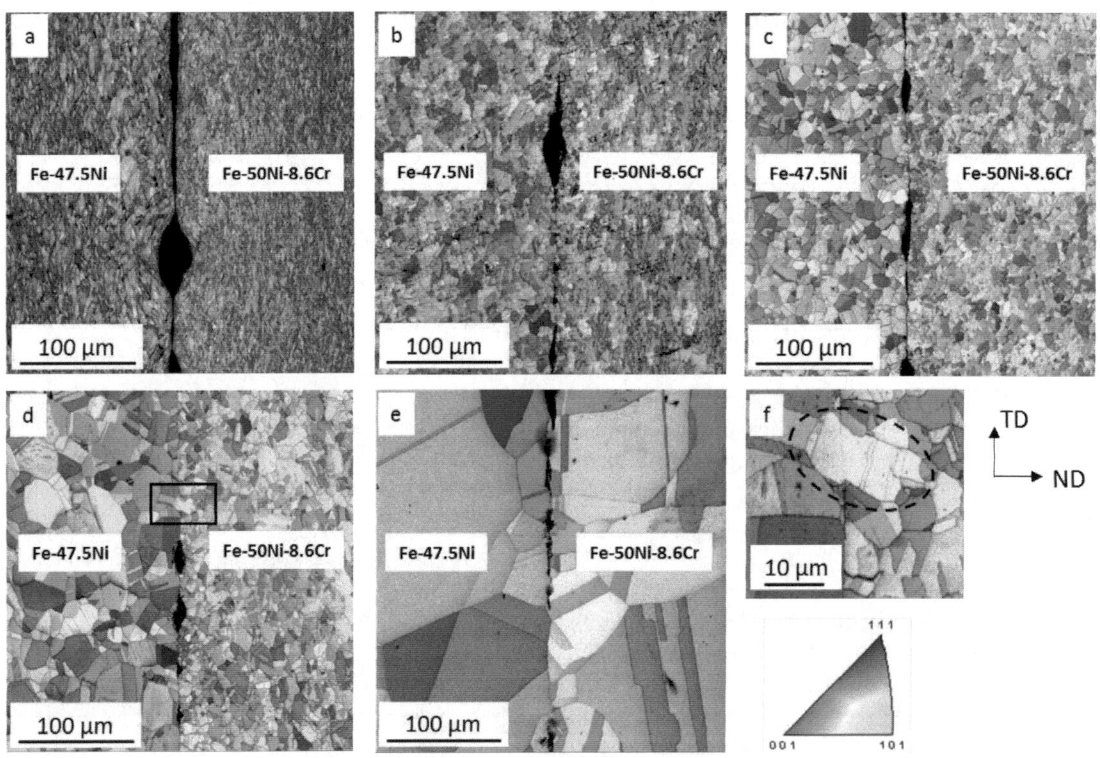

Figure 2. EBSD orientation maps showing the microstructure of the samples after 90 min of annealing at different temperatures: (a) deformed state; (b) 650°C; (c) 750°C; (d) 850°C; (e) 1200°C; (f) the box at the interface of the sample annealed at 850°C with higher magnification.

IOP Conf. Series: Journal of Physics: Conf. Series **1270** (2019) 012024 doi:10.1088/1742-6596/1270/1/012024

Figure 2f shows with higher magnification a part of the orientation map made at the interface of the sample annealed at 850°C (see in Figure 2d). The dotted circle points a grain which has grown and crossed the interface during annealing. Inside this grain, a ghost line [7] can be seen, marking the location of the interface before annealing.

Figure 3 displays EBSD orientation maps focused on the particles located along the interface of as-rolled and annealed samples. Similar particles arising from wire-brushing surface treatments were observed by Mishin et al. [8] at the interfaces of nickel processed by ARB. The authors found out that these work-hardened fragments affected the progress of recrystallization by providing highly refined deformation zones acting as preferential nucleation sites. This is in agreement with Figure 3a and Figure 3b, where it can be seen that these particles affect the rolled microstructure around them, creating deformation zones which contain finer grains than in the core.

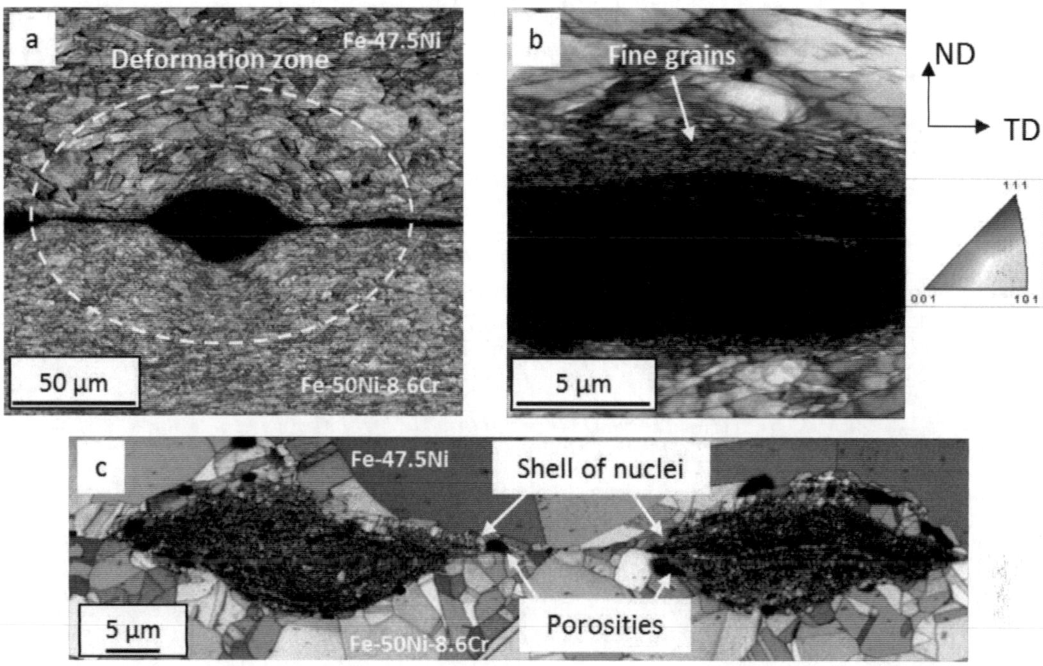

Figure 3. EBSD orientation maps showing particles at the interface of composite sheets: (a) before annealing; (b) before annealing with higher magnification; (c) after annealing at 850°C.

In Figure 3c, a "shell" made of nucleated recrystallized grains with various crystallographic orientations can be noticed around the fragments. It should be noted that regions containing fragments were recrystallized after annealing at 850°C but are composed of smaller grains than in the core of the sample. Recrystallization around fragments and interface contributed to the "interface closure" observed at larger scale in the maps of Figure 2 as annealing temperature increased (compare Figure 2a and 2d). Some porosities can be noticed around the particles after annealing at 850°C (see in Figure 3c). These voids were indeed observed along the interface of every annealed sample and seemed to become more frequent with increasing temperature.

Figure 4 shows EDS concentration profiles of major elements across the bonding interface for samples annealed at 450°C and 1200°C. As seen on the profiles, an interdiffusion layer is established at the interface and becomes larger with increasing temperature. The diffusion coefficient of Ni into Fe-52 %Ni was calculated with equation (1):

$$D(T) = D_0 \exp(-\frac{Q}{RT}) \quad (1)$$

IOP Conf. Series: Journal of Physics: Conf. Series **1270** (2019) 012024 doi:10.1088/1742-6596/1270/1/012024

where D_0 is the diffusion factor, Q is the activation energy for diffusion, R is the gas constant ($R = 8.314\ J.mol^{-1}.K^{-1}$) and T is the annealing temperature.

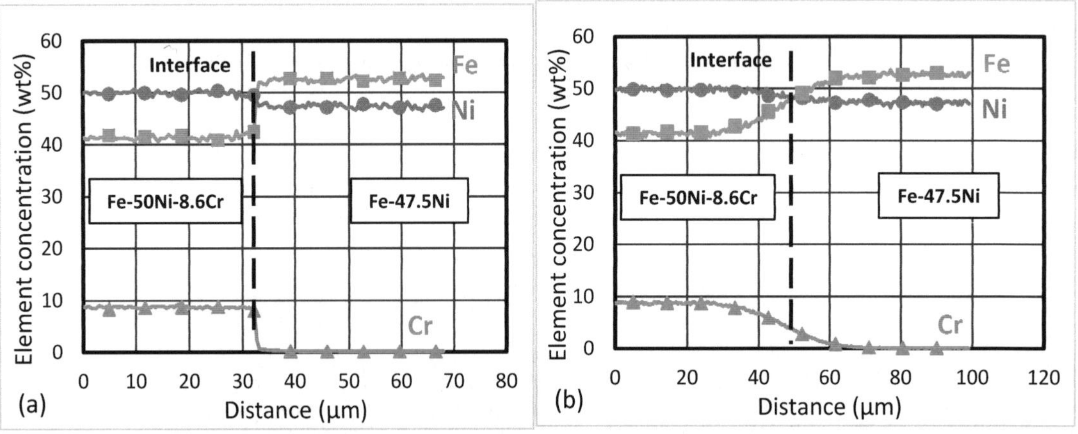

Figure 4. EDS line scanning curves of interfacial zone after annealing at: (a) 450°C; (b) 1200°C.

The theoretical variations of the interdiffusion layer thickness $L_D = \sqrt{D(T)t}$ (using equation (1)) and the bonding strength are plotted versus annealing temperature in Figure 5. The diffusion thickness is very weak at 450°C and is expected to remarkably expand at 1200°C (around 12 µm). This is in accordance with the experimental diffusion distance of Ni found from EDS measurements: 16 µm at 1200°C (Figure 4b).

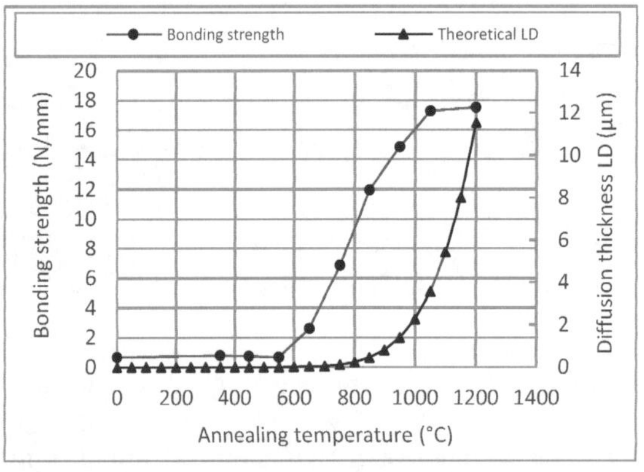

Figure 5. The effect of the annealing temperature on the bonding strength and the diffusion layer thickness of nickel at the interface of the composite sheet.

The annealing treatment performed after roll bonding involved several phenomena highlighted in this study, contributing to the bonding strength improvement of the composite with increasing temperature. Both recrystallization and grain growth driven by grain boundary migration took place during annealing. As seen previously in Figure 1b, grain growth may have promoted the increase in bonding strength from 650°C. One can assume that nucleation of recrystallized grains around coarse work-hardened fragments and at the interfacial regions contributed to enhance the bonding strength. Besides, at high annealing temperatures, some growing grains were able to cross the interface undoubtedly reinforcing bonding (see in Figure 2f).

Moreover, the observed atomic diffusion between the two sheets at high temperatures may have improved the metallurgical bonding at the interface, through a solid-solution strengthening effect. A similar process leading to an increase of mechanical bonding strength was described by Hwang et al.

[9] after annealing of an aluminum/stainless steel bimetal plate. It is suggested that volume diffusion during annealing not only resulted in strengthening of welded areas, but also led to welding of interfacial regions which may have been in contact during rolling without achieving preliminary metallurgical bonding or which may have come into contact during annealing through thermal expansion of metals. Thus, the bonding strength was assisted by diffusion phenomena at high temperatures (>850°C).

Finally, the growing porosities observed along the interface may also have led to reduction in bonding strength for the highest temperatures. These porosities may have been formed due to the Kirkendall effect [10] or due to the swelling of air bubbles trapped at the interface during rolling. This phenomenon is supposed to be a major reason for saturation of bonding strength beyond 1050°C. Besides, grain growth causing mechanical softening in welded areas can also have contributed to the bonding strength reduction and may have become critical compared to the beneficial effect of diffusion.

4. Conclusion

The effects of annealing treatments on the bonding strength of a bimetal strip consisting of two different iron-nickel alloys were investigated. The following conclusions were drawn from this work:

- As annealing temperature rose, the bonding strength was significantly improved, reaching its maximum at 1050 °C.
- It was suggested that the increase of bonding strength, first promoted by grain growth from 650°C, was also assisted at higher temperatures by interfacial diffusion, which helped to improve the metallurgical bonding in welded areas and expand the region of welding.
- Despite the continuous increase of diffusion, the saturation of bonding strength beyond 1050°C is likely due to porosities along the interface and strong grain growth causing mechanical softening in welded areas.
- Nucleation of new grains was observed in deformation zones around fragments and interface regions. Further investigation is required to understand the influence of recrystallization and grain growth mechanisms on the increase of interfacial bonding.

References

[1] Trostel H and Tiers J F 1996 *The Iron-Nickel Alloys* ed Springer (USA) chapter 18 pp 367-84
[2] Vaidyanath L R, Nicholas M C and Milner D R 1959 *Brit. Weld. J.* **6** 13-28
[3] Bay N 1983 *Weld. Res. Suppl.* **62** 137-42
[4] Li L, Nagai K and Yin F 2008 *Sci. Technol. Adv. Mater.* **9**(2) 023001
[5] Verstraete K 2017 *Étude du multi-colaminage de matériaux différents* PhD Thesis, Université Paris-Sud, Orsay, France
[6] Humphreys F J and Hatherly M 2004 *Recrystallization and Related Annealing Phenomena (2nd edition)* ed Elsevier (Oxford: Pergamon Press) chapter 13 pp 415-50
[7] Yin W, Wang W, Fang X, Qin C and Xing X 2015 *Mater. Charac.* **107** 134-8
[8] Mishin O V, Zhang Y B and Godfrey A 2017 *J. Mater. Sci.* **52**(5) 2730–45
[9] Hwang W S, Wu T I and Sung W C 2011 *J. Eng. Mater. Technol.* **134**(1) 014501
[10] Philibert J, Vignes A, Bréchet Y and Combrade P 2002 *Métallurgie - Du Minerai au Matériau (2ème édition)* ed Dunod (Paris) p 412

IOP Conf. Series: Journal of Physics: Conf. Series **1270** (2019) 012025 doi:10.1088/1742-6596/1270/1/012025

On the Recrystallization Characteristics and Kinetics of Two High-Si DQ&P steels

M C Somani[1], D A Porter [1], L P Karjalaien [1], P Kantanen [1], J Kömi [1], R D K Misra [2]

[1] Materials and Mechanical Engineering, Centre for Advanced Steels Research, University of Oulu, 90014 Oulun yliopisto, Finland
[2] Dept. of Metallurgical, Materials and Biomedical Engineering, University of Texas at El Paso, El Paso, TX 79968, USA

Email: mahesh.somani@oulu.fi

Abstract. In the direct quenching and partitioning (DQ&P) process to develop tough, ultrahigh-strength steel, the hot rolling stage involving deformation and recrystallization between successive passes affects the state of the austenite prior to the quenching and partitioning step. Therefore, in order to design appropriate thermo-mechanically controlled rolling processes, the recrystallization characteristics of two experimental high-Si DQ&P steels, with compositions (in wt.%) 0.3C-1Si-2Mn-1Cr and 0.25C-1.5Si-3Mn, were determined using the stress relaxation technique on a Gleeble thermomechanical simulator and modelled in terms of the effects of temperature, strain, strain rate and initial grain size. Data analysis resulted in the estimation of the powers of strain (-2.8 and -2.4) and strain rate (-0.23 and -0.14) and the recrystallization activation energies (303 and 289 kJ/mol) for the 0.3C-1Si-2Mn-1Cr and 0.25C-1.5Si-3Mn steels, respectively. This suggests that increasing the contents of Mn and Si made the recrystallization kinetics less sensitive to strain, strain rate and temperature. New regression equations derived to describe the recrystallization kinetics can be used in the design of the rough rolling part of thermomechanical processing.

1. Introduction

Conventionally, quenching and tempering is used to obtain high-strength structural steels with good impact toughness, but with limited strain hardening capacity. The novel concept of quenching and partitioning (Q&P) has been proposed as a potential processing route for achieving desired martensite-austenite microstructures and improving the balance of elongation to fracture and tensile strength for advanced high-strength steels [1, 2] Based on the principles of the Q&P process, a novel processing route comprising thermomechanical rolling followed by direct quenching and partitioning (TMR-DQP) has been developed with the aim of achieving yield strength ≥1100 MPa combined with good ductility and impact toughness [3, 4, 5].

In order to design appropriate thermo-mechanically controlled rolling processes of the DQ&P steels, it was considered important to characterize the static recrystallization (SRX) behavior in the form of semi-empirical relationships. SRX kinetics have been studied using a fractional softening approach to determine the time for 50% recrystallized fraction (t_{50}) in accord with the modeling performed by Somani et al. [6, 7] using the empirical relation:

$$t_{50} = A\ \varepsilon^{p}\ \varepsilon'^{q}\ d^{s} \exp(Q_{app}/RT) \quad (1)$$

where A is a material constant, ε is strain, ε′ is strain rate, Q_{app} is the apparent activation energy of recrystallization, d is grain size, R is the gas constant and T is absolute temperature. The material dependent constants p, q and s describe the powers of the strain, strain rate and the grain size, respectively. Using t_{50} in combination with an Avrami-type equation, the recrystallized fraction can be predicted as a function of temperature and time. Data for SRX of high-Mn TWIP type austenitic steels have been published [8, 9], but practically no SRX data are available for medium-Mn (3-10% Mn) steels, except a few observations made by Grajcar et al. [10, 11] for 3% and 5% Mn steels and hot deformation and the dynamic recrystallization study of binary Fe-Mn alloys by Cabanas et al. [12]. Somani et al. [6, 7, 13] developed a unique linear regression model that is able to predict satisfactorily the SRX kinetics of common carbon steel grades including microalloyed steels and also several special steel grades. In the course of the development of this model, the influence of alloying with Mn in the range 0.02–2% and Si up to 1.5% alloying was considered by including the instances of a couple of dual-phase and TRIP steels.

In this study, the SRX characteristics of two Si-alloyed Q&P steels with the compositions (in wt.%): 0.3C-2Mn-1Cr-1Si and 0.25C-3Mn-1.5Si were evaluated and kinetics modelled using a fractional softening approach together with metallography to determine the time for 50 % recrystallization as a function of strain, strain rate, grain size and temperature by applying SRX tests using a Gleeble 3800 thermomechanical simulator, similar to the regression modelling performed previously by Somani et al. [6, 7] for various steels.

2. Materials and experimental

The chemical compositions of the Si-alloyed C-Mn steels intended for DQ&P processing are given in table 1. Both were procured from OCAS, Belgium. Bars of 11x10 mm cross section cut from the hot rolled plates of the two steels were homogenized at 1250 °C for 2 hours prior to preparing the specimens for stress relaxation testing.

Table 1. Chemical Compositions of the High-Si DQ&P Steels

Steel code	C	Si	Mn	Al	Cr
2Mn-1Si-1Cr	0.3	0.99	1.9	0.007	1.0
3Mn-1.5Si	0.25	1.45	2.9	0.024	0.01

Cylindrical specimens of ϕ8 x 10 mm were machined for axisymmetric hot compression testing on a Gleeble 3800 thermomechanical simulator. The specimens were heated at a rate of 10 °C/s to the reheating temperatures of 1200 °C and 1250 °C for the 2Mn-1Si-1Cr and 3Mn-1.5Si steels, respectively, held for 2 minutes, followed by cooling at 2 °C/s to the deformation temperature and then compressed up to a prescribed strain after stabilizing the temperature for 15 s at the deformation temperature. Stress relaxation tests were carried out in the temperature range 850–1100 °C and 900–1200 °C for 2Mn-1Si-1Cr and 3Mn-1.5Si steels, respectively. Both the strain and strain rate were suitably varied at specific temperatures in order to achieve a broad range of deformation parameters for evaluating the SRX characteristics and kinetics of the two steels. For the 2Mn-1Si-1Cr steel, these were strain range 0.125-0.4 and strain rates 0.01-5 s^{-1} at 1000 °C, and for the 3Mn-1.5Si: 0.125–0.3, 0.001–2 s^{-1} at 1050 °C.

Austenite grain size was measured using the linear intercept method on specimens directly quenched from the reheating temperature with a water spray and etched in saturated picral/teepol solution. A few segregation bands, possibly Mn-rich regions, were noticed in the 3Mn-1.5Si steel, despite the homogenization treatment given prior to hot rolling and also the reheating at the higher temperature of 1250 °C in the Gleeble simulator. Some specimens were reheated at different temperatures prior to stress relaxation in order to produce relatively coarser or finer grain structure to understand the influence of grain size on SRX kinetics and to check the validity of the empirical equations for fractional softening.

3. Results and discussion

3.1. Flow stress behaviour

Typical examples of true stress – true strain curves obtained on the 2Mn-1Si-1Cr steel compressed to a true strain of ~0.2 at a constant true strain rate of 0.1 s^{-1} in the temperature range 850–1100 °C prior to stress relaxation is shown in figure 1. The flow stress behaviour displayed the presence of work hardening and dynamic recovery at all deformation temperatures. A similar flow stress behaviour is also observed in the case of 3Mn-1.5Si steel. Subsequent stress relaxation following hot compression should exhibit static recovery and recrystallization processes, except presumably at 1100 °C in the case of 2Mn-1Si-1Cr steel and at 1200 °C in the case of 3Mn-1.5Si steel, where the critical strain ε_c (≈ 0.8 times peak strain ε_p [14]) for the initiation of dynamic recrystallization (DRX) might have been exceeded, resulting in metadynamic (MDRX) recrystallization. The flow stress behavior of samples tested at higher strains and lower strain rates were also carefully examined in order to exclude possible MDRX cases from the modelling of the fractional softening equations for SRX. There are no significant differences between the two steels in respect of flow stress levels. Similar observations were made for 3%Mn and 5%Mn steels tested in the range 850–1150 °C by Grajcar et al. [10, 11].

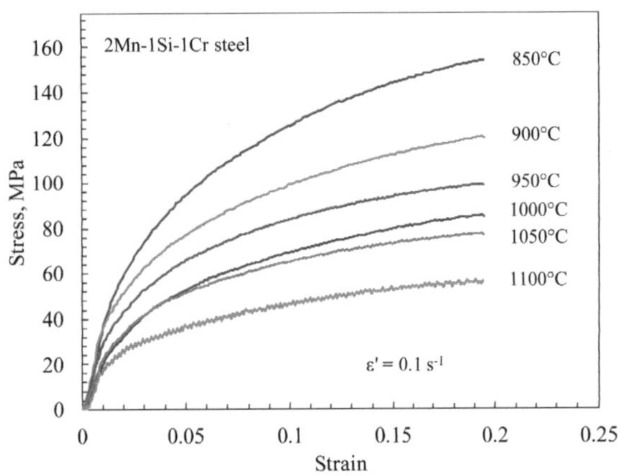

Figure 1. True stress – true strain curves obtained on 2Mn-1Si-1Cr steel following hot compression to 0.2 strain at 0.1 s^{-1} in the temperature range 850–1100 °C.

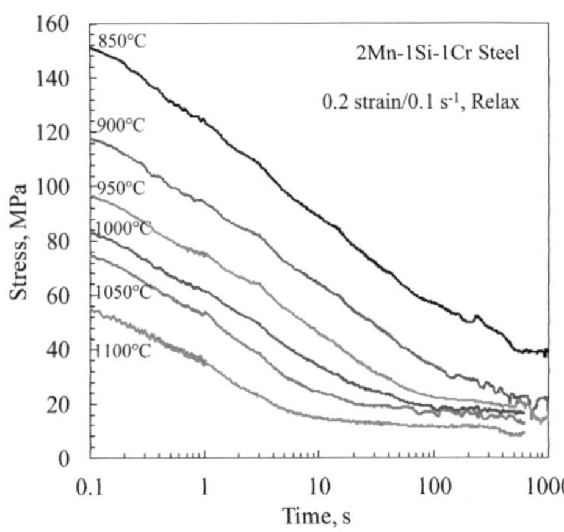

Figure 2. Typical stress relaxation curves obtained on 2Mn-1Si-1Cr steel specimens deformed to a true strain of 0.2 at 0.1 s^{-1}.

3.2. Stress relaxation behaviour

Typical stress relaxation curves for the 2Mn-1Si-1Cr steel following the compression tests shown in figure 1, are presented in figure 2. The initial and final linear stages on the curves correspond to the occurrence of static recovery (SRV) and the intermediate fast drop in the stress level is typical of SRX or MDRX process [15]. Stress relaxation curves were carefully analyzed to determine the kinetics of the SRX as a function of deformation parameters. Examples of recrystallized fraction vs. time curves fitted with the sigmoidal Avrami-type curves for the 3Mn-1.5Si steel are shown in figure 3, illustrating the effect of temperature on the kinetics of SRX.

3.3. SRX behaviour and apparent activation energy of recrystallization (Q_{app})

Referring to figure 2, the stress relaxation curves of the 2Mn-1Si-1Cr steel reveal the influence of temperature on the SRX kinetics following deformation to 0.2 strain at different temperatures in the range 900 – 1100 °C in steps of 50 °C. Complete softening was obtained in most cases, except at the lower temperatures (≤ 900 °C). The stress relaxation curves for the 3Mn-1.5 Si steel too showed that the softening was complete at most temperatures, except at 900 °C, where only partial recrystallization occurred. Similarly, the grain size effect on SRX rate was clearly revealed by the stress relaxation technique. For instance, in the case of 3Mn-1.5Si steel samples deformed at 1050 °C to 0.2 strain at 0.1 s^{-1}, the recrystallization rate was greatly accelerated in the case of fine grained structure (80 μm; t_{50} = 2.6 s) in comparison with that of the coarse-grained structure (480 μm; t_{50} = 12 s).

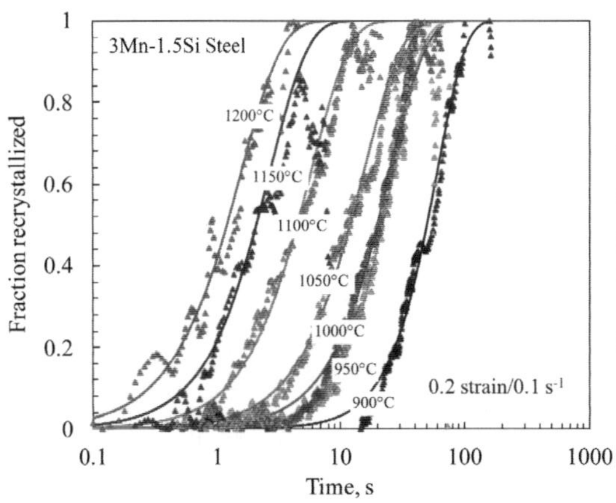

Figure 3. Fraction recrystallized vs. time data fitted with Avarmi type curves for the 3Mn-1.5 Si steel.

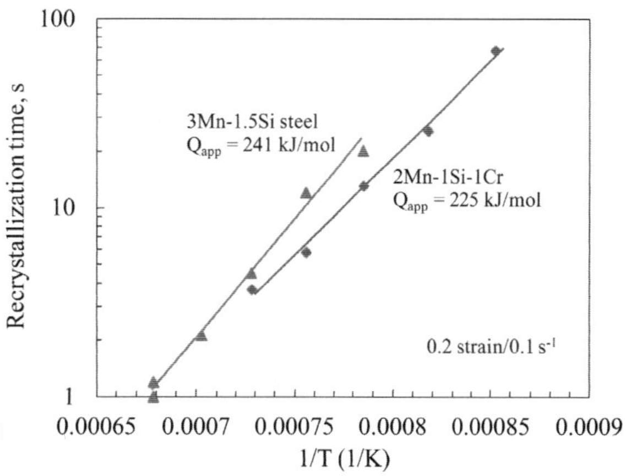

Figure 4. Estimation of Q_{app} for the SRX process for the two steels.

The temperature dependence of the SRX kinetics for the 2Mn-1Si-1Cr and 3Mn-1.5Si steels are shown in figure 4, where t_{50} times estimated at different temperatures for specimens deformed to 0.2 strain at 0.1 s^{-1} are plotted against the inverse absolute temperature. The slope of the plots was used to estimate the apparent activation energy for recrystallization (Q_{app}). When doing this, the cases of incomplete recrystallization (at low temperatures) and MDRX (at high temperatures) were carefully excluded. The estimated Q_{app} values for the 2Mn-1Si and 3Mn-1.5Si steels were about 225 and 241 kJ/mol, suggesting that an increase in Si and Mn contents increases Q_{app}. The observed Q_{app} for the two steels fall in the 177–283 kJ/mol range of values reported for other steels, for example, C-Mn 184 kJ/mol [6], 0.2C-2Mn-0.6Cr-(0-1.5)Si 193-227 kJ/mol [16], C-Mn-Nb 230 kJ/mol [6] and Type 304 stainless steel 283 kJ/mol [17].

Using a constant Q_{def} of 340 kJ/mol [6, 13] for both the 2Mn-1Si and 3Mn-1.5Si steels gives the activation energy of SRX (Q_{rex}) values of 303 and 289 kJ/mol, respectively. The low value of Q_{rex} for the 3Mn-1.5Si steel is due to its less negative q (-0.14 as opposed to -0.23), even though its Q_{app} was higher than that of the 2Mn-1Si steel, as expected from its higher alloy content.

3.4. Powers of strain and strain rate

The powers of strain (p) and strain rate (q) have been estimated based on the plots of (i) t_{50} vs. strain at 0.1 s^{-1} at either 1000 °C for the 2Mn-1Si-1Cr steel or 1050 °C for the 3Mn-1.5Si steel (figure 5), and (ii) t_{50} vs. strain rate following compression to 0.2 strain at the deformation temperature (figure 6). Referring to figure 5, the powers of strain (p) estimated from the slopes of the line fits were estimated to be about -2.8 and -2.35 for the 2Mn-1Si-1Cr and 3Mn-1.5 Si steels, respectively. The power of strain (p) for the 3Mn-1.5Si steel (-2.35) is slightly lower than that of 2Mn-1.5Si (-2.8). The strain exponent for C-Mn steels has been reported to be in the range -2.5 to -4 [6, 7, 13, 18, 19]. Values between -2 and -3 were measured by the stress relaxation technique for some microalloyed steels [6, 20]. Suikkanen et al. [16] obtained a value of -2.1 for p in the case of a 0.2C-2.0Mn-1.48Si-0.6Cr steel. Hence, no distinct dependence on alloying can be ascertained.

Similarly, the slopes of the linear fits in figure 6 show the powers of strain rate (q), estimated as -0.23 and -0.14 for the 2Mn-1Si-1Cr and 3Mn-1.5 Si steels, respectively. The power of strain rate for a 0.1C-1.2Mn-1.15 Si steel (-0.23) was comparable to that of the 2Mn-1.5Si steel [16]. Lang et al. [21]

determined the SRX kinetics for a 0.21C-1.48Si-2.04Mn-0.6Cr composition, also tested by Suikkanen et al. [16], and found q to be -0.18, i.e. between the values of -0.23 and -0.14. In comparison, relatively low values of q of -0.11 and -0.12 have been reported for C/C-Mn and Ti steels [6, 7, 22]. For Nb and Nb-Ti and also Mo-steels, q has been estimated to be of the order of -0.23 [6, 7]. In summary, all these values of q fall within a small range (-0.11 to -0.23) indicating a weak dependence of SRX on strain rate irrespective of the steel alloying.

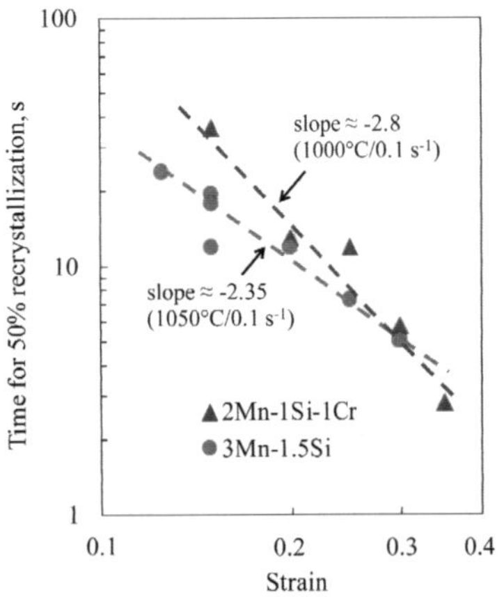

Figure 5. Dependence of t_{50} on strain for 2Mn-1Si-1Cr and 3Mn-1.5 Si steels deformed at 1000 °C/0.1 s^{-1} and 1050 °C/0.1 s^{-1}, respectively.

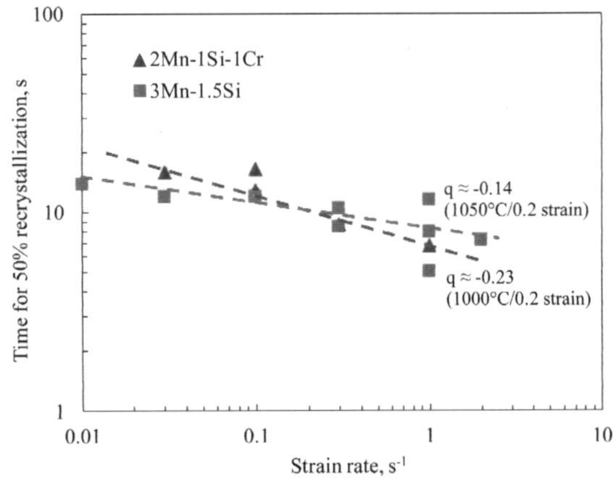

Figure 6. Estimation of the power of strain rate for the two steels deformed to 0.2 strain in the SRX regime.

3.5. Fractional softening equations for SRX

Taking the power of grain size to be described by the relation $s = 2.13d^{-0.105}$, together with the above values for the other parameters (Q_{app}, p and q) in Equation 1 gives the constant A for the two steels. The SRX rate, therefore, can be reasonably described by the following SRX equations:

$$\text{2Mn-1Si steel:} \quad t_{50} = 1.2\text{x}10^{-13}\,\varepsilon^{-2.8}\,\varepsilon'^{-0.23}\,d^{s}\exp(225000/RT) \tag{2}$$

$$\text{3Mn-1.5Si steel:} \quad t_{50} = 5.7\text{x}10^{-14}\,\varepsilon^{-2.35}\,\varepsilon'^{-0.14}\,d^{s}\exp(241000/RT) \tag{3}$$

In order to check the reliability of the equations, a few confirmation experiments were carried out by varying the deformation parameters and/or grain sizes for the two steels as described elsewhere [23]. The results confirmed that the SRX regression equations can be used with reasonable accuracy in designing the thermomechanical processing schedules for the two DQ&P steels.

The t_{50} times for the present steels, estimated with the aid of Equations 2 and 3, are compared with those of other steels elsewhere [23]. The SRX kinetics of the present high-Si steels is quite similar despite the different levels of Si and Mn alloying. The SRX kinetics of high-Si steels have been found to be much slower than C-Mn steels and marginally faster but close to that of 0.03%Nb-microalloyed C-Mn steel as well as an ordinary TWIP steel. An addition of 1–1.5% Si in the high-Si steels seems to have a potent effect, but weaker than that of Nb, in C-Mn steels.

4. Summary and conclusions

The recrystallization characteristics of two experimental high-Si DQ&P steels, viz., 2Mn-1Si-1Cr and 3Mn-1.5Si, have been evaluated using the stress relaxation technique on a Gleeble 3800 thermomechanical simulator. Data analysis resulted in the powers of strain (-2.8 and -2.35) and strain rate (-0.23 and -0.14) and the apparent (225 and 241 kJ/mol) and recrystallization activation energies (303 and 289 kJ/mol) for the 2Mn-1Si and 3Mn-1.5Si steels, respectively. This suggests that with the increase in the contents of Mn and Si, the SRX kinetics became less sensitive to strain, strain rate and temperature. The differences between the two steels appeared to be small and could be partly due to experimental scatter. Regression equations describing the SRX kinetics of the two steels fit well with the experimental data. Confirmation experiments indicated that the power of grain size suggested earlier for carbon steels was reasonably accurate for these steels, too.

Acknowledgement

The funding of this research activity through the RFCS Grant Agreement RFSR-CT-2014-00019 is gratefully acknowledged.

References

[1] Speer J G, Edmonds D V, Rizzo F C and Matlock D K 2004 *Curr. Opin. Solid State Mater. Sci.* **8** 219
[2] Li H Y, Lu X W, Li W J and Jin X J 2010 *Metall. Mater. Trans. A* **41A** 1284
[3] Somani M C, Porter D A, Karjalainen L P, Suikkanen P P and Misra D K 2014 *Mater. Sci. Forum* **783-786** 1009
[4] Somani M C, Porter D A, Karjalainen L P, Suikkanen P P and Misra R D K 2015 *J. Mater. Today: Proc.* **2S** S631
[5] Somani M C, Porter D A, Kömi J I, Karjalainen L P and Misra D K 2017 *Proc. Int. Symp. on New Developments in Advanced High-Strength Sheet Steels* (Keystone) (Warrendale: AIST) p 331
[6] Somani M C, Karjalainen L P, Porter D A and Morgridge R A 2003 *Proc. Int. Conf. on Thermomechanical Processing: Mechanisms, Microstructure and Control* (Sheffield) eds. E J Palmiere, M Mahfouf and C Pinna (Sheffield: The University of Sheffield) p 436
[7] Somani M C and Karjalainen L P 2012 *Mater. Sci. Forum* **715-716** 751
[8] Hamada A S, Somani M C and Karjalainen L P 2007 *ISIJ Int.* **47** 907
[9] Somani M C, Porter D A, Hamada A S and Karjalainen L P 2015 *Metall. Mater. Trans. A* **46A** 5329
[10] Grajcar A and Kuziak R 2011 *Adv. Mater. Res.* **287-290** 330
[11] Grajcar A and Kuziak R 2011 *Adv. Mater. Res.* **314-316** 119
[12] Cabañas N, Akdut N, Penning J and De Cooman B C 2006 *Metall. Mater. Trans. A* **37A** 3305
[13] Somani M C and Karjalainen L P 2004 *Mater. Sci. Forum* **467-470** 335
[14] Sellars C M 1980 *Proc. Int. Conf. on Hot Working and Forming Processes* (Sheffield) eds. C M Sellars and G J Davies (London: The Metals Society) p 3
[15] Karjalainen L P 1995 *Mater. Sci. Technol.* **11** 557
[16] Suikkanen P P, Lang V T E, Somani M C, Porter D A and Karjalainen L P 2012 *ISIJ Int.* **52** 471
[17] Karjalainen L P, Koskiniemi J A and Liu X D 1995 *Proc. 37th MWSP Conf. & Int. Symp. on Recovery and Recrystallization in Steel* (Hamilton) **XXXIII** p 861
[18] Hodgson P D and Gibbs R K 1992 *ISIJ Int.* **32** 1329
[19] Sellars C M 1990 *Mater. Sci. Technol.* **6** 1072
[20] Karjalainen L P, Perttula J, Xu Y and Niu J 1997 *Proc. 7th Int. Symp. on Physical Simulation* (Tsukuba) p 231
[21] Lang V T E, Suikkanen P P, Somani M C, Porter D A and Karjalainen L P 2011 *Proc. Int. Sci. Technol. Conf. on Advanced Metals, Materials and Technologies* (St. Petersburg) p 301
[22] Airaksinen K, Karjalainen L P, Porter D and Perttula J 1998 *Mater. Sci. Forum* **284-286** 119
[23] Somani M C, Porter D A, Karjalainen L P, Kantanen P K, Kömi J I and Misra D K 2019 *Int. J. Mater. Res.* **110** 3

IOP Conf. Series: Journal of Physics: Conf. Series **1270** (2019) 012026 doi:10.1088/1742-6596/1270/1/012026

The assessment of growth kinetics in the intermetallic layers of Al-Al_xNi_y-Ni laminate composites

M Azimi.[1,2,3*], M R Toroghinejad[1], M Shamanian[1], L A I Kestens[2,3]
[1]Dept. of Materials Engineering, Isfahan University of Technology, Isfahan, Iran;
[2]EEMMeCS Dept., Metals Science and Technology Group, Ghent University, Ghent, Belgium;
[3]Dept. of Materials Science Engineering, Delft University of Technology, Delft, The Netherlands

*Email: M.Azimi@tudelft.nl, M.Azimi@iut.ma.ac.ir; Monireh.Azimi@ugent.be

Abstract. Ni aluminides have technologically attracted much attention as oxidation protective layers on Ni-superalloys for high-temperature and harsh environments applications as well as for reinforcements in metal-matrix composites. Among the Ni aluminides, the AlNi compound exhibits the best combination of oxidation-protective characteristics and hardness; thus there is a progressive demand to produce it particularly through convenient in-situ fabrication processes. Therefore, the evaluation of growth kinetics in Al_xNi_y layers is of crucial importance in determining an optimum compound formation process. To this purpose, Al-Ni intermetallic laminate composites were produced through cold roll bonding and subsequent annealing of aluminum and nickel sheets. The microstructure of the intermetallic layers was investigated in order to specify the controlling mechanisms and subsequently the growth model of the different phases. The Al_3Ni layer was kinetically the first to appear but started to decompose at the expense of the AlNi compound when the direct source of Al disappeared for the reactive diffusion couples. The Al_3Ni layer growth was initially controlled by bulk diffusion, but then at $T \geq 525°C$ was modified as a function of competition between formation and consumption, whereas the AlNi growth was governed strongly by the interfacial reaction. The time dependence of the growth rate revealed different behaviors of linear and parabolic kinetics. The overall assessment revealed a bulk diffusion-controlled growth for all of the intermetallic layers. Arrhenius parameters could be derived for the Al_3Ni layer, while it was impossible for the AlNi phase because the formation of this layer was caused by a mixture of diffusion mechanism.

1. Introduction

Between nickel aluminides, the AlNi intermetallic phase exhibits the best combination of oxidation-protective characteristics and hardness, which provokes a great interest in Al-AlNi-Ni composites. Metal-intermetallic laminates (MIL) composites have the potential to combine various functions, which may be of use for high-temperature structural applications such as heat exchangers, which require appropriate thermal management [1,2].

There are some inconsistencies in the data on the growth kinetics of intermetallic phases. For instance, according to Michaelsen and Barmak [3] the growth of Al_3Ni phase indicates a parabolic dependence on the annealing time, whereas Jung et al [4] have reported that the growth kinetics of Al_3Ni phase does not obey the parabolic law. Moreover, researchers have reported conflicting results on Al and Ni asymmetric interdiffusion coefficient. Hence, the dominant diffusing element during the formation of intermetallic phases is still entirely unclear under similar process conditions [2,5–7]. However, Al_3Ni phase was recognized as the first kinetically favored intermetallic phase, independently of the concentration gradient [5] and imposed strain level [8].

Therefore, a more detailed examination of the formation mechanism of intermetallic phases is of crucial importance to better control the processing of these materials and the aim of this work is to investigate the formation mechanisms and growth kinetics of intermetallic phases in the Al-Ni diffusion couple fabricated by the roll bonding process.

2. Experimental procedure

Commercially pure aluminum and nickel were used as initial materials in sheet form. An Al-Ni-Al sandwich was subjected to cold roll bonding (CRB) by 90% reduction in thickness. The details of materials and CRB process were given elsewhere [8]. In order to form Al_xNi_y intermetallic compounds, the cold roll bonded composite with the dimensions of $200 \times 60 \times 0.5\ mm^3$ was annealed in the temperature range of 300 to 640 °C for 1 and 4 hours.

In addition to a conventional optical microscopy (OM, Olympus) the microstructural evolution was monitored with a field emission gun scanning electron microscope (FEG-SEM Quanta-450®, FEI) equipped with EBSD facility of type TSL-EDAX® and an energy dispersive spectrometer (EDS).

The microstructure of Al/Ni interface was identified by observing sites of size (100-400 μm) × (30-150 μm). Kikuchi band indexing in EBSD were acquired with OIM™ software and gathered at an accelerating voltage of 20 kV, a working distance of 14 to 16 mm with step sizes in the range of 50 to 80 nm. Samples were tilted 70° for acquiring the maximum yield of backscattered electrons. The thickness of the continuous or segmented layers of the intermetallic phases was determined at several points of the formed layers using image analysis software (ImageJ2)

3. Results and discussion

Figure 1 shows BSE images of intermetallic laminate composites obtained through annealing. The intermetallic layer(s) was formed along the Al/Ni original interface, visibly present except for the samples annealed at 300°C. Figure 2a and b shows the dimensions of the intermetallic layers as a function of temperature. The layers are mainly composed of Al_3Ni and AlNi compounds according to the phase analysis results listed in Table 1. This table also reveals that other compounds possess relatively negligible thickness.

Table 1. Various phase zones which were detected across Al-Ni interface in TD plane of samples using EDS and Kikuchi band indexing in EBSD.

Zone	Average of Spectrums in EDS		Probable phases	Indexed results in EBSD with confidence index (CI)	Confirmed phases	Ave. thickness of reacted zones (μm)
	Al at.%	Ni at.%				
(A)	1.3	98.7	Nickel	well-indexed nickel, CI=0.8	Nickel	Base metal
(B)	17	83	Diffusion zone	well-indexed (Ni), CI= 0.8	Ni-riched solid solution	~0.5
(C)	30.1	69.9	$AlNi_3$ or ($AlNi_3$+Al_3Ni_5)	well-indexed $AlNi_3$, CI=0.75 not indexed Al_3Ni_5, CI=0.03	$AlNi_3$	~0.5
(D)	55.3	44.7	(Al_3Ni_2+AlNi) or Al_3Ni_2	indexed grains as AlNi, CI= 0.3 indexed grains as $AlNi_3$, CI= 0.41 not indexed Al_3Ni_2, CI=0.01	Mixture of $AlNi_3$+AlNi	Various (0-1.6)
(E)	48.3	51.7	AlNi	well-indexed, CI= 0.49	AlNi	Various (0-20)
(F)	76.8	23.2	Al_3Ni	well-indexed, CI= 0.3	Al_3Ni	Various (0-9)
(G)	100	0	Al	well-indexed nickel, CI=0.7	Aluminum	Base metal

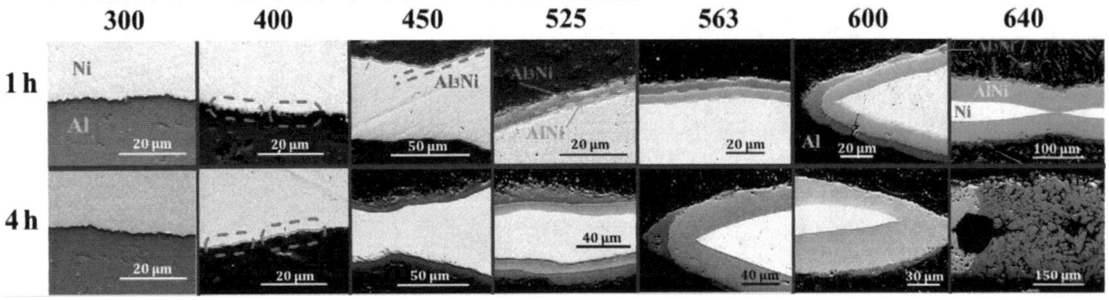

Figure 1. BSE-SEM images of interfaces on TD section of CRB-processed composites after annealing at various times and temperatures.

As shown in Figure 1, the Al phase was melted at 640°C and the interface reactions were no longer solid state. Hence, this temperature data are not given in next results.

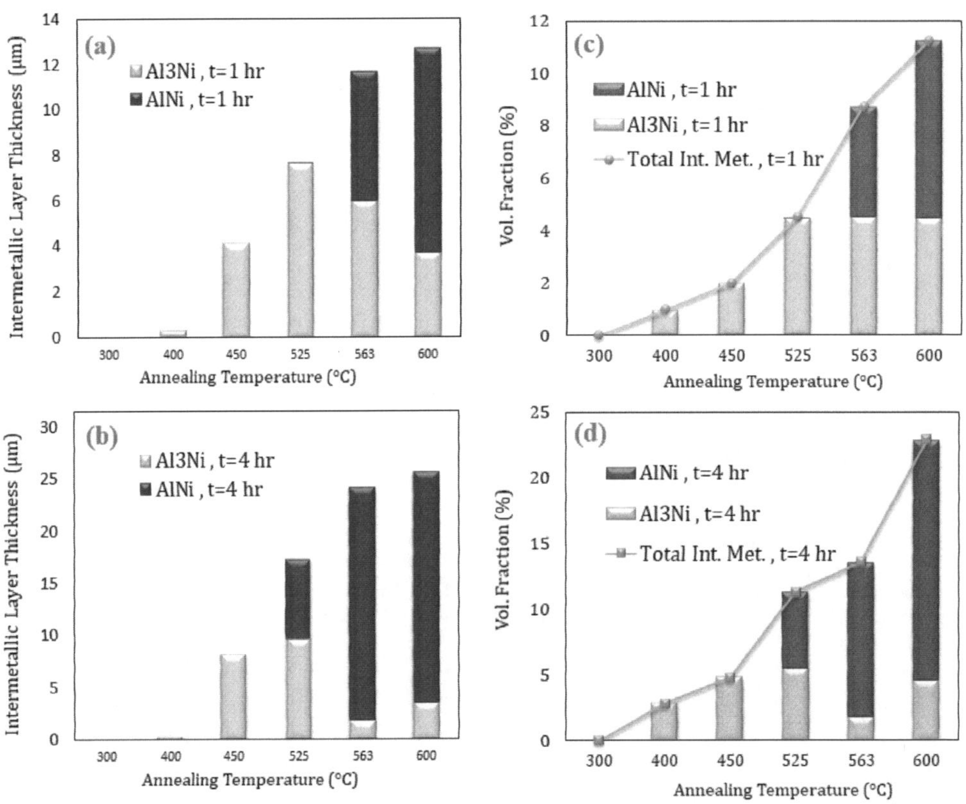

Figure 2. Intermetallic layer thickness/volume fraction vs. temperature during 1 h (a, c) and 4 h (b, d) annealing.

Based on the general empirical equation (1) proposed by Kidson [9], it is possible to model the growth of the thickness of a layer:

$$W_i = kt^n \tag{1}$$

$$ln(W)_i = n \,.\, ln(t) + ln(k) \tag{2}$$

whereby W_i represents the thickness of layer i; k= the growth rate constant; t= time [min] and n= kinetics exponent. From a double logarithmic plot, n and k values can be derived from a linear fit of the experimental data (cf. Figure 3). However, the accuracy of the results (cf. Table 2) may be influenced by the only two values of the time.

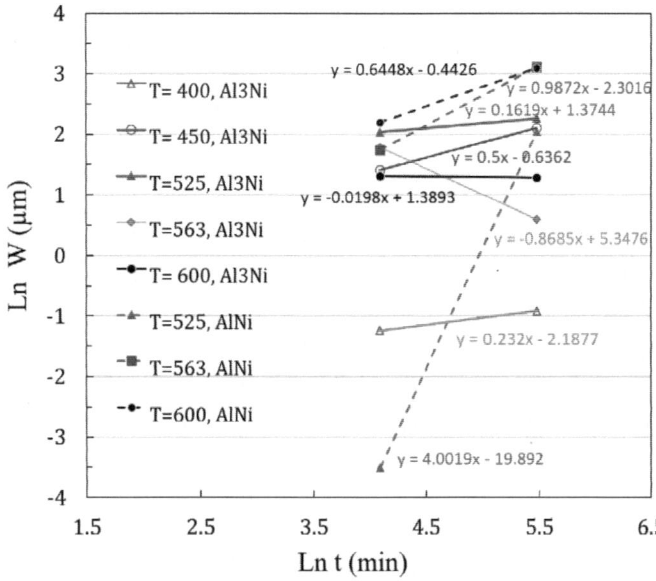

Figure 3. LnW_i versus Lnt for Al_3Ni and AlNi compounds.

Table 2. Growth kinetics parameters of n and k in Al_3Ni and AlNi layers.

Kinetics parameters \ *T(°C)*		400	450	525	563	600
n	Al_3Ni	0.38	0.5	0.16	-0.81	-0.02
	AlNi	-	-	4.00	0.99	0.65
Ln k	Al_3Ni	-2.19	-0.64	1.37	5.35	1.39
	AlNi	-	-	-19.89	-2.30	-0.44
k	Al_3Ni	0.12	0.53	3.95	180.10	4.01
	AlNi	-	-	2.30	1.00	0.64

The factor n is associated with the controlling mechanism of the intermetallic layer growth process. In terms of quantitative assessment, n=1 demonstrates interfacial reaction controlled growth; n=0.5: bulk diffusion controlled growth and n=0 corresponds to a balance between the layer formation and decomposition during a secondary reaction.

Intermetallic layers may exhibit different behaviors, i.e. sequential growth, concurrent growth and layer shrinkage through decomposition. The last one was observed from 563°C onward, owing to the fact that Al_3Ni (i.e. the Ni-poorest intermetallic) had already started to be consumed at expense of the AlNi compound when the Al direct source disappears from the reactive diffusion couples. Up to 450°C, only Al_3Ni layer was formed and afterwards followed the second stage of a two-stage growth process. The parabolic growth kinetics suggests bulk diffusion controlled growth until the growth kinetics is disturbed by the emerging new AlNi phase at $T \geq 525$°C (the volume fraction of intermetallic layers readily indicates a change in the trend due to this disturbance, cf. Figures 2 c, d). Indeed, a part of the Al_3Ni layer is consumed through the reaction at the Al_3Ni/Ni interface, cf. Equation 3. Eventually, the growth rate is a function of the competition between Al_3Ni direct formation by diffusion and consumption by the interfacial reaction, while the new phase (AlNi) growth is strongly controlled by the interfacial reaction.

$$Al_3Ni + 2Ni \rightarrow 3AlNi \tag{3}$$

Afterwards, increasing the temperature by only ~40° (T=563°C) results in a further increase of the relative thickness/fraction of the AlNi layer. In this stage, AlNi growth kinetics is linear, still controlled by the reaction (equation 3) at Al_3Ni/AlNi interface, but not as strong as that of at the lower temperature (525°C). In the meanwhile, the Al_3Ni decomposition rate prevailed the formation rate at the Al/Al_3Ni interface. After that, increasing the temperature close to the melting point of Al, the atomic mobility is promoted accelerating bulk diffusion. On the one hand, it results in a balance between Al_3Ni phase consumption and formation rate, giving rise to constant growth kinetics, while on the other hand, it leads to parabolic growth kinetics of the AlNi layer. This parabolic function implies that the AlNi growth dependency on the reaction has subsided.

The growth rate constant k is an exponential function of the temperature variable, cf. equation 4. Whereby Q is the activation energy for diffusion-limited growth and k_0 represents an independent factor of time and temperature and R= the global gas constant.

$$k = k_0 \exp(\frac{-Q}{RT}) \tag{4}$$

$$ln\, k = ln(k_0) - \frac{Q}{RT} \tag{5}$$

By combining equations (1) and (4), equation (6) is obtained as follows:

$$W = f(T, t, n) = k_0 t^n \exp(\frac{-Q}{RT}) \tag{6}$$

Defining the B parameter as equation (7), W is a function of temperature according to equation (8) (for constant values of n and t).

$$B = f(t, n) = k_0 t^n \tag{7}$$

$$W = f(T) = B\exp(\frac{-Q}{RT}) \tag{8}$$

Table 3. B function according to n values.

Layer - T(°C)	n	B	$\frac{\partial W}{\partial t}$, growth rate	Considerable remarks
Al_3Ni - 563	-1	$\frac{k_0}{t}$	<0	Layer consumption
Al_3Ni - 525, 600	0	k_0	0	Growth depends on temperature, independent of time
Al_3Ni - 400, 450 AlNi - 600	0.5	$k_0 t^{0.5}$	>0	Parabolic kinetics
AlNi - 563	1	$k_0 t$	>0	Linear kinetics
AlNi - 525	4	$k_0 t^4$	>0	Growth depends on interface reaction strongly

Based on the present values for the n variable, the B function appears in five forms which are listed in Table 3. In both layers, the dependency of B on the time is decreasing with temperature increase, independent of the function form and except at 600°C for B_{Al_3Ni}. At 600°C, this dependency is higher owing to the presence of the thick and compact layer of AlNi, which gives rise to a deceleration of the supply of Ni atoms by the diffusion process in order to form the Al_3Ni phase.

The Q value is calculated, cf. Figure 4, under the condition that growth is controlled by diffusion (n=0.5). Hence, this calculation is possible for the Al_3Ni layer in the 400-450°C range and estimation for AlNi layer in the 563-600°C range with the least possible error. The estimation here obtained is far from the finding by Urrutia et al [10] because of the mixture of bulk and grain boundary (GB) diffusion for AlNi formation in the present work. This can be explained by the wave-like morphology of the Al_3Ni/AlNi interface as one might expect from GB diffusion, cf. Figure 5.

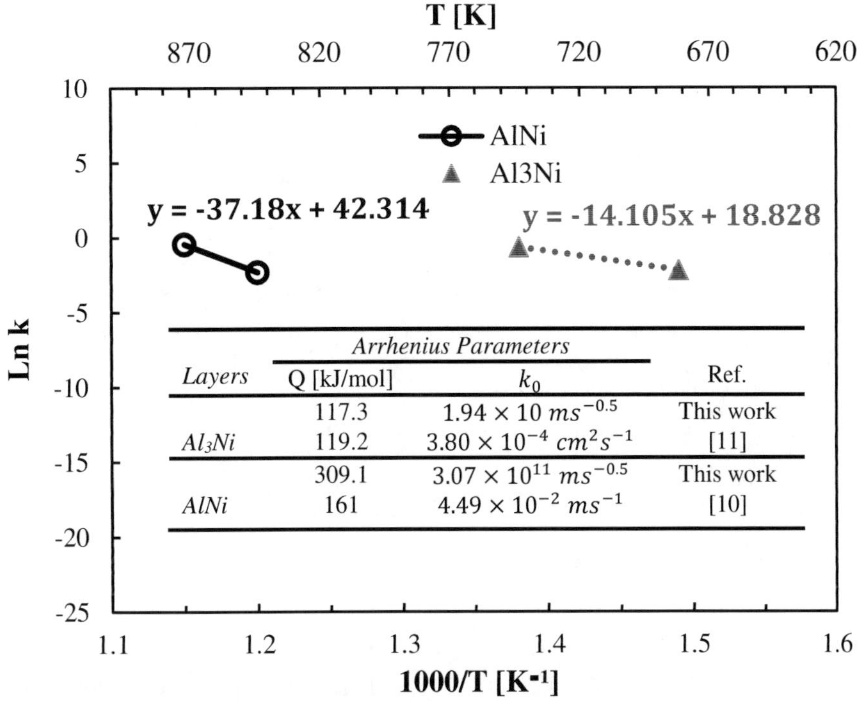

	Arrhenius Parameters		
Layers	Q [kJ/mol]	k_0	Ref.
	117.3	$1.94 \times 10\ ms^{-0.5}$	This work
Al₃Ni	119.2	$3.80 \times 10^{-4}\ cm^2s^{-1}$	[11]
	309.1	$3.07 \times 10^{11}\ ms^{-0.5}$	This work
AlNi	161	$4.49 \times 10^{-2}\ ms^{-1}$	[10]

Figure 4. ln k versus 1/T plot and the corresponding Arrhenius parameters.

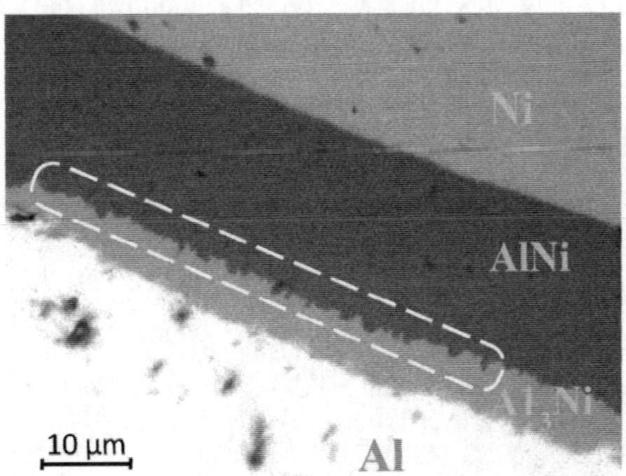

Figure 5. A typical wave-like interface of the Al_3Ni/AlNi in TD section (the sample was annealed at 600 for 4 h).

The overall growth assessment of intermetallic layers reveals an exponential trend as a function of temperatures, cf. Figure 6, indicating bulk diffusion-controlled growth of layers according to the Arrhenius model. Further, diffusion acceleration via base metals grain boundaries does not play a significant role during layers growth because the cold rolled grains of the base metals have already recrystallized and grown in the first stages of the annealing. After the annealing, Al and Ni grains exhibited the equivalent diameter of 60 and 102 µm from 2.3 and 50 µm in the cold roll-bonded stage, respectively.

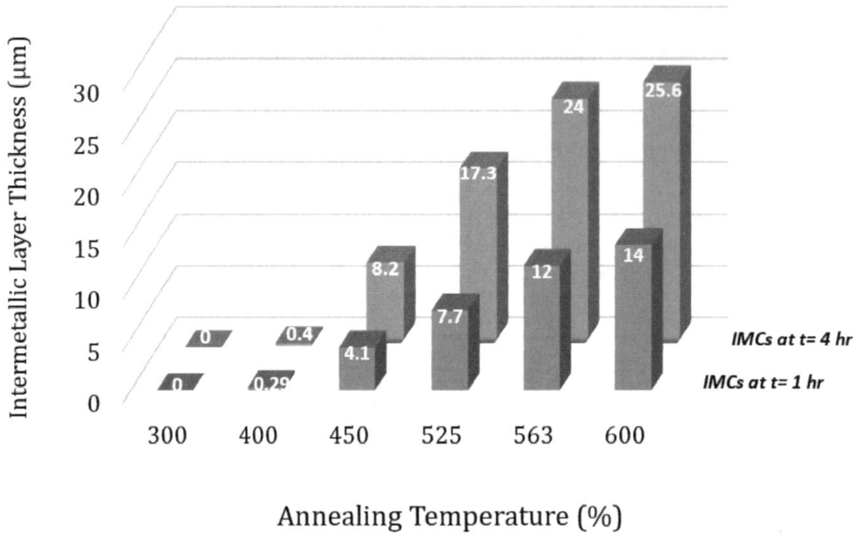

Figure 6. The overall Int. Met. layers thickness vs. annealing temperature during 1 and 4 hours.

4. Conclusion

An Al-AlNi-Ni laminate composite is relatively obtainable by annealing above 560°C longer than 4 h. In the term of growth kinetics and based on the overall assessment, intermetallic layers growth obeys Arrhenius model which indicates bulk diffusion controlled growth. However, internal layers exhibit different behaviors of sequential growth, concurrent growth and layer decreasing through decomposition. Up to 450°C, only Al_3Ni layer was formed and followed the second stage of a two-stage growth process. By emerging AlNi compound at higher temperatures, Al_3Ni growth behavior changes to consumption. The AlNi layer formation was controlled by a mixture of bulk and grain boundary diffusion while bulk diffusion was the only mechanism which controls the growth of Al_3Ni with activation energy similar to the previous studies.

References

[1] Konieczny M 2012 *Mater. Charact.* **70** 117–124
[2] Srivastava V C, Singh T, Ghosh Chowdhury S and Jindal V 2012 *J. Mater. Eng. Perform.* **21** 1912–1918
[3] Michaelsen C and K. Barmak K 1997 *J. Alloys Compd.* **257** 211–214
[4] Jung S, Minamino Y, Yamane T and Saji S 1993 *J. Mater. Sci. Lett.* **12** 1684–1686
[5] Sauvage X, Dinda G P and Wilde G 2007 *Scr. Mater.* **56**, 181–184
[6] Vojtěch D, Novák M, Zelinková M, Novák P, Michalcová A and Fabián T 2009 *Appl. Surf. Sci.* **255** 3745–3751
[7] Liu J C, Mayer J W and Barbour J C 1988 *J. Appl. Phys.* **64** 651–655
[8] Azimi M, Toroghinejad M R, Shamanian M and Kestens L A I 2017 *Metals* **7** 1–14
[9] Kidson G 1961 *J. Nucl. Mater.* **3**, 21–29
[10] Urrutia A, Tumminello S, Aricó S F and Sommadossi S 2014 *CALPHAD Comput. Coupling Phase Diagrams Thermochem.* **44** 108–113
[11] Ren X, Chen G, Zhou W, Wu C and Zhang J 2009 *J. Wuhan Univ. Technol. Sci. Ed.* **24** 787–790

Simulation of deformation and static recrystallization in the stress relaxation test

O Seppälä, A Pohjonen, J Ilmola, A Jokiranta, A Kaijalainen, M Somani and J Larkiola

Materials and Mechanical Engineering, Faculty of Technology, University of Oulu, PL4200, 90014, Oulu, Finland

e-mail: oskari.seppala@oulu.fi

Abstract. A finite element (FE) simulation model illustrating the stress relaxation test was established with the Abaqus™ software. The microstructural evolution of steel during relaxation includes the complex phenomena of recrystallization. While the compression introduces the planned deformation and stress into the test piece, subsequent softening relieves the stress and at the same time creates microstructural reconstitution and refinement. In this study, a model was developed to simulate the kinetics of static recrystallization taking place during holding, using a technique based on FE-simulation. The simulation results have been compared to the experimental stress relaxation data obtained on a Gleeble™ 3800 thermo-mechanical simulator. The model can be used to estimate the recrystallization kinetics throughout the test piece. In the future, these results can be used for estimating the required rolling forces for multi-pass roughing with reasonable accuracy, for instance. The modelling methodology can be extended to other steels too, with or without microalloying additions.

1. Introduction

Steel research traditionally requires a lot of time-consuming and expensive experiments. Experimental methods also require simplifications when trying to study the industrial-scale processes. For these reasons, simulation appears to be an appealing alternative for studying the constitutive flow and recrystallization behaviours of multiple steel grades and associated thermo-mechanical processes.

In the current study, a comprehensive comparison has been made between experimental and simulated stress relaxation data. The experimental data has been obtained with the Gleeble™ 3800 thermo-mechanical simulator and the simulated data with a model that has been established using the Abaqus™ FEM software. The model includes constitutive static recrystallization (SRX) kinetics that has been implemented using a well-established model, as described in the modelling and simulation session, as well as various models for estimating the grain size of steel after SRX. In the current version the implementation is uncoupled, which is not a problem with single compression experiments. In the future, the model will be developed into a coupled model, which will be used to simulate industrial multi-pass roughing rolling with high accuracy.

2. Thermomechanical treatments and material characterization

A Gleeble 3800 thermomechanical simulator, equipped with the Hydrawedge module, was used to carry out (1) stress relaxation and (2) interrupted stress relaxation tests on a low-alloy steel of the following composition (in wt.%): 0.16C-0.3Si-1.0Mn-0.6Cr-0.2Mo. Cylindrical test specimens of the dimensions ϕ 10 x 12 mm were first reheated at 10 °C/s to 1200 °C, held for 5 minutes, followed by cooling at 2 °C/s to a temperature at which the samples were deformed and relaxed for 200 s prior to subsequent quenching to room temperature (RT). The deformation temperatures were 950, 1000 and 1050 °C and the strain rates were 1, 10 and 40 /s and strains were 0.2 and 0.4. Interrupted relaxation experiments were performed at a constant true strain rate of 10 /s following compression to 0.2 and 0.4 strains to determine the prior austenite grain morphology. Holding times recorded for approx. 95 % recrystallization were 12 s at 950 °C and 6 s each at 1000 °C and 1050°C. The samples were quenched to RT following the stress relaxation / interrupted stress relaxation tests.

Prior austenite grain morphology was investigated using a laser scanning confocal microscope (LSCM) after picric acid etching and mean grain sizes were measured using mean linear intercept - procedure [1] from the centerline in the middle and quarter thickness of the deformed sample, as can be seen in figure 1.

3. Modeling and simulation

In order to calculate the non-uniform recrystallization kinetics and resulting average grain size after deformation, it is useful to apply the empirical equations describing the effect of temperature, strain and strain rate on these quantities [2].

3.1. SRX regression model

To estimate the time required for 50 % completion of static recrystallization, t_{50}, we used equation (1)

$$t_{50} = A'\varepsilon^p d^s \dot{\varepsilon}^q \exp(Q_{rex}/RT) \tag{1}$$

where ε is the strain, d the grain size prior to deformation and Q_{rex} the recrystallization activation energy, A' is a composition dependent constant, p and q are powers of strain and strain rate, respectively. The value for Q_{rex} is calculated using equation (2)

$$Q_{rex}(\mathrm{J/mol}) = 3803CF + 109418 \tag{2}$$

The composition factor CF = 2Cr + 10Cu + 15Mn + 50Mo + 60Si + 70V + 230Ti + 700Nb, where the concentrations of elements are, in wt.%; Si and Nb are considered to have maxima of 0.4% and 0.044%, respectively, for higher concentrations. Modelling of A' in equation (2) has been earlier fitted by regression, $\log A' = aCF + b$, where a and b are constants [3, 4]. The basis of selection of exponents p, q and Q_{def} has been given in detail elsewhere [3, 4]. The grain size exponent (s) has been found to be strongly dependent on grain size itself, decreasing nearly asymptotically with the grain size, varying from ~2.1 at the finest grain size (12 μm) to ~1.15 at large grain sizes (≈800 μm), in agreement with the values suggested by Fernandez et al. [5]. The dependence of s on grain size is described with a power function: $s = 2.13d^{-0.105}$. [6]

3.2. Grain size model

In order to estimate the average grain size D_r after the recrystallization has been completed, we applied equation (3), from [7, 8]

$$D_r = 25\left(14.925\ln\frac{10^{-9}Z}{8.5}\right)^{-0.67}\frac{\sqrt{d}}{\varepsilon} \tag{3}$$

where Z is the Zener-Hollomon parameter, which is calculated using equation (4)

$$Z = \dot{\varepsilon}\exp\left(\frac{Q_{def}}{RT}\right) \tag{4}$$

Another equation was also applied, from [9]

$$D_r = 0.5D^{0.67}\epsilon^{-0.67} \tag{5}$$

3.3. Abaqus model

The Gleeble upsettings (hot compression tests) were modelled by the FEM software Abaqus. The model utilizes non-linear facilities of a coupled temperature-displacement solver of the software package

Abaqus Explicit. Abaqus is a multifunctional FEM software capable of handling a multitude of different cases, including coupled temperature-dependent deformation of a solid piece, as in upsetting.

Upsetting is assumed to be an axisymmetric forming process. The specimen is isolated from the compressive jaws to reduce the effect of friction and heat conductance. Tantalum and graphite together with a nickel paste are used in the insulating layers and thus the friction coefficient was set to a value of 0.05 using Coulomb's friction theory. Heat conductance between the jaws and specimen was ignored. Due to vacuum used in upsetting tests, only heat radiation from the specimen surface is considered using emissivity of 0.8. Finely discretized geometry of the specimen is meshed with axisymmetric CAX4RT thermally coupled quadrilateral elements with enhanced hourglass control and reduced integration. Compressive jaws were modelled as rigid parts. Figure 1 shows the axisymmetric cross-section of a deformed test piece, with plastic equivalent elongation as field variable (blue is low, red is high) and includes the areas used for further simulations marked as black circles.

A previously fitted Hensel-Spittel (H-S) equation has been used to define the plastic stress-strain behavior of the Abaqus model [10]. The H-S equation was implemented into the FE-model using the VUHARD-subroutine. Comparison of upsetting forces between experimental and FEM results are depicted in figure 2. Temperature dependent material properties (conductivity, specific heat, density and heat expansion) are specified with Inter Dendritic Solidification (IDS) analysis package using chemical composition of a selected steel grade [11].

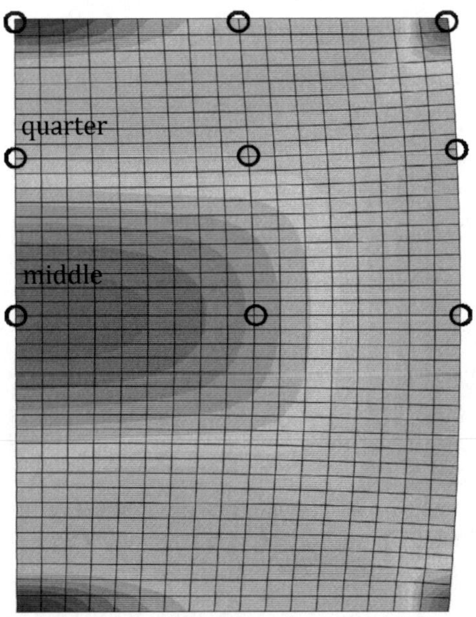

Figure 1. Cross-section of compressed test piece with PEEQ as variable

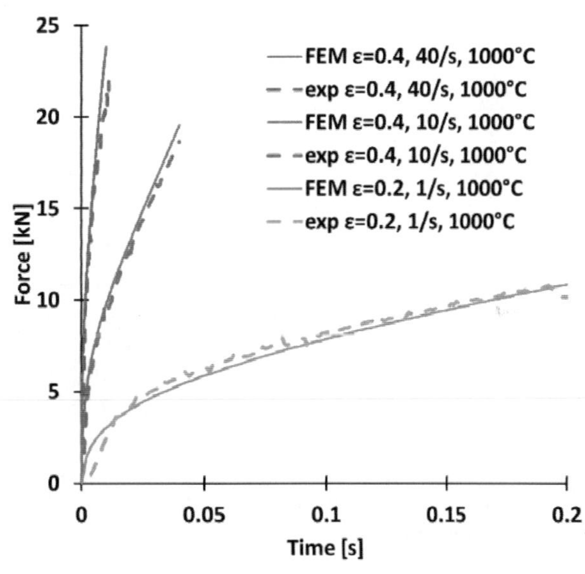

Figure 2. Comparison of extrusion forces

The validity of the applied Abaqus model was checked using reaction force data from various Gleeble tests. The experimental data was then compared with the reaction forces obtained from the Abaqus simulations. The simulated reaction forces seem to correspond with experimental data very nicely. Based on this comparison, the model is deemed accurate enough to be used as a base for further simulations.

Data from the Abaqus model has been used in the SRX regression model and grain size model as input data, thus tying the models together. Strain and strain rate data from key points throughout the test piece cross-section has been obtained from output files using a python script. t_{50} and grain size equations have then been applied to each point.

IOP Conf. Series: Journal of Physics: Conf. Series **1270** (2019) 012027 doi:10.1088/1742-6596/1270/1/012027

4. Results and discussion

4.1. SRX model

t_{50}-times estimated from the experimental data and the simulations were compared. Simulation data was chosen from test piece center and quarter thickness, since that area corresponds best to macroscopic relaxation data obtained experimentally. The results are shown in table 1 and figure 3.

Table 1. t_{50}-times [s] for simulated and experimental data

	T1000 e02 SR1	T1000 e02 SR10	T1000 e04 SR1	T1000 e04 SR10	T1050 e02 SR1	T1050 e02 SR10	T1050 e04 SR1	T1050 e04 SR10
sim	6.85	4.04	1.16	0.67	2.27	1.35	0.38	0.23
exp	4.10	2.90	1.50	1.74	1.80	1.80	1.50	1.80

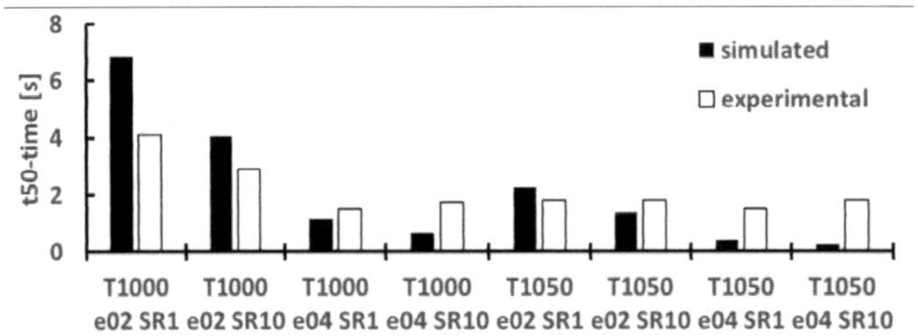

Figure 3. t_{50}-times for simulated and experimental data

From figure 3 it can be seen that the simulation and experimental data show the effect of strain and strain rate for temperature 1000 °C in a similar fashion, where increasing the strain and strain rate seems to decrease the t_{50}-time. The same effects cannot be seen from the experimental data for 1050 °C though. When taking to consideration the small absolute differences between the values and the fact that the model variables are calculated based on composition, it can be said that the model works reasonably well. For temperature 950 °C the difference between experimental and simulation data was 15 s.

Due to the time-consuming experiments and extensive data analysis, a single stress relaxation test is carried out for each set of test parameters to determine the activation energy of recrystallization and various exponents. However, some tests may need to be repeated, if the corresponding data distinctly fall as outliers during the fitting with related trendlines, leading to poor fits.

The simulated t_{50}-times have also been compared with varying strains, strain rates, and temperatures for different areas around the test piece cross-section. Center of the sample was chosen for closer inspection in figure 4.

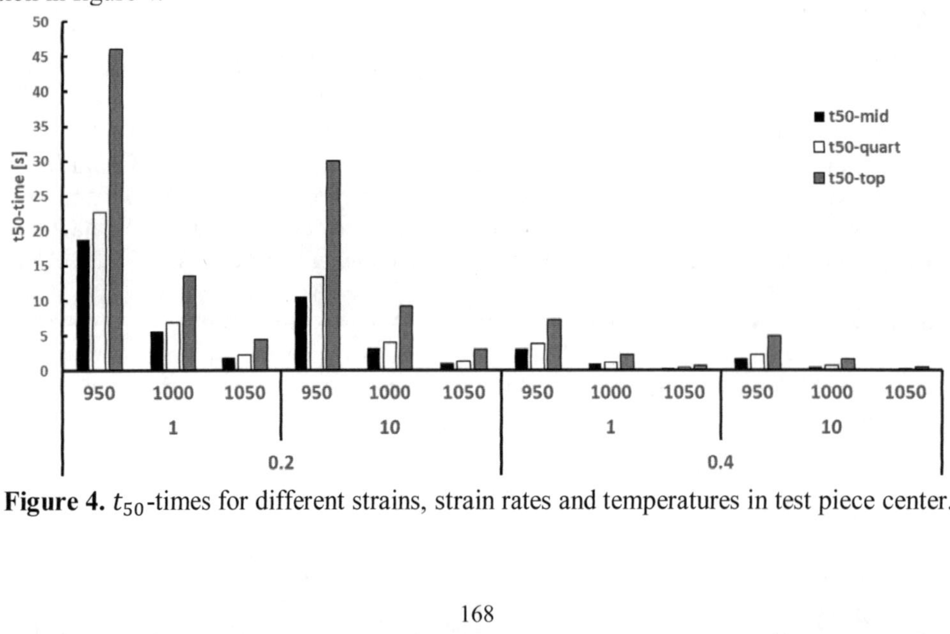

Figure 4. t_{50}-times for different strains, strain rates and temperatures in test piece center.

Large differences in t_{50}-times were observed for low strain, strain rate and temperature throughout the thickness of the test piece center. These differences seem to settle down when strain or temperature is increased. Strain rate has a small effect compared to the other variables.

4.2. Grain size model

Examples of etched austenite and mean grain size of experimental and simulated are shown in figures 5 and 6, respectively. Grain sizes with 0.2 strained are slightly bigger than 0.4 strain. Also increasing deformation temperature increased the experimentally defined mean grain size. Equation (3) is useful when comparing the grain sizes of 0.2 strain samples. However, equation (4) is more precise with the 0.4 straining samples.

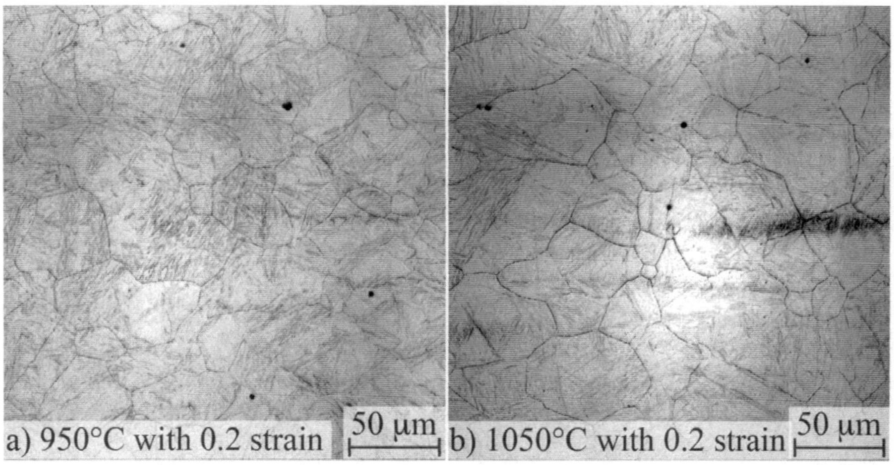

Figure 5. Examples of prior austenite morphologies of 0.2 strained at (a) 950 °C and (b) 1050 °C from middle of sample after picral etching.

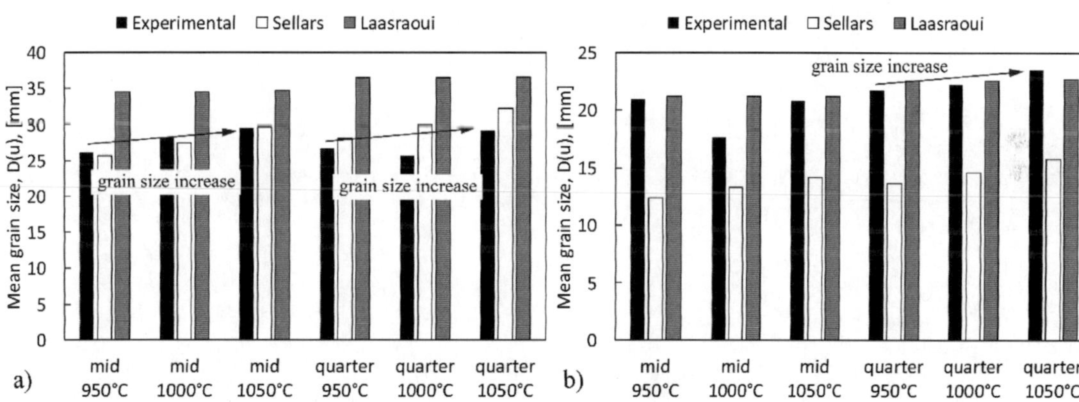

Figure 6. Mean grain sizes at middle and quarter of experimental Gleeble sample and calculated using Sellars and Laasraoui equation. Strained at 950, 1000 and 1050 °C with (a) 0.2 and (b) 0.4 strain.

5. Conclusions

An Abaqus FE model was successfully created for the Gleeble™ 3800 thermo-mechanical simulator and was shown to be accurate. The t_{50}-times and grain sizes were simulated in different points around the cross-section of the test piece and some significant differences were observed in t_{50}-times. The most significant factor affecting grain size was found to be strain, whereas temperature and strain rate had a negligible effect.

Grain size was simulated with two different equations, both of which seemed to correspond decently well to experimental data. The t_{50}-model was found to correspond well with experimental data in 1000 and 1050 °C, but this was not the case for 950 °C. There are several possible reasons for this, most likely the variables used were not ideal and the difference between the calculated value and the experiment may be related to variations in the sample material. In any case, further studying is required to find the cause for these issues.

Acknowledgements
Two of the Authors (A Pohjonen and M Somani) would like to acknowledge funding from the Academy of Finland through Genome of Steel project (Project #311934).

References

[1] Higginson R and Sellars C 2003 *Worked Examples in Quantitative Metallography* (London: Maney) p 116

[2] de Souza A L B, Balancin O 2019 In review

[3] Karjalainen L P, Somani M C, Porter D A and Morgridge R A 2002 *The 5th Int. ESAFORM Conf. (Krakow)*, ed M Pietrzyk et al. p 603.

[4] Somani M C, Karjalainen L P, Porter D A and Morgridge R A 2003 *Int. Conf. on Thermomechanical Processing: Mechanisms, Microstructure and Control (Sheffield)*, ed Palmiere E J et al. p 436.

[5] Fernández A I, Uranga P, Lopez B and Rodriguez-Ibabe 2000 J M *ISIJ Int.* **40** 893

[6] Somani M C and Karjalainen L P 2012 *Materials Science Forum* **715-716** 751-757

[7] Sellars C M 1979 *Physical Metallurgy of Hot Working, in: Hot Working and Forming Processes*, ed Sellars C M and Davies G J pp 3-15

[8] Lenard J G, Pietrzyk M and Cser L *Mathematical and Physical Simulation of the Properties of Hot Rolled Products* pp 151-158

[9] Laasraoui A and Jonas J J 1991 *ISIJ Int.* **31** 95-105

[10] Ilmola J, Seppälä O, Leinonen O, Pohjonen A, Larkiola J, Jokisaari J and Putaansuu E 2018 *AIP Conf. Proc.* vol 1960 090007

[11] Miettinen J, Louhenkilpi S, Kytönen H and Laine J 2010 *Mathematics and Computers in Simulation* **80** 1536-1550

IOP Conf. Series: Journal of Physics: Conf. Series **1270** (2019) 012028 doi:10.1088/1742-6596/1270/1/012028

Preferred orientation formation in surface layer of aluminum sheet subjected to friction roll surface processing and temperature gradient annealing

Yoshimasa Takayama[1], Ryuichi Hamano[2], Takuya Arakawa[2] and Hideo Watanabe[1]

[1]Department of Mechanical and Intelligent Engineering, Graduate School of Engineering, Utsunomiya University, Utsunomiya 321-8585, Japan

[2]Graduate student, Utsunomiya University

E-mail: takayama@cc.utsunomiya-u.ac.jp

Abstract. Preferred orientation formation in surface layer of aluminum sheet subjected to friction roll surface processing (FRSP) and temperature gradient annealing has been investigated by SEM/EBSD analysis. FRSP imposed severe shear strain in the surface layer of the aluminum sheet, where the indentation was selected as 0.1mm and the feeding speed was set as 0.3mm/s. The annealing with steep temperature gradient in longitudinal direction was applied to the small specimen of 12mm long, 10mm wide and 1mm thick, which made microstructural evolution concentrated to the nearest layer to the surface. Shear strain more than 10 was imposed at a site of 14μm beneath the surface. The temperature gradient more than 20K/mm was attained even for aluminum which has high thermal conductivity. The {100} pole of ND/TD (FRSP plane/FRSP direction) FRSPed specimen showed very weak intensity of <001>//ND while the {111} pole implies <110>//TD. In contrast, the shear texture component {111}<110> mainly evolved in the specimen subjected to FRSP and the ordinary annealing. Evolution of the component {111}<110> after the temperature gradient annealing were compared with that after the ordinary annealing in consideration of stored strain energy in each crystallographic orientation.

1. Introduction

Severe plastic deformation (SPD) has attracted great interest owing to not only formation of extremely fine microstructure but also improvement of related properties. Numerous studies have been performed to produce ultrafine grains and make clear their formation mechanism in various SPD processes such as equal channel angular pressing (ECAP) [1], accumulative roll bonding (ARB) [2] and high pressure torsion (HPT) [3]. The crystallographic textures often appeared in the SPDed materials, because extremely large strain should be accompanied with necessary directional priority in slip deformation. The texture evolution during ECAP by only the two shear systems {111}<110> and {001}<110> ({shear plane}<shear direction>) in an Al-Zn-Mg-Cu alloy having a strong initial texture was reported [4]. To an aluminum single crystal ECAP was applied with consistence of theoretical shear plane and direction to crystal slip system [5], and moreover, a crystal plasticity finite element method model showed the textural change with the lattice rotation around the transverse axis of ECAP [6]. For ARB processed aluminum the different local textures corresponding to the different microstructural regions were observed [7] and evolution of the textures with strong gradients was investigated in detail with experimental observations and numerical simulations [8]. However, attempts to the possibility of texture control by the SPD processes have been limited. On the other hand, it is well known that some of materials properties, such as corrosion, formability and wear, are strongly related to the surface microstructure and texture.

Friction roll surface processing (FRSP) has been attempted as another SPD process for the surface of the sheet materials in order to control microstructure and texture [9,10]. An attempt of FRSP to a commercial purity titanium sheet [9] revealed that a large plastic strain was imposed in the surface layer and caused grain refinement up to about 100nm. Characteristic shear textures were formed in the surface layer during FRSP and subsequent annealing, which was strongly related to the processing direction [9,10]. It should be noted that the shear texture formed in the FRSPed specimen remained preferential after annealing. Sakai et al. [11] already pointed out that the shear texture developed by the shear rolling had a broad one spreading around the typical shear components after annealing. It is, therefore, suggested that the shear texture formed by shear deformation tends to remain after annealing. Shear deformation and subsequent annealing could be a useful way for textural control.

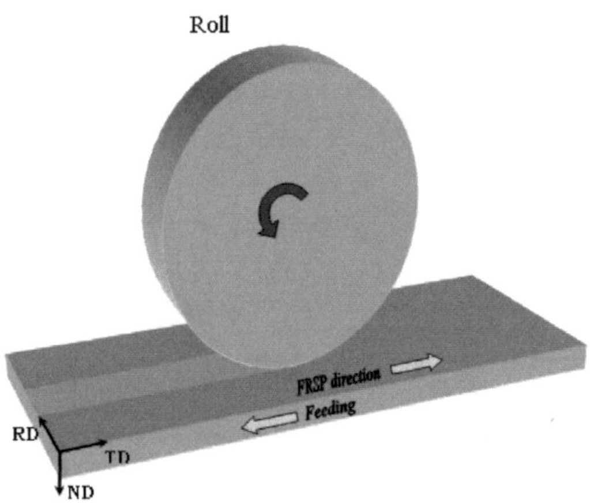

Figure 1. Schematic illustration of Friction Roll Surface Processing (FRSP). Rotating roll is fed on ND plane in feeding direction, which is parallel to TD (the width direction of rolled sheet sample) in the present figure. This processing is called ND/TD FRSP.

As it is needless to say dependence of temperature on microstructural evolution, restoration process proceeds for the shorter time at the higher temperature [12]. The annealing with high temperature gradient can lead to a local preferred evolution of microstructure intendedly by using gradient stored strain in FRSPed specimen.

The purpose of the present study is to review texture evolutions in specimens subjected to FRSP subsequent annealing to commercial purity aluminum [13]. Further, textural evolution in the specimen subjected to FRSP and temperature gradient annealing is investigated in comparison with that in the specimen subjected to FRSP and the ordinary annealing.

2. Experimental Procedure

2.1. Material

Commercial purity aluminum sheet (1050) 1mm thick was received as a starting material. It was produced by a process of casting, hot rolling, cold rolling with a reduction of 90% and annealing at 623K for 3.6ks.

2.2. Friction roll surface processing (FRSP)

Rotational direction of roll, feeding direction of sample and FRSP direction are illustrated in Figure 1. A fixed roll with a dimension of $\phi 70 \times 5$mm, which was made of tool steel SK3, was rotated at 240 rpm. The indentation, defined as the depth at which the rotating roll was pressed down into the specimen, was selected as 0.1mm. The feeding speed of specimen was set as 0.3 mm/s. FRSP was performed as described in our previous studies [9,10]: first, the interface where the roll and the specimen just touched was located, and this was regarded as the reference plane at which the indentation was zero. Then, the roll was placed to the given indentation without contacting the specimen. After rotating the roll, the specimen was moved forward until the entire surface was treated. This procedure resulted in making the working plane lower than the reference plane and led to a great amount of friction. The deformation mode in FRSP is similar to that in asymmetric rolling (ASR). Tangential speed of FRSP roll is faster than feeding speed on the surface layer of the material while a faster tangential speed of rolls in ASR is almost similar to a moving speed of surface material in ASR. After FRSP, the specimens were annealed at 573K, 623K or 673K for 3.6ks (1 h) in Ar gas and air cooled to study the texture evolution during annealing.

Here the plane and the direction of the specimen subjected to FRSP are expressed as plane/ direction sited before "FRSP", where the rolling, transverse and normal directions (RD, TD and ND, respectively) are used as the specimen axes. For example, ND/TD FRSP (See Figure 1) refers to the processing in which FRSP was performed on ND plane (the plane perpendicular to ND) of the specimen along the direction parallel to TD.

2.3. *Microstructural analysis*

Crystallographic textures of annealed specimens were evaluated using a scanning electron microscope/ electron back scatter diffraction pattern technique (SEM/EBSD; HITACHI S-3500H, TSL Orientation Image Microscopy system) after electropolishing. The electropolishing was carried out at 283K and 8V for 480s in a solution containing perchloric acid and ethanol with a ratio of 1/8. The scan areas were chosen on the cross sections of the specimen perpendicular to FRSP direction. Microstructure and texture evolution through the thickness of specimen was investigated from the processed surface to center of the specimen, the area was 400 μm×400 μm and the step size was 3μm. High magnification measurement was used with a smaller step size of 0.1μm for the microstructure in the vicinity of the processed surface.

2.4. *Temperature gradient annealing*

The annealing with steep temperature gradient in longitudinal direction was applied to the small specimen of 12mm long, 10mm wide and 1.5mm thick as follows. Figure 2 illustrates the sectioned drawing of temperature gradient annealing instrument.

An infrared furnace was employed as heating device because of its high energy density. As the top face of the specimen was sited to the concentrated part of the infrared rays in the center of the furnace, the stainless specimen holder was specially designed to cool the bottom face by combined water cooling device with copper tube.

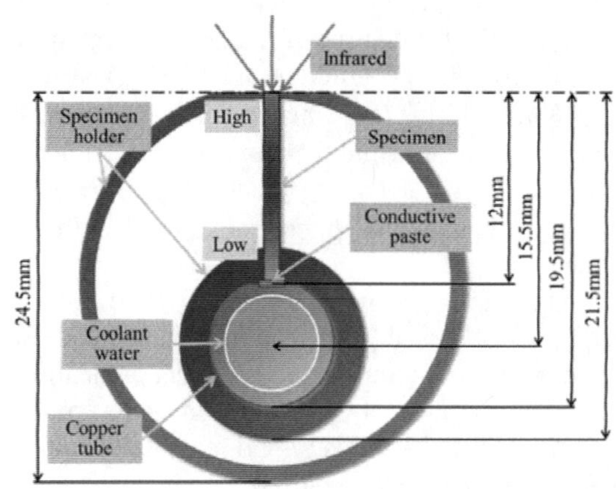

Figure 2. Schematic illustration of apparatus for temperature gradient annealing.

3. Experimental results and discussion

3.1. Texture of processed materials

Figure 3 shows the normalized pole figures of as-received, ND/TD FRSPed and subsequently annealed specimens measured by XRD. {100} and {111} pole figures of the as received specimen indicate typical cube texture (Figure 3(a)). As displayed in Figure 3(b), the {100} pole of ND/TD FRSPed specimen represents very weak intensity of <001>//ND while the {111} pole implies <110>//TD. Then, the main textural component is {001}<110> ({FRSP plane}<FRSP direction>). The central intensity of the {111} pole extended in RD likely means formation of the main component rotated about TD axis. After the annealing (Figure 3(c)), {001}<110> orientation is somewhat sharpened though the characteristically rotated spread of the main component remains after annealing.

These results probably show the texture formed mainly in the surface layer with thickness of 10μm order as they were obtained by X-ray diffraction with penetration depth ranging from 0.01 to 0.1mm. The texture after annealing is also mentioned below based on EBSD analysis on cross-section.

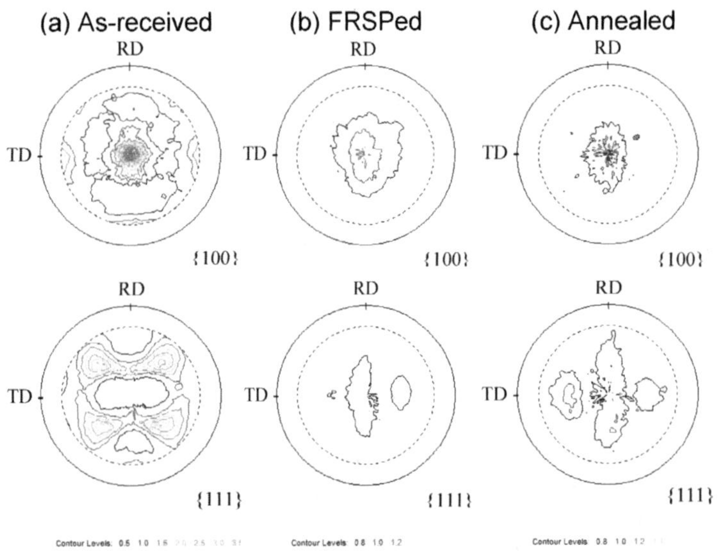

Figure 3. Pole figures showing difference in texture before and after FRSP and subsequent annealing in 1050 aluminum. (a) As received, (b) ND/RD FRSPed and (c) Annealed samples. Contour levels of pole figures in (a): 0.5, 1.0, 1.5, 2.0, 2.5, 3.0, 3.5, 4.0, 4.5, 5.0, 5.5, 6.0, 6.5, 7.0; Contour levels in (b): 0.8, 1.0, 1.2; Contour levels in (c): 0.8, 1.0, 1.2, 1.4 [13].

3.2. Microstructure after FRSP and temperature gradient annealing

Figure 4 shows inverse pole figure (IPF) and kernel average misorientation (KAM) maps of the sample subjected to RD/TD FRSP and temperature gradient annealing. The temperature gradient was kept almost as a mean value of 21.3K/mm during the annealing. The area from FRSP surface to 80μm depth is fine grained region without stored strain while the inner one to 150μm has high KAM more than 1.5 degrees and moreover, the deeper region seems as received grain structure. The fine grained region composed of ones of 4.6μm. Which is one third to the grain size of 13.9μm in the fine grained region of the ordinarily annealed sample.

The inner region with high KAM is found between the fine grained and the as-received ones. It is very characteristic manner, which is not observed in the ordinarily annealed sample. In addition, no coarse grained region was observed after the temperature gradient annealing. These facts suggest that restoration process takes place rapidly nearest the surface layer to compared with the deeper region. Such priority of the nearest surface layer in recovery and recrystallization probably leads to possibility of activation on nuclei related to the higher stored strain.

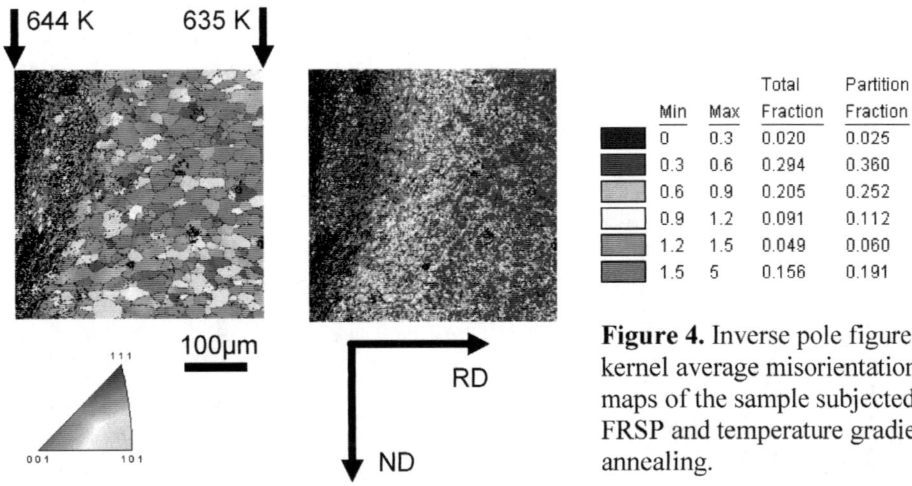

Figure 4. Inverse pole figure (IPF) and kernel average misorientation (KAM) maps of the sample subjected to RD/TD FRSP and temperature gradient annealing.

3.3. Texture after FRSP and temperature gradient annealing

Material deformation generated by FRSP is probably similar to that in asymmetric rolling (ASR). Preferred orientation components in asymmetric rolling were reported as H {001}<110>, E {111}<110> and F {111}<112> [14]. The texture after RD/TD FRSP and temperature gradient annealing (TGA) is summarized in Table 1. The 1050 aluminum used as a starting material had a considerably sharp Cube texture. The fraction of Cube within a tolerance of 15° decreased from 26.7% to 3.5% or 1.9% through FRSP and annealing. On the other hand, a shear texture component H changed from 3.6% to 4.8% or 2.9% according to the sort of annealing. The F orientation is defined as one of components with <111> parallel to RD for RD/TD FRSP. The orientation was found by 5.7% after the ordinary annealing and by 4.4% after TGA while it existed by only 0.4% before FRSP. It should be noted that E orientation evolved by three times from 3.3% to 10.3%, corresponding to higher strain near the surface.

Table 1. Fractions of textural components after RD/TD FRSP and two kinds of annealing (%)

Texture component (FRSP plane/ FRSP direction)	As received	Homogeneous Temperature at 623K	Temperature gradient at 623K		
		Fine grains	Fine grains	Strain stored	As received grains
FRSP_nd//<001>	44.2	15.9	11.4	12.3	35.8
FRSP_d//<110>	9.6	25.4	28.2	25.0	9.2
FRSP_nd//<111>	7.5	12.9	20.4	14.8	9.7
Cube {001}<100>	26.7	3.5	1.9	2.3	15.3
H {001}<110>	3.6	4.8	2.9	2.9	2.2
F {111}<112>	0.4	5.7	4.4	4.0	2.0
E {111}<110>	3.3	3.5	10.3	6.4	3.6

3.4. Microstructure and preferred orientation in the surface layer after FRSP

It was found that FRSP/TGA enables the high fraction of E{111}<110> in fine grained and strain stored regions compared with other orientations. This is attributed to restoration process limited in highly stored strain region. E orientation recovers rapidly during annealing, and/or hardly store strain through FRSP. Then, E orientation grows preferentially by strain induced boundary migration.

Figure 5 shows comparison between E orientation and the others in SEM/EBSD analyses for two fields of FRSPed sample without annealing. The KAM value of E orientation is lower than that of others. E orientation has higher IQ (image quality) and CI (confidence index) values than other orientations. The

IQ value was applied to analyze recrystallized volume fraction [15], while the stored strain energy was derived from the KAM value based on the relationship to dislocation density [16]. The facts on three values indicate that E orientation stores less strain than the other ones. It is, therefore, concluded that E orientation is not imposed strain during FRSP.

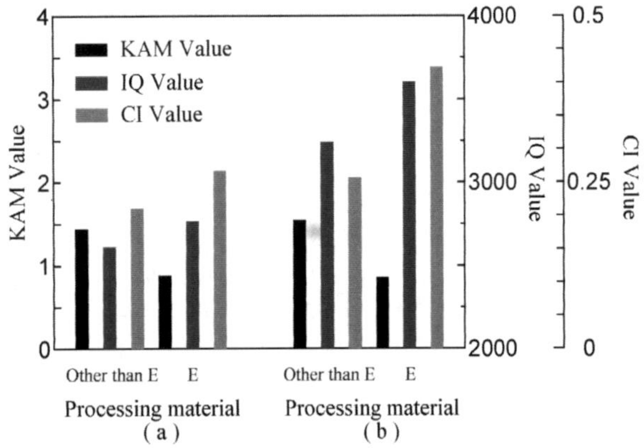

Figure 5. Comparisons of E and other orientations in KAM, IQ and CI values for (a) field A and (b) field B of FRSPed sample.

4. Summary

Preferred orientation formation in surface layer of aluminum sheet subjected to friction roll surface processing (FRSP) and temperature gradient annealing has been investigated by SEM/EBSD analysis. beneath the surface. The temperature gradient 21.3K/mm was attained even for aluminum with high thermal conductivity. The shear texture component {111}<110> mainly evolved in the specimen subjected to FRSP and the ordinary annealing. More evolution of the component {111}<110> after the temperature gradient annealing were compared with that after the ordinary annealing in consideration of stored strain.

References

[1] Iwahashi Y, Wang J, Horita Z, Nemoto M and Langdon T G 1996 *Scr. Mater.* **35** 143–6
[2] Saito Y, Tsuji N, Utsunomiya H, Sakai T and Hong R G 1998 *Scr. Mater.* **39** 1221–7
[3] Zhilyaev A P, Nurislamova G V, Kim B-K, Baro M D, Szpunar J A and Langdon T G 2003 *Acta Mater.* **51** 735–65
[4] Wang S C, Starink M J, Gao N, Qiao X G, Xu C and Langdon T G 2008 *Acta Mater.* **56** 3800–9
[5] Fukuda Y, Oh-ishi K, Furukawa M, Horita Z and Langdon T G 2004 *Acta Mater.* **52** 1387–95
[6] Lu C, Deng G Y, Tieu A K, Su L H, Zhu H T and Liu X H 2011 *Acta Mater.* **59** 3581–92
[7] Kamikawa N, Tsuji N, Huang X and Hansen N 2006 *Acta Mater.* **54** 3055–66
[8] Li S, Sun F and Li H 2010 Acta Mater. **58** 1317–31
[9] Shi M, Takayama Y, Umetsu T, Kato H, Watanabe H and Inoue H 2009 *Mater. Trans.* **50** 210–4
[10] Shi M, Takayama Y, Ma C, Watanabe H and Inoue H 2012 *Trans. Nonferrous Met. Soc. China* **22** 2616–27
[11] Sakai T, Hamada S and Saito Y 2001 *Scripta Mater.* **44** 2569–73
[12] Humphreys J F and Hatherley M 2004 *Recrystallization and Related Annealing Phenomena* 2nd Edition, (Oxford: Elsevier Science) pp 220–1
[13] Takayama Y, Hamano R, Arakawa T, Nonaka K, Watanabe H and Inoue H 2015 *Mater. Trans.* **56** 303–12
[14] Sidor J, Miroux A, Petrov R and Kestens L 2008 *Acta Mater.* **56** 2495–507
[15] Tarasiuk J, Gerber Ph and Bacroix B 2002 *Acta Mater.* **50** 1467–77
[16] Takayama Y and Szpunar J A 2004 *Mater. Trans.* **45** 2316–25

Observation of grain growth in U-10Mo alloy

K Kalia[1], W E Frazier[2], D P Field[1] and V V Joshi[2]

[1]School of Mech. and Matls. Engineering, Washington St. Univ, Pullman, WA, USA
[2]Pacific Northwest National Laboratory, Richland, WA, USA

dfield@wsu.edu

Abstract. Both as-cast and homogenized wedge specimens of U-10 wt%Mo (U-10Mo) alloy were thermomechanically processed in both cold and hot rolled conditions and annealed at 700 °C for periods lasting up to 2 h. The hot rolled samples had a net strain of 70% whereas the cold rolled ones were subjected to 30% strain. Annealed microstructures were examined using electron backscatter diffraction in order to estimate the completeness of recrystallization and grain size, and these were mapped as a function of strain. In as-cast samples, the stored energy with strain increases but in the homogenized samples the stored energy remains constant with strain. This is due to a difference in the initial microstructure (before deformation) of the homogenized samples.

1. Introduction

For several years, research to develop alternate fuels for high-performance research reactors (HPRR) has been under way to replace high enriched uranium fuel with low enriched uranium (LEU) fuel. To achieve similar U-235 content with LEU, given the reactor space constraints, a higher bulk density of fuel is needed, which necessitates the use of monolithic metal alloy fuel. To achieve this density, an alloy that offers a good combination of irradiation performance, oxidation resistance, strength, and ductility is required for HPRR applications. Various uranium alloys have been evaluated for their favorable mechanical properties and susceptibility to oxidation and corrosion [1] [2]. One of the broad challenges faced is the swelling kinetics of this metallic fuel during the irradiation process. The fuel plates start to swell due to emission of inert gases and effects the integrity and shape of the sample which further deteriorates the irradiation performance.

Uranium alloyed with 10 wt% molybdenum (U-10Mo) shows improvement in these characteristics compared to similar alloys. Another key factor to improve the performance is to have stable phase of U-10Mo during irradiation. The nominal fabrication process for U-Mo alloys is to cast and quench the material to stabilize the metastable gamma (γ) phase. Casting in an inert gas is recommended to prevent the formation of oxides in the alloy. Rapid cooling is necessary to keep the alloy in the γ phase. A common technique used to accomplish this includes an homogenization step in which the alloy is brought to an elevated temperature between the eutectoid temperature (560°C) and the melting temperature for the alloy (1135°C to 1200°C). The alloy is held at temperature for an extended period to ensure that the metal has completely transformed to the γ phase and to enable an even distribution of the molybdenum throughout the sample. Uranium α and β phase have orthorhombic and tetragonal lattice structure respectively whereas Uranium Υ phase has BCC crystalline structure. BCC crystalline structure possesses less anisotropic properties which somehow improves the irradiation performance of this metallic fuel [2] [3]. However, several efforts are now under way to optimize the U-10Mo

IOP Conf. Series: Journal of Physics: Conf. Series **1270** (2019) 012029 doi:10.1088/1742-6596/1270/1/012029

processing parameters and Pacific Northwest National Laboratory (PNNL) has supported these activities.

Research is being performed on the microstructure of this material with modeling and experimental simulation to reduce the defects caused during and after the irradiation process [1-3]. With a motive to that, microstructure evolution is studied with processing of this alloy. The first set of samples includes study of recrystallization texture in U10Mo alloy with hot rolling and annealing process. Rolling and annealing texture is studied with quantified microstructure analysis. The second set of samples includes study of recrystallization behavior in wedge shaped annealed samples. Microstructure evolution is studied with respect to strain.

2. Experimental Procedure

All samples were prepared using standard metallographic procedures and final polishing was performed on a vibro-polisher using 0.05 μm colloidal silica suspension before EBSD/SEM measurements. Recrystallization behavior was studied in U10Mo wedge-shaped samples with post annealing process after deformation. Two sets of cast samples were prepared. One set is homogenized at 900°C for 48 hours and another set is left as cast only. Each set includes two samples out of which one is deformed with cold rolling and another one with hot rolling (at 650^{0}C). Finally, each sample is post-annealed at 700°C for two hours. After deformation/ rolling process, the wedge-shaped samples were flattened with accumulation of specific uniform strains i.e. 0.1 to 0.7. All strains were calculated by using engineering strain formula along the length of the sample. On average, the cold roll deformed samples were 22 mm long, whereas, the hot roll deformed samples were 30 mm long. Strain up to 0.3 and 0.7 were calculated in cold rolled and hot rolled samples, respectively. At higher strains chances of cracking increases during cold rolling. Therefore, the total length of the cold rolled samples was kept shorter than the hot rolled samples.

Sample processing is explained with the flow chart shown below. The sample name with its processing terminology is given below:

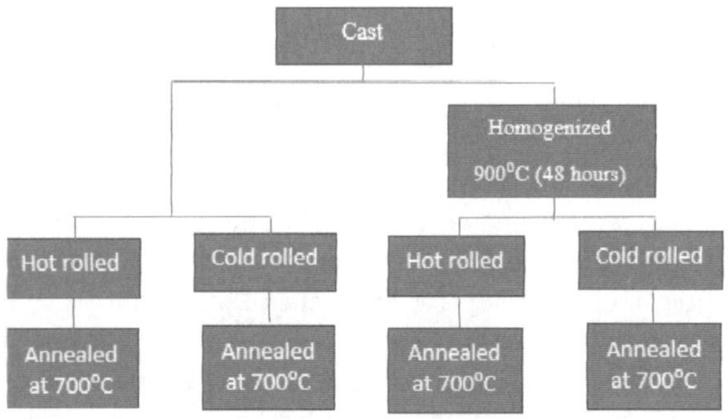

CH-700: Cast hold rolled post-annealed at 700°C

CC-700: Cast cold rolled post-annealed at 700°C

HH-700: Homogenized hot rolled post-annealed at 700°C

HC-700: Homogenized cold rolled post-annealed at 700°C

2.1. EBSD measurements

Sample preparation methods and EBSD parameters used were similar for the hot rolled and annealed samples. The total area scanned (for orientation maps) for EBSD analysis was approximately 300x900 μm^2 at each specific strain for all samples. Measurements were made over a regular hexagonal grid using a step size of four microns for cold rolled samples and two microns for hot rolled samples. Step size was decided as per the average grain size of the samples. Cold rolled samples have a larger mean grain size (>25μm) compared to hot rolled samples so the step size taken was larger (up to four microns). Specific strain positions were calculated by percentage deformation or engineering strain in all wedge-shaped samples.

In figure 1, dotted lines represent the initial shape of the sample before rolling. ϴ is the wedge angle, Δl is the change in height, h is the final height of the sample and b is the distance of each deformed region, from the 0% deformed region. For example: 50% deformation has an engineering strain of 0.5= $\frac{change\ in\ height}{original\ height}=\frac{2\Delta l}{h+2\Delta l}$. Δl calculated was finally used to calculate distance 'b' to find the X-coordinate of 50% deformation or 0.5 strain. This was done by assuming a right-angled triangle with Δl as the perpendicular distance and b as the base of the triangle. Distance b=$\frac{\Delta l}{tan^{\theta}/_{2}}$ was calculated and the sample (SEM stage) was moved manually in the X direction from 0% deformation or zero strain. The wedge angle ϴ is similar for all samples between 16~18^0 and is calculated by using SEM images in photo editing software.

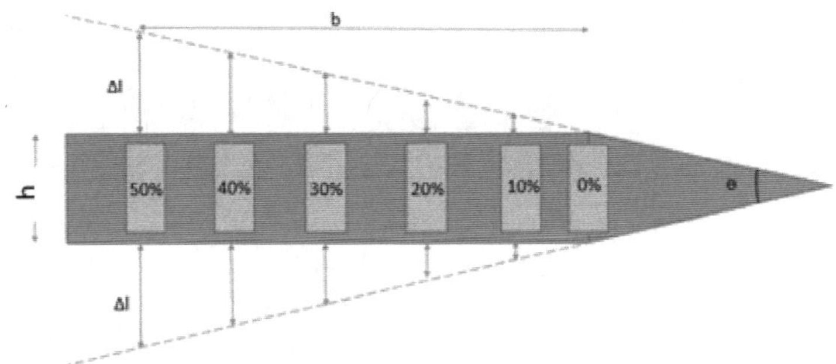

Figure 1: Schematic diagram of a wedge shaped sample representing dimensions for engineering strain calculation. Yellow areas denote the percentage of deformation

3. Results and Discussion

3.1. Microstructural analysis

Figure 2 shows the orientation maps (RD-ND planes) indexed as ϒ U-phase at different strain (0-0.7) of CC-700, HC-700, CH-700 and HH-700 samples. Black spots are either 2nd phase particles or points with CI<0.1. It was observed that the grain size drops with strain in all wedge-shaped post annealed 700^{0}C samples, and this drop is due to recrystallization behavior of grains. In both CC-700 and HC-700 samples, the grain size drops from 100 to 30 microns approximately at 0.3 strain. Whereas in CH-700 and HH-700 samples, the grain size drops from 100 to 20 microns approximately at 0.7 strain. The drastic drop in grain size is seen at 0.2 strain in all samples as shown in figure 3 (a). It can be stated that the critical deformation required for recrystallization to begin is at these temperatures is 0.2 strain. In sample CH-700 and CC-700, bimodal mean grain size charts were observed at 0.2 strain as shown in figure 3 (b) and (c). Bimodal mean grain size represents the presence of deformed and recrystallized grains in cast samples at 0.2 strain. This signifies that in cast samples (at 0.2 strain), the deformed grains start getting recrystallized. However, no such bi-modal grain size charts were observed in homogenized samples at 0.2 strain. Near random textures were observed at all strain levels in all samples.

During recrystallization, the fraction of high angle grain boundaries tends to increase because the low angle grain boundary fraction resulting from the dislocation structure is consumed by the growing grains. From figure 4, the fraction of HAGBs in all samples is above 0.9 signifying that a majority of the grains are recrystallized at all levels of strain.

3.2. Kernel average misorientation

Kernel average misorientation (KAM) is the average local misorientation of a single point in relation to orientations of neighboring points. Based on the assumption that local misorientation gradients are generated by dislocations, the KAM, is therefore a measure of the stored energy in terms of dislocation density in the microstructure. The measure of kernel average misorientation as a quantitative value is used to characterize the distribution of dislocation density [4] [5]. The KAM value in this work is calculated using a misorientation definition of less than 5° and the third nearest neighbors. As expected, this shows a similar trend compared to the GND (geometrically necessary dislocation) density in all samples (figure 5).

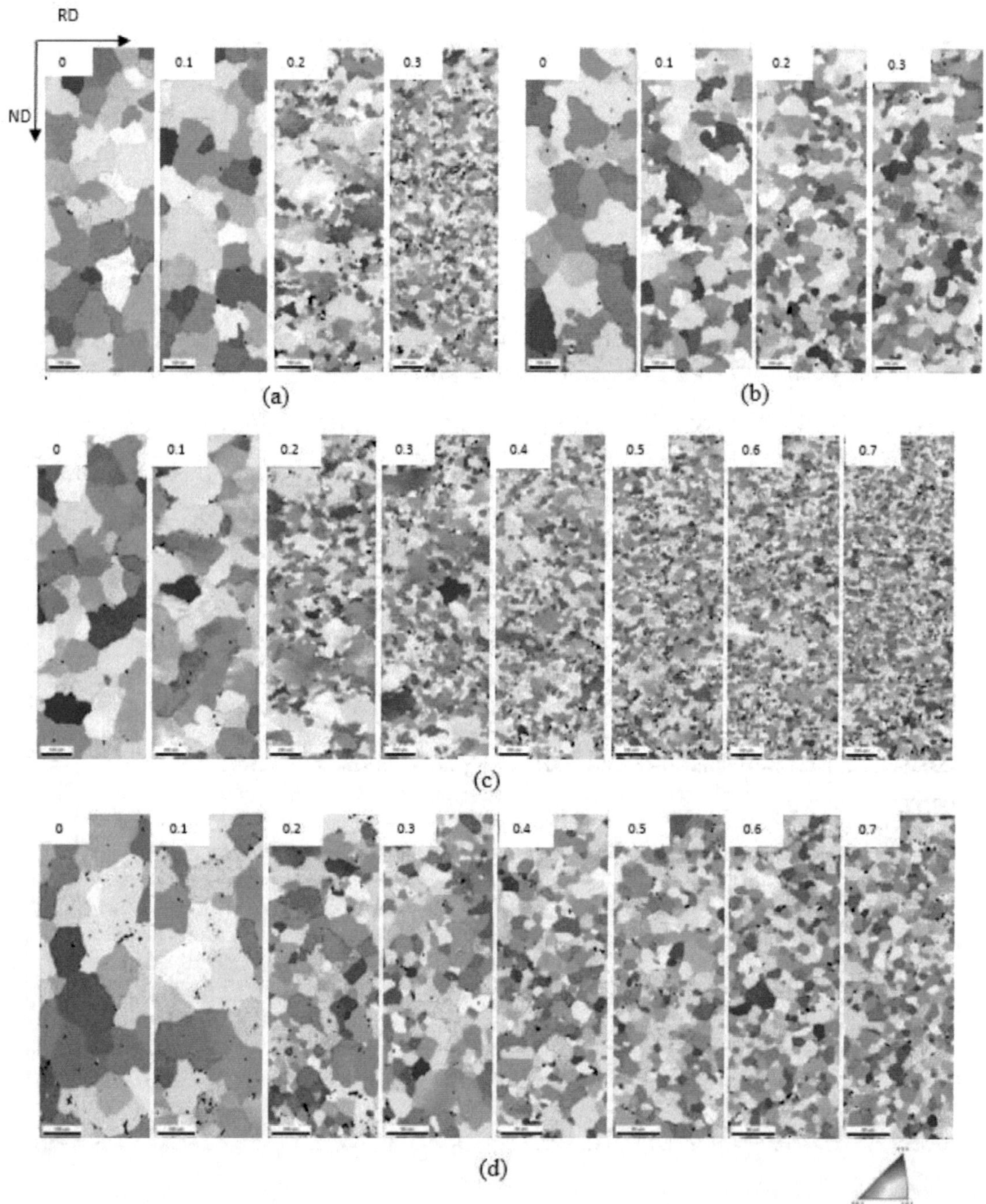

Figure 2: Orientation maps (RD-ND plane) indexed as U-ϒ phase at different strain (0-0.7) of (a) CC-700 (b) HC-700 (c) CH-700 (d) HH-700 samples. Black spots are either 2nd phase particles or points with CI<0.1. The micron bar is 100 μm for all images.

From figure 6(a) and (c), there is no certain change in stored energy of homogenized (HH-700 and HC-700) samples. In homogenized samples, little drop in fraction of GOS<2° at 0.2 strain in HH-700 and at 0.1 strain in HC-700 is considered an outlier. However, in KAM value there is no significant change as shown in figure 6(e). Whereas in cast samples as shown in figure 6(b) and (d), the stored energy increases with strain up to 0.4 in hot rolled and up to 0.3 in cold rolled samples. Due to shorter length of cold rolled samples, the strain accumulated is only up to 0.3 so we cannot determine beyond that strain level.

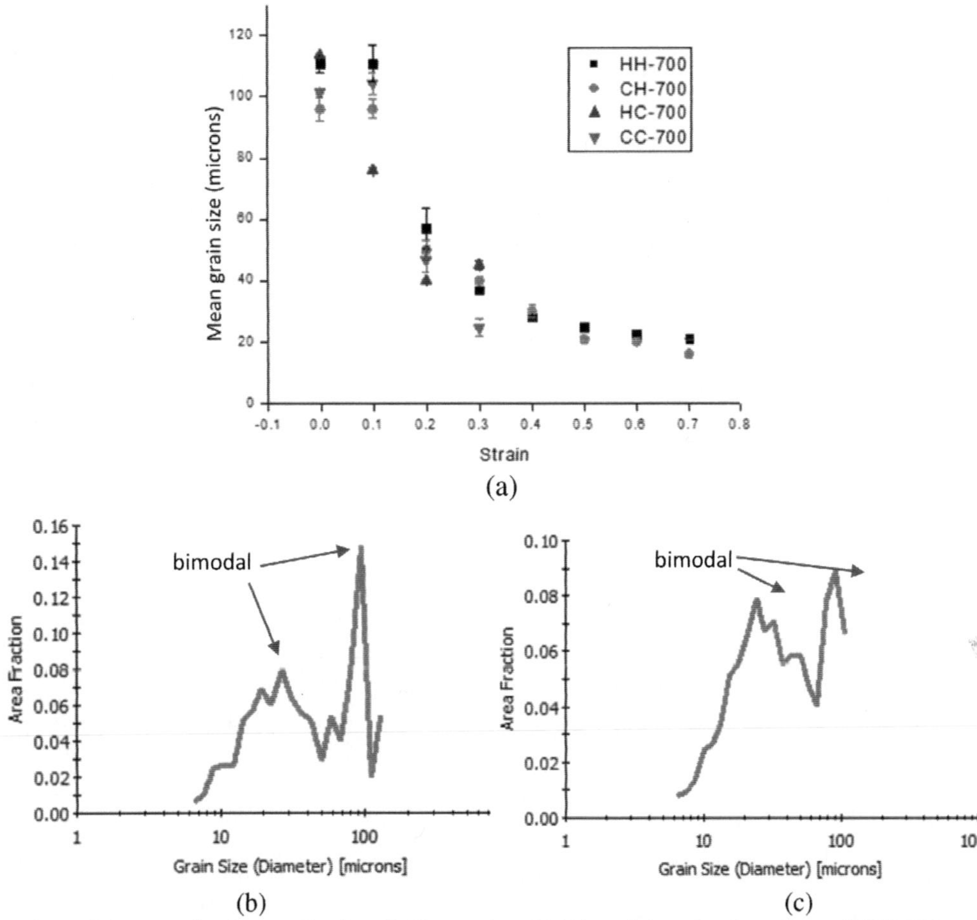

Figure 3: (a) Mean grain size vs Strain of all samples (b) Bimodal grain size chart of CH-700 and (c) CC-700 sample at 0.2 strain

From figure 7 (e) in sample CH-700 (cast hot rolled), the stored energy starts to drop from 0.4 and gets constant after 0.5 strain level. This study shows us that the critical deformation/ strain required for cast hot rolled samples is in between 0.4 and 0.5 for grains to get fully recrystallized with post-annealing process (at 700°C). After 0.5 strain, the deformation gets constant, which denotes that concurrent recovery at such high strains and temperature probably did not increase the stored energy (with deformation) [6]. In sample CC-700 (cast cold rolled), the stored energy at 0.3 strain is enough for grains to get fully recrystallized. Study of KAM maps concluded that the stored energy is higher in cast samples as compared to homogenized samples. Due to high stored energy, the grain refinement is more at higher strains in cast samples as the mean grain size is reduced to 16 microns (at 0.7 strain). Whereas, in homogenized samples the stored energy is little (compared to cast samples) so the mean grain size drops to 20 microns at a maximum strain of 0.7.

3.3. Grain orientation spread

A suitable value of GOS is required to differentiate the deformed from the recrystallized regions in the sample. The GOS value taken was less than 3^0 and grain size greater than $9\mu m^2$ [7-10]. In figure 8(a), the decrease in grain size with strain is due to recrystallization. Fraction of GOS (<2°) in sample HH-700 is up to 0.8 at all strain levels (except at 0.2 which is an outlier) whereas in sample CH-700, the fraction of GOS (<2°) reaches to 0.8 after 0.5 strain. The critical deformation required for recrystallization phenomenon to occur was studied. The minimum strain required is 0.5 for grains to get fully recrystallized in CH-700 (cast hot rolled) sample. Whereas in CC-700 (cast cold rolled) sample, the minimum strain required is 0.3 for grains to get fully recrystallized. It can also be concluded that the recrystallization behavior depends on the initial microstructure of the material. Before deformation, the homogenized samples had fully strain-free (recovered) microstructure but in cast samples the microstructure was not strain-free. So, in homogenized samples with post-annealing process (at 700°C), grains are fully recrystallized at all strains (except few strains which are outliers as explained earlier). In cast hot rolled sample, grains start to get fully recrystallized after 0.5 strain. This is probably because of concurrent recovery occurred during hot rolling at and beyond 0.5 strain [6]. Whereas in cast cold rolled sample the stored energy is too high that even at 0.3 strain the grains get fully recrystallized.

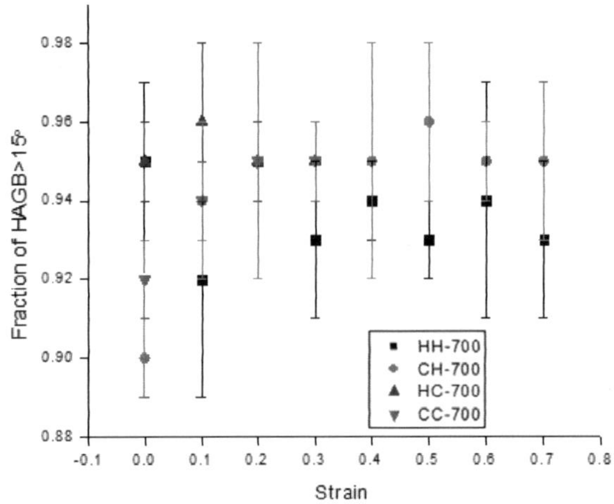

Figure 4: Fraction of HAGB>15° vs Strain of all samples

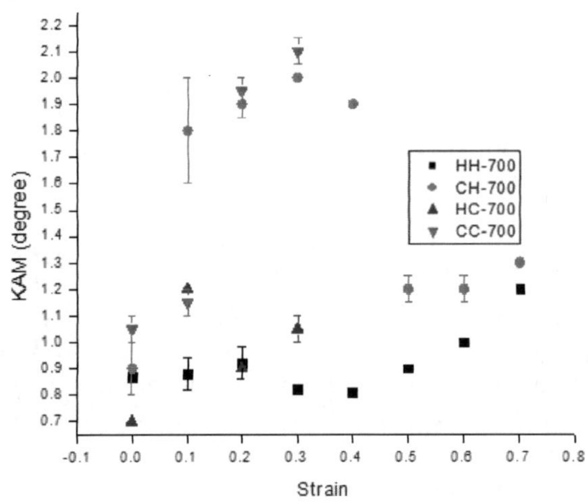

Figure 5: KAM vs Strain of all samples. The threshold value taken for misorientation is maximum of 5° with nearest 3rd neighbor.

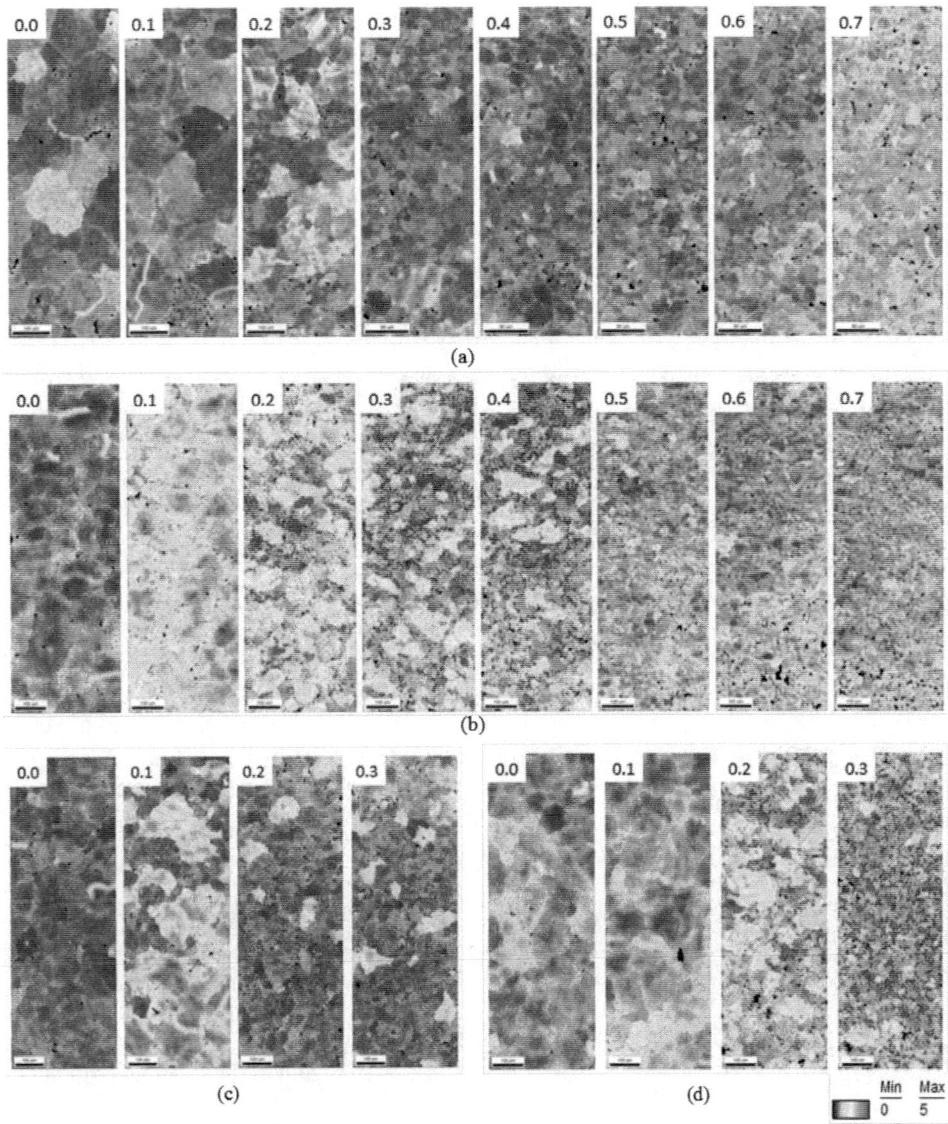

Figure 6: KAM maps of (a) HH-700 (b) CH-700 (c) HC-700 (d) CC-700 samples at all strains (0-0.7)

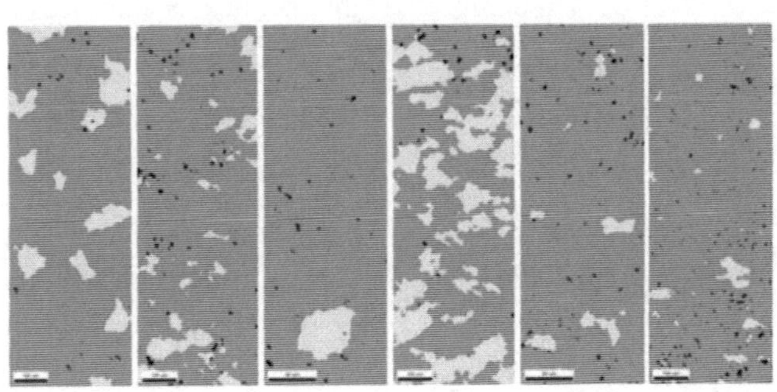

Figure 7: (a) GOS (<2°) maps of all samples at different strains

4. Conclusions

In CH-700 (cast and hot rolled) sample, the recovery and recrystallization phenomenon overlapped up to 0.4 strain wherein partial recrystallization occurred. It is anticipated that concurrent recovery, at and above 0.5 strain, diminished the increase in stored energy with deformation in sample CH-700. The fraction of GOS (<2°) increases to 0.8 after 0.5 strain and remains constant with further deformation. Values of KAM and the fraction of GOS (<2°) remain constant after 0.5 strain and this denotes that there was no further stored energy in the structure after 0.5 strain. Whereas in cast cold rolled samples, the fraction of GOS (<2°) is greater than 0.8 at 0.3 strain. The stored energy in the cast cold rolled sample at 0.3 strain is about double compared to the cast hot rolled sample. In the homogenized hot rolled and cold rolled samples, the recrystallization fraction is greater than 0.8 at all strain levels, except one, which is considered as an outlier. There is no strong evidence for the small drop in the fraction of GOS (<2°). However, as explained it could be anticipated that due to shear deformation during hot rolling (at that strain), the critical deformation (required for recrystallization phenomenon) did not occur.

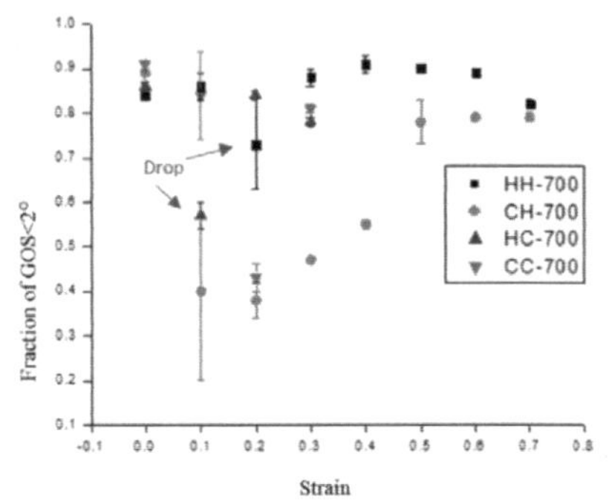

Figure 7: (b) Fraction of GOS (<2°) vs Strain. The threshold value taken is maximum misorientation of 2° with grain size >9μm^2. Orange color represent the recrystallized fraction of grains

References

[1] Nyberg EA, Burkes DE, Joshi VV, and Lavender CA 2015 'The microstructure of rolled plates from cast billets of U10Mo alloy' PNNL-24160 (March 2015) Page 15-16

[2] Meyer MK, Hofman GL, Ray AE 1998 'Design of high density gamma-phase uranium alloys for LEU dispersion fuel applications' Presented at: The 1998 International Reduced Enrichment for Test Reactor Conference Sao Paulo, Brazil, Page 1-3

[3] M.K. Meyer, G.L. Hofman, Jong-Man Park, Yusong-Ku 'Observation on the Irradiation behavior of U-Mo Alloy dispersion fuel' Presented at: The 2000 International Meeting on Reduced Enrichment for Research and Test Reactors (October 2000) Page 1-3

[4] Moussa C, M Bernacki, R Besnard and N Bozzolo 2015 *IOP Conf. Series: Materials Science and Engineering* **89** 012038

[5] Kamaya M 2011 *Ultramicroscopy* **111** 1189-99

[6] Humphreys FJ and M. Hatherly 2004 'Recrystallization and related annealing phenomena' Elsevier, Pages 76, 389, 415-27

[7] Alvi MH, SW Cheong, H Weiland and AD Rollett 2003 'Microstructural evolution during recrystallization in hot rolled Aluminum Alloy 1050' Metallurgical Modeling For Aluminum Alloys - Proceedings From Materials Solutions Conference: 1st International Symposium On Metallurgical Modeling For Aluminum Alloys. 191-197

[8] Field DP, LT Bradford, MM Nowell, and TM Lillo 2007 *Acta mater* **55** 4233-41

[9] Field DP 1995 *Matls. Sci Eng.* A **190** 241-6

[10] Schneider CA, Rasband WS, and Eliceiri KW 2012 Nat Methods 9 671-5

IOP Conf. Series: Journal of Physics: Conf. Series **1270** (2019) 012030 doi:10.1088/1742-6596/1270/1/012030

The effect of Niobium on Austenite Evolution during Hot Rolling of Advanced High Strength Steel

A Smith[1], F Vercruysse [2], R Petrov [2] and P Verleysen [2]

[1]Rina Consulting – Centro Sviluppo Materiali SpA, Rome, Italy
[2]Ghent University, Ghent, Belgium

ali.smith@rina.org

Abstract. Two AHSS steels (0.2C-2.4Mn-1.5Si with/without Nb+Cr addition) were subjected to pilot reversing mill hot rolling in 5 passes varying the finish rolling and coiling temperature. After coiling simulation, the final microstructures were characterised by the EBSD technique measurements. One sample was analysed to determine the presence and the level of Nb precipitated by TEM and chemical extraction. The austenite evolution was inferred from the measured transformation textures and reconstructed final austenite texture and microstructure. Further insights were obtained by comparing the results to a semi-empirical austenite evolution model, which predicted recrystallization behaviour and level of Nb precipitation. From the combined experimental and modelling work it was shown that Nb addition led to recrystallization inhibition via partial strain induced precipitation for this high Mn and high Si steel. Applying the austenite evolution model to a more realistic industrial hot strip rolling schedule revealed earlier Nb precipitation during hot rolling and higher strain accumulation with a finer grain size. The final Nb level in solution was lower for the industrial schedule, although a significant amount was not precipitated (two thirds of the total).

1. Introduction

Austenite evolution during hot rolling has been intensively studied and modelled in the last decades for plain carbon, (C-Mn) and low carbon microalloyed (Nb,V,Ti) steels [1,2]. Recently there has been explosive growth in the development of advanced high strength steels (AHSS) for the automotive sector [3]. Since these steels are mainly produced as thin sheets following cold rolling and heat treatment, the microstructural changes occurring during hot rolling have received much less attention. Nevertheless, understanding and control of austenite evolution in AHSS during hot rolling is of importance, to ensure e.g. cold rollability and to optimise the microstructure for processing after cold rolling [3]. The latest AHSS chemistries employ higher carbon contents and significant levels of Mn (e.g. up to 3 wt%) and Si (e.g. 1.5 wt%), with or without microalloying [4]. These rich alloying chemistries may significantly affect austenite evolution in terms of recrystallization behavior and Nb strain induced precipitation.

The present work explores the role of niobium addition on austenite evolution in laboratory rolling of AHSS steels. The Nb state, (i.e. as solute or as strain induced precipitate) and austenite microstructure were inferred from a combination of texture measurements, and prior austenite reconstruction, precipitation characterization, and microstructure evolution modelling. Finally, based on the results obtained, the validated austenite evolution model was applied to an example industrial strip rolling schedule.

2. Experimental methods and models used

Two advanced high strength steel chemistries were cast into 50 kg ingots using vacuum induction melting. The measured compositions are shown in table 1.

Table 1. AHSS steel chemistries. Element amounts in wt% with balancing Fe.

	C	Mn	Si	Cr	Nb	Al	N
Base	0.20	2.4	1.5	0	0	0.02	0.002
Nb	0.20	2.4	1.25	0.3	0.025	0.02	0.002

The ingots were first rough rolled into slabs of 20 mm thickness. After reheating at 1250°C for 1 hour, hot rolling was performed down to a strip thickness of 5 mm in 5 passes (percent reductions of 30-25-25-20-20). The temperature of the final pass was varied i.e. finishing temperatures (FT) of 920°C and 860°C were studied. Temperature was measured during rolling via a thermocouple inserted at the mid thickness of the slab. Since a reversing mill was used, the interpass times were relatively long i.e. in the range 10-20s. After finish rolling, the strips were subjected to air or water cooling to the target coiling temperature (CT) i.e. 620°C or 540°C. The coiling was simulated by furnace cooling the strips at a rate of 20°C/h.

Samples were taken parallel to the rolling direction and prepared for EBSD examination via grinding and electro polishing. Orientation distribution functions were determined to obtain the bulk texture. In addition, samples were analysed after mechanical polishing via optical microscopy (Nital/Le Pera etch). For the Nb steel, analysis of precipitates was carried out for one condition only (FT = 920°C and CT = 540°C), using both TEM replicas and chemical extraction followed by inductively coupled plasma optical emission spectroscopy (ICP-MS).

For modelling studies, the EBSD measurements were used to reconstruct the prior austenite grain structure and its texture, using the APRGE software [5]. In addition, the austenite microstructure evolution during hot rolling was simulated using an in-house model [6]. This model is based on a combination of classical semi-empirical recrystallization models [1,2]. Strain induced precipitation was treated using the original physical model of Dutta and Sellars [7], extended it to predict the fraction of niobium in solution according the approach of Pereda et al. [8]. The interaction of recrystallization with precipitation was treated in the most simple fashion i.e. assuming that once precipitation had started that no further softening (recrystallization) could take place during the remaining passes [1]. Finally, the austenite evolution model should be adapted consider the effect of steel chemistry, since the AHSS compositions are richer than those used to develop the original equations. This was taken into account in the following way. Firstly, static recrystallization was modelled using the activation energy (composition dependent) i.e. in a linear fashion for Mn and Si solute drag according to Medina [2]. To extrapolate this model to high Si levels, a modification was made considering the non-linear effect of Si seen in TRIP steels by Suikkanen et al. [9]. For Mn no modification to the Medina equation was made i.e. linear behavior was assumed. This is considered justified, considering the much weaker effect of Mn on recrystallization kinetics in microalloyed steels and the smaller extrapolation i.e. 2.4 wt% Mn studied here, versus 1.2-1.4 wt% Mn steels mainly used to generate the Medina model equation. Another influence of Mn and Si is on the Nb strain induced precipitation kinetics [1]. Thus Si can accelerate precipitation and Mn retards it. Although the Siciliano model takes this into account by correction factors, the Si and Mn levels used were significantly different to those of AHSS chemistries. Thus for the present study, the original Dutta Sellars model was used, adjusting the pre factor for precipitation start time if necessary.

3. Results and Discussion

3.1. Optical microscopy and TEM

Microstructures obtained for air/water cooling with coiling at 620°C, were ferrite-pearlite structures for both steels and finishing temperatures. For coiling at 540°C, a bainitic structure was obtained for the Nb steel with air/water cooling. The base steel with reduced hardenability (no Cr or Nb) led to a partially bainitic structure with water cooling. Figure 1 shows example microstructures for the steels.

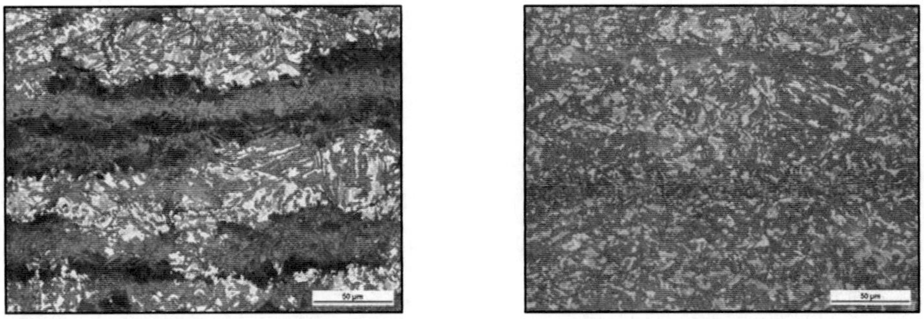

Figure 1 Microstructures for base (left) and Nb (right) steels for finish rolling temperature of 860°C followed by water cooling and coiling at 540°C.

For the Nb steel an example TEM replica micrograph is shown by figure 2 below.

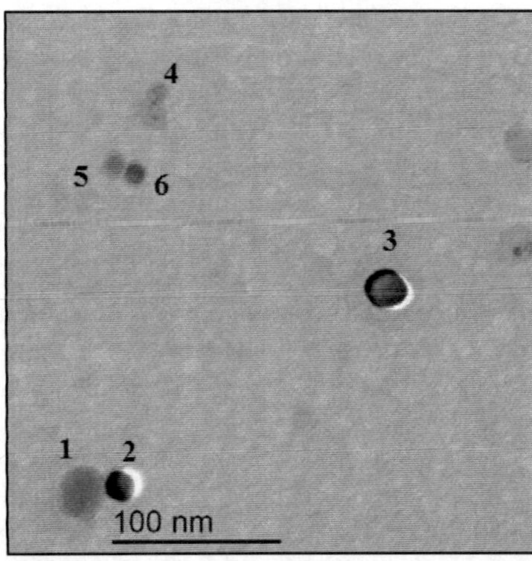

Particle	Size (nm)	Nb wt% from EDS
1	32	91
2	18	91
3	23	88
4	11	83
5	10	87
6	11	88

Figure 2 Nb steel TEM replica after hot rolling with finish rolling temperature of 920°C followed by air cooling and coiling at 540°C.

As can be seen in figure 2, TEM analysis revealed the presence of Nb (C,N) precipitates with a size range of 10-32 nm. For the conditions given in figure 2 the chemical extraction results revealed that 0.020 wt% Nb remained in solution i.e. 0.005 wt% or 20% of the Nb was present as Nb(C,N) precipitates.

3.2. EBSD results and simulations

EBSD scans are shown by the orientation distribution functions (ODFs) in figure 3. The results were compared to the theoretical textures expected for different austenite states [10,11].

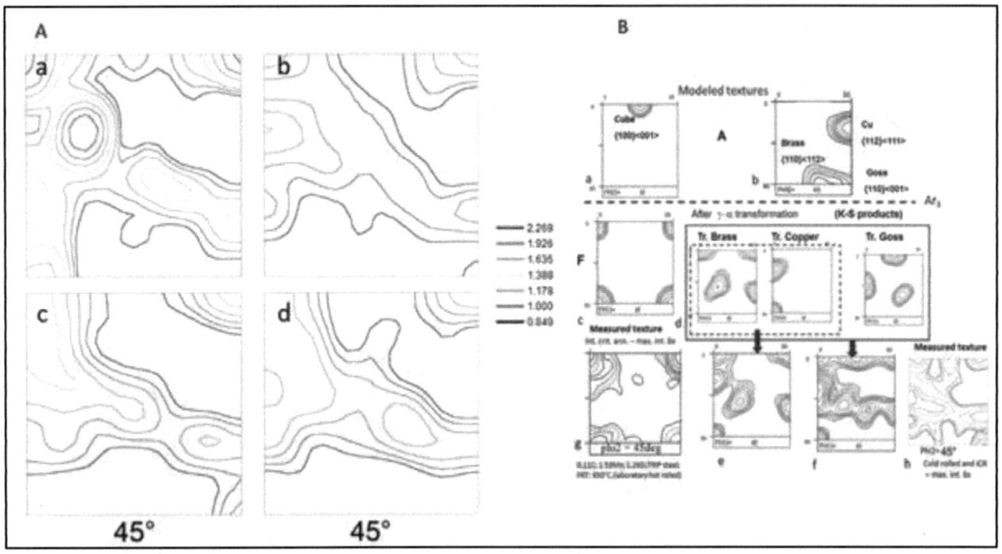

Figure 3 (A) ODF's of hot rolled samples of the base alloy with CT 620°C and FT 920°C (a) and FT 860°C (b) as well as similarly for the Nb alloy FT 920°C (c) and FT 860°C (d). (B) Theoretical formation of transformation texture in steels [11].

As can be seen from figure 3, for both steels and finishing temperatures, a clear transformation fiber texture is observed. This texture would suggest that before transformation deformed austenite was present i.e. having a texture with copper and brass orientations. In addition, the rotated cube texture component is present in all cases (strong and weak). This component may indicate the presence of recrystallized austenite i.e. having a texture with cube orientation. Unfortunately, the rotated cube component can also be formed from deformed austenite with a brass orientation. However, since the measured rotated cube component for the base steel at 920°C finishing temperature was significantly higher than that for the Nb steel, at the same finishing temperature (maximum intensity of 2.3 versus 1.4), one can infer a greater final recrystallized fraction for the base steel. For the lower finishing temperature, the situation is more complex since the rotated cube component for both steels is weak. Rather, it can be seen that the intensity of the general deformation fiber is more intense for the Nb steel (for example transformed Goss component). Thus it can be tentatively concluded that the final austenite recrystallized fraction in the base steel was also higher for finishing at 860°C.

To gain further insights, the austenite texture was reconstructed using the APRGE software, applied to the measured textures at the lower finish rolling temperature. Since this software requires the presence of diffusionless transformation products, the measured EBSD textures from samples water cooled and coiled at 540°C were used as the input. The results are shown in figure 4.

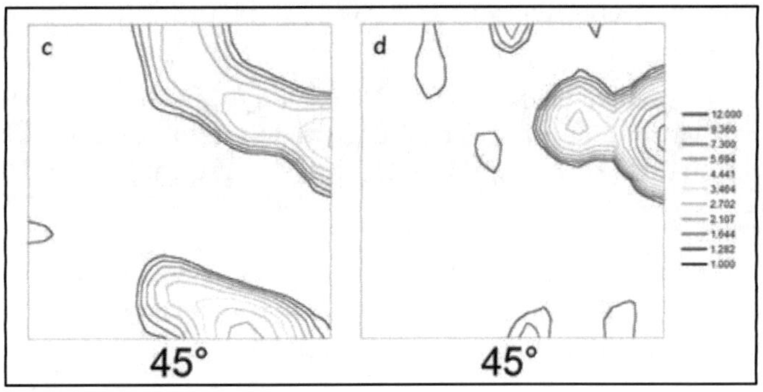

Figure 4 ODF's of reconstructed prior austenite structure just before transformation of the Base (c) alloy and Nb alloy (d), both in FT860°C-water cooling-CT540°C condition.

As can be seen, the base steel austenite texture exhibits copper, Goss and cube components, indicating a partially recrystallized state, whilst the Nb steel reveals a strong copper component i.e. indicating deformed austenite. Combining these results with those from figure 3 clearly illustrate the role of Nb in the rich AHSS chemistry, namely that for both finishing temperatures, Nb addition has led to more significant inhibition of recrystallization.

3.3. Recrystallization modelling of lab scale hot rolled strips

From the previous sections, several important questions arise i.e. is the Nb effect observed due to solute drag or to strain induced precipitation? Are the observed precipitates formed before or during/after transformation? Can recrystallization modelling of lab rolling reproduce the same trends as those inferred from the texture results?

Figure 5 below shows the results of the modelling activity for the cases of air cooling after finishing to the coiling temperature of 620°C. Only a summary is shown i.e. final total retained strain (indicating the degree of austenite deformation retained after 5 passes), and the final recrystallized fraction (X). Note that similar trends were obtained if water cooling after finishing to coiling at 540°C was used.

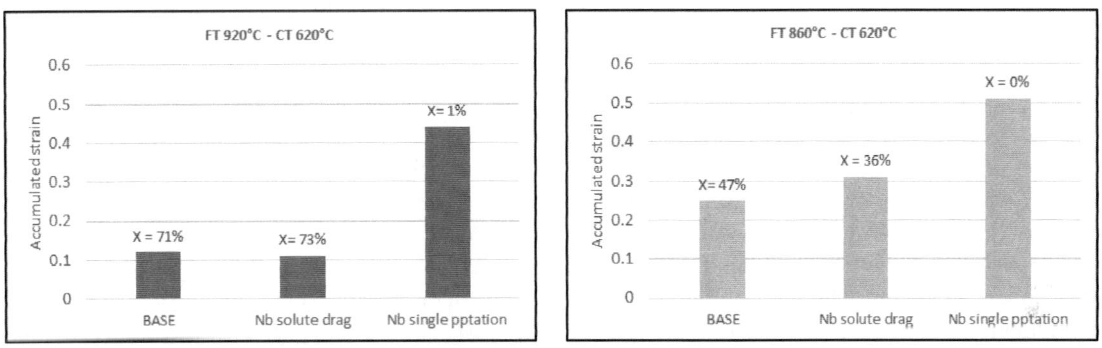

Figure 5 Recrystallization modelling results summary for base (left) and Nb steel (right).

As shown in figure 5, for both finishing temperatures, the accumulated strain values between base and Nb steels are very similar when the Nb effect on recrystallization inhibition is considered only as solute drag. (Note that the slightly lower strain for Nb solute drag at the higher temperature is due to the actual temperature of the last passes being slightly higher than for the base steel). On the other hand, when Nb strain induced precipitation was allowed to occur, (started between the last two passes F4-F5), the accumulated strain was significantly higher, i.e. recrystallization was completely supressed for both finishing temperatures starting between F4 and F5. In addition, figure 5 also shows the effect of finishing temperature on recrystallization kinetics i.e. for all three cases the accumulated strain in austenite was higher for the lower temperature. Comparing these simulations to the trends from the texture analysis reveals good qualitative agreement, i.e. partial recrystallization of the base steel for both finishing temperatures, and a much higher accumulated strain in the Nb steel, (considering strain induced precipitation rather than a solute drag effect).

Concerning the predicted final amount of Nb in solution for when strain induced precipitation was active, values of 0.018-0.019 wt% were obtained for all schedules, including the schedule where TEM and chemical extraction analysis was made afterwards. Thus, good agreement was obtained with chemical extraction results with no modification to the original Dutta Sellars model coefficients. This suggests that the opposite influence of Si and Mn on precipitation kinetics for the current steel chemistry would tend cancel each other out.

3.4. Recrystallization modelling of 0.025wt% Nb steel for an industrial hot strip schedule

The model was extended to an industrial strip rolling schedule of 7 finishing passes (with final pass at 900°C) taken from literature [12]. Table 2 shows the comparison of results with the lab schedule.

Table 2. Recrystallization modelling results for 0.2C-2.4Mn-1.25Si-0.025Nb-0.3Cr steel.

Schedule	Total reduction ratio	Interpass times (sec)	Final Nb level in solution (wt%)	Final residual strain	Final fraction recrystallized	Final austenite grain size (microns)
Lab	4	10-20	0.019	0.44-0.51	0	15
Industrial	9	1-4	0.016	1.18	0	7

For the industrial schedule with much higher total reduction, Nb precipitation occurred earlier (F2-F3 compared to F4-F5 for lab rolling), this led to greater strain accumulation, a finer final austenite grain size, and a lower amount of Nb remaining in solution. Interestingly the model predictions suggest that a significant amount of Nb remains in solution (two thirds of 0.025 wt%) for the industrial schedule. This information is of great interest when considering optimization of AHSS mechanical properties. Thus Nb remaining in solution is available for downstream processes e.g. assisting steel hardenability during transformation or for precipitation hardening after cold rolling and annealing.

4. Conclusions

From a combination of precipitate analysis, texture analysis, and recrystallization modelling adapted to AHSS chemistries, a picture was built up of the austenite evolution during laboratory hot rolling of two AHSS steel compositions (with and without 0.025 wt% Nb). The base steel final austenite microstructure was partially recrystallized, whilst for the Nb steel a mainly deformed austenite structure was obtained. For both steels, a lower finish rolling temperature increased the amount of retained strain. For the Nb steel, results were best explained by strain induced precipitation rather than by solute drag. Kinetics of strain induced precipitation were in agreement with classical model predictions, suggesting that Mn and Si effects cancelled each other out. Finally, extending the validated model to an industrial strip rolling schedule revealed earlier Nb strain induced precipitation i.e. giving much more strain accumulation and a finer grain size. Nevertheless, a significant amount of Nb remained in solution, potentially available for downstream processes.

Acknowledgments

This research has received funding from the European Union's Research Fund for Coal and Steel under grant agreement N°709755, OptiQPAP project.

References

[1] Siciliano F and Jonas J 2000 *Metall. Mater. Trans. A* **31A** 511
[2] Medina S and Quispe A 2001 *ISIJ Int.* **41** 774
[3] Poliak E and Bhattacharya D 2014 *Mater. Sci. Forum* **783-786** 3
[4] Smith A, Vercruysse F, Petrov R, Verleysen P and Linke B 2018 *Mater. Sci. Forum* **941** 106
[5] Cayron C 2007 *J. Appl. Cryst.* **40** 1183
[6] Smith A, Lubrano M, Di Schino A and Guindani A 2014 Metall. Ital. **n. 3/2014** 43
[7] Dutta B and Sellars C 1987 *Mater. Sci. Technol.* **3** 197
[8] Pereda B, Rodriguez-Ibabe J and Lopez B 2008 *ISIJ Int.* **48** 1457
[9] Suikkanen P, Lang V, Somani M, Porter D and Karjalainen L 2012 *ISIJ Int.* **52** 471
[10] Ray R, Jonas J, Butron-Guillen and Savoie J 1994 *ISIJ Int.* **34** 927
[11] Petrov R, Sidor J and Kestens L 2016 *Encyclopedia of Iron, Steel, and Their Alloys* ed Calos R and Totten G (New York: Taylor and Francis) pp 70-99.
[12] Poliak E 2012 *Mater. Sci. Forum* **715-716** 782

IOP Conf. Series: Journal of Physics: Conf. Series **1270** (2019) 012031 doi:10.1088/1742-6596/1270/1/012031

Effect of Thermomechanical Parameters on Grain Growth and Recrystallization during Grain Boundary Engineering of Austenitic Stainless Steel

Shun Tokita[1], Hiroyuki Kokawa[2] and Yutaka S. Sato[3]

[1] Joining and Welding Research Institute, Osaka University, Osaka, Japan
[2] Department of Materials Processing, Graduate School of Engineering, Tohoku Univeristy, Sendai, Japan (Now at Welding and Laser Processing Institute, Shanghai Jiao Tong Univeristy, Shanghai, China)
[3] Department of Materials Processing, Graduate School of Engineering, Tohoku University, Sendai, Japan

E-mail: tokita@jwri.osaka-u.ac.jp

Abstract. Grain boundary engineering has been attracting attention as an effective method to prevent intergranular corrosion of austenitic stainless steel and Ni based alloys. It has been considered that grain growth and recrystallization play important role in evolution of grain boundary character distribution during thermomechanical process. However, systematic researches to examine the effect of thermomechanical process parameters on grain growth and recrystallization have not been performed. In this study, grain boundary character distribution of 304 austenitic stainless steel after thermomechanical process with various parameters was analysed by electron backscatter diffraction (EBSD). Grain boundary character distribution was mainly affected by the reduction ratio of cold rolling. Abnormal grain growth was observed in the specimen with small reduction ratio (3%). Length ratio of coincident site lattice (CSL) boundaries was drastically increased to 86% in the 3% cold rolled and annealed specimens from 67% in the base material. On the other hand, normal grain growth was observed in thermomechanical processed specimens with slightly higher reduction of cold rolling (5%). In these specimens length ratio of CSL boundaries did not exceed 80%, which is required to disconnect the random boundary networks and to improve intergranular corrosion resistance effectively. Detailed analysis of grain boundary character distribution has shown that disconnection of random boundary networks was achieved by formation of annealing twins during abnormal grain growth.

1. Introduction

Austenitic stainless steels have been widely used in many industrial applications because they have good mechanical properties and high resistance to general corrosion. However, weld decay, which is severe intergranular corrosion in weld heat affected zone, often occurs at grain boundary during the welding process. Intergranular corrosion is caused by the formation of chromium carbides at grain boundaries.

Grain boundary engineering [1, 2] has been attracting attention as an effective method to prevent grain boundary degradation such as intergranular corrosion. Grain boundaries can be categorized into coincidence site lattice (CSL) boundaries and the other random boundaries. It has been reported that the CSL boundaries have higher intergranular corrosion resistance than random boundaries because precipitation of chromium carbide was prevented in the CSL boundaries [3, 4]. Utilizing these properties, resistance to grain boundary degradation can be improved by an increase in the frequency of CSL boundaries and disconnection of random boundary networks. Recent researches has reported an increase in the frequency of CSL boundaries and an improvement of intergranular corrosion resistance of 304 austenitic stainless steel by application of grain boundary engineering [5-9]. Suppression of grain boundary degradations such as grain boundary cracking, segregation, and intergranular stress corrosion cracking has been reported in the previous study [10].

The thermomechanical process is often used to optimize the grain boundary character distribution (GBCD). Considering the practical use of the thermomechanical process for GBE, the optimization mechanism of GBCD needs to be revealed to improve the energy efficiency and reliability as a microstructural control process. However, the mechanism of the evolution of grain boundary character distribution during the thermomechanical process is not fully understood. In this study, systematic examination of thermomechanical process parameters and measurement of intergranular corrosion resistance were conducted to reveal the effect of thermomechanical parameters on the evolution of GBCD and intergranular corrosion resistance.

2. Experiments

2.1. Specimen preparation

The material used in this study was 304 austenitic stainless steel with the chemical composition of 18.36 Cr, 8.15 Ni, 0.58 Si, 1.27 Mn, 0.040 C, 0.032 P, 0.001 S (wt%). The solution heat treatment was conducted in electric furnace at air atmosphere in 1323 K for 0.5 h and subsequently quenched in cold water. The solution heat treated material will be denoted as the base material in this study. The thermomechanical process for grain boundary engineering can be divided into strain annealing and strain recrystallization based on their parameters, such as reduction ratio of cold rolling and annealing time [11, 12]. In this study, strain annealing process was conducted because the drastic increase of the frequency of CSL boundaries of 304 and 316 austenitic steel was reported in the previous study [5, 6]. Specimens were deformed by cold rolling, resulting in the reduction of the thickness of 3, 5 and 10%. The cold rolled specimens were cut into small chips, 5W × 4L × 2T mm3 in size. The annealing process was performed in an electric furnace with an annealing temperature of 1220, 1280, 1340 K and specimens were subsequently quenched in cold water.

Thermomechanical processed specimens were mechanically polished by the water-abrasive paper and subsequently polished with 1 μm Al2O3 particles. Electropolishing was conducted in 10% perchloric acid ethanol at 273 K at a potential of 25 V to obtain the suitable surface conditions for EBSD observation.

2.2. EBSD observation

The GBCD analysis was conducted with a field emission scanning electron microscopy HITACHI S-4300SE equipped with TSL OIM system. EBSD observation with a step size 5 μm was conducted in the 2 mm x 2 mm observed area, which was perpendicular to the rolling direction. The cleanup procedure in the OIM analysis software was performed using the neighbor confidence index (CI) correlation method with a minimum CI value of 0.1. Grain definition with a minimum grain size of 5 analysis pixels and a minimum misorientation angle of 15 degree was set for the grain dilation clean up method and other grain-based analysis such as grain orientation spread analysis.

Brandon's criteria [13] was adopted to define the maximum tolerance angle of CSL boundaries. The relationship between grain boundary character and intergranular corrosion resistance have been examined in recent studies. [5, 14, 15] Σ3 CSL boundary has been widely recognized as excellent corrosion resistant boundary while other Σ CSL boundaries do not always show the high corrosion resistance. On the other hand, other Σ CSL boundaries sometimes prevent propagation of intergranular corrosion, because the previous study showed that non-corroded or non-sensitized Σ9, 13b, 17a, 29a boundaries disconnected corroded random boundaries [5, 15]. In this study, the grain boundaries in EBSD data were roughly categorized into CSL boundaries with low-Σ value (Σ ≦29) and the other random boundaries, because this study focused on the relationship between thermomechanical process parameters and the general tendency of the grain boundary character distribution. The frequency of CSL boundaries was measured as a length ratio of the CSL boundaries to a total length of all low angle and high angle boundaries.

3. Results and discussions

3.1. Effect of thermomechanical parameters on the grain boundary character distributions

The GBCD maps of the thermomechanical processed specimens with 3, 5, 10% of cold rolling and annealing in 1220, 1280, 1340 K are shown in figure 1. Random boundary and CSL boundary ware depicted as black and green lines in the GBCD maps. The GBCDs of the thermomechanical processed specimen was mainly affected by the reduction ratio in cold rolling. A significant increase of CSL boundaries during grain growth and annealing twinning was observed in the 3% cold rolled and annealed specimens. The CSL boundaries that replaced and disconnected random boundaries were observed. The random boundary networks were effectively disconnected in the specimen.

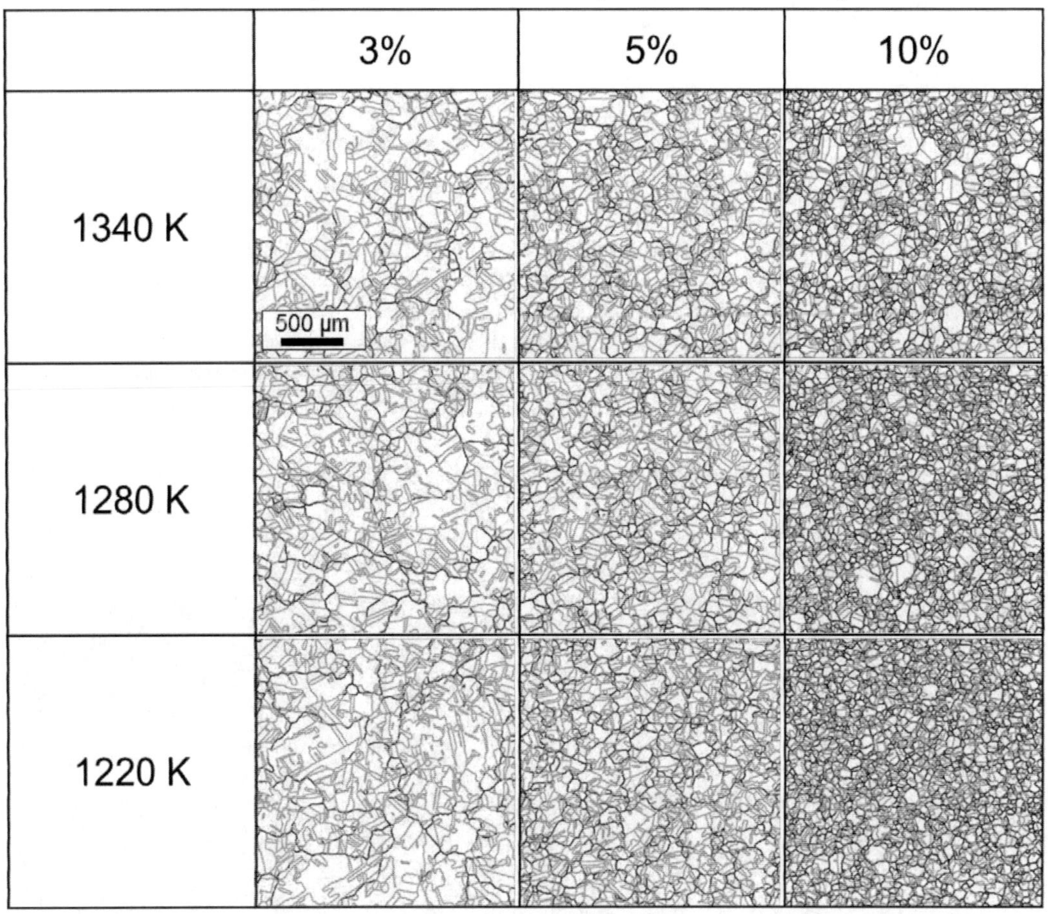

Figure 1 Grain boundary character distribution map of the thermomechanical processed materials.

The 5% cold rolled and annealed specimen also showed an increase of CSL boundaries, but most of CSL boundaries were distributed inside of random boundary grains and significant disconnection of random boundary networks was not observed. The GBCDs of the specimens with 10% cold rolling did not change significantly from that of the base material. Although the grain size of the specimens with 10% cold rolling was slightly increased in higher annealing temperature condition, a significant effect of the annealing temperature on the GBCD was not observed in figure 1. Annealing time for GBCD change in high annealing temperature conditions was shorter than that in low annealing temperature conditions. These results showed that the processing time for the thermomechanical process can be shortened by increasing annealing temperature in this system.

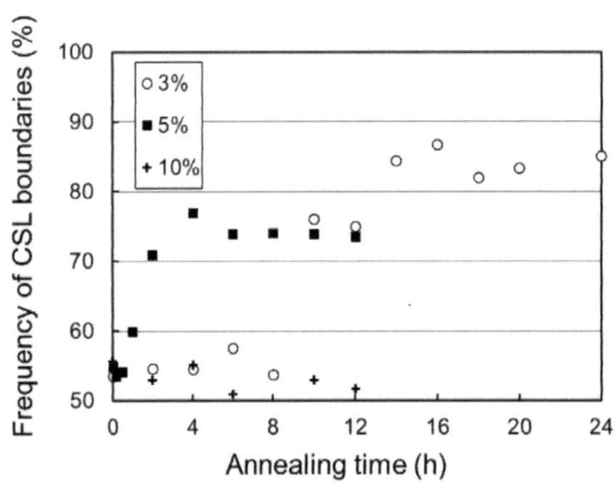

Figure 2 Frequency of CSL boundaries of the thermomechanical processed materials.

Figure 2 shows the frequency of CSL boundaries of thermomechanical processed specimens with different annealing times. The frequency of the CSL boundaries in the base material was 51%. The frequency of the CSL boundaries of the 3% cold rolled specimen increased until 16 h of annealing and then stabilized above 80%. The frequency of CSL boundaries of the 5% cold rolled specimen increased within 2 h and the highest frequency was 78%, which was lower than that in the 3% cold rolled specimen. The 10% cold rolled specimen did not show the increase in the frequency of CSL boundaries. It has been reported that the frequency of CSL boundaries higher than 80% is effective to disconnect three-dimensional random boundary networks [16] and to improve intergranular corrosion resistance [6]. Therefore, the improvement of intergranular corrosion resistance was expected in the thermomechanical processed specimens with 3% cold rolling.

In order to discuss the reason why the GBCDs of the thermomechanical processed specimen were mainly affected by reduction ratio in cold rolling, grain orientation spread (GOS) maps of the

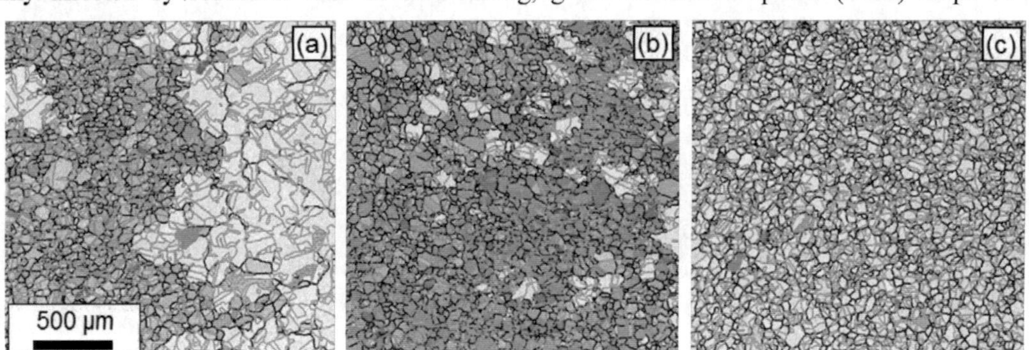

Figure 3 Grain orientation spread maps of (a) 3%, (b) 5% and (c) 10% cold rolled and annealed materials.

specimens are shown in figure 3. GOS represents the average deviation between the orientation of each analysis pixel in a grain and the average orientation of the grain. Grains with higher GOS value, those are depicted as darker red in figure 3, are considered to have larger strain. The grains with lighter red were considered as growing grains by strain induced grain growth. Large growing grains with lower GOS value were observed in the 3% cold rolled specimen, as shown in right side of the observed area in figure 3 (a). Previous research has shown the frequency of CSL boundaries increased through the annealing twinning during abnormal grain growth [17]. It is considered from this result that microstructural change for optimization of the GBCD could be successfully obtained in the 3% cold rolled and annealed specimen. In the 5% cold rolled and annealed specimen, size of the growing grains was relatively smaller than that in the 3% cold rolled specimen. It is estimated that impingement of growing grains should occur more frequently in this condition. Therefore, abnormal grain growth could not occur efficiently in this condition because impinged grains stopped growing. This might be the reason why the frequency of CSL boundaries could not be increased adequately in the 5% cold rolled and annealed specimen in figure 3. In the 10% cold rolled specimen, homogeneous distribution of small growing grains with low strain was observed. This result shows recrystallization occurred in this condition. Optimized GBCD could not be obtained in this condition because recrystallized grains were mainly surrounded by random boundaries as shown in figure 1.

4. Conclusion

The GBCDs of thermomechanical processed specimens were mainly affected by the reduction ratio of cold rolling. The 3% cold rolled and annealed specimen showed the abnormal grain growth and drastic increase of CSL boundaries, while the 5% cold rolled and annealed specimen showed normal grain growth. It is suggested that increase in length ratio of CSL boundaries was achieved by the abnormal grain growth and frequently formation of annealing twins.

References

[1] Watanabe T 1984 *Res. Mech.* **11** 47
[2] Palumbo G, Lehockey E M and Lin P 1998 *JOM* **50** 40
[3] Palumbo G and Aust K T 1990 *Acta Met. Mater.* **38** 2343
[4] Kokawa H, Shimada M and Sato Y S 2000 *JOM* **52 7** 34
[5] Shimada M, Kokawa H, Wang Z J, Sato Y S and Karibe I 2002 *Acta Mater.* **50** 2331
[6] Michiuchi M, Kokawa H, Wang Z J, Sato Y, S and Sakai K 2006 *Acta Mater.* **54** 5179
[7] Kokawa H 2005 *J. Mater. Sci.* **40** 927
[8] Tokita S, Ikeshoji M, Yamada G, Kurihara K, Kokawa H, Sato Y S and Fujii H T 2015 *Sci. Technol. Weld. Join.* **20** 702
[9] Yamada G, Kokawa H, Yasuda Y, Tokita S, Yokoyama T, Sato Y S, Fujii H T and Tsurekawa S 2013 *Philos. Mag.* **93** 1443
[10] Watanabe T 2011 *J. Mater. Sci.*, **46 12** 4095
[11] Lin P, Palumbo G, Erb U and Aust K T 1995 *Scripta Met. Mater.* **33 9** 1387
[12] King W E and Schwartz A J 1998 *Scripta Mater.* **38 3** 449
[13] Brandon D G 1966 *Acta. Metall.* **14** 1479
[14] Jones R and Randle V 2010 *Mater. Sci. Eng.* **A527** 4275
[15] Bi H Y, Kokawa H, Wang Z J, Shimada M and Sato Y S 2003 *Scripta Mater.* **49** 219
[16] Frary M and Schuh C A 2005 *Philos. Mag.* **85** 1123
[17] Tokita S, Kokawa H, Sato Y S, Fujii H T 2017 *Mater. Charact.* **131** 31

Effect of Cr and C on microstructure evolution of medium carbon steels during friction stir welding process and their mechanical property

C Cheng [1], K Kadoi [1], S Tokita[1], H Fujii[1], K Ushioda[1] and H Inoue[1]

[1] Joining and Welding Research Institute, Osaka university, Osaka, Japan

Email: kadoi@jwri.osaka-u.ac.jp

Abstract. The aim of the present study was to investigate the effect of chromium and carbon in medium carbon steels on the microstructure evolution during FSW process and their mechanical property. The materials are dynamically recrystallized in austenite during FSW followed by phase transformation from austenite to ferrite, pearlite, bainite, and martensite. In the case of 0.2%C, the microstructure of the stir zone transited from pearlite and ferrite to bainite and martensite with the increase in chromium content. On the other hand, in the case of 0.4%C, the microstructure transited from pearlite to martensite with the increase in chromium content. The specimens of 0.2%C-4%Cr, 0.4%C-2%Cr and 0.4%C-4%Cr showed high tensile strength of over 1.5 GPa. In addition, even though the microstructures in the stir zones of the specimens are martensite, the elongation of 0.2%C-4%Cr was 31%while those of the others were less than 5%. TEM observation revealed that many tiny carbides precipitated in martensite of the stir zone of 0.2%C-4%Cr and micro twin formed in martensite in those of 0.4%C-2%Cr and 0.4%C-4%Cr. Thus, it was considered that auto-tempering occurred in the stir zone of 0.2%C-4%Cr during FSW process. The carbide precipitation in martensite is more effective to improve the tensile elongation compared with micro twin formation.

1. Introduction

Medium carbon steels have advantages of high strength with low cost. Therefore, they have a wide range of applications in industries. However, the weldability and the mechanical properties of the fusion-welded joint of the steels deteriorate with increasing in C content due to the formation of hardened phase such as martensite. Martensite with high carbon content is brittle and susceptible to cracking. Furthermore, the ductility of carbon steel also decreases due to the formation of brittle martensite during welding process [1].

It is known that chemical composition (alloying elements) and thermal cycle during welding corresponding to weld heat input mainly influence on the microstructure evolution and the morphology. In the case of the chemical composition, the alloying elements owing high affinity to carbon affect the phase transformation from austenite during welding. Chromium is one of the candidate elements since chromium enhances high hardenability and an attractive interaction with carbon (formation of chromium carbide) [2]. The ductility of martensite should also modify by the formation of the carbide [3,4]. On the other hand, in the case of the thermal cycle during welding, it is required to apply the proper welding process and to control the heat input and the peak temperature. Friction stir welding (FSW) process is a solid-state joining process with low peak temperature and low heat input [5]. Thus, the application of FSW process on medium carbon steels induces the modification of microstructure such as martensite phase compared to the fusion welding process. Besides, FSW process has also an advantage of refining austenite grains by dynamic recrystallization corresponding to work strain.

Even though there are considerable numbers of study on the application of FSW process to carbon steels, the influence of alloying elements in medium carbon steels on microstructural evolution during FSW is still under the discussion due to complicate phenomena such as the dynamic recrystallization of austenite and the phase transformation [6-8]. The aim of this study is to investigate the effect of chromium and carbon in medium carbon steels on the microstructure evolution during FSW process and their mechanical properties (tensile strength and hardness) and the microstructure.

2. Experimental method

Table 1 shows chemical compositions of specimens. Carbon and chromium contents were varied as 0.2 and 0.4 mass% and 2 and 4 mass%, respectively. The specimens were fabricated through vacuum induction melting under argon atmosphere and hot rolling. The specimen surface was machined after the hot rolling. The specimen geometry for FSW test was 180 - $200^{W} \times 45^{L} \times 3.3^{t}$ mm. The chemical compositions were measured with an optical emission spectrometer. Hereafter, the base composition of Fe-0.2%C-2%Cr is called 0.2C-2Cr for simplicity. The other specimens are also called 0.2C, 0.2C-4Cr, 0.4C, 0.4C-2Cr, and 0.4C-4Cr, respectively.

Stir in plate FSW were carried out on the base metal plates by using a loading controlled FSW machine. The tool was made of WC and the geometry was the shoulder diameter of 15 mm, the probe diameter of 6 mm, and the probe length of 3.1 mm as depicted in figure 1. FSW was performed under welding speed of 400 rpm and the rotation speed was 150 mm/min. The load was adjusted according to the base metal. The tool was tilted 3 degrees. The peak temperature during FSW under the conditions is considered above Ac3 temperature in order to investigate the phase transformation after fully austenitizing [6].

The cross-sectional microstructures of the specimens were analyzed through an optical microscope (OM), a field emission type scanning electron microscope (SEM, JEOL JSM-7001FA) and a transmission electron microscope (TEM, JEOL 2100F). The specimens were prepared for OM and SEM observations by chemical etching with 3% nital solution after final polishing with colloidal silica. The prior austenite grain size in the stir zone was measured by liner intercept method using the specimen etched by 1.5% picric acid. The specimens for TEM observation were prepared thinning by mechanical polishing and further thinned by electropolishing in a twin-jet polisher (Struers TenuPol-5).

Table 1 Chemical compositions of base metal (mass%)

	C	Cr	Si	P	S	Mn
0.2C	0.16	0.0	0.010	<0.01	<0.01	0.01
0.2C-2Cr	0.19	1.97	0.014	<0.01	<0.01	<0.01
0.2C-4Cr	0.20	4.01	0.017	<0.01	<0.01	<0.01
0.4C	0.39	0.0	0.010	<0.01	<0.01	<0.01
0.4C-2Cr	0.39	1.97	0.014	<0.01	<0.01	<0.01
0.4C-4Cr	0.39	4.02	0.017	<0.01	<0.01	<0.01

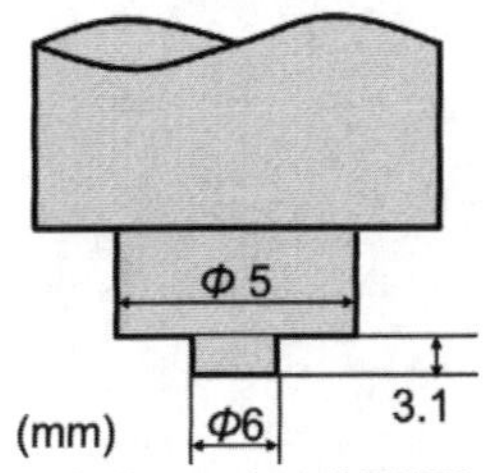

Fig. 1 Schematic of FSW tool.

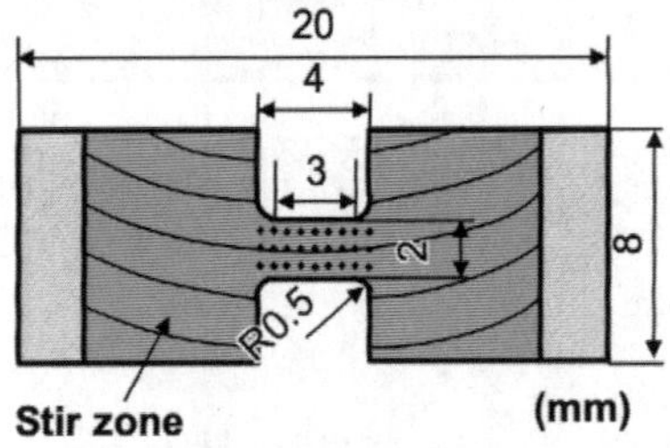

Fig. 2 Schematic of specimen geometry for tensile test

The Vickers hardness and the tensile property in the stir zones were determined as the mechanical property. The small special test specimen for tensile test was paralleled length of 3 mm and width of 2 mm as shown in figure 2. The entire paralleled area was covered with the stir zone to investigate the tensile property of the stir zone. The tensile test was carried out under 0.04 mm/min of a cross-head speed at room temperature. The strain was measured by tracing gauge marks (Vickers indents) on the specimen using a high-speed camera.

3. Results and discussion

3.1. Microstructure of stir zone

Figure 3 shows SEM images of microstructure in the stir zone of the specimens with different carbon and chromium contents. In the case of 0.2 % C, the microstructure of the stir zone changed from pearlite and ferrite to bainite and martensite with the increase in chromium content. In contrast, in the case of the carbon content of 0.4%, the microstructures are pearlite in 0.4C and martensite in both 0.4C-2Cr and 0.4C-4Cr. The martensite found in the specimens of 0.2C-0.4Cr, 0.4C-2Cr and 0.4C-4Cr is mainly lath martensite. However, the detailed morphologies are different.

Figure 4 represents prior austenite grains in stir zones of each specimen. The stir zones in the specimens containing chromium consist of fine grains which is around 20 μm. The average grain sizes

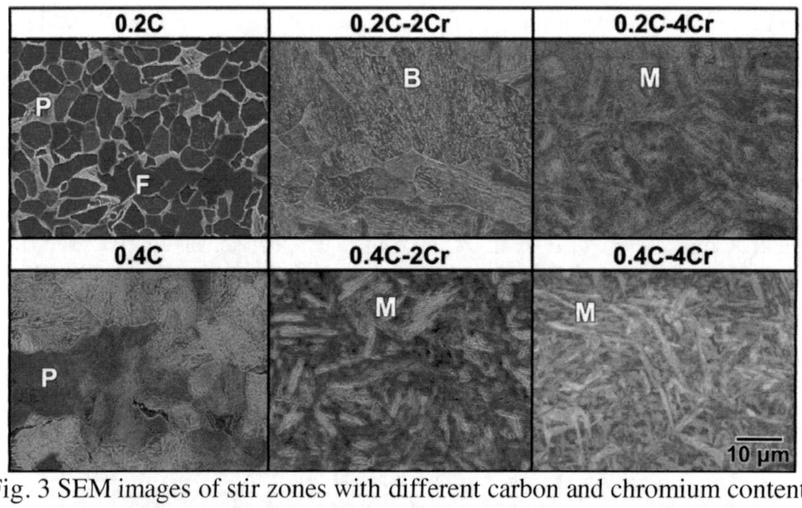

Fig. 3 SEM images of stir zones with different carbon and chromium content.

Fig. 4 Prior austenite grains in stir zones with different carbon and chromium content.

of the prior austenite in the stir zones of 0.2C-2Cr, 0.2C-4Cr, 0.4C-2Cr, and 0.4C-4Cr are 25.5, 21.8, 24.1, and 19.4 μm, respectively. It was impossible to determine the prior austenite grain size in 0.2C and 0.4C specimens (no chromium containing) because the grain boundaries were not clear. The prior austenite grains tend to be refined with the increase of chromium content regardless of carbon content. Addition of carbon also induces grain refining. Thus, the recrystallization of austenite by intense plastic deformation and heat generation during FSW attributes to refine the grains.

In order to analyze the detailed morphology of the martensite which was observed in figure 3, TEM-EDS analysis was carried out on the specimens of 0.2C-4Cr, 0.4C-2Cr and 0.4C-4Cr. Figure 5 shows TEM images of the stir zone of the specimens. Tiny precipitates (< 100 μm) distribute in the martensite laths in 0.2C-4Cr. The dislocations are also observed around the precipitates. Dense micro twins are mainly observed in 0.4C-2Cr and 0.4C-4Cr. Additionally, there is little precipitation in spite of high carbon content compared to 0.2C-4Cr. TEM-EDS spectroscopy maps in the stir zone of 0.2C-4Cr and 0.4C-4Cr are shown in figure 6. Carbon and iron are detected at the tiny precipitates in 0.2C-4Cr. Thus, the precipitates must be identified as iron carbide. There is no partitioning of chromium because the diffusion of chromium atoms after martensitic transformation is difficult under high cooling rate by FSW. On the other hand, in the case of 0.4C-4Cr, the distributions of carbon and chromium are uniform. This is caused by relatively low martensitic start temperature (Ms) as compared with that in 0.2C-4Cr. Thus, carbide is hardly precipitated in the stir zone of 0.4C-4Cr.

The stir zones of the specimens were firstly austenitized since the peak temperature was considered to be above Ac3 temperature during FSW process and then transformed into ferrite, pearlite, bainite and martensite depending on chemical compositions and cooling rate. Besides, chromium is known to induce a separation of CCT curve into ferrite formation field and bainite formation field and simultaneously retards the kinetics of phase transformation. Carbon is austenite stabilizing element, the incubation period of austenite to transform to pearlite and bainite increases with the increase of carbon content [9]. Thus, pearlite and ferrite were transformed from austenite in 0.2C and 0.4C without chromium. Bainite transformation also occurred in 0.2C-2Cr during the FSW process. The further increase in chromium content delays the bainite transformation [10]. Thus, martensite formed in 0.2C-4Cr, 0.4C-2Cr and 0.4C-4Cr since the super-cooled austenite in the stir zone of 0.4C with chromium transform to martensite at further lower Ms temperature compared to that of 0.2C-4Cr. The carbide precipitation in 0.2C-4Cr is considered by the auto-temper process after martensitic transformation since Ms temperature is relatively high. In order to investigate the carbide precipitation in auto-tempered martensite, the heat treatment at 400 °C for 1 hour was applied for the specimen of 0.4C-4Cr. The TEM image and TEM-EDS mappings are shown in figure 7. The carbide precipitations were a size of 200 - 300 nm. The precipitates are rich in carbon, chromium and iron. Moreover, it should be noticed that micro-twins in 0.4C-4Cr still remained even after the heat treatment.

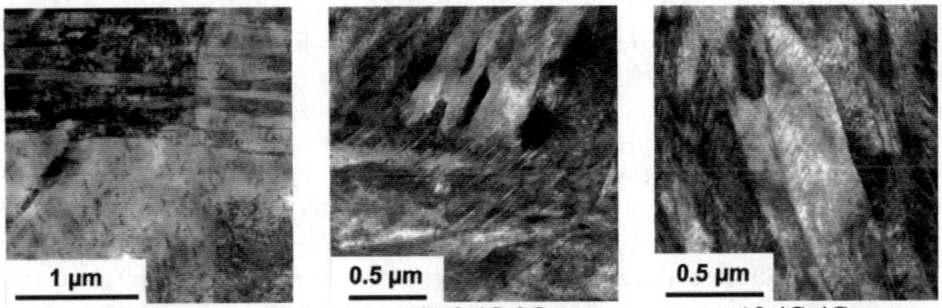

a) 0.2C-4Cr b) 0.4C-2Cr c)0.4C-4Cr

Fig. 5 TEM images of stir zone.

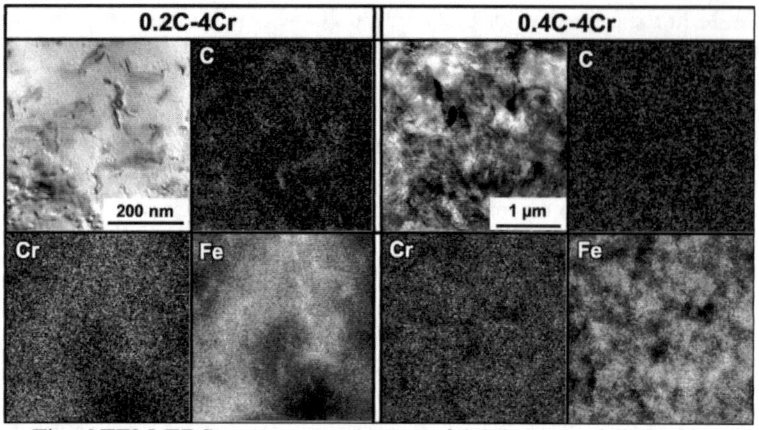

Fig. 6 TEM-EDS spectroscopy maps of 0.2C-4Cr and 0.4C-4Cr.

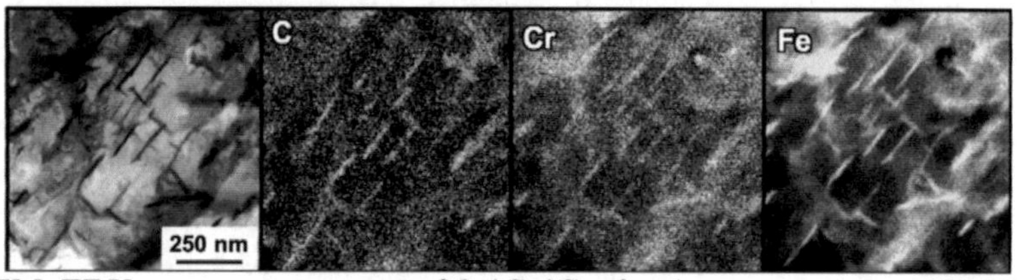

Fig. 7 TEM–EDX spectroscopy maps of 0.4C-4Cr after heat treatment at 400 °C for 1 h.

3.2. Mechanical properties

The hardness of the stir zones of 0.2C–2Cr, 0.2C–4Cr, 0.4C–2Cr, and 0.4C–4Cr are 148, 365, 484, 266, 669, and 707 HV, respectively. In both cases of the carbon content of 0.2C and 0.4C, the hardness increases with the chromium content. This tendency corresponds to the microstructural transition with carbon and chromium contents as shown in figure 3. Even though the microstructures of stir zone in 0.2C-4Cr, 0.4C-2Cr, and 0.4C-4Cr are commonly martensite, the hardness of 0.2C-4Cr is the much lower than those of the other martensite in 0.4C-2Cr and 0.4C-4Cr. This should be caused by auto-tempering and hardness of martensite owing to carbon content.

Figure 8 shows the stress-strain curves of the specimens with different carbon and chromium contents. The ultimate strength of the stir zone in 0.4C, 0.2C-4Cr, 0.4C-2Cr and 0.4C-4Cr are 852 MPa, 1601 MPa, 2181MPa and 2479 MPa, respectively. The total elongation of the stir zone in 0.4C, 0.2C-4Cr, 0.4C-2Cr and 0.4C-4Cr are 35.1%, 29.6%, 4.3% and 4.1%, respectively. It is remarkable that material still keeps high ductility with high strength, although martensite formed in 0.2C-4Cr. On the other hand, 0.4C-2Cr and 0.4C-4Cr showed very low elongation and brittle fracture.

The strength and the elongation of tempered 0.4C-4Cr are 1715 MPa and 19.1%. The tempering heat treatment could modify the elongation from 4.2 to 19.1% and the strength keeps high. Besides, the strength and the elongation of tempered 0.4C-4Cr are close to those of the 0.2C-4Cr auto-tempered martensite. It is consistent with that martensite is softened after tempering in 0.2-4Cr. The carbide precipitation observed in the stir zone of 0.2C-4Cr indicated that auto-temper process occurred during the FSW process. In contrast, high carbon content attributes to the ultrahigh ultimate strength and low elongation by the formation of martensite with micro twins. It is worth noting that auto-tempered lath-martensitic structure of 0.2C-4Cr exhibited an excellent balance of strength and ductility among the steels with the commonly martensitic structure in the stir zone of 0.2C-4Cr, 0.4C-2Cr and 0.4C-4Cr.

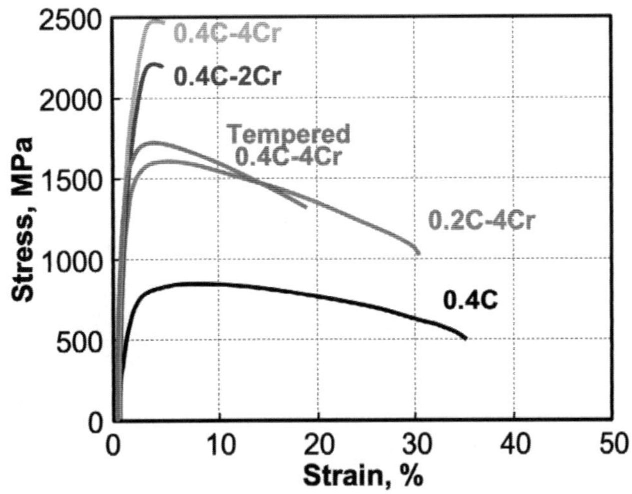

Fig. 8 Stress-strain curves.

4. Summary

The microstructure of the stir zone of the specimens containing 0.2 % C, changed from pearlite and ferrite to bainite and martensite with the increasing chromium content. In the case of the carbon content of 0.4%, the microstructures are pearlite in 0.4C and martensite in both 0.4C-2Cr and 0.4C-4Cr. The martensite found in the specimens of 0.2C-0.4Cr, 0.4C-2Cr and 0.4C-4Cr is mainly lath martensite.

With the increase of chromium and carbon contents, the strength of the stir zone increased and the elongation decreased. The specimens of 0.2%C-4%Cr, 0.4%C-2%Cr and 0.4%C-4%Cr showed high tensile strength of over 1.5 GPa. In addition, even though the microstructures in the stir zones of the specimens are martensite, the elongation of 0.2%C-4%Cr was 31%while those of the others were less than 5%. The auto tempering process induced to tiny carbide formation within lath martensite in 0.2C-4Cr specimen. Micro twins mainly formed in 0.4C-2Cr and 0.4C-4Cr specimens. Thus, the tiny carbide precipitation in the martensite has a significant effect on the balance between high strength and good elongation.

References

[1] Chen H, Kikuta Y, Araki T, Yoneda M and Matsuda Y 1984 *Acta Metall.* **32** 779
[2] Lee H J and Lee H W 2015 *Int. J. Electrochem. Sci.* **10** 8028
[3] Kostryzhev A, Singh N, Chen L, Killmore C and Pereloma E 2018 *Metals* **8** 134
[4] Wang Z, Rahka K, Nenonen P and Laird C 1985 *Acta Metall.* **33** 2129
[5] Mishra R and Ma Z 2005 *Mater. Sci. Eng. R* **50** 1
[6] Fujii H, Cui L, Tsuji N, Nakata K, Nogi K, Ikeda R and Matsushita M 2006 *Mater. Sci. Eng. A* **429** 50
[7] Cui L, Fujii H, Tsuji N and Nogi K 2007 *Scripta Mater.* **56** 637
[8] Chung Y, Fujii H, Ueji R and Tsuji N 2010 *Scripta Mater.* **63** 223
[9] Krauss G and Thompson S 1995 *ISIJ Int.* **35** 937
[10] You W, Xu W, Liu Y, Bai B and Fang H 2007 *J. Iron Steel Res. Int.*, **14** 39

IOP Conf. Series: Journal of Physics: Conf. Series **1270** (2019) 012033 doi:10.1088/1742-6596/1270/1/012033

Grain orientation stability and residual elastic energy induced by intergranular elastic interaction in rolled Al sheet

W Mao[1,2], Y Li[1] and H Ren[1]

[1] School of Materials and Metallurgy, Inner Mongolia University of Science and Technology, Arding Street 7, 014010 Baotou, China
[2] State Key Laboratory of advanced Metals and Materials, University of Science and Technology Beijing, Xueyuan Road 30, 100083 Beijing, China

wmmao@ustb.edu.cn

Abstract. Plastic behaviors of grains in polycrystalline metals are induced by external loading and affected by intergranular elastic interactions during deformation. A proposed reaction stress (RS) model including the effects of external and intergranular interaction stresses can simulate the behaviours, whereas the necessary equilibria of intergranular stress and strain are reached reasonably. Rolling texture development in roughly elastic isotropic Al as an example is simulated based on the RS model, whereas main texture components can be predicted accurately. The level of intergranular interactions may fluctuate depending on many factors and become stabile as texture forms at higher deformation degree, whereas all grains encounter similar deformation environment. Cube orientation indicates certain stability under the effect of RSs. A portion of elastic strain energy remains around the grains beside normal stored energy after deformation. The energy is orientation dependent and could not be predicted accurately by currently predominant crystallographic deformation theories and models based on Taylor strain principles. The elastic strain energy might affect the recrystallization process.

1. Introduction

Orientations of nuclei and recrystallization grains are related closely to the orientations of deformed grains. Therefore, the crystallographic deformation theories have attracted wide attention. It is known, for example, that cube oriented grains are not stable during rolling of FCC metals. However, cube texture appears very frequently in recrystallized FCC metal sheets. It is not yet entirely clear how those cube oriented nuclei could appear, although they may grow rapidly during recrystallization [1-3].

Recrystallization nuclei should somehow form in deformed matrix at early annealing stage. The deformed matrix offers not only an appropriate environment for rapid growth of grains, but also substructures with some special orientations for nucleation. Stress and strain equilibria are always maintained among the grains during deformation. However, distributions of stress and strain inside a rolling sheet are fluctuant in penetrating the polycrystalline body. It has been indicated that plastic deformation is conducted by combined activation of slips penetrating grains and some non-penetration slips around boundary areas [4]. Different combination of the slips in different grains helps to reach the fluctuant stress and strain equilibria in a nature way, and the fluctuation levels depend on grain orientations. After plastic deformation, some elastic stresses and strains will remain between grains in deformation matrix, which may provide extra recrystallization driving force in addition to ordinary stored energy. The extra driving forces may be grain orientation-dependent and affect the orientations of recrystallization grains.

Strain equilibrium in current theories and models including the viscoplastic self-consistent (VPSC) model [5], the advanced Lamel model [6], and the grain–interaction model [7], has been reached. However, the strains predicted by the models are not so fluctuant as they should be in penetrating the polycrystalline body, since the models are based on Taylor principles and prescribe in advance that strain tensor of grains or grain clusters is basically identical to the macroscopic strain tensor during plastic deformation. The strain prescription does not agree with the experimental observations very well [8]. On the other hand, for example for steel and Al alloy, the plastic deformation in VPSC model is accomplished mainly by penetrating slips. Therefore, the models can not predict the residual elastic stress and strain between grains after deformation accurately. In this work, we attempt to calculate the stability of different substructures of Al during rolling by means of a reaction stress (RS) model [4, 8], and to analyze the relationship between residual elastic energy and grain orientation after deformation.

2. Experimental observation

A 7.8 mm thick hot rolled and annealed band of industrial pure Al, as experimental material, was cold rolled for 10% reduction. Grains before and after the rolling are shown in Figure 1a and 1b. Many traces of active slip systems in different grains can be observed in Figure 1b. Figure 1c indicates the grain structure (full lines) determined by EBSD after rolling. The grain shapes should be similar to those indicated by dashed lines in Figure 1c, if they were deformed according to many current models based on Taylor principles [5-7], which are different from the experimental observations. Although the differences are not so drastic, the phenomenon implies that the multiple activation of slip systems does not agree with those predicted by many theoretical models above. Different slip systems may lead to different orientation evolutions and texture formation. The traces of slip systems indicate also, that many penetration slips occurred in the grains (Figure 1b) during deformation to achieve the necessary rolling strain. At the same time, there appear also many non-penetrating slips, including some slips closely localized around grain boundary (e.g. Figure 1d grain *) and some half penetrating slips (e.g. Figure 1d grain #). The non-penetrating slips help to reach the necessary intergranular strain and stress equilibria.

IOP Conf. Series: Journal of Physics: Conf. Series **1270** (2019) 012033 doi:10.1088/1742-6596/1270/1/012033

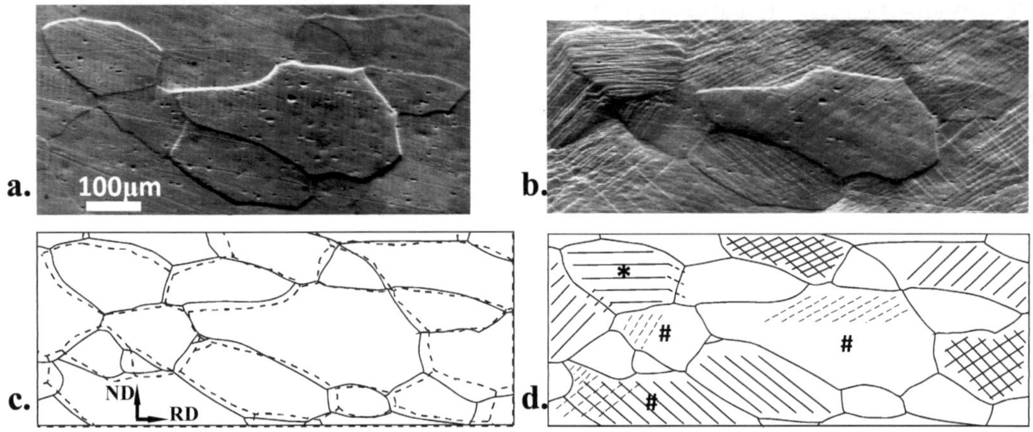

Figure 1. Grains of the 10% rolled Al sheet. (a) grains before rolling; (b) slip traces in the grains after rolling; (c) shape comparison of deformed grains (full lines) and those according to Taylor principles (dashed lines); (d) the penetrating slip (parallel full lines) and non-penetrating local slips (parallel dashed lines, including localized around boundary in grain * and half penetrating in grain #).

3. Reaction stress (RS) model

If a slip system in a grain is activated by external loading, the plastic strain tensor produced by the slip will induce stress and strain incompatibility. The slip must encounter resistance from the neighboring grains first in the form of elastic RSs. Plastic strain components produced will be partially reduced in an elastic manner, whereas the neighboring grains must also be strained elastically. Therefore, the incompatible plastic strains are relieved by the elastic strains of the concerned grain and its neighboring grains, between which stress and strain equilibria are reached elastically. The rolling stress tensor $[\sigma_{ij}]$ bore by the concerned grain, according to [4,8], becomes:

$$[\sigma_{ij}] = \sigma_y \begin{bmatrix} \frac{1}{2} & 0 & 0 \\ 0 & 0 & 0 \\ 0 & 0 & -\frac{1}{2} \end{bmatrix} - \sigma_y \begin{bmatrix} -\mu\varepsilon_{22}\frac{d}{b} & \mu\varepsilon_{12}\frac{d}{b} & \mu\varepsilon_{13}\frac{d}{b} \\ \mu\varepsilon_{21}\frac{d}{b} & \mu\varepsilon_{22}\frac{d}{b} & \mu\varepsilon_{23}\frac{d}{b} \\ \mu\varepsilon_{31}\frac{d}{b} & \mu\varepsilon_{32}\frac{d}{b} & 0 \end{bmatrix} \quad (1)$$

where the first and second terms refer to the external rolling stress tensor and internal RS tensor, respectively; ε_{ij} denotes the plastic strain produced by the slip; 1, 2, and 3 represent the rolling direction (RD), transverse direction (TD), and normal direction (ND) of the rolling sheet, respectively; σ_y indicates the macroscopic flow yield stress during rolling; μ specifies the generalized Schmid factor of the activated slip system; d corresponds to the effective distance between dislocations, and b is the length of the Burgers vector. The approximate relationship between σ_y and yield shear stress τ_y and the d can be expressed according to Frank–Read theory as follows [9]:

$$\tau_y = \frac{\sigma_y}{2} = \frac{Gb}{d} \text{ or } d = \frac{2Gb}{\sigma_y} \quad (2)$$

where G, in approximately elastically isotropic metals (such as Al), is shear modulus. The flow yield stress σ_y of an annealed metal, i.e., σ_{y0} denotes the yield strength, and its top limit in extremely hardened state, e.g., at ε_{33}=−4.0 (98% rolling reduction) can be the tensile strength σ_b. According to the conventional flow stress of most deformation metals, the flow yield stress σ_y for the present simulation can be assumed to develop with rolling reduction ε_{33}<0 as follows:

$$\sigma_y = \sigma_{y0} + \left(\sigma_b - \sigma_{y0}\right) \cdot \sqrt[n]{\frac{-\varepsilon_{33}}{4}} \tag{3}$$

where σ_{y0}=20 MPa and σ_b=50 MPa are valid for Al, and n=8 is used in the present work. The RSs indicated in equation (1) accumulate during rolling until the Schmid factor of another slip system sufficiently increases and is instantaneously activated. The multiple slips can be alternatively followed step by step, and any requested combination of slip systems can be implemented.

However, the accumulated RSs should feature a limitation and a yield point that must not be exceeded. The limitations of deformation stresses should be $[\sigma_{ij}]_{\text{lim}} \leq \alpha_{ij}\sigma_y/2$, where α_{ij} represents the effective coefficient of maximal RSs at a range of 0 to 1 [8]. α_{ij}=0 means no RS, whereas α_{ij}=1 implies the highest level of RS. If any of the RSs acting on a grain reaches its up-limit, certain slip systems in the neighboring grains may be additionally activated locally in addition to the normal penetration slips. In this case, the stress and strain equilibria are also maintained by the local plastic behavior around the boundary areas, as induced by several non-penetration slips, on which shear stress has reached the critical value for activation. The non-penetration characteristics of the slips imply that stress and plastic strain in equilibrium run from one grain over the boundary to another grain in a fluctuant way (Figure 1d) and cannot be similar to that predicted by Taylor principles. On the other hand, the non-penetration slips will induce also random orientation evolution, i.e. random texture.

It is necessary to prove whether the model can correctly predict deformation texture. An Al sample in commercial purity (99.9% Al) with 8.6 mm thickness and random initial texture was prepared. The sample was cold rolled to reductions of 70%, 90%, and 98% (true strain ε_{33}: −1.2, −2.1 and −3.8, respectively). Figure 2a shows the rolling textures in ODF sections, in which the copper and brass textures in φ_2=45° section and S texture in φ_2=65° section were observed [10].

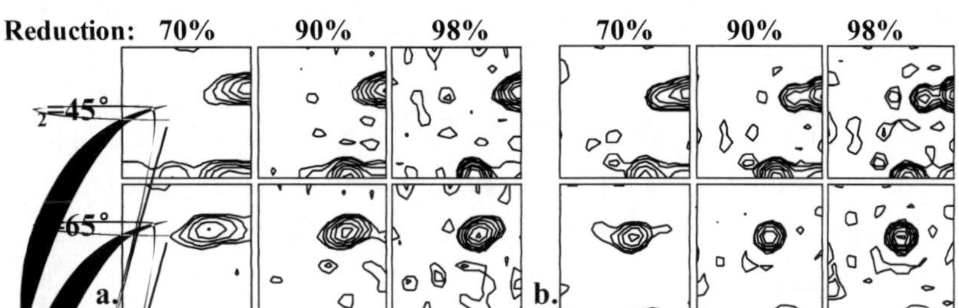

Figure 2. Rolling texture of Al sheets. a. experimental observation; b. simulated based on the RS model including 20% random texture (936 initial random orientations; steps $\Delta\varepsilon_{33}$=0.001, α_{12}=0.7, α_{23}=0.01, α_{31}=0.001, α_{22}=0.08; ODF φ_2=45° and φ_2=65° sections, density levels: 1, 2, 4, 7, 11, 16, 22).

All slip systems bear shear stress during deformation no matter they are active or not. The shear stress may reduce the necessary RS for additional activation of non-penetration slips, i.e., the effective coefficients α_{ij} are commonly lower than 1. The α_{ij} must be determined for simulating the rolling texture and are influenced by grain orientations and those of the neighboring grains. The α_{ij} of grains may be various because of different orientations of their neighboring grains, but will become stable during the rolling, since all grains fall into the same statistical environment with the formation of rolling texture. An appropriate values for the α_{ij} are found to be α_{12}=0.7, α_{23}=0.01, α_{31}=0.001, and α_{22}=0.08. Figure 2b shows the simulation results, including 20% random texture resulting from the non-penetration slips. This outcome is closely similar to the experimental observations (Figure 2a) showing the central positions of copper, brass, and S textures and their peak densities. Therefore, the RS model can predict and reproduce the formation of rolling texture in Al correctly.

4. Stability of deformation substructure and residual elastic energy of grains based on RS model

The main texture components after heavy rolling of Al commonly include copper texture {90°, 30°, 45°}, brass texture {35°, 45°, 0°} and S texture {55°, 35°, 65°}(Figure 2). Besides, the cube texture {45°, 0°, 45°} can be predicted in a relative wide range of higher α_{ij} values (Figure 3).

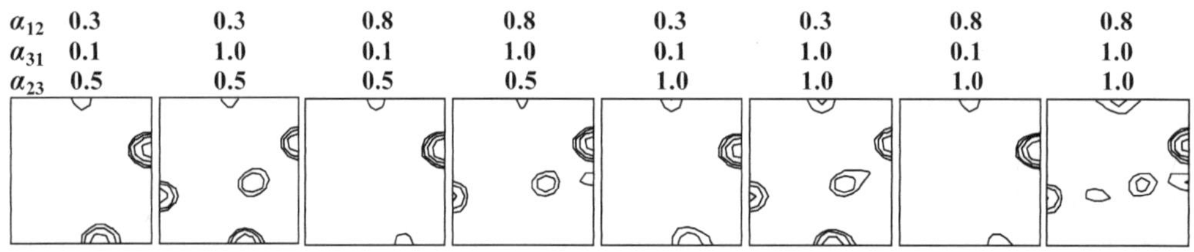

Figure 3. Cube orientation after 95% rolling simulated based on the RS model under different α_{ij} values (α_{22}=0), (ODF φ_2=45° sections, steps $\Delta\varepsilon_{33}$=0.001, density levels: 5, 10, 20, and 40).

It is obvious that cube orientation becomes stable locally in heavy rolling matrix if the RSs of grains reaches higher levels, which is unavoidable in an intricate rolling process. The corresponding stable cube substructure should be the source of cube nuclei of cube recrystallization texture.

The elastic stresses and strains induced by intergranular interactions will remain if external loading is removed. A part of the elastic energy of different grains remains in addition to normally stored energy. RSs σ_{33}=0 and σ_{11}=$-\sigma_{22}$ are valid, whereas deformation volume remains basically constant by the removal of external loading. The elastic energy W around the boundary area can then be calculated according to the elastic theory and the remaining RS σ_{ij} after deformation of differently oriented grains:

$$W = \frac{1}{2G}\left(\sigma_{12}^2 + \sigma_{23}^2 + \sigma_{31}^2 + \sigma_{22}^2\right) \quad (4)$$

where G=26 GPa is valid for Al. The orientations based on RS simulation are classified into the main rolling texture components of copper, brass or S if their misorientations to the main components are less than 5°. The elastic energy of the orientations in the main texture components is calculated according to equation (4). Table 1 shows the average elastic energy of orientations in different textures in comparison with that of all orientations. The average elastic energy of cube orientation is also calculated in a similar manner under different α_{ij} values. Notably, the orientations in different texture components indicate different levels of elastic energy under the same simulation conditions. Brass texture demonstrates much higher elastic energy in comparison with those of copper and S texture, and the elastic energy of cube texture could be very low or very high depending on α_{ij} values (Table 1).

Table 1. Average elastic energy induced by intergranular mechanical interactions in different texture components during rolling simulation by the RS model

Texture component	Effective coefficient α_{ij} of maximal RSs				Rolling reduction		
	α_{12}	α_{23}	α_{31}	α_{22}	70%	90%	98%
Brass	0.7	0.01	0.001	0.08	5009 Pa	5499 Pa	5922 Pa
S					2749 Pa	2865 Pa	2642 Pa
Copper					2420 Pa	2933 Pa	2844 Pa
All orientations					2596 Pa	3375 Pa	3213 Pa
Cube	0.8	1.0	0.0	0.0	2526 Pa	2066 Pa	2311 Pa
	0.3	1.0	1.0	0.0	8300 Pa	9991 Pa	7883 Pa

Normally, the stored energy of deformed Al approximates 2MPa, and it reaches 0.3MPa in recovery state [9]. In comparison, the elastic stain energy (Table 1) is much lower. However, the level of elastic stain energy in differently oriented deformation grains can still exhibit certain effects during the following thermal treatment, including recrystallization nucleation and grain growth. The RS model indicates that the elastic strain energy retained around the cube structure may increases with increasing α_{ij} levels (Table 1). The lower elastic energy may increase the stability of cube structure.

5. Summary

Real deformation behaviors of grains in Al do not agree with that predicted under Taylor principles. The RS model based on combination of external and intergranular interaction stresses can accurately predict textures in Al rolling sheets both in terms of orientation positions of the textures and their peak densities. Furthermore, according to the RS principles, cube orientation can achieve frequently certain stability in rolling process, which is of great significance for the formation of recrystallization cube texture. The residual elastic strain energy around the deformation grains is orientation-dependent, but is much lower than ordinary stored energy. The energy can lead to different orientation stabilities after deformation and may extend its influence to the subsequent recovery and recrystallization process.

Acknowledgement

This work is financed by the National Natural Science Foundation of China (grant no. 51571024).

References

[1] Ibe G, Dietz W, Fraher A C and Lücke K 1970 *Zeitschrift für Metallkunde*, **61** 498
[2] Mao W and Yang P 2014 *Trans. Nonferrous Met. Soc. China*, **24** 1635
[3] Mao W 1999 *J. Mater. Eng. Perf.* **8** 556
[4] Mao W 2016 *Mater. Sci. Eng.* **A 672** 129
[5] Lebensohn R A, Tomé C N and Castaňeda P P 2007 *Philoso. Mag.* **87** 4287
[6] Van Houtte P, Li S, Seefeldt M and Delannay L 2005 *Int. J. Plast.* **21** 589
[7] Mu S, Tang F and Gottstein G 2014 *Acta Mater.* **68** 310
[8] Mao W 2016 *Front. Mater. Sci.* **10** 335
[9] Hornbogen E and Warlimont H 1991 *Metallkunde*, (New York: Springer) p102 and p 128
[10] Mao W 1998 *Mater. Sci. Eng.* **A257** 171

IOP Conf. Series: Journal of Physics: Conf. Series **1270** (2019) 012034 doi:10.1088/1742-6596/1270/1/012034

Evaluation of the kinetics of recrystallization based on scaling 1D cellular automaton by DSC measurements

Tamás Bubonyi[1], Péter barkóczy[1], Szilvia Gyöngyösi[2]

[1] University of Miskolc, 3515 Miskolc, Miskolc Egyetemváros, Hungary
[2] University of Debrecen, 4032 Debrecen, Egyetem tér 1., Hungary

fembubo@uni-miskolc.hu

Abstract. There are several methods to study the kinetics of the recrystallization process. DSC measurement is one of the most effective measurement methods. Cellular automata simulations are frequently used in computational studies. Both methods give and use kinetic constants. The aim of this study to find the relation between the kinetic constants. Additionally an evaluation method is introduced based on the DSC evaluation and cellular automaton which can compute a more detailed kinetic description of recrystallization.

1. Introduction

The kinetic analysis of recrystallization is one of the most important steps in the simulation studies [1]. There are several methods which give a proportional measure to recrystallized volume fraction [2]. Optical microscopy is one of them and can show a detailed picture about the recrystallization, but ineffective when the annealing caused mostly by recovery. Some other method is to the evaluate the kinetics of nucleation and growth [3]. The results of mechanical testing or resistivity measurement can also provide the recrystallization volume fraction and can additionally show the effect of recovery too [4]. In several cases calorimetry, especially DSC measurement is used to evaluate the changes of recrystallized volume fraction as a function of time and temperature [5]. The analysis of DSC measurement is well studied and developed. The disadvantage of this methods is it doesn't show information about the microstructure and the recovery. But in several cases the kinetic constants from DSC evaluation are enough for simulating property changes.

Cellular automaton is one of the oldest non-conventional computation methods used in simulation of phase transformation processes [6]. This method is successfully applied in the simulation of recrystallization [7,8]. Cellular automaton is a beloved simulation technique in this field, because this can reproduce the stochastic nature of the nucleation and growth processes [9]. The major problem in the application of these simulations is the scaling. The automata work with arbitrary units, but we need natural terms. It is possible to simplify the simulation to 1D as much as possible [10]. The big advantage of 1D is that the number of kinetic constants decreases to the minimal number, but the simulation computes all necessary aspects of the recrystallization [11]. The results in this case is not as sophisticated as a 2D [12] or 3D [13,14] computation and cannot cover all detailed phenomena of course. But the computation become extremely fast.

The 1D cellular automata simulations of recrystallization also use the kinetic constants of nucleation and growth. Using these values, the scaling of the 1D automata becomes an easy optimization process

[15]. Due to the arbitrary units of the automaton the question on the meaning of these numbers is important. These values on one side could be only parameters of the simulation and are only used by the automata. On the other side, the scaling provides physical meaning for the arbitrary units. Hence the used kinetic constants can have a relation to that kinetic equations which are evaluated from DSC measurement. The main aim of the present study is to examine this question.

Pure copper (OFHC Cu) is used to study the recrystallization process to avoid the disturbing effect of recovery. DSC measurements were performed, and a 1D recrystallization automaton was scaled to the measured results. The kinetic constants from the DSC evaluation and from the automaton will be compared.

2. Materials and methods

Continuously casted OFHC copper slab was heat treated at 400°C for 1h. The thickness of the slab was 10mm. The slab was rolled at room temperature with a Von-Roll laboratory mill to 2mm thickness. DSC samples were machined from the cold rolled sheet. The samples were heated up in a Netsch DSC 204 heat flux DSC equipment at 10K/min, 15K/min, 25K/min and 30K/min.

The DSC curves were analyzed with a self-developed software. The transformed volume fraction was calculated from the peaks according to [15]. The curves were evaluated with Kissinger [16] and Takhor [17] method. The Takhor method calculates the activation energy and rate constant of the recrystallization process. The Ozawa method was used to evaluate the value of the Avrami exponent.

The 1D cellular automaton used was published in [10]. The scaling was performed according to [15]. The scaling was made in the same time unit as the rate constant was calculated. The linear size in this case is an independent variable, and the grain size was not calculated in the present study. The scaling procedure gives the activation energy of nucleation and growth during the simulated recrystallization process.

3. Results and discussion

Figure 1 shows the DSC curves and the recrystallized volume fraction of the cold rolled OFHC Cu samples. On the curves it is clearly seen that the recrystallization process started around 200°C and finished at 300°C. As the heating rate increases, the start, finish, as well as the peak temperature shifts toward higher temperatures. Around the peaks it is possible to apply a linear base line to determine the peaks. Based on the integral of the peaks it is possible to determine the transformed volume fraction.

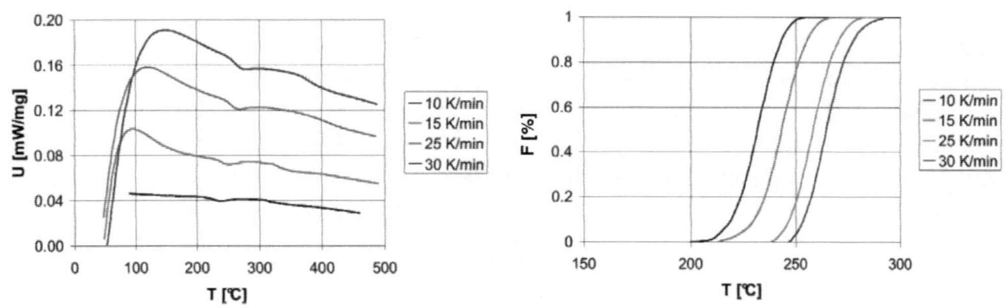

(a) (b)

Figure 1. The DSC curves (a) and the transformed volume fraction (b) as a function of temperature at different heating rates.

There are several thermo-kinetic methods to evaluate the DSC curves. The Kissinger and Takhor methods are based on the shift of peak temperature (T_{peak}) as a function of heating rate (v). The basic equations of both methods are shown as eq. 1 and 2. Both methods calculate the activation energy of recrystallization (Q_{rex}) and the rate constant (B_{0rex}).

IOP Conf. Series: Journal of Physics: Conf. Series **1270** (2019) 012034 doi:10.1088/1742-6596/1270/1/012034

$$ln\left(\frac{v}{T_{peak}^2}\right) = -\frac{Q_{rex}}{R}\frac{1}{T_{peak}} + ln\left(\frac{B_{0rex}R}{Q_{rex}}\right) \quad (1)$$

$$ln\left(\frac{v}{T_{peak}-T_{start}}\right) = -\frac{Q_{rex}}{R}\frac{1}{T_{peak}} + ln(B_{0rex}) \quad (2)$$

The symbols R and T_{start} denote the gas constant and the starting temperature. Figure 2 shows the left values of the eq. 1 and 2 as a function of reciprocal temperature. The points are laying on a straight line with slope $-Q_{rex}/T_{peak}$. The rate constant can be computed from the intersepts of the lines.

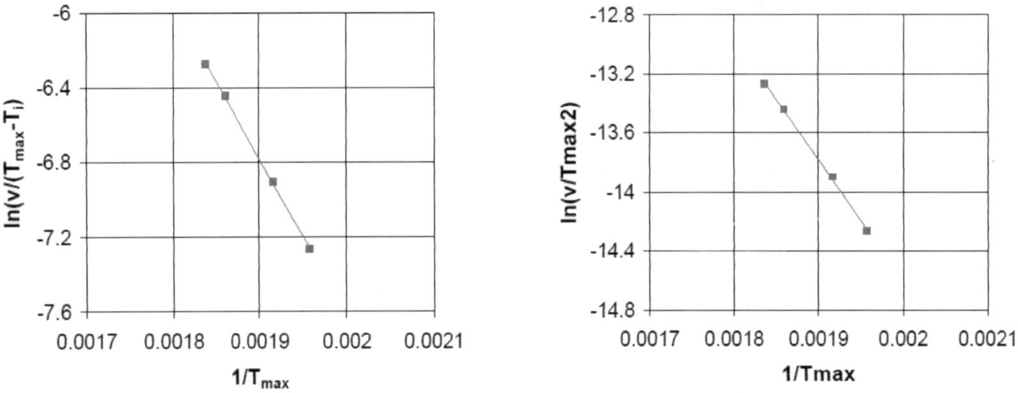

a) b)

Figure 2. The determination of the kinetic constants at 80% reduce based on eq. 1 and 2. a): Takhor method, b) Kissinger method.

The analysis gives 58589J/mol and 58523J/mol to the activation energy of recrystallization based on the Kissinger and Takhor method. A small difference exists between the two number. This small difference is enough to cause a large difference in the rate constant. The value of this is 3183.9 1/s and 486.6 1/s based on the Kissinger and Takhor method. Examining the properties of the JMAK equation it becomes evident that both parameters describe the same process.

The Avrami exponent is the missing value. Neither the Kissinger nor the Takhor method can determine this value. There are several approximations to calculate the Avrami exponent. Based on Kissinger's theory the half width of the peaks are proportional to the Avrami exponent. Equation 3 shows this relation. For the calculation of the Avrami exponent the value of the activation energy is necessary. The value from the Kissinger's analysis is used. All peaks give an exponent, the average is calculated. The Avrami exponent based on this analysis is 3.26.

$$n = \frac{2.5T_{peak}^2}{\Delta T_{1/2}\frac{Q_{rex}}{R}} \quad (3)$$

After the detailed kinetic analysis of the curves the [18] cellular automata simulation was scaled with the transformed volume fraction curves based on the method [15]. During the scaling a difference between the curves was defined, and the activation energy of the nucleation and growth were changed by a simplex method. The minimum of the difference between the measured and simulated recrystallized volume fraction was determined based on the mentioned kinetic constants. Figure 3 shows the results of the scaling.

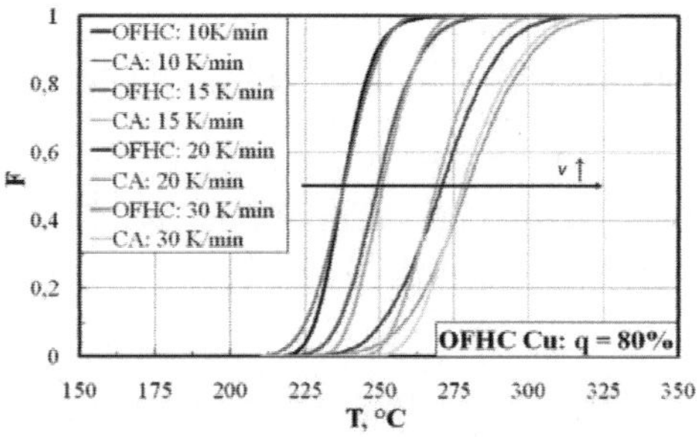

Figure 3. The recrystallized volume fraction as a function of temperature based on the DSC measurement and scaled cellular automata computation.

The value of activation energy of nucleation (Q_n) is 36627.3 J/mol, and growth (Q_g) is 9890.5 J/mol. Originally these values are used in the automata for simulations studying annealing. But the well-known equation 4 [9,19] provides a connection between the activation energies.

$$Q_{rex} = Q_n + (n-1)Q_g \quad (4)$$

Using the values from the scaling and the Avrami exponent from the analysis above, the computed activation energy of recrystallization is 58980 J/mol, which is a bit higher than the Kissinger's or Takhor's value. But these values are suitable to make a detailed study, or simulation without metallographic sample and measurement.

4. Summary

Recrystallization of cold rolled OFHC Cu samples were studied by DSC measurement. The DSC curves were analyzed using Kissinger's and Takhor's theory and the activation energy and rate constant were computed. The Avrami exponent is calculated from the half width of the DSC peaks. Additionally, a recrystallization 1D cellular automaton was scaled with the DSC results. During the scaling the value of the activation energy of nucleation and growth were changed to find the best fit of the measured and simulated transformed volume fraction curves. The activation energies from the kinetic analysis and from the scaled simulation are comparable based on the Avrami exponent. Completing the DSC measurements with these fittings provide a more in-depth kinetic description of the process without information about the microstructure. Of course, the introduced method cannot provide information on grain size, for this reason metallographic examinations are still necessary.

5. Acknowledgements

The described article was carried out as part of the GINOP-2.3.2-15-2016-00027 "Sustainable operation of the workshop of excellence for the research and development of crystalline and amorphous nanostructured materials" project implemented in the framework of the Szechenyi 2020 program. The realization of this project is supported by the European Union.

The described research was also supported through the National Research, Development and Innovation Office – NKFIH K119566 project.

References

[1] Raabe D 1998 *Computational materials science: the simulation of materials microstructures and properties* Wiley-VCH
[2] Cotterill P and Mould P R 1982 *Recrystallization and Grain Growth in Metals* Surrey University

Press London.

[3] Chen H, Yao Y, Warner J A, Qu J, Yun F, Ye Z, Ringere S P and R Zheng, 2017 Grain size quantification by optical microscopy, electron backscatter diffraction, and magnetic force microscopy *Micron* 'ed F. Braet, R.F. Egerton p. 41–7

[4] Mittmeijer E J, Van Gent A and Van Der Schaaf P J 1986. Analysis of transformation kinetics by nonisothermal dilatometry *Metallurgical and Materials Transactions* **17A** (Springer US), p. 1441-45.

[5] Mandal D and Baker I 1995 Determination of the stored energy and recrystallization temperature as a function of depth after rolling of polycrystalline copper *Scripta Metallurgica Et Materiala* **33** (Elseiver) 645-50

[6] Chopard B and Droz M 1998 *Cellular automata modeling of physical systems*. University of Geneva.

[7] Midownik M A 2002 A review of microstructural computer models used to simulate grain growth and recrystallisation in aluminium alloys *Journal of Light Metals*, **2** (Elseiver), p. 125-35.

[8] Barkóczy P, Roósz A and Geiger J 2005, Simulation of Dynamic and Meta-dynamic Recrystallization by Cellular Automaton EUROMAT 2005, European Congress on Advanced Materials and Processes

[9] Molnar D, Niedermeier C, Mora A, Binkele P and Schmauder S 2012 Activation energies for nucleation and growth and critical cluster size dependence in JMAK analyses of kinetic Monte-Carlo simulations of precipitation *Continuum Mechanics and Thermodynamics*, **24** (Springer), pp 607–17

[10] Gyöngyösi Sz and Barkóczy P, 2013 Simulation of grain coarsening using one-dimensional cellular automaton *Materials Science Forum* **752** 'ed A. Roósz and K. Tomolya pp. 217-22.

[11] Bubonyi T and Barkóczy P 2018 Simulation of crystallographic changes during recrystallization by one-dimensional cellular automaton, *IOP Conf. Series: Materials Science and Engineering* **426**

[12] Zheng C, Raabe D and Li D 2012 Prediction of post-dynamic austenite-to-ferrite transformation and reverse transformation in a low-carbon steel by cellular automaton modeling, *Acta Materialia* **60** (Elseiver), pp 4768-79

[13] Radhakrishnan B, Sarma G B and Zacharia T 1998 Modeling the kinetics and microstructural evolution during static recrystallization—Monte Carlo simulation of recrystallization *Acta Materialia* **46** (Elseiver) pp 4415-33

[14] Bubonyi T Bánóczy Zs Barkóczy P and Buza G 2019 Local Annealing of Cold Rolled Aluminum Sheets by LASER Treatment, *International Journal of Engineering and Management Sciences* **4** 'ed E Szűcs pp 21-6

[15] Barkóczy P and Gyöngyösi Sz 2013 Scaling cellular automaton simulations of short-range diffusion processes, *Materials Science Forum* **729** 'ed Berecz T et al pp 150-55

[16] Liu F, Liu X N and Wang Q 2008 Examination of Kissinger's equation for solid-state transformation *Journal of Alloys and Compounds* **473** (Elseiver) p 152-6

[17] Takhor R L 1972 Advances in nucleation and crystallization of glasses *Journal of the American Ceramic Society* 'ed William G Fahrenholtz p. 166.

[18] Benke M Tranta F, Barkóczy P, Mertinger V and Daróczi L 2008 Effects of heat flux features on the differential scanning calorimetry curve of a thermoelastic martensitic transformation, *Materials Science and Engineering A* **481** 'ed Gunther Eggeler, Gernot Kostorz pp. 522-5.

[19] Lu K and Wang J T 1991 Activation energies for crystal nucleation and growth in amorphous alloys *Materials Science and Engineering A* **133** 'ed H. Herman, pp. 500-3

Comparison between experimental data and a cellular automata simulation of martensite formation during cooling

A Kaijalainen, O Seppälä, V Javaheri, A Pohjonen, D Porter and J Kömi

Materials and Mechanical Engineering, Centre for Advanced Steels Research, University of Oulu, Oulu, Finland

Corresponding author email: antti.kaijalainen@oulu.fi

Abstract. Computer simulations of steel microstructural development provide a powerful tool, which can form the basis of mechanical property predictions. However, in order to create detailed understanding of the factors affecting the properties, the model should predict microstructural evolution during cooling. The present study compares the results of cellular automata simulations with experimental data for two distinct austenite conditions, recrystallized and deformed. Detailed microstructural features were studied using a laser scanning confocal microscope, FESEM and FESEM-EBSD. The two-dimensional cellular automata (CA) model for simulating the formation of lath martensite was parameterized using fitted Johnson-Mehl-Avrami-Kolmogorov and Koistinen-Marburger equations. The parent austenite microstructure for the CA model was determined from the final martensitic microstructure using austenite grain reconstructions based on the use of MATLAB software and the MTEX toolbox. The results of this cellular automata simulation can be used to estimate the shapes and sizes of martensite blocks, which offers new possibilities for the qualitative estimation of the mechanical properties of high-strength steels formed from recrystallized or deformed austenite.

1. Introduction

The development of novel ultra-high strength steels (UHSS) requires metallurgical knowledge together with much experimentation. Especially, for direct-quenched steels, where the cooling process after the last hot rolling pass dictates the final mechanical properties, a good understanding of on-cooling phase transformation is needed [1]. Achieving a fully martensitic microstructure is the key factor for having high strength and high toughness in these steel grades [2]. However, the process parameters along with the chemical composition affect the phase transformations and thus the microstructure. The evolution of microstructure with the aid of cellular automata (CA) has been studied by several authors [3–6]. CA has been successfully used to simulate the recrystallization and grain growth of steel by Sieradzki and Madej [3] and annealing following cold deformation by Madej et al. [4]. Also, austenite-ferrite phase transformations and the resulting carbon diffusion during isothermal heating and continuous cooling have been simulated by An et al. [5] and austenite-ferrite-pearlite transformations during the thermal cycle of dual-phase steel by Bos et al. [6]. The current study employs the model of Seppälä et al. [7], a novel CA method used to simulate bainite and martensite transformation. The model is based on the well-established Koistinen-Marburger (K-M) equation [8].

Prior austenite state and grain size also plays a crucial role regarding the final microstructure. Hence, to evaluate and to estimate the on-cooling phase transformations, the prior austenite grain structure should also be taken into consideration. To reveal the prior austenite grain structure, a reconstruction method is commonly used for martensitic and bainitic microstructures that have been formed from equiaxed [9,10] and highly pancaked prior austenite [11,12] with a Kurdjumov–Sachs orientation relationship (K–S OR) [13]. In the near future, microstructural computation will be a powerful tool able to reduce the amount of experimental work needed in steel development. The novel CA model for simulating the formation of martensite in steels is one step towards this objective. The aim of this study is the verification of the CA simulation proposed in [7] for martensite formation with the help of experimental EBSD data.

2. Experimental material and methods

2.1. Thermomechanical treatments and material characterization

The studied steel composition in wt.% was 0.14C-0.2Si-1.1Mn-0.7Cr-0.2Mo-0.03Ti-0.002B. Dilatation data for recrystallized and pancaked austenite were obtained using a Gleeble 3800 thermomechanical simulator. Cylindrical samples made from the steel were initially solution treated at 1250 °C for 2 hours followed by water quenching. The specimens were subsequently heated at 10 °C/s to 1100 °C, held for 2 min, cooled to 1050 °C or 850 °C, held 15 s, and then compressed with three hits each having a strain of ~0.2 at a strain rate of 1/s. Then the specimens were held 5 s before cooling at 70 °C/s. This gave two distinct austenite conditions: one recrystallized (1050 °C) and the other pancaked (850 °C) for comparison with the CA simulations. A general characterization of the transformation microstructures was performed with a field emission scanning electron microscope (FESEM) (Sigma, Zeiss) on specimens etched with nital.

Electron backscatter diffraction (EBSD) measurements and analysis were performed using the EDAX-TSL 7.0 acquisition and analysis software. The FESEM for the EBSD measurements was operated at 15 kV with a step size of 0.15 μm when studying areas ~110 x 110 μm and 0.4 μm for the larger area ~220 x 220 μm. Samples were subsequently re-polished to remove the contaminated layer after EBSD and the prior austenite grain structure was investigated using a laser scanning confocal microscope (LSCM) after picric acid etching from the same area.

2.2. Parent austenite reconstruction method

In order to have a good initial assumption of the parent austenite grain structure for the CA simulations, the parent austenite grain structure was revealed using a reconstruction technique. The EBSD data of the dilatometry samples were subjected to the reconstruction using Matlab software supplemented with the MTEX texture and crystallography analysis toolbox [14]. Initially, grain maps were created from the data sets with a grain boundary tolerance of 3-5 degrees. Then, the parent austenite orientation map was reconstructed from the assembled grain maps with a two-step reconstruction algorithm. In the first step of the process, the orientation relationship between austenite and martensite was determined using K-S orientation relationship i.e. $\{111\}_\gamma//\{110\}_\alpha$, $<110>_\gamma//<111>_\alpha$ where there are 4 $\{111\}_\gamma$ planes which are parallel to a $\{110\}_\alpha$ plane and each $\{111\}_\gamma$ plane contains 3 $<110>_\gamma$ directions, each of which is parallel to 2 $<111>_\alpha$ directions. Consequently, 24 different crystallographic variants of the bcc structure (bainite/martensite) can form within one former fcc (parent austenite) grain. According to Morito et al. [15], the 24 variants generated by the K-S OR are categorized into four packet groups. In a packet, all variants share the same close-packed planes. Each packet then contains six variants depending on the parallel directions. Although the exact parallelism of planes and directions of the K-S orientation relation cannot be observed in actual crystallographic measurements of martensite, the consistent variant labeling system by Morito et al. [15] has been used to successfully identify a number of martensitic/bainitic microstructures with varying orientation relationships between the fcc and bcc structures. Afterwards, in the second step, the grain map was divided into distinct clusters using the Markov clustering method [16] proposed by Gomes and Kestens [17]. The parent austenite orientation was then calculated for each cluster separately, resulting in a reconstructed orientation map. The average misorientation between the reconstructed orientation for each cluster and the best fit for each individual grain was approximately 2 degrees, indicating a good fit for the reconstructed result. The full details for the reconstructed procedure are available in Ref. [18].

2.3. Cellular automata algorithm

Cellular automata [19] is a method in which the modelled area is divided into equal-sized elements, i.e. cells. These cells have various properties which can be altered by a user-defined set of rules. In this case the only property needed for a cell is a color code that defines its current phase and separates instances of the same phase divided from each other by a high-angle grain boundary. Each cell represents a finite area of a single phase and several adjacent cells with the same color code represent a single grain (austenite) or block (martensite).

The CA-algorithm used in the current study starts from a situation where the microstructure is fully austenitic. At the end of every step the output created by the simulation is saved and an end condition check is performed. If end condition requirements are satisfied the simulation ends, otherwise it continues. Algorithms calculate the amount of new martensite nuclei and their growth speeds. As the simulation advances, new nuclei are placed into the simulation area and start growing with respect to their growth rules. Lath martensite grows as blocks, which consist of thin disc-like laths which grow next to each other. A single lath grows lengthwise at a very high speed with a constant width. The lath grows until it collides with a high-angle grain boundary. Widthwise growth of a block occurs gradually as new nuclei appear next to an existing block and grow in the same direction, adding to the width of a block [20].

3. Results and discussion

Figures 1-4 illustrate the microstructures of samples strained at 1050 °C (recrystallized) and 850 °C (pancaked) with two acquisition areas including EBSD image quality (IQ) maps with high-angle grain boundaries (HAGB, >15°), i.e. prior austenite, block and packet boundaries (figures 1-4a). A smaller area with a finer step size can produce better IQ and therefore more detailed grain size information. However, only few prior austenite grains are covered in the 110 x 110 μm area (figures 1a and 3a). Thus, a larger area (220 x 220 μm) was also examined, producing quantitative data of prior austenite grains, which are reconstructed with Matlab MTEX code as can be seen in figures 2b and 4b. In the CA simulated images (figures 1-4c), the black lines are parent austenite grain boundaries from the MTEX reconstructions and red tones are martensite blocks with their boundaries as blue lines. MTEX reconstructed parent grains were used as the initial austenitic microstructure in the CA simulations. MTEX reconstructed, recrystallized parent austenite grains were found to match extremely well with the picral etched optical microscope images even though a thin layer was polished away after the EBSD acquisitions. In the pancaked austenite, the grain boundaries are not so clear since some new small grains have nucleated on the grain boundaries after strain at 850 °C. For this reason, the MTEX reconstruction does not work as well as with equiaxed grains, but the information is still suitable for the CA simulations.

The EBSD grain sizes were determined using the equivalent circle diameters (ECD) of grains surrounded by HAGB. CA simulations show block boundaries. Grain size distributions and area fractions of the experimental data with two levels of data clean-up together with the CA simulations are shown in figures 5-6. The CA simulations are done with a 1 μm step size to prevent formation of artefacts during the simulations and the finest ECD grain size is ~1.13 μm. EBSD ECD grain size distributions, also filtered to remove grains smaller than ~1.0 μm, are comparable to those obtained from the CA simulations. In both cases mean ECD grain sizes are at the same level. However, there are more differences when comparing the area fractions. A few large grains might be artefacts due to undetected grain boundaries in the EBSD acquisitions. Nevertheless, experimental and simulated grain sizes and morphologies are seen to correspond well for the martensitic microstructures studied.

IOP Conf. Series: Journal of Physics: Conf. Series **1270** (2019) 012035 doi:10.1088/1742-6596/1270/1/012035

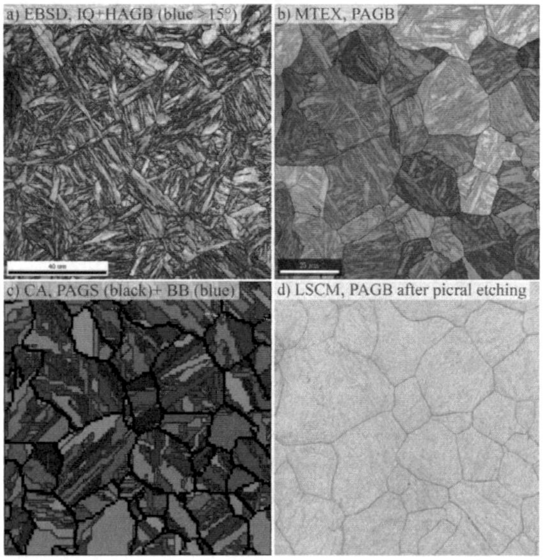

Figure 1. 3×0.2 strain at 1050 °C, small scanned area. (a) Original EBSD image quality map with high-angle boundaries (HAGB, >15°, blue), (b) MTEX reconstructed prior austenite grains, (c) CA simulated martensite blocks (red) with block boundaries (BB, blue) and (d) prior austenite grain boundaries after picral etching.

Figure 2. 3×0.2 strain at 1050 °C, large area. (a) Original EBSD image quality map with high-angle boundaries (HAGB, >15°, blue), (b) MTEX reconstructed prior austenite grains, (c) CA simulated martensite blocks (red) with block boundaries (BB, blue) and (d) prior austenite grain boundaries after picral etching.

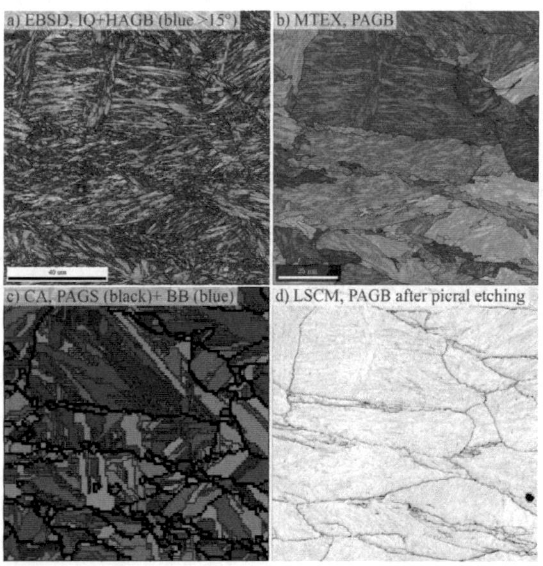

Figure 3. 3×0.2 strain at 850 °C, small area. (a) Original EBSD image quality map with high-angle boundaries (HAGB, >15°, blue), (b) MTEX reconstructed prior austenite grains, (c) CA simulated martensite blocks (red) with block boundaries (BB, blue) and (d) prior austenite grain boundaries after picral etching.

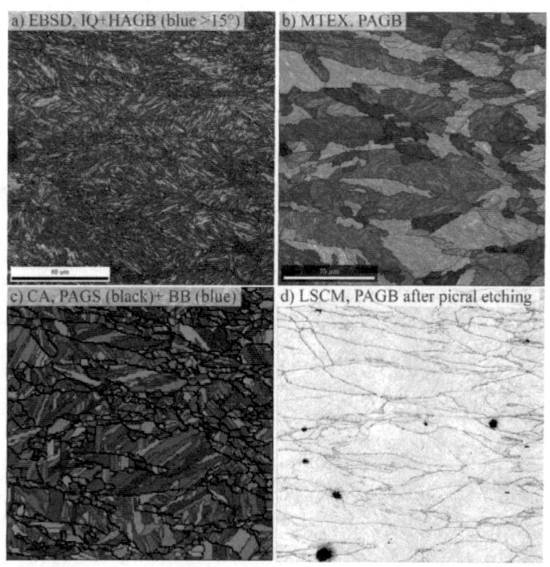

Figure 4. 3×0.2 strain at 850 °C, large area. (a) Original EBSD image quality map with high-angle boundaries (HAGB, >15°, blue), (b) MTEX reconstructed prior austenite grains, (c) CA simulated martensite blocks (red) with block boundaries (BB, blue) and (d) prior austenite grain boundaries after picral etching.

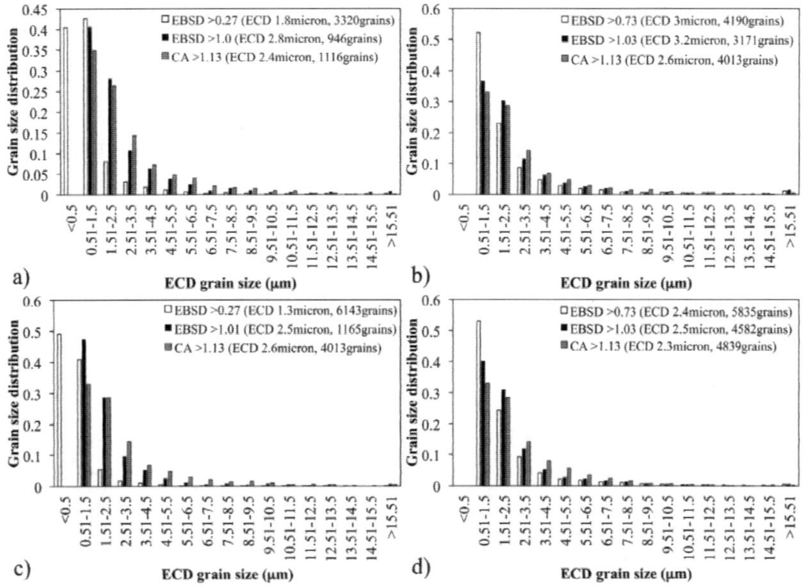

Figure 5. Grain size distribution of specimen strained 3×0.2 at 1050 °C. a) smaller and b) larger area and specimen strained 3×0.2 at 850 °C c) smaller and d) larger area.

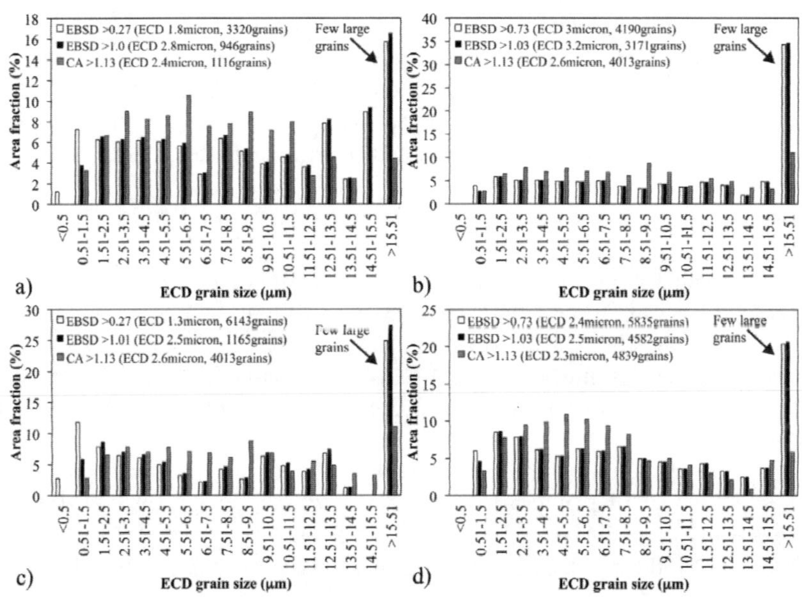

Figure 6. Grain area fractions of specimen strained 3×0.2 at 1050 °C a) smaller and b) larger area and specimen strained 3×0.2 at 850 °C c) smaller and d) larger area.

4. Conclusions

The aim of this study was to verify a detailed model for martensite transformations by using a novel cellular automaton that simulates martensite transformation graphically in a given austenite grain structure. The simulation results need to be linked to a certain material to make them comparable to real life. The simulation and experimental results were compared by analyzing microstructural images and grain sizes. The shape and size of martensite blocks was found to be in good correspondence when similar grain size filtering was used.

In its current form, the CA simulation can be used to simulate the resulting morphology of the martensite blocks in the thermomechanically processed steel in question to a higher degree of detail than is possible using the macroscopic K-M model. The detailed morphology can be used to estimate the mechanical properties of the steel and it could also be used as input information into other simulation tools, for example fatigue models. Hopefully, in the future, the orientations of austenite grains can be implemented into the CA model to produce the texture of the transformed bcc phase on the basis of a K-S orientation relationship as in authors' previous study [11].

References

[1] Kaijalainen A J, Liimatainen M, Kesti V, Heikkala J, Liimatainen T and Porter D A 2016 Influence of Composition and Hot Rolling on the Subsurface Microstructure and Bendability of Ultrahigh-Strength Strip *Metall. Mater. Trans. A* **47** 4175–88
[2] Kaijalainen A J, Suikkanen P P, Limnell T J, Karjalainen L P, Kömi J I and Porter D A 2013 Effect of austenite grain structure on the strength and toughness of direct-quenched martensite *J. Alloys Compd.* **577** S642–8
[3] Sieradzki L and Madej L 2013 A perceptive comparison of the cellular automata and Monte Carlo techniques in application to static recrystallization modeling in polycrystalline materials *Comput. Mater. Sci.* **67** 156–73
[4] Madej L, Sieradzki L, Sitko M, Perzynski K, Radwanski K and Kuziak R 2013 Multi scale cellular automata and finite element based model for cold deformation and annealing of a ferritic–pearlitic microstructure *Comput. Mater. Sci.* **77** 172–81
[5] An D, Pan S, Huang L, Dai T, Krakauer B and Zhu M 2014 Modeling of Ferrite-Austenite Phase Transformation Using a Cellular Automaton Model *ISIJ Int.* **54** 422–9
[6] Bos C, Mecozzi M G and Sietsma J 2010 A microstructure model for recrystallisation and phase transformation during the dual-phase steel annealing cycle *Comput. Mater. Sci.* **48** 692–9
[7] Seppälä O, Pohjonen A, Kaijalainen A, Larkiola J and Porter D 2018 Simulation of bainite and martensite formation using a novel cellular automata method *Procedia Manuf.* **15** 1856–63
[8] Koistinen D P and Marburger R E 1959 A general equation prescribing the extent of the austenite-martensite transformation in pure iron-carbon alloys and plain carbon steels *Acta Metall.* **7** 59–60
[9] Cayron C, Artaud B and Briottet L 2006 Reconstruction of parent grains from EBSD data *Mater. Charact.* **57** 386–401
[10] Miyamoto G, Iwata N, Takayama N and Furuhara T 2011 Reconstruction of Parent Austenite Grain Structure Based on Crystal Orientation Map of Bainite with and without Ausforming *ISIJ Int.* **51** 1174–8
[11] Kaijalainen A, Javaheri V, Lindell D and Porter D A 2018 Development of crystallographic texture under plane and shear strain in ultrahigh-strength strip steels *IOP Conf. Ser. Mater. Sci. Eng.* **375** 012026
[12] Javaheri V, Khodaie N, Kaijalainen A and Porter D 2018 Effect of niobium and phase transformation temperature on the microstructure and texture of a novel 0.40% C thermomechanically processed steel *Mater. Charact.* **142** 295–308
[13] Kurdjumow V G and Sachs G 1930 Über den Mechanismus der Stahlhärtung *Zeitschrift für Phys.* **64** 325–43
[14] Nyyssönen T, Isakov M, Peura P and Kuokkala V-T 2016 Iterative Determination of the Orientation Relationship Between Austenite and Martensite from a Large Amount of Grain Pair Misorientations *Metall. Mater. Trans. A* **47** 2587–90
[15] Morito S, Huang X, Furuhara T, Maki T and Hansen N 2006 The morphology and crystallography of lath martensite in alloy steels *Acta Mater.* **54** 5323–31
[16] Dongen S van 2000 *Graph Clustering by Flow Simulation* (University of Utrecht)
[17] Gomes E and Kestens L A I 2015 Fully automated orientation relationship calculation and prior austenite reconstruction by random walk clustering *IOP Conf. Ser. Mater. Sci. Eng.* **82** 012059
[18] Nyyssönen T 2017 *Quenching and Partitioning of High-Aluminum Steels* (Tampere University of Technology)
[19] Wolfram S 2002 *A New Kind of Science* (Champaign: Wolfram Research)
[20] Morito S, Adachi Y and Ohba T 2009 Morphology and Crystallography of Sub-Blocks in Ultra-Low Carbon Lath Martensite Steel *Mater. Trans.* **50** 1919–23

Limit conditions for the onset of abnormal grain growth in a homogeneous microstructure: Theory and Experiments

GC Abbruzzese[1,2]
[1] CEO , Spires Srl- Terni, IT
[2] Strategic Advisor, Rina Consulting-CSM, Rome, IT

Email:giuseppeabbr1@virgilio.it, giuseppe.abbruzzese@externals.rina.org

Abstract. In previous papers it has been demonstrated that abnormal grain growth in a polycrystalline microstructure is a natural phenomenon even in the presence of an homogeneous microstructure. Namely, in a statistical sense, each portion of the material has the same grain size distribution and no local heterogeneities are involved. The only necessary condition is the presence of a grain growth inhibitor which can be represented by second phase particles (Zener Drag), homogeneously distributed as well, or by segregating atoms on the grain boundary (Atoms drag).
Considering specifically the Zener Drag effect, it can be shown that in the presence of an evolution of the restraining force (even if extremely slow), due to Ostwald ripening or to dissolution of the particles, the corresponding grain growth process allows the grain size distribution to continuously evolve reaching a critical shape. Such peculiar shape corresponds to a "structural" instability which irreversibly determines the onset of an abnormal growth phase with a drastic change of grain growth kinetics (kinetics exponents for the mean radius largely above the canonical 0.5).
In this context the mechanism of Abnormal Grain Growth onset is discussed as a general physical phenomenon and a set of experimental data on several steels will be presented for supporting such conclusion.

1. Introduction

The presence of second phase particles or solute atoms segregating along the grain boundaries affects the kinetics of grain boundaries motion resulting in a restraining force ($P_{Z,S}$) which reduces the kinetics of normal grain growth (mean radius $R=k\,t^n$ with $n < 0.5$) [1-4].

Particles cause an inhibition of grain boundary motion which can be considered similar to a retarding force P_Z acting along the moving boundary; there are thus two possible conditions:

- The external driving force P has the same magnitude as P_Z but opposite sign (the net force on the boundary is zero).
- The external driving force P is greater than the maximum force P_{Z0}, which can be exerted by the particles, than the boundary can move.

The maximum force P_{Z0} can be expressed as follows (γ = boundary energy):

$$P_{Z0} = \gamma \cdot I_{Z0} \quad ; \quad I_{Z0} = \beta \cdot \frac{f_v}{r_p}. \tag{1}$$

with f_v (volume fraction of particles), r_p (mean radius) and β a constant in the range of 0.75 ÷ 2 according to particle geometry. The grain boundary velocity can be calculated as follows (m = boundary mobility):

$$v = m \cdot \Delta P = m \cdot (P - P_{Z0}) \tag{2}$$

In figure 1 the net forces ΔP according to Eq.2 are plotted as a function of boundary velocity pointing out three ranges for such a function [2,3].

If we apply Eq. 1 and 2 to the ranges described above we obtain the values of ΔP_{ij} as a function of the different boundary curvatures (R_i, R_j):

$$\Delta P_{ij}(I) = \gamma \cdot \left(\frac{1}{R_j} - \frac{1}{R_i} - I_{Z0} \right) \quad ; \quad \Delta P_{ij}(II) = 0 \quad ; \quad \Delta P_{ij}(III) = \gamma \cdot \left(\frac{1}{R_j} - \frac{1}{R_i} + I_{Z0} \right)$$

(3)

Under the homogeneity assumption valid in any statistical system [2-4] it is possible to use the following kinetics equation for grain i, by summing the motion of ij grain boundary subjected to the net driving force ΔP_{ij} and moving with v_{ij} velocity

$$\frac{dR_i}{dt} = \sum_j \Delta P_{ij} = \sum_j w_{ij} v_{ij} = M \sum_j w_{ij} \cdot \left(\frac{1}{R_j} - \frac{1}{R_i} - I_{ij} \right) \quad ; \quad w_{ij} = \frac{\varphi_j R_j^2}{\sum_j \varphi_j R_j^2}$$

(4)

where the expression for the probability function w_{ij} is derived from basic topological laws in [4] and $M = 2m\gamma$ is the boundary diffusivity.

The growth rate for each class i can be derived specializing Eq.4, according to Eq.3, as :

$$\frac{dR_i}{dt} = M \left[\psi_I \cdot \left(\frac{1}{R_C^I} - \frac{1}{R_i} \right) + \psi_{III} \cdot \left(\frac{1}{R_C^{III}} - \frac{1}{R_i} \right) - (\psi_I - \psi_{III}) \cdot I_{Z0} \right]$$

(5)

$$\text{with} \quad \psi_\lambda = \frac{\sum_\lambda \varphi_j R_j^2}{\varphi_j R_j^2} \quad ; \quad \frac{1}{R_c^\lambda} = \frac{\sum_\lambda \varphi_j R_j}{\sum_\lambda \varphi_j R_j^2} \tag{6}$$

and λ=I, II, III interval [3]. By analyzing, in respect to a grain R_i, the GSD we can divide it into three zones, as pictured in figure 2: one containing the grains which tend to shrink, one containing the grains which tend to stagnate and the last one containing the bigger grains which are allowed to grow when in contact with the grain R_i. After the initial period of restricted grain growth, in the presence of stable grain growth inhibition, the mean radius tends to stabilize around a constant value [1-3]. It has also been shown that, starting from stagnation condition, for a decreasing value of the Inhibition factor I_z the mean radius tends to reach a stagnation condition (n=0) characterized by a GSD, converging to a very well defined shape, independent of the initial conditions and the characteristics of which are [2]:

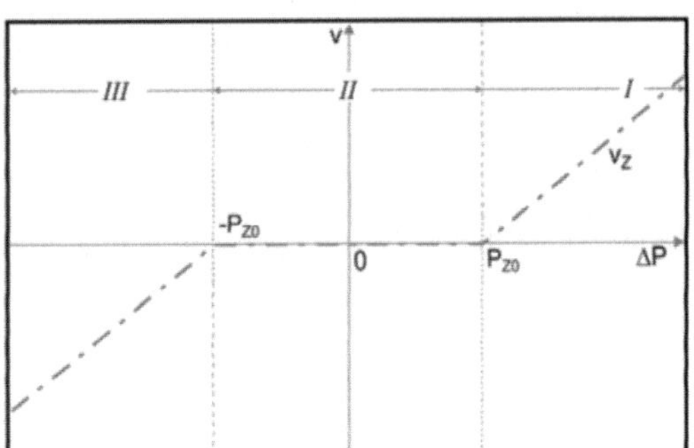

Figure 1. GB velocity vs. driving force ΔP in the presence of Zener drag.

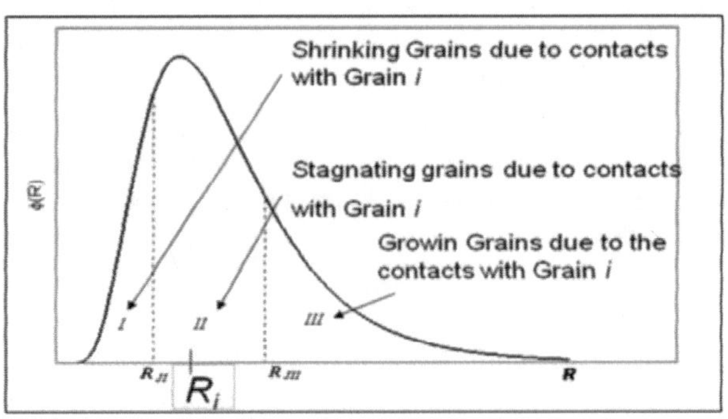

Figure 2. Grain Size Distribution (GSD) and relative behavior in correspondence to grain R_i.

IOP Conf. Series: Journal of Physics: Conf. Series **1270** (2019) 012036 doi:10.1088/1742-6596/1270/1/012036

- A GSD characterized by the absence of small grains, which disappeared during the restrained grain growth process, up to a value $u_{min} = R_{min}/R_c$ which is in equilibrium with the largest grains in the system $u_{max} = R_{max}/R_c$ (R_c is the critical radius in the classical sense and calculated in other papers [1,2]).
- A peaked GSD (density function) with a maximum in correspondence of the smallest grains present in the system ($R_{max}/R_{min}=3$, $p(u_{min})/p(u_{max}) = 3^4$, $p(u) = (12/13) * 1/u^4$).

The stability analysis of the limit GSD at any, even infinitesimal, drop of the Zener Drag show that in this case the boundary energy minimization process reaches maximum instability [2].

Such approach to a limit GSD as a precondition for the onset of abnormal grain growth (AGG) can be more clearly appreciated in the presence of homogeneous Zener Drag instability.

In figure 3, in a double logarithmic plot, the case of normal grain growth ($I_z=0$), starting from a log-normal distribution with mean radius of 5 micron, is shown. In this case the kinetics exponent of the mean radius is n=0.5 and the steady state distribution is the Hillert type for 3-D case [3] .
Moreover in figure3 a set of simulations are also shown where the Inhibition factor I_z is allowed to decrease continuously with a simple linear function but with different kinetics parameters (dI_z/dt) according to the law $I_z = I_{zo} - (dI_z/dt)\ t$ with $I_{zo} = 1000\ cm^{-1}$.

The effects, although of the same nature, are different in intensity according to the kinetics of the allowed inhibition drop. Namely, through a faster growth of the mean radius, larger abnormal grain sizes are obtained the slower the inhibition drop kinetics is (see the maximum coefficients n in figure 3; the apparent grain growth exponents for the mean radius are much higher than 1/2 while the maximum radius grows with $n=1$ according to the physics of the phenomena). After the fast growth of the mean radius (size of millimeters can even be reached, as it is well known in practice) , proceeding the decrease of I_z until zero, the growth process converges to the normal grain growth "route". However at this point, the growth process proceeds with much slower kinetics after the dramatic and fast change of grain size (and correspondingly to the change of the total boundary energy in the system). If the Inhibition drop is arrested during or after the discontinuous process of grain growth the system will reach a new stagnation condition, finding a new equilibrium between the value of the residual Inhibition and the grain size reached by the abnormal growth (even in this case the final result will be strongly influenced by the drop rate of the Inhibition value during the unstable period).

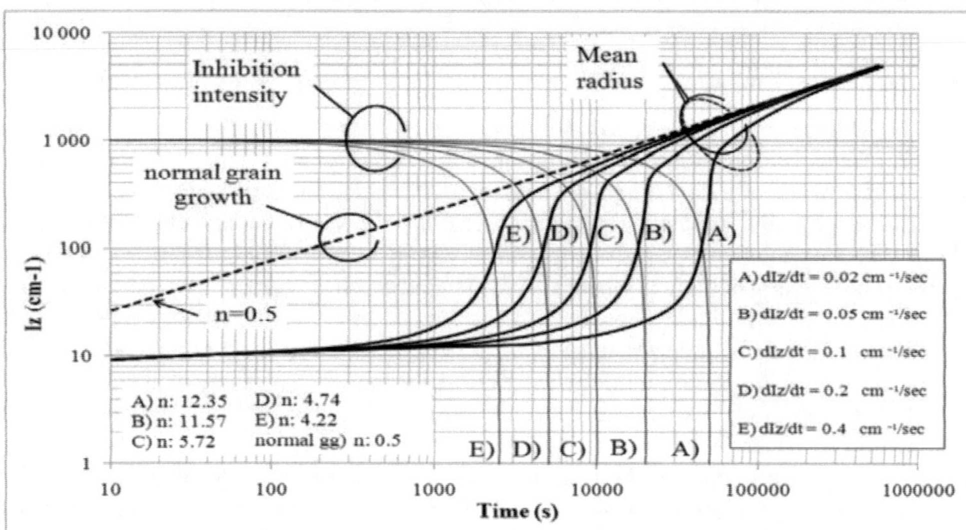

Figure 3. Mean radius evolution as a function of different GG Inhibition drop rate.

In figure 4, starting from the B GSD of figure 3, the evolution of the GSD is shown for a select case of Inhibition drop kinetics. Here it is possible to see how the instability in the Zener Drag

restraining force, through a continuous change of the GSD, leads to an unstable GSD which presents a high peak concentrated around relatively small grains in the system.
For any further drop of the inhibition intensity this unstable condition develops in fast grain growth kinetics, characterized by a broader GSD (even double peaked if the drop rate is sufficiently slow) and ending up with large average grain size (from the initial grain radius of a few microns to the final one in the order of a millimetre). When the inhibition fades ($I_z \rightarrow 0$), a steady state grain growth tends to establish(t_8 case in figure 6) which is asymptotically approaching the theoretical "route" for the normal grain growth shown in figure 3 (*n* coefficient = 0.5). In figure 5 and 6 the shape of the grain size density function is compared at different stages of the process.

The evolution of the GSD suggests an "apparent" strategy adopted by the grain system to efficiently reduce the free energy (grain boundary energy) in the presence of a restraining force to a free grain growth (here specifically of Zener Drag type, but similar behavior can be expected also for the atoms drag [3]).

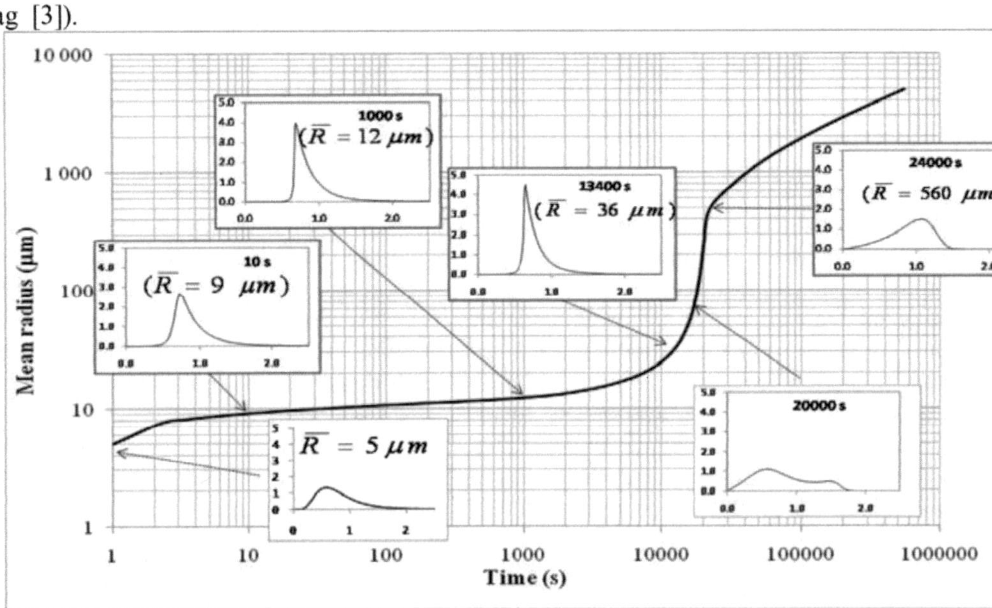

Figure 4. Evolution of GSD during Inhibition drop at different stage of the process (case B of figure 3).

In fact the change of the shape of the GSD, as previously described, suggests a kind of "preparation" of special topological conditions for the larger grains, namely the realization of a surrounding grain neighborhood formed preferably of the smallest grains available in the system, producing an energetically "unstable" equilibrium condition. For any further, even little, variation of the inhibition intensity the few relatively large grains will grow at the expenses of the smallest grains in the system with the most efficient kinetics [4,5]. In figure 7 the variation coefficient and the ratio between maximum and mean radius (of the simulated cases in figure 3) are shown as a function of time. It can be seen that the initial narrowing of the distributions is followed by a peak which becomes higher the slower the inhibition drop rate is during the abnormal process (coincident with max n values in figure 3) but never overtakes the theoretical limit value of $p(u_{min}) = 4.67$ [2]. As a matter of fact, the limit distribution can never be exactly reached during the grain growth process if the presence of a certain thermally activated decay of the restraining force is present (particles dissolution or Ostwald ripening, as normally is the case). Moreover, such distribution should be seen as the final stage by which the microstructure loses "memory" of the initial GSD completely and therefore its shape becomes an invariant to any "path" the initial microstructure goes through.

2. Comparison between experiments and theoretical predictions

2.1.Structural Carbon Steel

Experiments on a set of Cr-Mn micro-alloyed steel, for quenching and tempering, either alloyed with Ti or Nb, have been carried out in laboratory. After ingot casting and hot rolling, samples of steel plates have been prepared to perform thermal treatments to study austenitic grain growth. The steels behave very similarly although the typical critical temperature of AGG changes with the chemical composition. For sake of brevity, only a Nb alloyed steel is hereby considered, with the following chemical composition (wt %):

C	Mn	Cr	Alsol	Nb	N	Si	S	P	Ti
0,7	1,28	0,52	0,031	0,088	0,012	0,3	0,01	0,02	–

The thermal treatments (1050 °C, for 8,16 and 24 hours) are carried out in a laboratory furnace with controlled atmosphere. Grain boundaries, after martensitic quenching and tempering, are evidenced by a picric etchant with special additives, capable to reveal the prior Austenite grains. Grain size has been obtained by measuring the area of about 1000 grains for each sample by using a quantitative metallographic system. From the grain area an equivalent radius (diameter) of each grain can be defined and the 2-D grain size distribution so obtained has been transformed in 3-D by numerical procedure using the Schwarz-Saltykov method [7]. With this method the approximation of a spherical shape for the grains can induce to an overestimation of small grains in 3D, which requires a correction , typically involving less than 5% of total grain population . An analysis of the statistical significance of the histogram, by the numerosity of each grain class and in relation with the total number of the grain population, led to the choice of a size class width (radius) for the histogram equal to 2 micron.

Because of this procedure a series of possible sources of errors are concurrently operating and must be counteracted as much as possible by adopting accurate procedures, for instance:

- creating a sufficient homogeneous material along the steel fabrication
- adequately singling out the stage at which abnormal grain growth starts (Temperature and Time)
- controlling the quality of micrographs obtainable to measure the prior austenitic grain (quality of etchant, possibly needed correction for evident incomplete boundary patterns, use of a magnetic table to contour the grains,...)
- adopting a sufficient numerosity of the sample and good quality of transformation from 2-D to 3-D

In figure 5 the size distributions, relative to the three different treatment times, are reported and compared (grain frequency in % vs diameter in μm). It is possible to clearly see that there is an evolution of the shape of the size distribution and of other average parameters (figures. 7a,b,c).

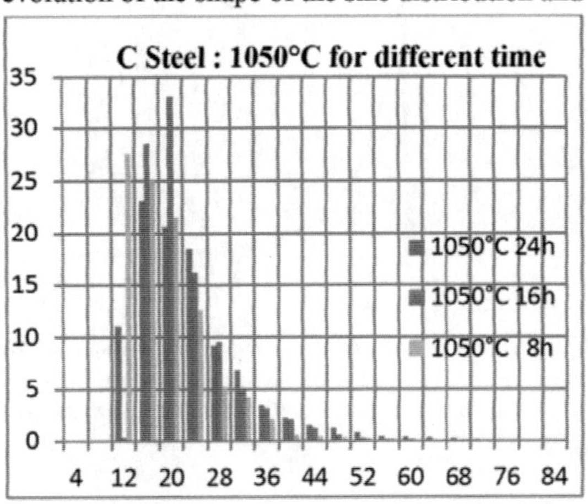

Figure 5. GSD (% vs diameter μm) for 8,16, 24h at 1050°C.

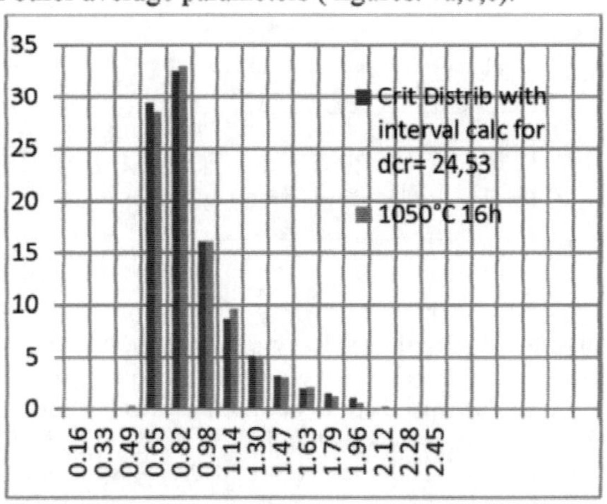

Figure 6. GSD at 16h vs the limit distribution $u \in \left(\frac{2}{3}, 2\right)$, with the interval width Δu normalized by the experimental critical diameter (d_{cr} 24,53 μm for sample 1050°C 16h).

Such a behavior is in line with the previous discussion of having an initial distribution narrowing in the presence of strong inhibitors of grain growth by the disappearance of very small grains in GSD (8h GSD in figure 5a) [1]. During the coarsening of (NbC,N) precipitates (continuous inhibition drop), the GSD evolves into a more peaked distribution (16h GSD), with an even narrower shape (see the variation coefficient in figure 5b), a ratio between maximum and critical radius around 2,1) and a ratio between the Mean radius and critical radius which reaches a value near the theoretical one for the limit distribution (12/13=0,923, figure 7a). The process is then followed by an initial phase of abnormal growth (24h GSD) in which a certain number of very large grains (5-10% in volume) appear in the microstructure. This can be statistically "measured" by: a shift of the peak of the 16h GSD to the left (smaller grains), a smaller ratio between the Mean radius and Critical radius - after having reached a value near the theoretical one for the limit distribution shown in figure 7a, and a variation coefficient that increases (also in correspondence of the creation of a tail of large grains in the GSD). The whole process is described in the simulation reported in figure 7d (peaking, shifting and tail creation in GSD) which corresponds to the evolution of distributions in first part of figure 4 up to the onset of AGG.

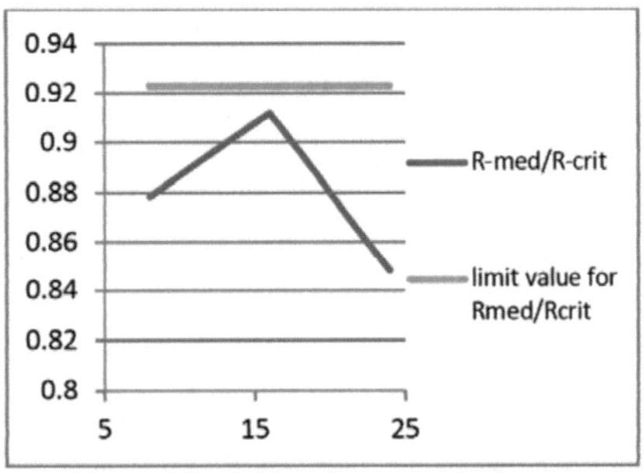

Figure 7a. Comparison as a function of time (h), between the experimental R_m/R_{cr} ratio with the limit value 0,923.

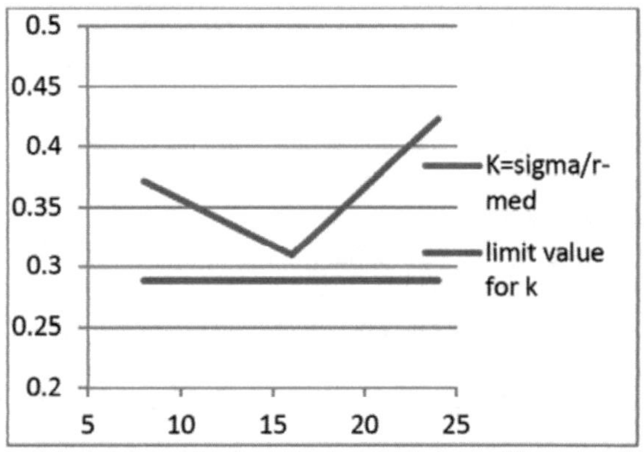

Figure 7b. Comparison of the experimental values for the variation coefficient k with the limit value: $k = 0.288 = 1/\sqrt{12}$.

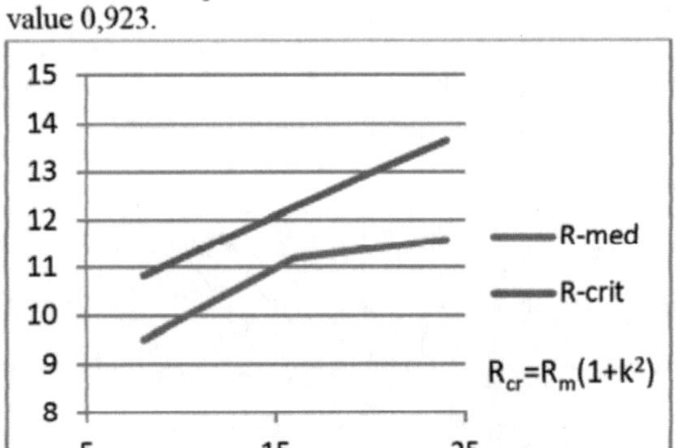

Figure 7c. Evolution of R_m and R_{cr} parameters as a function of time. First converging and then diverging. It means that the variation coefficient first is pointing toward the limit condition k=0,288 , narrowing the GSD and then, after the onset of AGG, starts to increase.

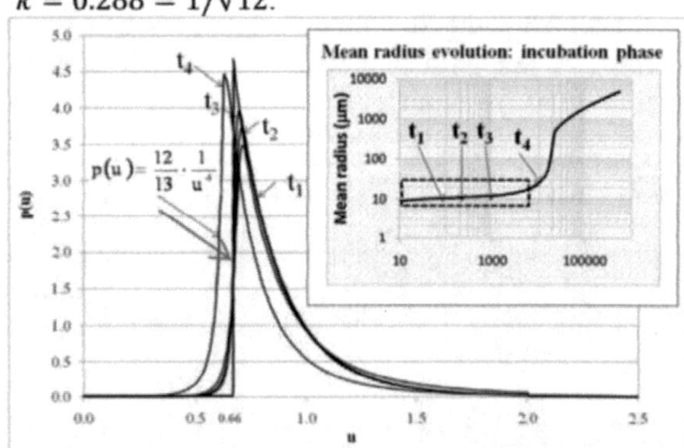

Figure 7d. Evolution of the GSD in the simulation of Figure 3-case B, showing the approaching to the limit GSD and the deviation as soon the AGG starts (see GSD at t_4).

2.2. *Grain Oriented Electrical Steel*

Two grades of Conventional Grain Oriented Electrical Steels of the following composition (wt%) have been investigated for grain growth process up to abnormal growth condition.

Steel	C	Si	Mn	S	Alsol	N	Cu
1	0,025	3,15	0,07	0,026	-	0,005	0,06
2	0,027	3,20	0,065	0,020	0,009	0,006	0,05

Steel 1 contains only MnS, Cu_xS precipitates and Steel 2 almost only contains AlN as second phase grain growth inhibitors (the latter by adding Aluminum and reducing the slab reheating temperature to a value in which the dissolution of Manganese Sulfide is negligible). The outcome of the growth process in these materials (Goss grains), is also influenced by the crystallographic texture in the primary matrix and is sensible to differences in the microstructure along the sheet thickness. However, the general grain growth process, before the establishment of the selective growth of the minor Goss Texture Component, follows with good approximation the rules of uniform grain growth in the presence of Zener Drag. The particle dissolution for Ostwald Ripening then induces microstructure evolution and instability.

Samples (30mm wide, 300 mm long // RD) from a cold rolled and decarburized strip (0.30 mm thick) have been treated in a special Thermal Gradient Furnace to simulate the box annealing treatment of secondary recrystallization (30°C/h up to 920°C, as maximum temperature on one end of the long side of the sample, with soaking treatment of 1 hour). Such treatment allows to single out the position (and the associated soaking Temperature) where the secondary recrystallization starts along the sample (in this experiment it was about 875 °C). It means that on the side of the strip sample above this temperature the secondary recrystallization is fully accomplished (secondary grains of millimeters size are visible to the naked eye) whereas on the side where the temperature is below 875 °C the grains are, in the stage of incipient AGG, continuously growing (typically in the range of 10-20 micron size). A metallographic sample just in front of the position in which the abnormal grains were visible has been taken (in the center of sheet thickness to avoid subsurface layer effects), namely where the continuous grain growth process was still operating, and then investigated for the GSD by measuring the area of more than 1000 grains. The 2-D grain size distribution has been transformed in 3-D by the method already mentioned [7]. The resulting experimental GSD shown in figure 8a refers to the case of Steel 1 at 870°C, which is very close to the temperature corresponding to the large secondary grains in the strip (875 °C in the temperature gradient). 875 °C here represents the case of onset of abnormal growth, in fact it's possible to see in the distribution the appearance of a tail of large grains and correspondingly the presence of a class of small grains. The difference between the two histograms, in figure 8b, shows this process even better at the two ends of the GSDs, as well as the variation in the body of the histograms, with the increase of presence of grains roughly above the conventional critical radius (9,51μm), and a decrease of grains just below the R_c. The coincidence of the conventional critical radius valid for normal grain growth R_c is valid, as a parameter for the whole distribution, only for the limit distribution[2]. The equations (5) and (6) are those that regulate the growth in the general case, with the consequence of the fall of the concept of a unique critical radius valid for the whole distribution. The difference in the shape of the two GSDs (figure 8a) is again of the type described by the simulations in the figure 7d, namely the GSD in t_4 represents the sample at 875°C, with the typical shift in respect to the one near the critical distribution (sample at 870 °C). In figure 9a the GSDs of Steel 1 and 2 at 870°C (namely just before the onset of the secondary recrystallization) are compared. The important fact shown here is that notwithstanding the grain growth inhibitors being completely different in the two cases (AlN and (Mn,Cu)S), the GSDs are substantially identical showing again that the process is rather generalized and does not depend on the nature of the material or on the type of precipitates (see also the case of structural steel previously discussed). In figure 9b the comparison of Steel 2 at 870 °C with the limit distribution, transformed in a frequency histogram with the same class width and normalized with the same experimental critical radius, shows a very good agreement.

The above results and other similar ones for silicon iron already reported [3] support the evidence that the onset of abnormal grain growth here is preceded by a transformation of the GSD providing the topological condition for a fast discharge of the grain boundary energy.

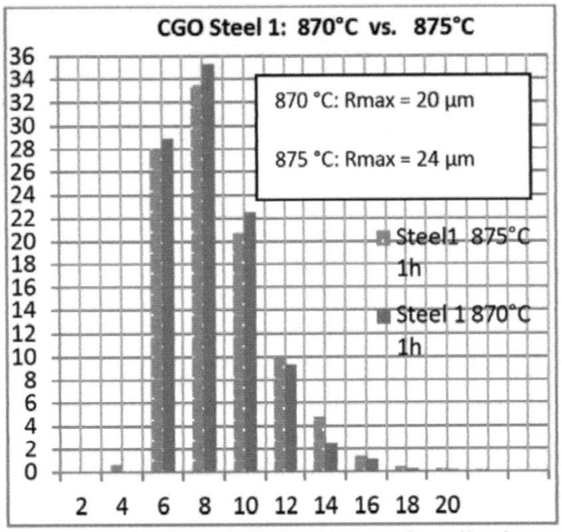

Figure 8a. CGO steel GSDs (% vs grain radius in μm), just before and at the onset of AGG.

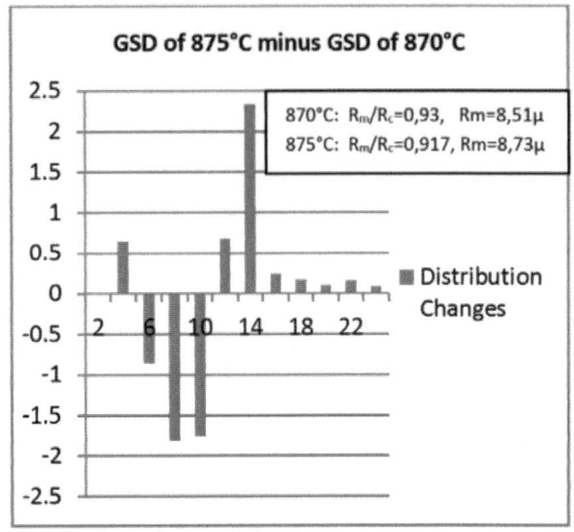

Figure 8b. Difference between the two histograms (Δ%) showing the relative shifting of the grains frequency.

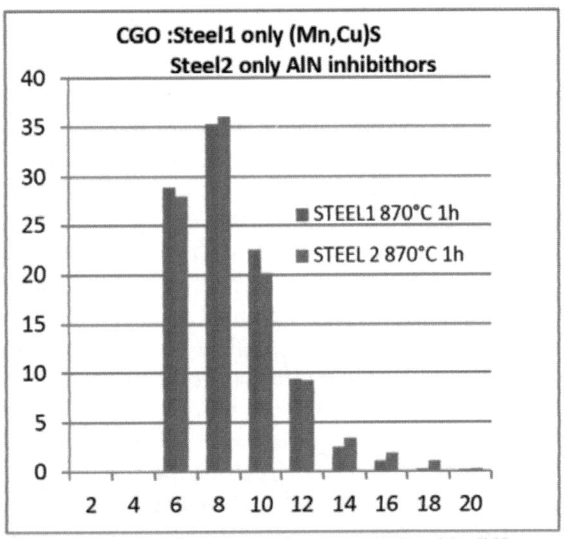

Figure 9a. Comparison of the two CGO with different grain growth inhibitors.

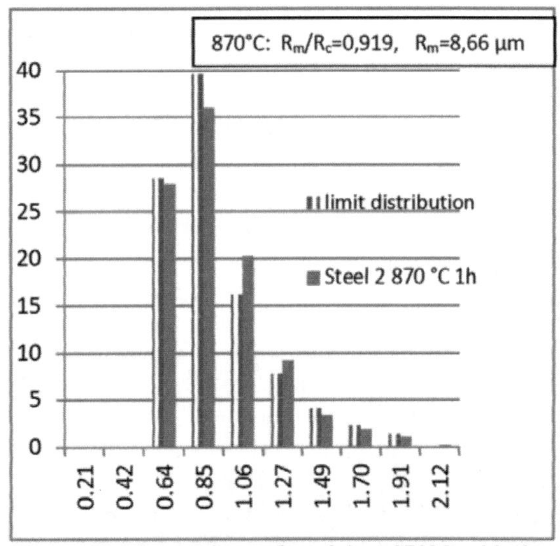

Figure 9b. Comparison of steel 2 at 870°C with the theoretical limit curve defined in the interval $u \in (\frac{2}{3}, 2)$: the class width $\Delta u = \Delta R / R_c$ is normalized with the experimental critical R_c.

3. Conclusion

In the framework of the Statistical Theory it has been shown that the abnormal grain growth in the presence of unstable inhibitors (Zener or Atom Drag, ..) is a continuous and homogeneous process and it is independent of local phenomena (as those possibly induced by topological, texture, precipitation or drag inhomogeneity) . The main characteristic of the grain growth process in these conditions is that the evolution of the GSD, according to the specific inhibition drop rate, passes close to a limit equilibrium GSD which, in the presence of a restraining force to grain growth, plays the role of *"microstructure organization for efficiently reducing the boundary energy"*. The formation of a microstructure with very few grains of relatively large size (3 times the smallest grains in the system) embedded in a large population of small grains (coincident with a high peak in GSD) produces an acceleration phase for the grain growth kinetics with features of a discontinuous process. The decay rate of the inhibition is responsible for the growth kinetics; in fact a lower drop rate increases the slope of the mean radius kinetics, enhances the presence of a two peaks distribution in correspondence of the maximum growth speed and enlarges the size of the final abnormal grains. The comparison of the experimental GSD of steels just before the abnormal grain growth induced by Zener drag shows a good agreement with the theoretical prediction of the limit GSD (in particular with the samples of Micro-alloyed steels for quenching and tempering and of Grain Oriented silicon iron).

Accurate grain size measurements on GO Silicon Iron (Ferritic Phase) and Construction Steels (Austenitic Phase) show that this basic mechanism is very generalized and, in substance, is dominating any other possible growth enhancing effect in the microstructure (like for example texture effect in Grain Oriented steel for which the controlled drag evolution becomes a "necessary condition" for a proper texture selection).

References

[1] Abbruzzese G 1985 Computer simulated grain growth stagnation *Acta Metall.* **33** 1329–37
[2] Abbruzzese G C, Brandt R and Lucke K 1998 Limit conditions for abnormal grain growth in the presence of Zener drag *Grain Growth in Polycrystalline Materials III* ed H Weiland, B Adams and A D Rollett (Pittsburg: The Minerals, Metals and Materials Society) p 263
[3] Abbruzzese G C and Forzanti C 2012 Homogeneous Abnormal Grain Growth in Polycrystalline Materials *Mater. Sci. Forum* **706–709** 1355–60
[4] Abbruzzese G C and Buccioni M 2007 Theory of Grain Growth in the Presence of Atoms Drag Effects *Mater. Sci. Forum* **558–559** 1005–12
[5] Abbruzzese G C 2012 Derivation of Statistical Model of Grain Growth Based on 3-D Von Neumann Equation *Mater. Sci. Forum* **715–716** 427–36
[6] Barteri M, Rossi A and Mascanzoni A 1983 *CSM internal report n° 4548R*
[7] Underwood E E 1968 Particle Size Distribution *Quantitative Microscopy* ed R T DeHoff and F N Rhines (New York: Mcgraw-Hill) p 162

IOP Conf. Series: Journal of Physics: Conf. Series **1270** (2019) 012037 doi:10.1088/1742-6596/1270/1/012037

Influences of cold rolling, recrystallization and surface effect on the transformation textures in a TA10 titanium alloy

Yang P[1], Wei Z G[1], Gu X F[1], Cui F E[2] and Mao W M[1]

[1]School of Mater. Sci.& Eng., Uni. of Sci. and Tech. Beijing, Beijing, 100083, China
[2]Institute of Advanced Mater. and Tech., Uni. of Sci. and Tech. Beijing, Beijing, 100083, China

Abstract. A transformation treatment based on surface effect induced transformation is applied to cold rolled TA10 titanium alloy sheets to change and control transformation texture. This kind of treatment can promote the α-phase nucleation at sheet surface region during cooling and control the growth of surface α-grains into center layer of sheet. The results indicate that as the heating temperature increased, the transformation textures in the final HCP-structured TA10 sheets changed in helium atmosphere from the recrystallization type texture of <11-20>//RD to the transformation type texture of <11-20>//ND. The textures in sheets treated at low transformation temperature of 1000°C reveal texture memory effect, while those treated at high transformation temperature of 1100°C illustrate the effects of both transformation strain and anisotropy in elastic modules of HCP titanium crystals. The surface effect is promoted by cooling gas of helium, which leads to layered structures along sheet thickness direction. The surface region often shows small equiaxed grains with recrystallization orientations, while the center layer reveals coarse grains with either plate-like or equiaxed grain morphology corresponding to transformation orientation or recrystallization orientation, respectively. The special grain boundaries in plate–like α variants are different from those in equiaxed grains. These differences in morphologies, sizes and orientations are explained in terms of the mechanism change of the factors affecting variant selection.

1. Introduction

Transformation textures in high symmetrical metallic materials obeying orientation relationships (OR) such as the K-S or Burgers OR are generally weak due to the more crystallographically equivalent variants. In special cases, however, transformation texture may be sharper because of strong variant selections. Many models are suggested for the variant selection leading to stronger transformation texture as summarized in [1], such as the special dislocation configurations related with deformation, the special anisotropic elastic strains or transformation strain. The occurrence of strong transformation texture is of both theoretical and practical significances because transformation texture is normally different to deformation or recrystallization textures, thus to demonstrate special physical or mechanical properties. The so-called surface effect induced transformation texture present in electrical steels, namely differently oriented grains nucleate firstly at sheet surface region due to either the anisotropies in strain energy or in surface energy under H_2 atmosphere, can be an example of producing favorite {100} texture for a magnetic property improvement [2,3]. Our previous work indicates that the surface effect induced transformation texture of <11-20>//RD with columnar grained structure can be produced in pure titanium sheets in Ar atmosphere [4], this work reports the evolution of microstructure and tetxures in transformation-treated TA10 sheets in helium atmosphere which has a 10 times higher thermal conductivity than that of Ar atmosphere.

2. Materials and processing

The materials used for this study is a commercial HCP-structured TA10 titanium sheets with nominal composition of Ti-0.3Mo-0.8Ni in mass percent. Starting material is a hot rolled and annealed sheet with thickness of roughly 2mm. Then it was cold rolled for 60%-80% reductions and sample was heated quickly to different temperatures above or below the critical transformation temperature (T_c=910°C which was measured using DSC method at a heating and cooling rate of 500°C/h). Helium gas (He) was used during transformation treatment and compared with those in Ar [4]. The samples for EBSD measurement were electropolished with a polishing solution consisting of 5% perchloric acid and 95% ethyl alcohol with a voltage of 30 V for 30s at room temperature. The EBSD system of Oxford-instrument mounted on an SEM Zeiss-FEG under 20KV is applied to reveal microstructure and texture. The allowance of 15° deviation angle is set to determine the area fractions of each texture component.

3. Results and analyses

3.1. Recrystallization texture

For the purpose of comparison the recrystallization texture is firstly determined and shown in figure 1 of TA10 sheet after 80% cold rolling and 800°C 15min annealing. As seen in most cold and warm rolled and annealed Ti sheets, a typical TD-rotated basal texture {01-13}<-2110> is present with an average grain size of 46.6μm and uniform equiaxed morphology. The misorientation angle distribution shown in figure 1c doesn't reveal any special peaks related to variants due to Burgers OR.

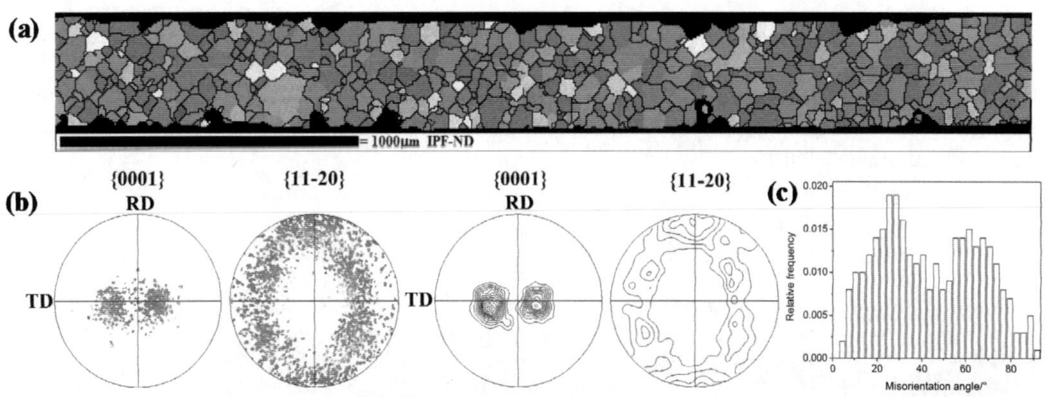

Figure 1. EBSD maps of 80% cold-rolled TA10 sheet with 0.4 mm thickness following recrystallization annealing at 800°C for 15min, **(a)** orientation map shown in IPF-ND; levels:1,2,3…**(b)** pole figures; **(c)** misorientation angle distribution

3.2. Influence of heating temperature

Figures 2-4 show the EBSD maps of TA10 sheets subjected to 80% cold rolling and subsequent heating in He gas for 1-3min to 920°C, 1000°C and 1100°C (T_c=910°C) respectively. The cooling rate during transformation is 300°C/h and final thickness is 0.4mm. It is seen that at all temperatures grain sizes are un-uniform, small grains formed at surface region and coarse grains developed in center layer. The grain morphologies are different. At 1100°C (figure 4) plate-like grains/variants (green color) with <11-20>//ND orientation are easily formed and their grain boundaries tend to be parallel to ND. Concerning grain boundary types it is seen that at low heating temperature (figure 2) the 5 special misorientation angles and axes among 12 variants (they are listed in the up-right of figure 2) which are related to variants for Burgers OR are less, whereas at 1100°C, the <11-20>60° grain boundaries increased from 2.1% to 15.3%. This indicates that the formation of corresponding variants can reduce transformation strain very effectively. 15.8% grains are in <11-20>//ND oriented (green color) in area fractions and 9.53% are <10-10>//ND oriented at 920°C, while 40.8% and 20.6% are present for two kinds of orientations at 1100°C totally amounting to 61.4%. Rotated basal recrystallization texture is still the main texture component as seen in figure 2c and figure 2d and they can be both small grains at surface region and coarse grains in center layer. The arrows in figure 2a, 3a, 4a reveal that a surface small grain with either of two kinds of orientations can grow into center layer to become a coarse grain. It is noted that the rotated basal texture here is regarded as transformation texture rather than recrystallization texture because grains are subjected to transformation, it is a texture memory effect.

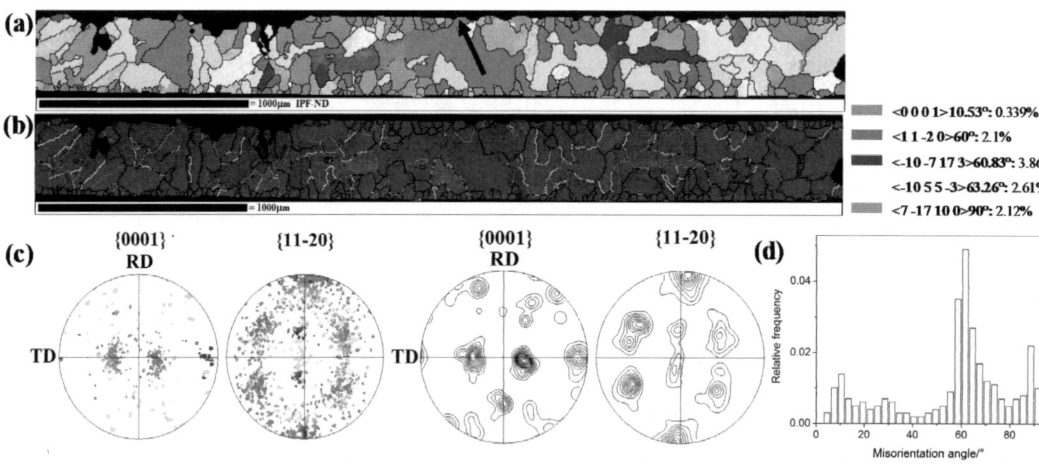

Figure 2. EBSD maps of 80% cold-rolled TA10 sheet with 0.4 mm thickness and heated to 920°C (T_c=910°C) for 3 min, and then cooled with a rate of 300°C /h, **(a)** orientation map shown in IPF-ND; **(b)** Kikuchi band quality map with 5 special grain boundaries in figure 2a; **(c)** pole figures; **(d)** misorientation angle distribution

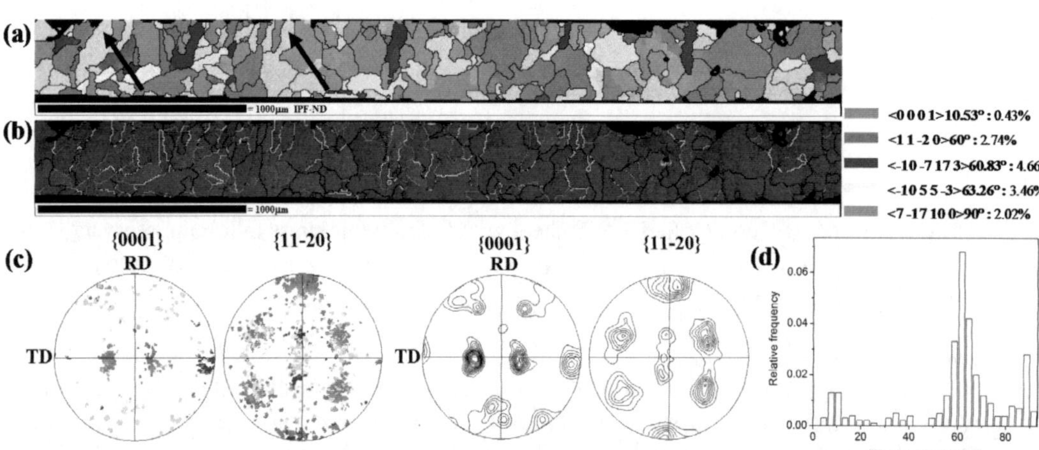

Figure 3. EBSD maps of 80% cold-rolled TA10 sheet with 0.4 mm thickness annealed at 1000°C for 1min, **(a)** orientation map shown in IPF-ND; **(b)** Kikuchi band quality map with 5 special grain boundaries in figure 3a; **(c)** pole figures; **(d)** misorientation angle distribution

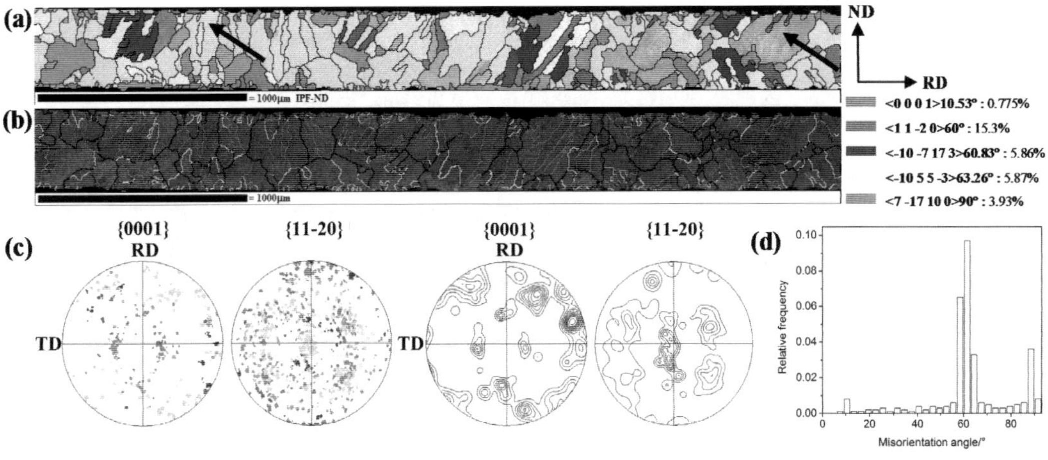

Figure 4. EBSD maps of 80% cold-rolled TA10 sheet with 0.4 mm thickness annealed at 1100°C for 1min, **(a)** orientation map shown in IPF-ND; **(b)** Kikuchi band quality map with 5 special grain boundaries in figure 4a; **(c)** pole figures; **(d)** misorientation angle distribution

3.3. Influence of rolling reduction

Figure 5 and figure 6 show the EBSD maps of TA10 sheet subjected to 60% cold rolling and subsequent 1000°C and 1100°C (T_c=910°C) annealing in He for 1min. The final thickness is 0.4mm (initial thickness is 1mm). By comparison to figure 3 and figure 4 of 80% rolled and transformation treated sheets, it is seen that there is no large difference between them and the main difference in grain morphologies and orientation distribution are due to heating temperature. At medium heating temperature of 1000°C, the <11-20>//ND texture is slightly stronger than the rotated basal texture and 28.5% grains in area fractions are <11-20>//ND oriented (green color) and 9.37% are <10-10>//ND oriented.

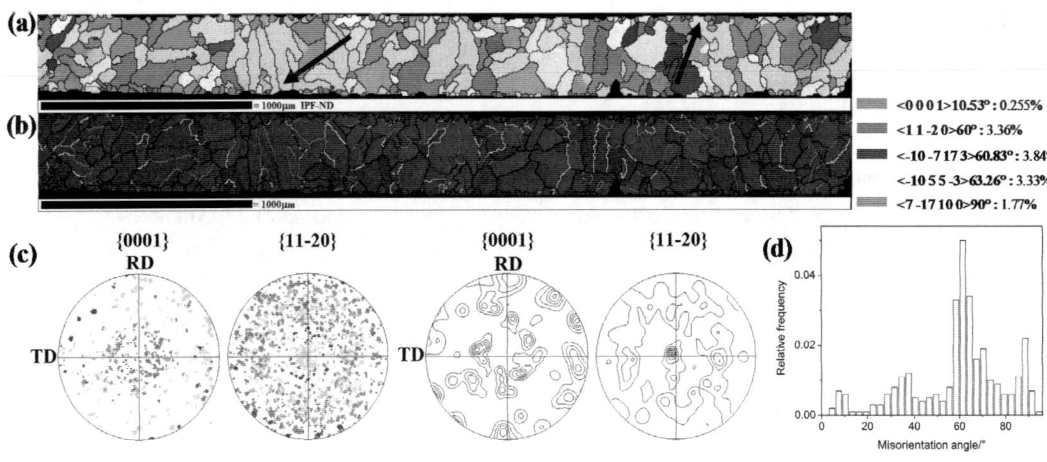

Figure 5. EBSD maps of 60% cold-rolled TA10 sheet with 0.4 mm thickness annealed at 1000°C for 1min, **(a)** orientation map shown in IPF-ND; **(b)** Kikuchi band quality map with 5 special grain boundaries in figure 5a; **(c)** pole figures; **(d)** misorientation angle distribution

In figure 6 of transformed sheet at 1100°C, 44.1% grains are in <11-20>//ND orientation (green color) in area fractions and 15.9% are <10-10>//ND oriented totally amounting to 57.05%. The <11-20>60° grain boundaries (red lines in figure 6b) increased to 18.6%. Figure 6e exhibits the 3 variants orientations in pole figures within a round large grain boundary as shown in an arrow in figure 6a and it shows that these 3 variants have a common [11-20] direction parallel to ND.

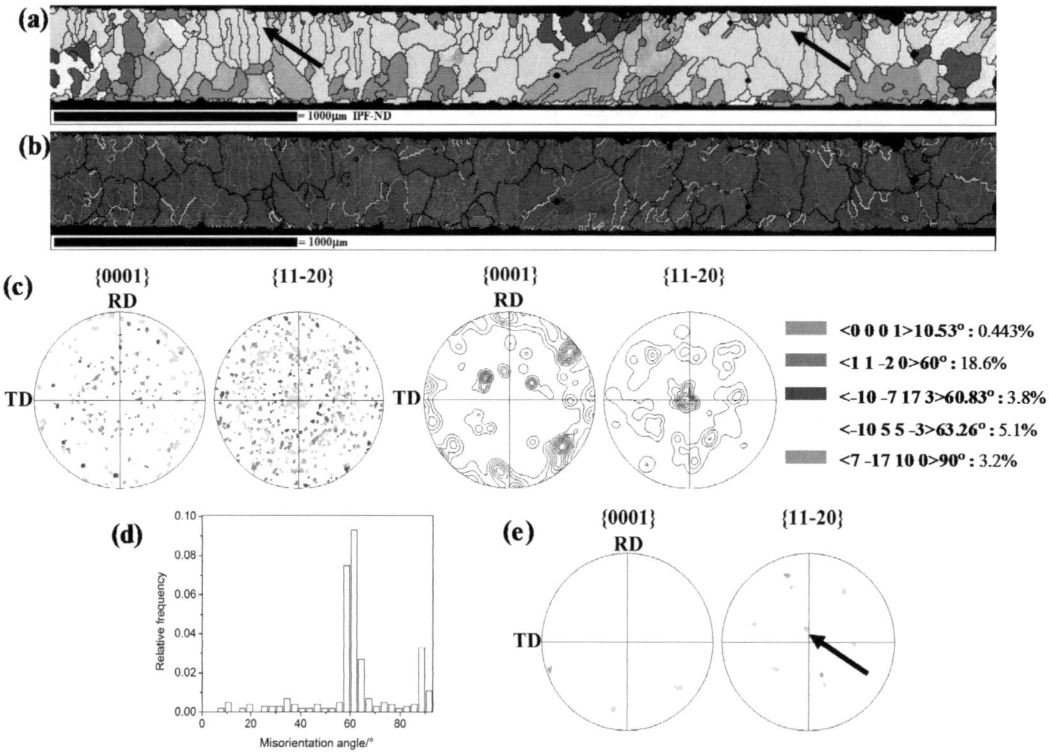

Figure 6. EBSD maps of 60% cold-rolled TA10 sheet with 0.4 mm thickness annealed at 1100°C for 1min, **(a)** orientation map shown in IPF-ND; **(b)** Kikuchi band quality map with special grain boundaries in figure 6a; **(c)** pole figures; **(d)** misorientation angle distribution; **(e)** pole figure of 3 variants within a large round grain boundary as shown by the left arrow in figure 6a

4. Discussion

Based on the above presented EBSD data it is clearly seen that surface effect plays important role by changing transformation texture from recrystallization type to transformation type. Heating temperature is also a key factor for controlling transformation texture because it changes the grain sizes of β-phase. If the grain size in β-phase is small at low heating temperature, namely slightly above critical transformation temperature, strong variant selection and less variant, possibly only one variant, may nucleate leading to texture memory effects and retaining of recrystallization texture as transformation texture. At high heating temperature, however, β-phase grows to large sizes and residual stress or crystal defects are released, therefore more variants are possible to nucleate at grain boundaries obeying Burgers OR, at the same time the high flow rate of helium gas of 8 Litre/min with high thermal conductivity maintains the nucleation in surface region, thus promoting <11-20>//ND grain formation and moreover the 3 variants with same <11-20> parallel to ND can more effectively reduce transformation strain which can be easily confirmed by calculation and comparison of transformation strain tensors between one variant and three variants. The preferred nucleation of <11-20>//ND is ascribed to the lowest elastic module in HCP titanium, the hardest direction is <0001> in Ti crystals. It is noted that the <11-20>//ND grains are rather less before the transformation from α to β phase during heating.

Protecting atmosphere is also an important factor. In our previous work [4] we used the less thermally conductive Ar and only obtained surface induced microstructure of two layers of columnar grains, but the resulted texture is a retained recrystallization texture of <11-20>//RD, namely only texture memory effect is activated. The higher thermal conductivity of He gas can induce 3 plate-like variants building a round normal grain boundary in which a lot of <11-20>60° special grain boundaries are present. This means that He gas promotes <11-20>//ND texture more effectively than Ar gas does.

Finally it should be noted that in our previous work [5], the similar <11-20>//CA (compression axis) as the <11-20>//ND texture in this work is observed during compression of β-phase at high temperature with subsequent water quenching. The same transformation texture of <11-20>//CA or ND is observed in two cases, but the processing routes are different, one is the compression of β-phase and let it transform into α-phase and the other is the cold rolling of α-phase and heat it quickly and let it transform according to $\alpha \rightarrow \beta \rightarrow \alpha$ sequence. In the first case there is no surface effect and the <11-20> texture is explained in terms of the release of transformation strain/stress during compression because the basal plane spacing of {0001} in α-phase is larger than that of {110} in BCC β-phase. In the second case of this work surface effect is activated by blowing He gas on the sheet surface to promote the preferred nucleation at surface region and further the preferred growth into center layer, thus the underlying mechanisms for two cases are different.

5. Conclusions

1) The TA10 titanium sheet shows a typical TD-rotated basal recrystallization texture {01-13}<-2110> below transformation temperature. At low heating temperature above T_c the transformation texture is similar to the recrystallization texture, showing the so-called texture memory effect, although microstructure is totally changed from uniform equiaxed grains to bi-grain sized microstructure with small grains locating at surface region.

2) At high heating temperature of 1100°C dominant <11-20>//ND texture is present with plate-like variants and their boundaries tending to be parallel to ND. The typical 3 <11-20>//ND variants show a common [11-20] axis towards ND. This feature is ascribed to the surface effect-induced transformation and the soft <11-20> direction with lowest elastic module. Over 50% grains in area fraction can be <11-20> or <1-100>//ND oriented.

3) At medium heating temperature of 1000°C, there exists a competition between rotated basal texture and prismatic texture of <11-20>//ND. The resulted texture is in general weak and microstructure is rather inhomogeneous consisting in the small grains at surface region with recrystallization texture orientation, the coarse round grains grown from surface region with recrystallization texture orientation and the plate-like grains also grown from surface region with prismatic texture orientations.

4) Rolling reduction didn't play larger role than heating temperature. High thermal conductive helium flow promotes plate-like variant formation more effectively in comparison to Ar atmosphere.

Acknowledgments

This work is finally supported by the National Natural Science Foundation of China (No. 51771024 and 51571024).

References

[1] Tomida T 2018 Variant selection mechanism by elastic anisotropy and double K-S relation for transformation texture in steel; difference between martensite and ferrite *Acta Materialia*. **146** 25–41

[2] Sung J K, Lee D N, Wang D H and Koo Y M 2011 Efficient generation of cube on-face crystallographic texture in iron and its alloys *ISIJ Int*. **51** 284–90

[3] Zhang L W, Yang P, Wang J H and Mao W M 2016 Transformation of {100} texture induced by surface effect in ultra-low carbon electrical steel *J Mater Sci*. **51** 8087–97

[4] Li K and Yang P 2018 Texture control of pure titanium sheet by the surface effect during phase transformation *Metals*. **8(5)** 358

[5] Li K and Yang P 2016 Interaction among deformation, recrystallization and phase transformation of TA2 pure titanium during hot compression *Trans. Nonferrous Met. Soc. China*. **26** 1863–70

Evolution of microstructure, texture and grain boundary character distribution of potassium doped tungsten fibers annealed at variable temperatures

L Tanure[1, 2*], D Terentyev[3], J Riesch[4] and K Verbeken[1]

[1]*Department of Materials, Textiles and Chemical Engineering, Ghent University (UGent), Technologiepark 46, B-9052 Ghent, Belgium*
[2]*Dutch Institute for Fundamental Energy Research, DIFFER, De Zaale 20, 5612 AJ Eindhoven, The Netherlands*
[3]*Nuclear Materials Science Institute, SCK.CEN, Boeretang 200, 2400 Mol, Belgium*
[4]*Max-Planck-Institut für Plasmaphysik, Garching, Germany*

E-Mail: leandro.tanure@ugent.be

Abstract: The effect of the annealing temperature on the microstructure and grain boundary character distribution of potassium doped tungsten fibers made of drawn wire was investigated by Electron Backscatter Diffraction. Samples, with a diameter of 148.7 μm, in the as-received condition and annealed at 1300, 1600, 1900, 2100 and 2300 °C were analyzed at the center of the transversal sections. Up to 1900 °C, a uniform microstructural coarsening and primary recrystallization followed by normal grain growth was observed. Between 1900 and 2100 °C abnormal grain growth took place. The strong texture (<110> parallel to the drawing axis) remained present in all conditions. With increasing the annealing temperature, the low angle grain boundary fraction increased at the expense of high angle grain boundaries while the amount of coincidence site lattice boundaries reached its maximum at 1600 °C. At this temperature, the most resistant configuration of triple junctions against intergranular crack propagation was obtained.

1. Introduction

Recently, several studies have been carried out to evaluate the use of tungsten (W) based materials as plasma facing components for future nuclear fusion reactors (i. e. ITER and DEMO) [1]. Despite its suitability for such application, due to its refractory nature, low sputtering erosion, good thermal conductivity, low tritium retention and high strength, W possesses intrinsic brittleness below the so-called ductile-to-brittle transition temperature and can be even further embrittled by neutron irradiation and overheating [2]. A possible way to mitigate these issues is to reinforce W with W fibers, made of W wires doped with potassium to improve their temperature stability. The fiber-reinforcement promotes a toughening mechanism analogous to that of a fiber-reinforced ceramic matrix composite which relies on energy dissipation by controlled mechanisms at the fiber/matrix interfaces [2]. Additionally, the dopant effect of potassium results in pinning of the grain boundaries as such suppressing recrystallization [3].

Given the concept of reinforcement, the behavior of the composite will be very much determined by the properties of the W-fibers [4]. Given that plasma-facing components will be subjected to significant heat loads, the resistance of the fibers against the high temperature annealing is important aspect to be addressed and has been extensively studied in the past years [5–11]. Potassium has already proven to restrict grain boundary mobility and keep ductile fracture even after annealing at very high temperatures (1600-1900 °C), since the decrease in fracture toughness is not as pronounced as in pure W-fibers, but a better understanding on the role of triple junctions is still missing. The goal of the present study is to assess the effect of the annealing temperature on microstructure, texture, grain boundary character distribution and evolution of the triple junctions of K-doped W-fiber via Electron Backscatter Diffraction (EBSD) analysis.

2. Materials and Methods

Drawn potassium doped wires (60 – 75 ppm K), with a diameter of 148.7 ± 0.2 μm, were provided by OSRAM GmbH, Schwabmünchen. Segments of 100 mm long were annealed in a tube furnace, under hydrogen atmosphere, at 1300, 1600, 1900, 2100 and 2300 °C for 30 minutes [6]. Samples with 5 mm length were cut, mounted, grinded and polished up to 1 μm diamond paste suspension following the standard metallographic procedures. An additional step of polishing with OP-U suspension was required for EBSD analysis. The specimens were scanned at the center of the transversal cross-section using a Field Emission Gun Quanta-450 FEI Scanning Electron Microscope in areas of 40x40 μm^2 with step-sizes of 0.05 and 0.10 μm. Image post-processing were carried out with the OIM® Analysis software. The size of the areas were chosen taking into account a balance between the number of grains (as much as possible) and the effect that the edges of a rectangular area could have on the analysis of a round cross-section.

3. Results and Discussion

Figures 1 (a-f) show Inverse Pole Figure (IPF) maps overlapped with Image Quality (IQ) maps of the K-doped W-fibers in all conditions. Figures 1(g) and (h) display the zoomed-in IQ maps (in smaller areas) of the samples in the as-received and annealed at 1300 °C conditions.

Figure 1. Inverse Pole Figure overlapped with Image Quality maps of K-doped tungsten samples in the (a) as-received condition and annealed at (b) 1300 °C, (c) 1600 °C, (d) 1900 °C, (e) 2100 °C and (f) 2300 °C. Color code with respect to the Drawing Axis (DA). (g) and (h) detail of the IQ maps of the as-received and annealed at 1300 °C conditions, respectively.

Up to 1900 °C, i.e. figures 1(a), (b), (c) and (d), one can notice a uniform and progressive change in the grain size. From 2100 °C on, figures 1(e) and (f), abnormal grain growth takes place and the maps are dominated by a few very large grains. Therefore, these conditions were excluded from further analysis due to the dominance of these grains which implies a lack of statistics in the analyzed parameters (texture, grain boundary character distribution and average grain size).

The average grain sizes, calculated by the software, are 0.45 ± 0.30, 0.63 ± 0.37, 0.90 ± 0.59 and 1.25 ± 0.71 μm for the as-received and annealed at 1300, 1600 and 1900 °C conditions, respectively. Despite the large standard deviation, a gradual and continuous growth can be noticed. These samples present unimodal grain size distribution and no exaggerated grain growth was observed.

In studies carried out by Jansen [3] and, recently, by Zhao et al. [8] on microstructure of W-fibers after different heat treatments, recrystallization was ascribed to small morphological changes in the transversal cross-section. According to Zhao, when a curly shape, also referred as Van Gogh sky structure (due to the resemblance with the painting style of the famous painter), is no longer observed and, according to Jansen, when ribbon-shaped fiber-crystals transform quickly into stem-like crystals with polygonal cross-section, recrystallization has taken place. Figures 1(g) and (h) show IQ maps of a 5 x 5 μm^2 area where clear differences between microstructures of the as-received and annealed at 1300 °C conditions can be seen. Grains slightly larger of the annealed condition replace the smaller grains of the as-received condition suggesting a recrystallized microstructure.

The grain growth is kept under control up to 1900 °C. This behavior agrees with observations made by Jansen and by Terentyev et al. [5] on K-doped W-fibers where abnormal grain growth occurred only above 2000 °C. Annealing at such high temperature is required to achieve the secondary recrystallization stage thanks to the very effective dragging of the grain boundary migration by the potassium dopant. Potassium remains entrapped inside some residual pores that are stretched along the axis direction during wire drawing forming narrow ellipsoids that break up into a single row of spherical and nano-sized bubbles upon annealing [12].

The samples presented a very strong <110> // DA texture as can be seen in both figure 1, due to the dominance of the green color on the IPF maps, and figure 2 with peaks of 15.7, 16.7, 17.0 and 16.6 times random for the as-received, and annealed at 1300, 1600 and 1900 °C conditions, respectively. This texture could be expected for body-centered cubic materials that underwent wire drawing [13]. Therefore, the annealing temperature did not have a significant effect on texture intensities since all conditions have similar texture intensities.

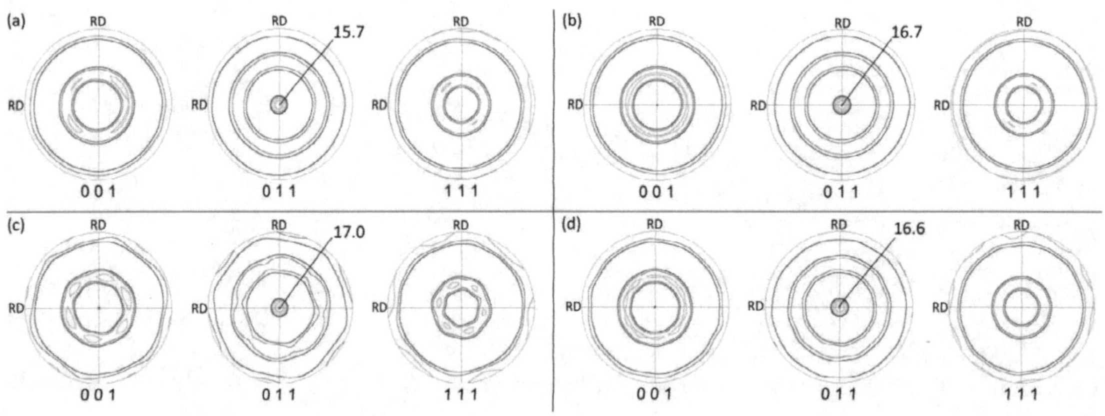

Figure 2. Pole figures of the samples with unimodal grain size distribution in the (a) as-received and annealed at (b) 1300, (c) 1600 and (d) 1900 °C conditions. DA perpendicular to the paper. RD: radial direction. Intensity levels: 1-2-4-6-8-11-16 times random.

The analysis of the Coincidence Site Lattices (CSL – from Σ3 to Σ29) followed the Brandon's Criterion for a maximum permissible deviation given by $\Delta\theta=\theta_0(\Sigma)^{-n}$ where θ=15 and n=0.5. By raising the annealing temperature the number fraction of low angle grain boundaries (LAGBs – misorientation angles between 2 and 15°) has gradually increased from 0.292 to 0.345, 0.375 and 0.394 for the as-received and annealed at 1300, 1600 and 1900 °C conditions, respectively, whereas the fraction of high angle grain boundaries (HAGBs – misorientation angles >15°) has reduced from 0.478 to 0.371, 0.341 and 0.357, respectively, for the same annealing conditions. The CSL fraction is 0.229 (as-received), 0.283 (annealed at 1300 °C), 0.284 (annealed at 1600 °C) and 0.249 (annealed at 1900 °C). This remarkable trend, expressed in the increase of LAGBs fraction at expense of HAGBs, as well as the large fraction of HAGBs in the as-received condition can be explained by the grain subdivision mechanism. It occurs when dislocation accumulation processes take place surrounding deformed crystallites during continuous deformation resulting in the increase of the misorientation angle. With the increase of annealing temperature, the reduction of HAGBs is ascribed to the strong texture developed during wire drawing (as shown in figure 2 and reported in [5,10]) that is responsible for the orientation pinning effect (when a growing grain meets a deformed crystallite with similar orientation causing the replacement of the mobile HAGB by a less mobile LAGB) [14].

Some studies have already pointed out that specific features of the materials such as stress corrosion crack propagation resistance can be enhanced with the increase of CSL boundary fraction due to particular properties related to the enhanced atomic diffusion or dislocation glide [15]. Figure 3(a) shows the distribution of the CSL clearly dominated by Σ3, Σ9, Σ11, Σ19a and Σ27a types. The same boundaries were also reported by Watanabe [16] and related to strong <110> texture in Fe-6.5%Si ribbons. The increase in Σ3 fraction at expense of both Σ9 and Σ27a is a well-known feature described in detail by the *"Σ3 regeneration model"* proposed by Randle in order to explain how the Σ3 boundary fraction can be enhanced in the network of low stacking-fault energy materials [17]. Although tungsten is a body-centered cubic metal with high stacking-fault energy, there is a slight similar but inconclusive trend observed in this work that Randle's prediction was followed.

The grain boundary connectivity has an important role regarding the control of a dominant fracture mode in polycrystalline materials, since triple junctions (TJ) can act as either a crack arrester or a crack assister (to accumulate stress concentration) depending on the distribution and interconnectivity of the boundaries prone to crack propagation[18]. Figure 3(b) shows the four types of TJs that can occur in a material, combinations of random boundaries (R) with special (S) CSL boundaries, i.e. 0-CSL (RRR), 1-CSL (RRS), 2-CSL (RSS) and 3-CSL (SSS), whereas figure 3(c) shows the TJ Distribution of the fibers with unimodal grain size distribution. If a TJ comprises at least 2-CSL (RSS) boundaries, the random boundary network connectivity will be broken, as well as the pathway to intergranular degradation. Therefore, the number of special CSL boundaries in a TJ provides an indication of failure resistance.

According to figure 3(c), up to 1600 °C, there is a reduction of RRR and RRS TJ types and an increase in both RSS and SSS (with the CSL fractions increase) which are critical components in predicting the breaking-up of the grain boundary network connectivity [19]. This behavior was observed and detailed by Fortier et al. [19] in aluminum and copper and can be explained based on the Σ3 regeneration model. When a Σ3 boundary interacts with another Σ3, the third boundary will probably be a Σ9 rather than a LAGB. If the interaction is between Σ3 and $\Sigma3^n$, the third boundary will most likely be $\Sigma3^{n-1}$ rather than a $\Sigma3^{n+1}$. Both interactions reduce the availability of random boundaries at TJ, therefore, leading to an increase of TJ fractions where special boundaries are more abundant (RSS and SSS) at the expense of the other TJs types. Finally, for the sake of completeness, if the Σ3 interacts with a random boundary, the newly created boundary at the TJ will have four out of five degrees of freedom fixed. When this happens, this implies that the indices of the grain boundary plane must be affected [17].

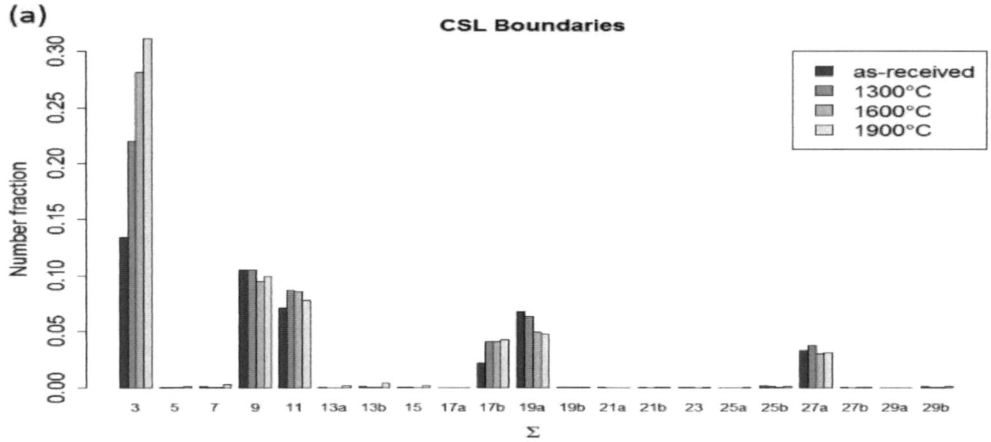

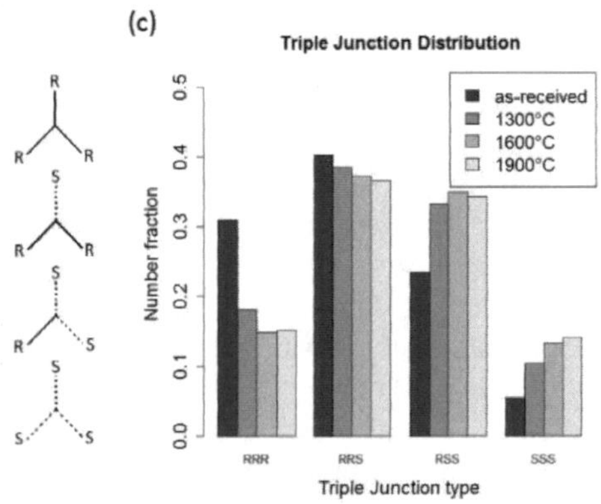

Figure 3. (a) Coincidence Site Lattice Boundaries distribution of the samples with unimodal grain size distribution. (b) Representation of the different types of triple junctions. R: Random boundary and S: Special boundary. (c) triple junction distributions of the as-received and annealed at 1300, 1600 and 1900 °C conditions.

Due to the large fraction of CSL boundaries in all conditions (around 0.25-0.30) the resistance of the microstructure must be assessed in terms of TJs since the connectivity of the grain boundaries plays a more important role on crack propagation than the CSL fraction itself [19]. For this evaluation, one should consider the distribution of TJs that comprise at least one boundary susceptible to cracking, therefore, the SSS type will be considered as non-active entity regarding crack propagation. If the ratio $f_{RSS}/(1-f_{SSS}) \geq 0.35$ (where f_{RSS} and f_{SSS} are the number fractions of TJs composed by two and three special boundaries, respectively) percolative paths in the microstructure, which could act as a crack assister, will be broken reducing the connectivity of random boundaries. This ratio of percolative-resistant TJs gives an estimate of the fraction of TJs capable of arresting a propagating crack [18] and are equal to 0.25, 0.37, 0.40 and 0.39 for the as-received and annealed at 1300, 1600 and 1900 °C conditions, respectively. Although these values are very close to each other, according to this evaluation, annealing promotes the increase of resistant TJ distribution regarding intergranular crack propagation indicating that another fracture mechanism could prevail. Depending on the TJ distribution, if an initially formed intergranular crack cannot continue to propagate because of meeting a resistant TJ, RSS, the crack proceeds as a mixture of intergranular and transgranular fracture. If there are no resistant TJs, the crack will propagate connecting random weak boundaries resulting in typical intergranular fracture showing severe brittleness. These results are consistent with our previous study where surface fracture analysis were carried out on tensile tested W-pure and K-doped fibers that were annealed in the same conditions [5]. The K-doped fibers were more resistant to intergranular fracture in comparison with W-pure fibers [7] due to the elongated grains that remained in the microstructure up to 1900 °C. Future comparative studies on the effect of annealing temperature on TJ character of both W-pure and K-doped fibers will help to clarify the role of TJs on mechanical behavior.

4. Conclusions

The effect of annealing temperature on microstructure, texture, grain boundary character and triple junction character distributions of K-doped W-fibers was evaluated and discussed in terms of EBSD analysis. Primary recrystallization was observed at 1300 °C followed by normal grain growth up to 1900 °C. Secondary recrystallization took place between 1900 and 2100 °C confirming the strong effect of potassium on suppressing the abnormal grain growth. The annealing temperature did not impact the texture development that remained very strong, around 16 times random, with <110> components parallel to the Drawing Axis. As the annealing temperature increased, there was a reduction on HAGBs and an increase of LAGBs fractions. The fraction of CSL has increased up to 1600 °C and could be linked with the increase of percolation-resistant TJ fractions as well as the increase of TJs where 2 and 3 CSL boundaries were present (RSS and SSS).

Acknowledgement

This work has been carried out within the framework of Erasmus Mundus Fusion-DC and the EUROfusion Consortium. It has received funding from the Euratom research and training programme 2014-2018 and 2019-2020 under grant agreement No 633053. The views and opinions expressed herein do not necessarily reflect those of the European Commission.

References

[1] Hirai T, Escourbiac F, Barabash V, Durocher A, et al. 2015 *J. Nucl. Mater.* **463** 1248–51
[2] Riesch J, Aumann M, Coenen J W, et al. 2016 *Nucl. Mater. Energy* **9** 75–83
[3] Jansen H H R 1987 *Philips J. Res.* **42** 3–14
[4] Gietl H, Riesch J, Coenen J W, et al. 2017 *Fusion Eng. Des.* **124** 396–400
[5] Terentyev D, Tanure L, Bakaeva A, et al. 2019 *Int. J. Refract. Met. Hard Mater.* **81** 253–71
[6] Terentyev D, Van Renterghem W, Tanure L, et al. 2019 *Int. J. Refract. Met. Hard Mater.* **79** 204–16
[7] Terentyev D, Riesch J, Dubinko A, Khvan T and Zhurkin E E 2019 *Fusion Eng. Des.* 1–4
[8] Zhao P, Riesch J, Höschen T, et al. 2017 *Int. J. Refract. Met. Hard Mater.* **68** 29–40
[9] Riesch J, Han Y, Almanstötter J, et al. 2016 *Phys. Scr.* **T167** 14006
[10] Nikolić V, Riesch J and Pippan R 2018 *Mater. Sci. Eng. A* **737** 422–33
[11] Nikolić V, Riesch J, Pfeifenberger M J and Pippan R 2018 *Mater. Sci. Eng. A* **737** 434–47
[12] Horacsek O, Briant C L and Horacsek K 1997 *High Temp. Process.* **16** 15–27
[13] Engler O and Randle V 2010 *Introduction to Texture Analysis: Macrotexture, Microtexture and Orientation Mapping* (CRC Press Taylor & Francis Group)
[14] Hughes D A and Hansen N 1997 *Acta Mater.* **45** 3871–86
[15] Lehockey E M, Palumbo G, Lin P and Brennenstuhl A M 1997 On the relationship between grain boundary character distribution and intergranular corrosion *Scr. Mater.* **36** 1211–8
[16] Watanabe T, Fujii H, Oikawa H and Arai K I 1989 *Acta Metall.* **37** 941–52
[17] Randle V 2004 *Acta Mater.* **52** 4067–81
[18] King W E, Stolken J S, Kumar M and Schwartz A J 2000 *Electron Backscatter Diffraction in Materials Science* ed A J Schwartz, M Kumar and B L Adams (New York)
[19] Fortier P, Miller W a. and Aust K T 1997 *Acta Mater.* **45** 3459–67
[20] Randle V 2010 *Mater. Sci. Tech.* **26** 774-80

IOP Conf. Series: Journal of Physics: Conf. Series **1270** (2019) 012039 doi:10.1088/1742-6596/1270/1/012039

Grain sizes and dislocation densities in fcc-metallic materials processed by warm to hot working

V Torganchuk, A Morozova, M Tikhonova, R Kaibyshev and A Belyakov

Belgorod State University, Belgorod 308015, Russia

E-mail: belyakov@bsu.edu.ru

Abstract. Our just received results on the deformation microstructures evolved in a medium-Mn austenitic steel during large strain warm to hot rolling were analyzed with a reference to other alloys with face centred cubic lattices and relatively low stacking fault energies. The structural changes were characterized by the development of dynamic recovery (DRV) and dynamic recrystallization (DRX). The deformation grain size decreased while the dislocation density increased with a decrease in deformation temperature. Both the grain size and the dislocation density could be expressed by power law functions of temperature-compensated strain rate, i.e., Zener-Hollomon parameter, with exponents of -0.33 and 0.2, respectively, in the case of discontinuous DRX. On the other hand, the exponents of -0.1 and 0.06 were obtained for the grain size and dislocation density dependencies on the temperature-compensated strain rate, respectively, under conditions of warm deformation accompanied by DRV and continuous DRX. Therefore, a power law function could be obtained between the grain size and the dislocation density with a grain size exponent of -0.6, irrespective of the mechanisms of microstructure evolution in a wide range of deformation conditions.

1. Introduction

Structural steels and alloys are commonly subjected to various thermo-mechanical treatments, which result in required combinations of mechanical properties of semi-products owing to appropriate microstructures. The most important microstructural parameters affecting the mechanical properties of metallic materials are the phase content including the distribution of dispersed particles, the grain size, and the dislocation density [1-4]. The phase content depends on the alloying design and can be realized for improving the properties in multiphase systems, whereas a beneficial combination of the grain size and the dislocation density enhances mechanical properties of almost all structural steels and alloys. Therefore, studying the mechanisms of microstructure evolution in metals and alloys during thermo-mechanical processing is of great practical importance.

Warm to hot working of metallic materials is frequently accompanied by dynamic recrystallization (DRX) [5-7]. Two DRX mechanisms are discussed in relevant literature [6]. Those are discontinuous DRX and continuous one. The former takes place in metals and alloys with relatively low stacking fault energy (SFE) subjected to hot working. This is a classical DRX mechanism, which has been revealed several ten years ago and clarified in numerous studies [5]. In contrast, the mechanism of continuous DRX has not been well documented and it is still arousing some debates. It is currently agreed that continuous DRX may occur in metals and alloys with relatively high SFE under hot working conditions as well as in almost all metallic materials irrespective of SFE under conditions of warm or even cold working [7]. The DRX grain size (D) depends sensitively on the deformation condition, which can be adequately represented by temperature-compensated strain rate, i.e., Zener-

Hollomon parameter, $Z = \dot{\varepsilon} \exp(Q/RT)$, where $\dot{\varepsilon}$ is the strain rate, Q is the activation energy, R and T are the universal gas constant and absolute temperature, respectively. A power law function, $D \sim Z^{-n}$, with an exponent of approx. 0.4 has been reported for discontinuous DRX during hot working of face centred cubic (fcc) materials with low SFE [8]. This strong temperature/strain rate dependence of the DRX grain size becomes much weaker as deformation temperature decreases leading to a change in DRX mechanism from discontinuous to continuous [8]. Correspondingly, an exponent in the power law function tends to decrease below 0.1. In contrast to the DRX grain size, the effect of deformation conditions and DRX mechanisms on the dislocation substructures in warm to hot worked steels and alloys has not been studied in detail. An inverse proportional relationship has been suggested between the DRX flow stress and the subgrain/cell size in DRX microstructures [1, 9]. However, the variation of dislocation density in DRX microstructures has not been explained satisfactorily. The aim of the present study, therefore, is to elaborate the relationship between the deformation conditions within the warm to hot working domain and the microstructural parameters, i.e., the deformation grain size and dislocation density, and their effect on the strength of a medium-Mn austenitic steel as a typical representative of fcc-metallic materials with low SFE.

2. Experimental

The medium-Mn steel, Fe-12%Mn-0.6%C-1.5%Al (all in wt.%) was produced in the Central Research Institute for Machine-Building Technology, Russia. The steel was characterised by a uniform austenite microstructure with an average grain size of 80 μm that developed by hot rolling at 1423 K. The steel samples were rolled to a total reduction of 60% at different temperatures ranging from 773 K to 1373 K. The rolling reduction in each rolling pass was 10%. The samples were reheated to rolling temperature after each rolling pass. The structural investigation were carried out by an orientation imaging microscopy (OIM) based on electron backscattering electron diffraction (EBSD) analyser set at a Quanta Nova NanoSEM 450 scanning electron microscope (SEM). The mean grain size and the kernel average misorientation were evaluated using TSL OIM Analysis software, ver. 6.2. The grain size was evaluated by a linear intercept method along the normal direction (ND). The dislocation density was calculated as $\rho = 1.15\ \theta_{KAM} / (b\ h)$, where θ_{KAM} is the kernel average misorientation, b is the Burgers vector, and h is the step size in the OIM map [10]. The tensile tests were carried out at ambient temperature using an Instron 5882 universal testing machine and dog-bone specimens cut along the rolling direction (RD).

3. Results and discussion

3.1. Deformation microstructures

The effect of rolling temperature on the deformation microstructures evolved in the steel samples is illustrated in figure 1. Warm rolling at 773 K to 973 K brings about the elongation of original grains along the rolling direction and the development of numerous microshear bands, which is accompanied by the formation of strain-induced grain boundaries (figures 1a and 1b). In contrast, the deformation microstructures evolved during hot rolling at 1073 K to 1373 K consist of almost equiaxed grains containing annealing twins (figures 1c and 1d). Such microstructures are typical of discontinues DRX that takes place in low SFE metals during hot working [5]. It can be concluded, therefore, that the studied range of rolling temperatures covers both the hot deformation domain of discontinuous DRX and warm deformation domain of dynamic recovery (DRV) or continuous DRX.

The transverse grain size (D) evolved in the present steel samples after rolling at different temperatures is shown in figure 2 as a function of temperature–compensated strain rate (Z) along with data from other studies on DRX microstructures [11-15]. The grain size decreases with a decrease in temperature and/or an increase in strain rate. Within the hot deformation domain with $Z < 12^{12}\ s^{-1}$, the grain size exhibit a rather strong power law relationship with Z with an exponent of 0.33. On the other hand, a power law function with much smaller exponent of about 0.1 holds between the grain size and the deformation conditions, i.e., Z. This change in the relationship between D and Z has been discussed in terms of DRX mechanisms [8, 14]. Namely, transition from discontinuous DRX to continuous one, when DRV serves as a softening controlling mechanism, with a decrease in deformation temperature occurs along with such a change in the DRX grain size dependence.

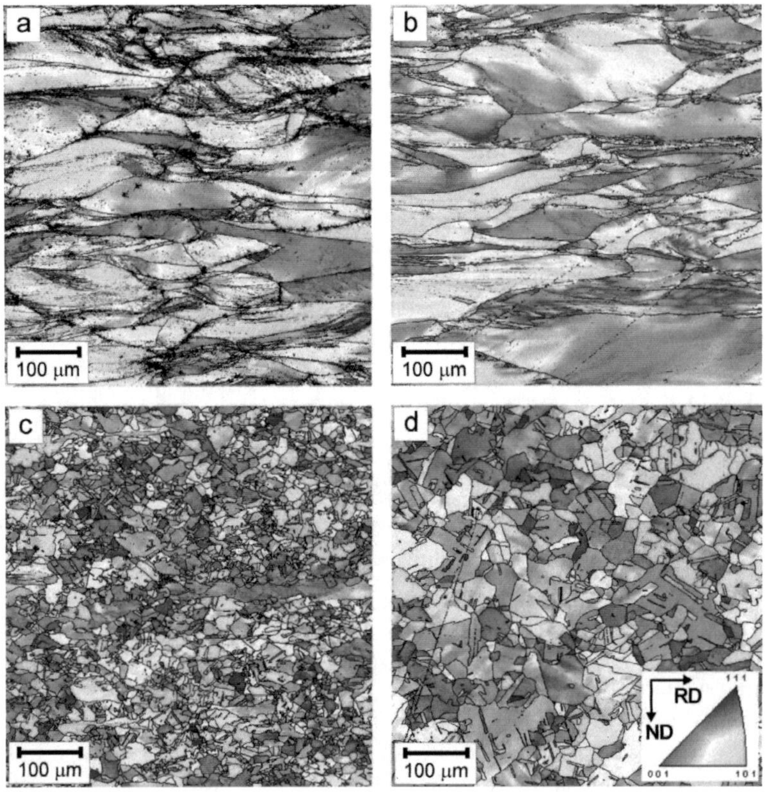

Figure 1. OIM micrographs of the deformation microstructures evolved in a medium-Mn steel processed by warm to hot rolling at 873 K (a), 973 K (b), 1073 K (c), 1173 K (d). The colours indicate the crystallographic direction along the normal direction (ND).

A reduction of the grain size with a decrease in deformation temperature is accompanied by an increase in the internal stresses within the warm to hot worked microstructures. The distribution of internal distortions can be mapped by the kernel average misorientation (figure 3). An average value of the kernel average misorientation increases from 0.5° to 2.9° as the rolling temperature decreases from 1373 K to 773 K. It should be noted that the deformation microstructures are characterized by inhomogeneous distribution of kernel average misorientations. The large kernel average misorientations are located near the grain boundaries and along the microshear bands in the warm rolled microstructures (figures 3a and 3b). The distribution of kernel average misorientations becomes more uniform in the discontinuous DRX microstructures, although different ultrafine grains evolved by hot rolling at 1073 K exhibit different levels of internal distortions (figure 3c). The non-uniform distribution of internal distortions in the ultrafine grained microstructures that develop during hot working is associated with discontinuous DRX mechanism, when the just nucleated stress-free DRX grains cyclically replace the previously grown work hardened grains. The kernel average misorientation decreases with an increase in rolling temperature; remarkable distortions can be observed only near the grain boundaries in the DRX microstructures evolved during hot rolling at a relatively high temperature (figure 3d). Such distribution of the internal distortions indicates that any post-DRX scarcely affected the deformation microstructures after the final rolling pass.

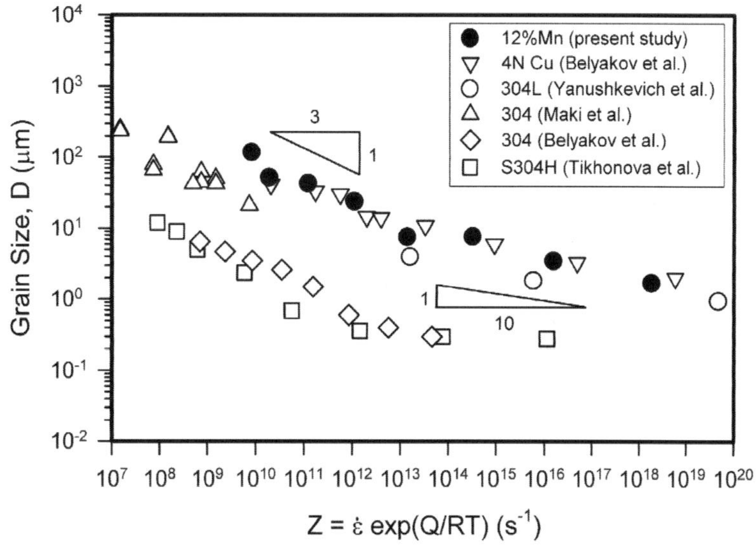

Figure 2. Effect of deformation conditions on the deformation grain size in a medium-Mn steel and other fcc-alloys with low to medium SFE [11-15].

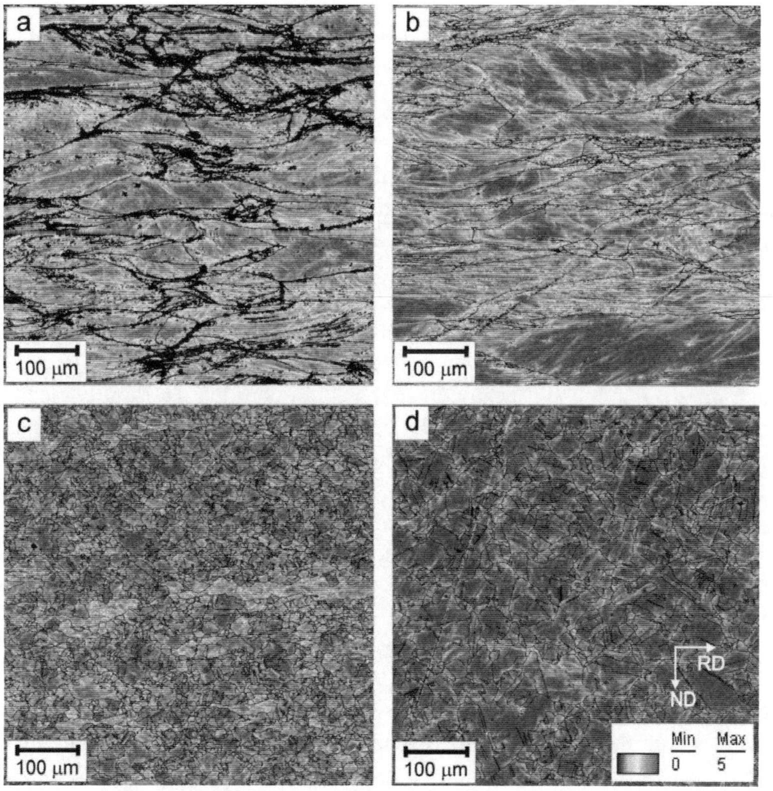

Figure 3. Kernel average misorientations in the deformation microstructures evolved in a medium-Mn steel processed by warm to hot rolling at 873 K (a), 973 K (b), 1073 K (c), 1173 K (d).

The effect of deformation conditions on the corresponding dislocation density is shown in figure 4. Some available date for the dislocation densities measured by direct transmission electron microscopy observations in fcc-metallic materials with low-to-medium SFE [12, 15, 16] are also represented in figure 4. It is clearly seen that the change in the DRX mechanism at Z of approx. 10^{12} s^{-1} is accompanied by the change in the temperature/strain rate dependence of dislocation density. Namely, an exponent of about 0.2 in a power law function of $\rho \sim Z^q$ decreases to 0.06 as deformation conditions change from hot working accompanied by discontinuous DRX to warm working with continuous DRX at sufficiently large strains.

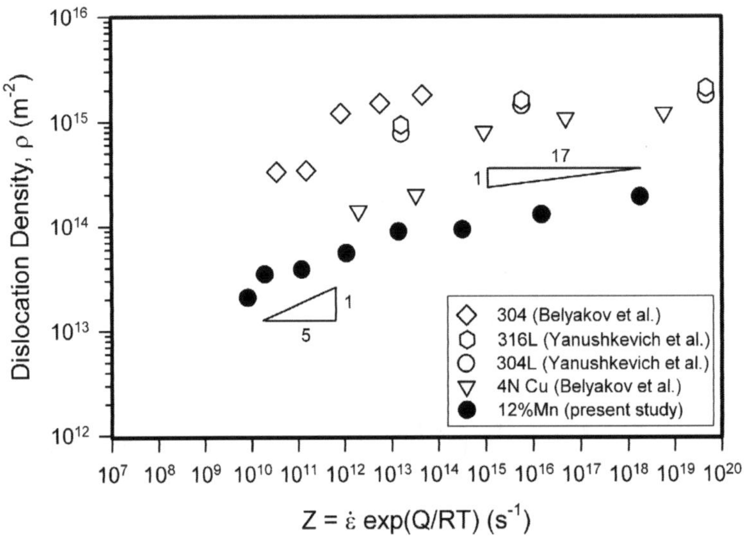

Figure 4. Effect of deformation conditions on the dislocation density in the deformation microstructures in a medium-Mn steel and other fcc-alloys with low to medium SFE [12, 15, 16].

3.2. Tensile properties

A series of the stress-strain curves obtained by tensile tests of the warm to hot rolled steel samples is shown in figure 5. In spite of the different microstructures evolved by warm or hot rolling, the steel samples demonstrate similar tensile behaviour. The strain hardening rapidly decreases at an early plastic deformation resulting in pronounced stage of uniform elongation, when the strain hardening slightly decreases with straining, followed by necking and rapid failure. Such tensile behaviour is typical of high-Mn austenitic steels exhibiting the effects of twinning/transformation induced plasticity [17]. A decrease in the rolling temperature significantly strengthens the steel samples. The yield strength increases from 340 MPa to 950 MPa as the rolling temperature decreases from 1373 K to 773 K. It should be noted that the strengthening is not accompanied by remarkable change in the strain hardening, although total elongation decreases.

3.3. Structure-property relationship

The results obtained in the present study and those reported in other papers for austenite and copper testify to unique power law relationships held between the deformation microstructures the deformation conditions. The most important that these relationships suggest a unique power law function between the deformation grain size and dislocation density. Indeed the power law functions of $D \sim Z^{-0.33}$ and $\rho \sim Z^{0.2}$ obtained for hot working conditions results in a power law function between ρ and D with a grain size exponent of -0.6. On the other hand, the grain size and dislocation density can be related to temperature-compensated strain rate with exponents of -0.1 and 0.06, respectively, in the case of warm working. These dependencies result in the same power law relationship between the dislocation density and grain size with an exponent of -0.6. Figure 6 represents the variation of dislocation density with grain size for the present steel samples and other austenitic steels and copper alloys [12, 13, 15, 18] and confirms the statement above. The same relationship between the grain size and dislocation density has been reported for continuous DRX microstructures evolved in austenitic stainless steels during warm rolling to large total strains [15]. The linear plots in figure 6 suggest that a unique relationship between the grain size and dislocation density may held in DRX microstructures evolved in wide range of materials and deformation conditions irrespective of DRX mechanisms.

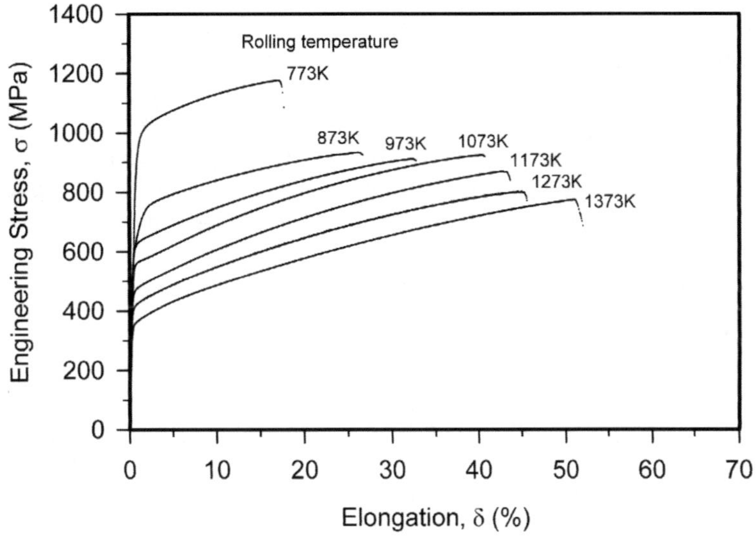

Figure 5. Tensile stress-strain curves for a medium-Mn steel subjected to warm to hot rolling at the indicated temperatures.

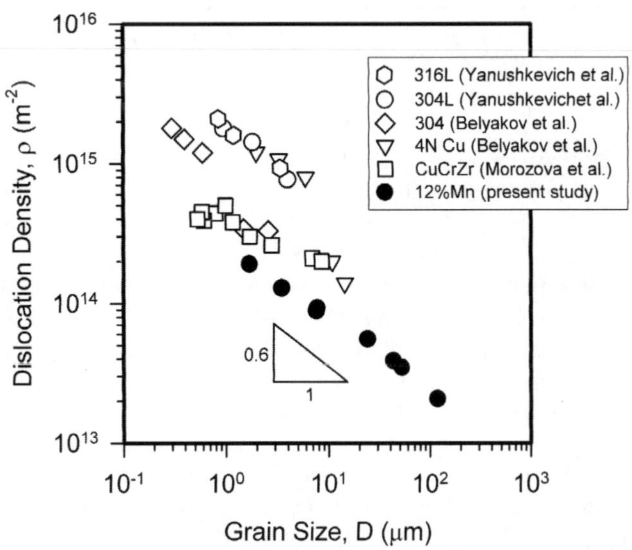

Figure 6. Relationship between the dislocation density and the grain size in a medium-Mn steel and other fcc alloys with low SFE subjected to warm to hot working [12, 13, 15, 18].

The yield strength of work hardened metallic materials processed by large strain deformation can be expressed by a modified Hall-Petch-type relationship, $\sigma_{0.2} = \sigma_0 + k\ D^{-0.5} + \Delta\sigma(\rho)$, where σ_0 is the strength of the same dislocation-free material with infinite grain size, k is the grain size strengthening factor, and the last term represents the dislocation (substructural) strengthening [15, 19]. The latter is commonly calculated by Tailor-type equation, $\Delta\sigma(\rho) = \alpha G b \sqrt{\rho}$, where α is a numerical factor, and G is the shear modulus. The revealed unique relationship between the deformation grain size and dislocation density, e.g., $\rho = 1.36E13 + 2.5E14\ D^{-0.6}$ for the present steel samples in figure 6, enables one to predict the strength of warm to hot worked materials by using either D or ρ. It should be noted that attempts to evaluate the strength by using either grain size or dislocation density have been discussed in other studies [15, 20, 21]. Setting $\alpha = 0.82$ [22], the Hall-Petch-type plot for the present steel samples is shown in figure 7a. Note here that the values of the first term of 172 MPa and the grain boundary strengthening factor of 648 MPa $\mu m^{0.5}$ in figure 7a are similar to those reported in other studies on the grain size strengthening [20, 22, 23]. The relationships between the experimental yield strength of the present steel samples and those calculated by using either grain size, $\sigma_{0.2}(D)$, or dislocation density $\sigma_{0.2}(\rho)$ as a unique variable are presented in figure 7b. A good correspondence between the experimental and both calculated values validates the presented treatment.

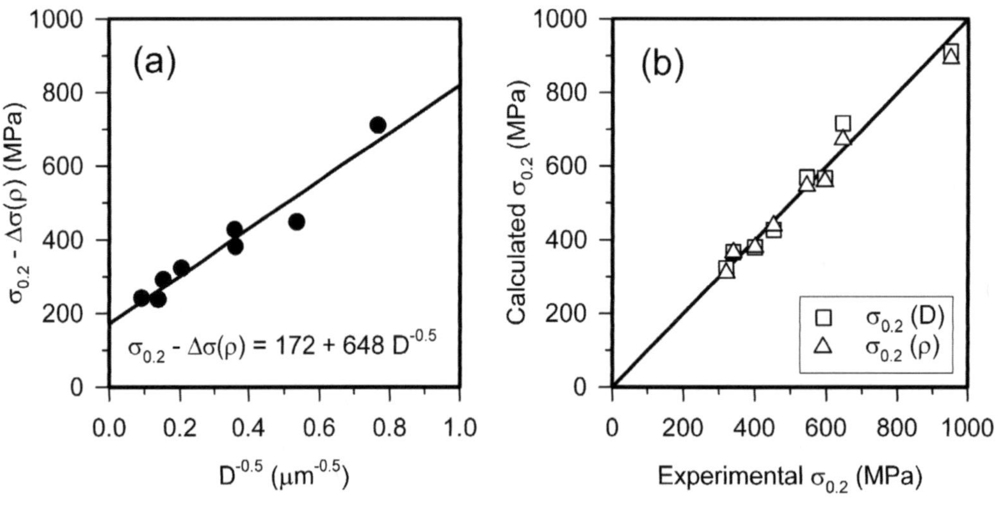

Figure 7. Grain size strengthening (a) and relationship between the experimental and calculated yield strengths (b) for a medium-Mn steel subjected to warm to hot rolling.

4. Summary

The warm to hot rolling of a medium-Mn austenitic steel at different temperatures of 773 K to 1373 K was accompanied by the development of dynamic recovery (DRV) and dynamic recrystallization (DRX). The latter readily developed at temperatures above 1073 K by the discontinuous DRX mechanism. The transverse grain size measured along the normal direction decreased with a decrease in deformation temperature and could be expressed by power law functions of temperature-compensated strain rate, i.e., Zener-Hollomon parameter (Z), with exponents of -0.33 in the range of $Z < 10^{12}\ s^{-1}$ corresponding to hot working accompanied by discontinuous DRX and -0.1in the range of $Z > 10^{12}\ s^{-1}$ corresponding to warm working with DRV as the main restoration process. The change in the dislocation density clearly correlated with that of the deformation grain size. Namely, the dislocation density increased with Z from power law with an exponent of 0.2 during hot working or 0.06 under warm working conditions. Correspondingly, a unique power law relationship with a grain size exponent of -0.6 was obtained between the deformation grain size and the dislocation density in a wide range of warm to hot rolling conditions, irrespective of the mechanisms of microstructure evolution. Therefore, the yield strength of the rolled steel samples could be expressed by a modified Hall-Petch-type relationship, which includes the dislocation strengthening, by using the grain size or dislocation density as a unique variable.

Acknowledgements
Authors acknowledge with gratitude the financial support received from the Ministry of Education and Science, Russia, under Grant No. 11.3719.2017/PCh (11.3719.2017/4.6).

References

[1] Young CM and Sherby OD 1973 *J. Iron Steel Inst.* **211** 640
[2] Mecking H and Kocks UF 1981 *Acta Metall.* **29** 1865
[3] Charleux M, Poole WJ, Militzer M and Deschamps A 2001 *Metall. Mater. Trans.* A **32A** 1635
[4] Di Leo CV and Rimoli JJ 2019 *Scripta Mater.* **166** 149
[5] Sakai T and Jonas JJ 1984 *Acta Metall.* **32** 189
[6] Sakai T, Belyakov A, Kaibyshev R, Miura H and Jonas JJ 2014 *Prog. Mater. Sci.* **60** 130
[7] Huang K and Loge RE 2016 *Mater. Des.* **111** 548
[8] Yanushkevich Z, Lugovskaya A, Belyakov A and Kaibyshev R 2016 *Mater. Sci. Eng. A* **A667** 279
[9] Tikhonova M, Dolzhenko P, Kaibyshev R and Belyakov A 2016 *Metals* **6** 268.
[10] Kusakin P, Kalinenko A, Tsuzaki K, Belyakov A and Kaibyshev R 2017 *Kovove Mater.* **55** 161
[11] Maki T, Akasaka K, Okuno K and Tamura I 1982 *Trans. Iron Steel Inst. Jpn. Int.* **22** 253
[12] Belyakov A, Sakai T, Miura H and Kaibyshev R 1999 *Iron Steel Inst. Jpn. Int.* **39** 593
[13] Belyakov A, Sakai T, Miura H and Tsuzaki K 2001 *Philos. Mag.* A **81** 2629
[14] Tikhonova M, Belyakov A and Kaibyshev R 2013 *Mater. Sci. Eng. A* **A564** 413
[15] Yanushkevich Z, Dobatkin SV, Belyakov A and Kaibyshev R 2017 *Acta Mater.* **136** 39
[16] Belyakov A, Miura H and Sakai T 1998 *Iron Steel Inst. Jpn. Int.* **38** 595
[17] Kusakin PS and Kaibyshev RO 2016 *Rev. Adv. Mater. Sci.* **44** 326
[18] Morozova A, Mishnev R, Belyakov A and Kaibyshev R 2018 *Rev. Adv. Mater. Sci.* **54** 56
[19] Hansen N 2004 *Scr. Mater.* **51** 801
[20] Shakhova I, Belyakov A, Yanushkevich Z, Tsuzaki K and Kaibyshev R 2016 *ISIJ Int.* **56** 1289
[21] Starink MJ 2017 *Mater. Sci. Eng. A* **A705** 42
[22] Yanushkevich Z, Belyakov A, Kaibyshev R, Haase C and Molodov DA 2016 *Mater. Charact.* **112** 180
[23] Odnobokova M, Belyakov A, Enikeev N, Molodov DA and Kaibyshev R 2017 *Mater. Sci. Eng. A* **A689** 370

IOP Conf. Series: Journal of Physics: Conf. Series **1270** (2019) 012040 doi:10.1088/1742-6596/1270/1/012040

Static recrystallization of pure titanium after cryo-deformation

C K Yan[1,2], G R Cui[3*], S J Qu[2], A H Feng[2], J Shen[2] and D L Chen[4]**

[1] School of Materials Engineering, Jiangsu University of Technology, Changzhou, China

[2] School of Materials Science and Engineering, Tongji University, Shanghai, China

[3] School of Materials Science and Engineering, Harbin Institute of Technology at Weihai, Weihai 264209, China

[4] Department of Mechanical and Industrial Engineering, Ryerson University, Toronto, Ontario M5B 2K3, Canada

E-mail address:
cuiguorong@hitwh.edu.cn (G.R. Cui); dchen@ryerson.ca (D.L. Chen).

Abstract. Commercially-pure (CP) titanium was first processed via cryogenic deformation to activate high-density deformation twins, and was subsequently annealed at 500°C to induce static recrystallization (SRX). Two types of twins were mainly activated during cryogenic deformation, *i.e.,* $\{11\bar{2}2\}$ contraction twins and $\{10\bar{1}2\}$ extension twins. Multiple point-like twins were also present because the growth of twins was remarkably impeded at the low temperature. During annealing the recrystallization mechanisms were identified to be $\{11\bar{2}2\}$ and $\{10\bar{1}2\}$ twinning induced SRX. The point-like twins could effectively promote the occurrence of SRX. The grains were significantly refined from ~40 to ~2.7 μm, while a few $\{11\bar{2}2\}$ and $\{10\bar{1}2\}$ TBs still retained with the same twinning angle/axis relationship during annealing.

1. Introduction

The Commercially pure titanium (CP-Ti) is a promising hexagonal close-packed (hcp) metals for medical devices and shipping equipment thanks to its excellent corrosion resistant performance [1, 2]. In addition to dislocation slip, deformation twinning is also a major deformation mechanism for hcp metals due to their limited slip systems [3]. The commonly-observed twinning systems in CP-Ti are $\{10\bar{1}2\}$ and $\{11\bar{2}1\}$ extension twinning, and $\{11\bar{2}2\}$ contraction twinning [4, 5]. The extension twinning is activated under tension parallel to *c* axis or under compression perpendicular to *c* axis of the hcp unit cell, while the contraction twinning is activated under tension perpendicular to *c* axis or under compression parallel to *c* axis [4, 5].

Except for coordinating plastic deformation, twinning in CP-Ti can promote dynamic recrystallization during hot deformation [6, 7]. Yan *et al.* [6] applied pre-cryo-deformation to activate twinning and hot deformation to induce dynamic recrystallization (DRX). Three types of twins were active in inducing DRX. It was found that the DRX stage included twin-active stage and discontinuous

DRX stage and the DRX mechanisms were twinning-induced DRX and discontinuous DRX. Yan *et al.* [7] also applied three-directional compression at cryogenic temperature to activate homogeneous deformation twins, followed by hot deformation to induce homogeneous fine grains. As a result, homogeneous and equiaxed grains with an average grain size of ~1.6 μm was achieved. The DRX mechanism also involved twinning-induced DRX stage and continuous DRX stage. However, till now, few researchers have focused on the effect of twinning activated at cryogenic temperature on static recrystallization (SRX).

Therefore, in this paper CP-Ti was firstly deformed at cryogenic temperature to activate twins, then was annealed at hot temperature to induce SRX. The study on twinning activated at cryogenic temperature and twinning-induced static recrystallization would enrich the theory on static recrystallization.

2. Experimental

A CP-Ti rolled plate with a thickness of 30 mm was utilized in our experiment. The cylindrical samples (Φ8×12) were machined from the as-received plate, where the axis of the cylinder was parallel to the normal direction (ND). The cylindrical samples were immersed in liquid nitrogen for 30 min, and then were compressed to a true strain of 0.30 at a strain rate of 0.01 s^{-1} on CMT 5205 testing machine to activate twins. After cryo-deformation, the samples were annealed in a muffle furnace at a temperature of 500°C for a series of time of 3 min and 14 min. After annealing, the samples were cooled in the air.

The microstructure was examined on the RD-TD plane in the central section of the samples by optical microscopy (OM, Zeiss Axio 5.m) and scanning electron microscopy (SEM) equipped with electron backscatter diffraction (EBSD). The samples for OM and EBSD characterization were electro-polished in a solution consisting of perchloric acid, butanol and methanol (3:17:30 in volume ratio) with a voltage of 30 V at a temperature of -35°C. The EBSD data were analysed via HKL Channel 5 software package.

3. Results and discussion

3.1. *Twinning activated at cryogenic temperature*

The microstructure of the as-received CP-Ti plate, characterized via EBSD, is shown in Figure 1 [6]. A homogeneous and equiaxial microstructure without deformation twins can be seen Figure 1(a). As shown in Figure 1(b), the grain size of the equiaxial grains was ~40.7 μm. The {0001} pole figure revealed a typical split basal texture, where the peak pole intensity of *c*-axis was split by about ±(25-35°) from ND toward TD (Figure1(c)).

Figure 2 shows the microstructure of CP-Ti after deformation at cryogenic temperature to a strain amount of 0.30 at a strain rate of 0.01 s^{-1}. The low magnification (100×) optical micrograph shows that twins were activated in almost all grains, however, the distribution of twins was inhomogeneous (Figure 2(a)). The morphology of twins could be divided into parallel type, intersectional type, and densely-distributed type, which are denoted by arrows in Figure 2(a). The parallel type twins present a parallel morphology, while the intersectional type twins are inter-crossed each other. The morphology of some high-density twins was not clear in the low magnification observations. Figure 2(b) is high magnification (2500×) OM image, where the widths of most twins were seen to be smaller than 1 μm. The densely-distributed type twins (in Figure 2(b)) exhibit an accumulative morphology. The twin morphology was further characterized via EBSD method. The $\{11\bar{2}2\}$ twin boundaries (64°<$10\bar{1}0$>) were colored green and the $\{10\bar{1}2\}$ twin boundaries (85°<$11\bar{2}0$>) were colored red in our study. It is seen that the parallel type twins and

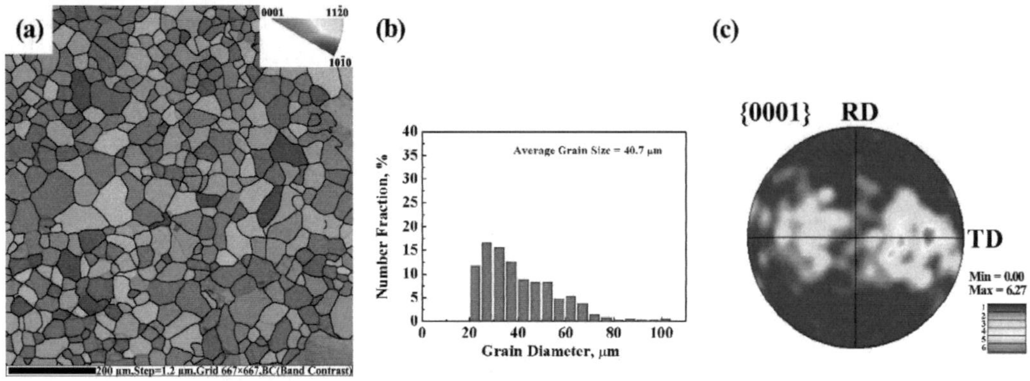

Figure 1 – Microstructure of the as-received CP-Ti plate: (a) inverse pole figure (IPF) map with respect to ND; (b) distribution of grain sizes; (c) {0001} pole figure [6].

intersectional type twins were $\{11\bar{2}2\}$ twins while the densely-distributed twins were $\{11\bar{2}2\}$ and $\{10\bar{1}2\}$ twins (Figure 2(c)). Figure 2(d) is the misorientation angle distribution, where three peaks fell in the ranges of 2-10°, 60-70° and 80-90°, respectively (The peak in the ranges of 2-10° reflects low angle grain boundaries). Inverse pole figures shown in Figure 2(e) were utilized to decide the rotation axis corresponding to the peak in Figure 2(d). The peak fell in the ranges of 60-70° was around the $\langle 10\bar{1}0\rangle$ axis and the peaks fell in the ranges of 80-90° was around the $\langle 11\bar{2}0\rangle$ axis, which confirmed that $\{11\bar{2}2\}$ and $\{10\bar{1}2\}$ twins were activated, respectively. Typical grains named as A and B in Figure 2(c) are depicted in Figure 2(f) together with their corresponding pole figures for a more detailed analysis. It is seen that in grain A shown in Figure 2(f), multiple intersectional $\{11\bar{2}2\}$ twins were activated, indicating the presence of $\{11\bar{2}2\}$ twin variants. Three twin variants were activated in the pole figure of Figure 2(f). The *c* axis of grain A matrix was nearly parallel to ND, then the compressive direction was along *c* axis, leading to the activation of $\{11\bar{2}2\}$ contraction twins. In grain B shown in Figure 2(f), two sets of parallel twins met so as to form new HAGBs, as revealed by the relevant pole figure, where two twin variants were seen to be activated. The reason for the activation of $\{11\bar{2}2\}$ in grain B is the same as that for the activation of $\{11\bar{2}2\}$ twins in grain A. In addition, multiple point-like $\{11\bar{2}2\}$ twins and $\{10\bar{1}2\}$ twins were activated in grain C in Figure 2(c). Twinning process is generally characterized by twin nucleation and growth, where twin nucleation is a process of formation of point-like nucleus [8]. During compression at cryogenic temperature, the growth of twins was remarkably impeded, thus numerous twin nuclei retained the point-like morphology.

3.2. Twinning activated at cryogenic temperature

Recrystallization mode of software Channel 5 was utilized to identify SRX grains. In the recrystallization mode, grains with an average misorientation of smaller than 3° is deformed microstructure, grains with an average misorientation of more than 15° is recrystallized microstructure, and grains with an average misorientation ranging from 3° to 15° is substructure, which means grains with crystal defects. IPF map and recrystallization map of the cryo-deformed CP-Ti after annealing at 500°C for 6 min are shown in Figure 3(a) and Figure 3(b), respectively. The blue-colored subset (with an average misorientation of more than 15°) in Figure 3(b) represents the static recrystallization grains. SRX mechanisms are commonly classified into continuous SRX (CSRX) and discontinuous SRX (DSRX) [9]. CSRX is an evolution process of transforming low angle grain boundaries (LAGBs) into high angle grain boundaries (HAGBs), while DSRX is a process of the nucleation of novel grains by bulging of HAGBs [9]. In addition, twinning-induced static recrystallization (TSRX) is regarded as a novel static recrystallization mechanism. In our study, the SRX grains frequently occurred in the interior of twins or twin boundaries, indicating that the SRX mechanism may be TSRX. During annealing, the

SRX grains resided on the TBs, and TBs evolved into HAGBs. This would confirm the TSRX mechanism, which will be discussed.

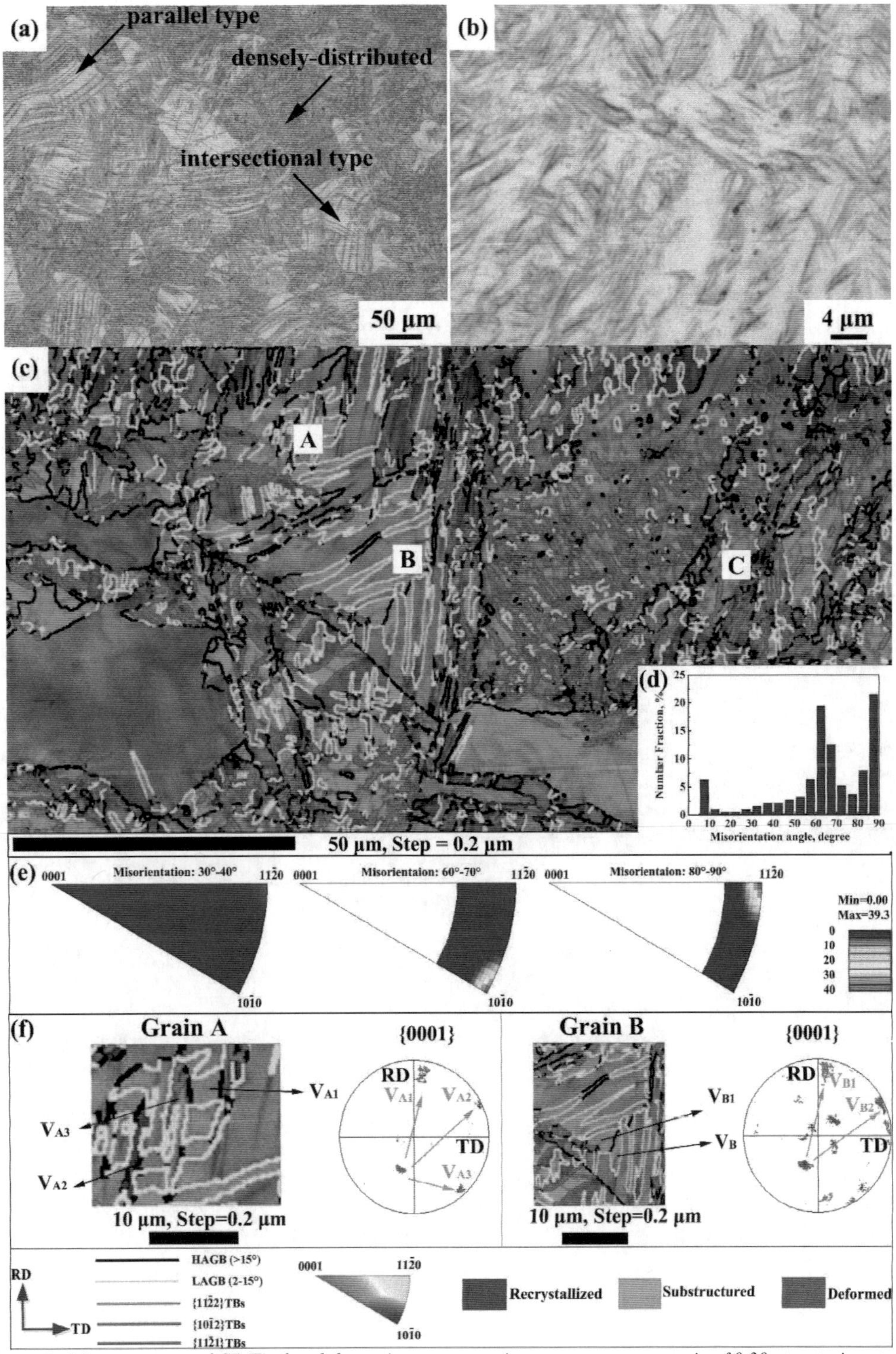

Figure 2 –Microstructure of CP-Ti after deformation at cryogenic temperature to a strain of 0.30 at a strain rate of 0.01 s^{-1}: (a)-(b) OM; (c) IPF map; (d) distribution of misorientation angles; (e) IPF; (f) local IPF maps.

Three typical zones named A, B and C are selected from Figure 3(a) for more detailed discussion. SRX grains in zone A evolved from grain A or B in Figure 2(f) since the SRX grains still had $\{11\bar{2}2\}$ TBs. In fact, this was the initial stage of SRX since $\{11\bar{2}2\}$ GBs had not evolved into ordinary GBs. The recrystallization mechanism was $\{11\bar{2}2\}$ twinning induced SRX. By comparing grain C in Figure 2(c) and zone B in Figure 3(a), it is deduced that the SRX grains in Figure 3(a) were induced by the point-like twins in Figure 2(c) for the SRX grains mainly resided on the TBs of the point-like $\{11\bar{2}2\}$ and $\{10\bar{1}2\}$ twins. Hence, the SRX mechanism was $\{11\bar{2}2\}$ and $\{10\bar{1}2\}$ twinning induced SRX. Moreover, the recrystallization rate of zone B was significantly higher that of other zones, suggesting that the high density point-like twins could predominantly promote the occurrence of SRX. The grain size in zone C was ~10 μm, which was significantly smaller than the average grain size of the as-received CP-Ti, meaning that the incubation period of the grain was short and the grain was SRX grain in the stage of grain growth. The grain was surrounded by $\{11\bar{2}2\}$ and $\{10\bar{1}2\}$ TBs, therefore it could be inferred that the recrystallization mechanism was $\{11\bar{2}2\}$ and $\{10\bar{1}2\}$ twinning induced SRX.

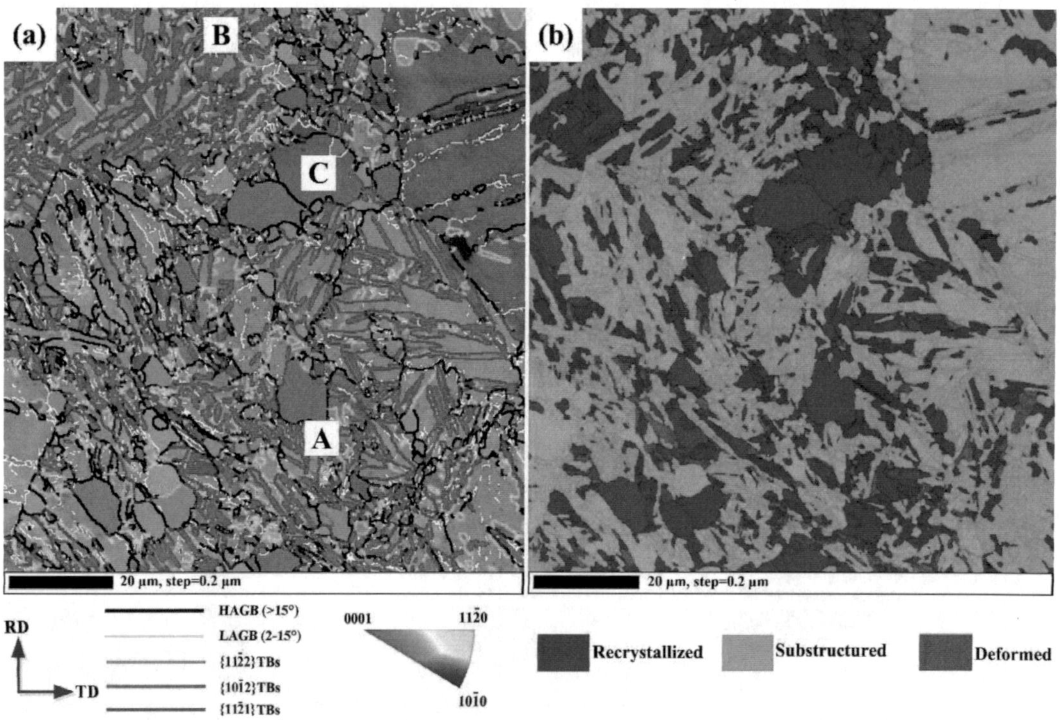

Figure 3 – Microstructure of cryo-deformed CP-Ti after annealing at 500°C for 6 min: (a) IPF map; (b) recrystallization map.

IPF map and recrystallization map of the cryo-deformed CP-Ti after annealing at 500°C for 14 min are shown in Figure 4(a) and Figure 4(b), respectively. The grains were significantly refined in comparison with the as-received microstructure. However, a few $\{11\bar{2}2\}$ and $\{10\bar{1}2\}$ TBs still retained the same twinning angle/axis relation during annealing process. The average grain size was ~2.7 μm, and the fraction of high angle grain boundaries was ~83% (Figure 3(c-d)).

IOP Conf. Series: Journal of Physics: Conf. Series **1270** (2019) 012040 doi:10.1088/1742-6596/1270/1/012040

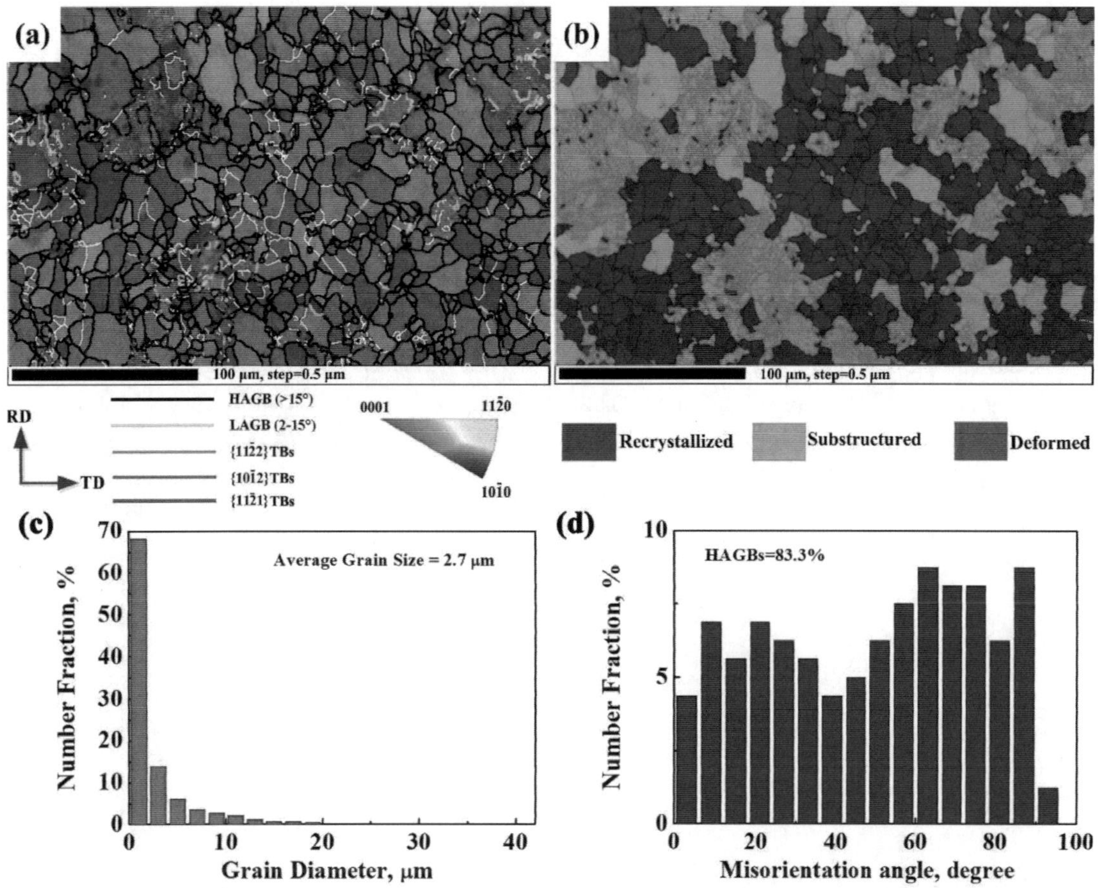

Figure 4 –Microstructure of cryo-deformed CP-Ti after annealing at 500°C for 14 min: (a) IPF map; (b) recrystallization map; (c) grain size distribution; (d) misorientation angle distribution.

4. Conclusions

The microstructure evolution of CP-Ti during cryo-deformation and subsequent annealing at 500°C was characterized. The following conclusions can be drawn:

(1) During cryogenic deformation, high-density deformation twins were activated, mainly $\{11\bar{2}2\}$ contraction twins and $\{10\bar{1}2\}$ extension twins. Multiple point-like twins were activated because the growth of twins was remarkably impeded at the low temperature.

(2) During annealing, the recrystallization mechanisms were $\{11\bar{2}2\}$ and $\{10\bar{1}2\}$ twinning induced SRX. The point-like twins could effectively promote the occurrence of SRX.

(3) The grains were significantly refined from ~40 to ~2.7 μm during annealing. A few $\{11\bar{2}2\}$ and $\{10\bar{1}2\}$ TBs still retained the same twinning angle/axis relation in the annealing process.

Acknowledgements

The authors are grateful to the financial support from the National Natural Science Foundation of China (NSFC) (Grant Nos 51871168) and the Natural Science and Engineering Research Council of Canada (NSERC) in the form of international research collaboration, Fundamental Research Funds for the Central Universities, and Major Science and Technology Project "High-end CNC Machine Tools and Basic Manufacturing Equipment" (2013ZX04011061).

References

[1] Ren J Q, Wang Q, Lu X F, Liu WF, Zhang P L and Zhang XB 2018 *Mater. Sci. Eng.* A. **731** 530-38.

[2] Wang Q, Xin C, Sun Q, Xiao L and Sun J 2017 *Mater. Sci. Eng.* A. **702** 125-32.

[3] Wronski S, Wierzbanowski K, Jędrychowski M, Tarasiuk J, Wronski M, Baczmanski A and Bacroix B 2016 *Mater. Sci. Eng.* A **656** 1-11.

[4] Xu S, Gong M, Jiang Y, Schuman C, Lecomte J-S and Wang J 2018 *Acta. Mater.* **152** 58-76.

[5] Qin H and Jonas J J. 2014 *Acta. Mater.* **75** 198-211.

[6] Yan C K, Feng A H, Qu S J, Cao G J, Sun J L, Shen J and Chen D L 2018 *Acta. Mater.* **154** 311-24.

[7] Yan C K, Feng A H, Qu S J, Sun J L and Shen J. 2018 *Mater. Sci. Eng* A **731** 266-77.

[8] Wu P D, Guo X Q, Qiao H L and loyd DJ 2015 *Mater. Sci. Eng* A **625** 140-45.

[9] Takayama A, Yang X, Miura H and Sakai T 2008 *Mater. Sci. Eng* A **478** 221-28.

IOP Conf. Series: Journal of Physics: Conf. Series **1270** (2019) 012041 doi:10.1088/1742-6596/1270/1/012041

Recrystallization of cladding tubes from Zr-based alloys for nuclear reactors

M G Isaenkova[1,*], Yu A Perlovich[1], V A Fesenko[1] and O A Krymskaya[1]

[1]National Research Nuclear University MEPhI, Kashirskoe shosse, 31, Moscow, 115409, Russia

[*]Corresponding author's e-mail: isamarg@mail.ru

Abstract. This paper is devoted to the regularities of change in the crystallographic texture and the structure of cladding tubes with a total degree of deformation by their cross-section of 70-80% as a result of polygonization or recrystallization during heat treatment in the temperature range from 480 to 600 °C. It is shown, that there are two competing processes of crystal lattice perfection in cladding tubes at annealing temperatures over 540 °C. The oriented growth of grain nuclei, formed at the boundaries of grains and disoriented with respect to the deformed matrix by an angle of 20-30° about basal axes, is predominant. At the recrystallization of cladding tubes from E110 alloy, a significant change in their crystallographic texture of cold rolling is observed, i.e. an increase of the integral texture parameter f_R increases, and a decrease of f_T. The grains, formed at the stage of plastic deformation by twinning, prove to be unstable even at the stage of polygonization, due to the mobility of highly disoriented boundaries under conditions of the anisotropic thermal expansion of neighboring regions.

1. Introduction

Heat treatment is the final stage in the fabrication of cladding tubes from zirconium-based alloys for nuclear reactors. Such treatment is accompanied by partial or complete recrystallization of the alloy, the structure of which is determined by the degree of the preceding deformation, the features of the original texture, the heating rate and the annealing temperature, and also the stressed state at the final stage of deformation [1-3]. The degree of recrystallization of the material is an important factor, which controls the level of operational properties and their anisotropy, because the crystallographic texture is significant changes during its recrystallization [4-7]. In some case at heat treatment of cladding tubes, a substantial reorientation of the basal axes are observed [8-9]. This reorientation is associated with an inhomogeneous distribution of a work hardening in grains of different orientations relative to the unstable position of texture maxima [10-12]. However, the mechanisms of the recrystallization of cladding tubes are very diverse and have not yet been finally established. Bozzollo's paper [13] states that the primary nuclei of new grains grow from perfect regions located within deformed grains, and therefore they are characterized by the same orientation as the deformed matrix. But, during the experiments the samples of the deformed metal are heated with high rate and then they are aged for 15-60 minutes. At high heating rates, the mechanism of the annealing texture formation in the deformed material is fundamentally different from the case of industrial heating of tubes in the furnace. In addition, short-term anneals do not allow to create a structure stable under the conditions of exploitation.

Thus, the aim of this paper is to identify recrystallization mechanisms in rolled cladding tubes by results of texture and structure changes in the course of their industrial annealing. At that, the proposed interpretation of the process is in accordance with the regularities of structure and texture formation by recrystallization described in the literature for alloys based on hexagonal metals such as Ti, Zr and Hf.

IOP Conf. Series: Journal of Physics: Conf. Series **1270** (2019) 012041 doi:10.1088/1742-6596/1270/1/012041

2. Studied specimens and used methods

Studied cladding tubes from Zr-1%Nb alloy of two final sizes, described at the table 1, have been annealed during 3 hours at the different temperatures from 480 to 600 °C. The annealing time has been chosen in accordance with the results of [14], showing that textural changes cease, when a duration of annealing exceeds 3 hours.

The outer diameter and wall thickness of the tubes studied, the Q-factor and the degree of deformation, calculated from the cross-sectional area of the tube at the final stage of fabrication are presented in the table 1. Also, the table 1 contents the texture integral f-parameters, the full width on half maximum (FWHM) of X-ray peak ($21\bar{3}3$), and the degree of recrystallization for some annealing temperatures. Q-factor is equal the ratio of variations of the tube wall thikness to its average diameter under plastic deformation and characterizes the stress state of tube at the process of its deformation.

Table 1. Texture and structure changes during increasing of annealing temperature.

Number	D x t, mm	Q	ε_s, %	T_{anneal}, °C	*f*-parameters R	*f*-parameters T	*f*-parameters L	r	FWHM $(21\bar{3}3)_R$, deg.
1	13.6 x 0.9	2.5	76.5	-	0.469	0.465	0.066	0.04	1.704
				480	0.459	0.476	0.065	0.06	0.421
				540	0.523	0.424	0.053	1.67	0.432
				600	0.517	0.431	0.053	3.83	0.355
2	9.13 x 0.6	1.6	80.7	-	0.485	0.462	0.053	0.01	1.319
				480	0.494	0.461	0.045	0.06	0.369
				540	0.558	0.399	0.043	1.85	0.313
				600	0.548	0.405	0.047	3.66	0.335

The scheme of specimen preparation, used for X-ray study of the structure and crystallographic texture of tubes, is shown in figure 1. The cladding tubes (a) are cut into separate segments (b), which are glued to a flat plate so that a composite sample is obtained. The width of the segment depends on the diameter of the tube. The composite sample is grinded to a flat surface and then is chemically etched on 30 μm for removing the layer of work hardening.

For texture analysis, the studied surface of samples was perpendicular to the radial direction (R) (figure 1-c). In order to estimate the degree of recrystallization using the ratio of intensities of X-ray peaks ($11\bar{2}0$) and ($10\bar{1}0$), a cross section perpendicular to the axial direction of the tube (L) was prepared (figure 1-a).

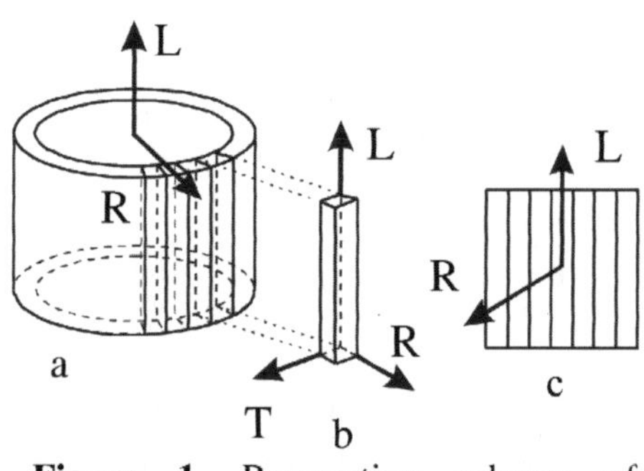

Figure 1. Preparation scheme of specimens:
a - the initial tube and specimen for study of its L-section; b - the segment of tube; c - the composite R-specimen.

X-ray analysis was carried out using DRON-3 and DRON-3M X-ray diffractometers of Russia manufacture, as well as D8 DISCOVER of Bruker manufacture using filtered copper or chromium radiations. Diffraction spectra of the investigated tubes were used for estimation of the recrystallization degree, parameters and stress state of crystalline structure. The texture direct pole figures (DPFs) were recorded in accordance with the standard method of sample tilting [15-16]. On the basis of the (0002) reflection measurement up to 80° of the sample tilt and using extrapolation to the undetermined region of the stereographic projection, the complete DPF (0001) was constructed and used for calculation of the Kearns integral texture parameters: f_R, f_T and f_L, where T is the tangential direction in tube [17]. Orientation distribution functions (ODF) were calculated by incomplete DPFs (0001), $\{10\bar{1}0\}$, $\{11\bar{2}0\}$ and $\{10\bar{1}2\}$ by using standard software LABOTEX [18-19].

The grain size was estimated by the method of X-ray intensity fluctuations in the process of shooting the DPF {11$\bar{2}$0}, described in detail in the work [20]. The degree of recrystallization was estimated from the ratio of the intensities of the X-ray peaks (11$\bar{2}$0) and (10$\bar{1}$0), taking into account their ratio in the textureless sample. Lattice parameters were calculated by peak positions of X-ray peaks (21$\bar{3}$3) and (30$\bar{3}$2).

3. Experimental results

3.1. Structural state of samples

Changes in lattice parameters of the crystal structure as a result of the annealing of the studied tubes are shown in Fig. 2. These plots prove, that already at a temperature of 480 °C the crystalline structure undergoes recovery to the equilibrium state of α-Zr, owing to the anomalously high self-diffusion coefficient of Zr [21]. The parameters of the structure are practically preserved at all other annealing temperatures (figure 2).

The increase of the annealing temperature causes a perfection of the crystalline structure of the deformed grains, about which we are judging by the FWHM of the X-ray lines (see table 1, figure 3). It is known [1-3, 9, 13], that as a result of recrystallization of Zr-based alloys, a crystal lattice of recrystallized α-Zr grains rotates about the basal normal with respect to the deformed matrix. By the study of considered processes, X-ray peaks (11$\bar{2}$0) and (10$\bar{1}$0) were recorded for the cross section perpendicular to the tube axis L and the ratio of their intensities r=I(11$\bar{2}$0)/I(10$\bar{1}$0). The last parameter allows to estimate the rotation of the hexagonal lattice about the basal direction. Table 1 contains the values r for some annealing temperatures, and figure 4 shows the change in the recrystallization degree r depending on the annealing temperature of the investigated cladding tubes.

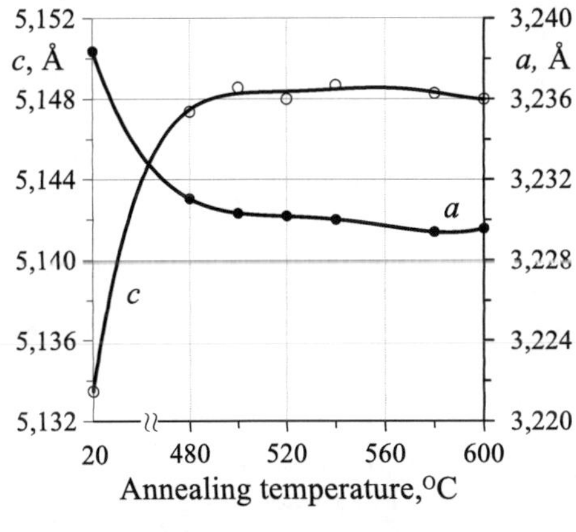

Figure 2. Changes of parameters of α-Zr crystal structure by increase of the annealing temperature.

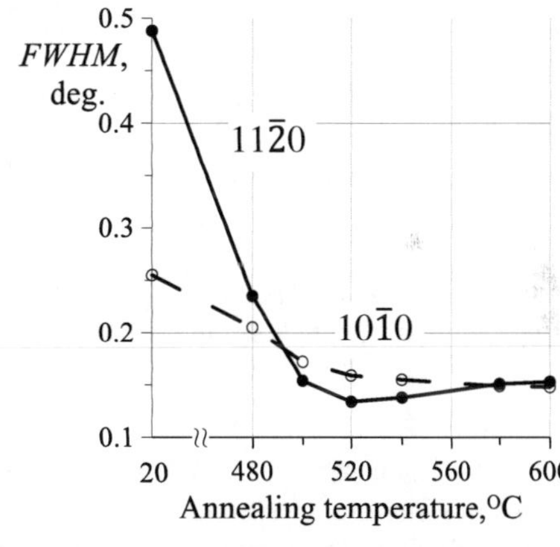

Figure 3. Changes of X-ray peaks' FWHMs by increase of annealing temperature (by an example of tube with external diameter 13.6 mm)

Figure 5 illustrates the dependence of the grain size on the degree of recrystallization of a zirconium alloy. From the given graphs it follows, when an annealing temperature is higher than 540 °C and r > 2, recrystallization by rotation of the grains lattice about basal normals of the deformed matrix by 30° is observed and a stable grain size is varied in a narrow range from 2.5 to 3.5 μm.

3.2. Crystallographic texture of annealed tubes

A typical changes in the ODF section with an Euler angle $\varphi_1 = 0$ at an increasing of the annealing temperature of cold-rolled tubes, having an external diameter of 13.6 mm, are shown in figure 6. Figure 7 illustrates the dependence of the change in the ratio of f-parameters on the annealing temperature of the studied tubes, differing by the technological characteristics of their deformation at the final stage.

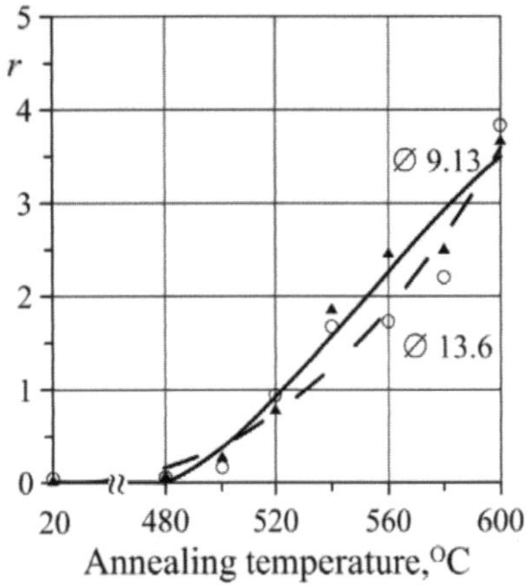

Figure 4. Rise of recrystallization degree *r* by increase of annealing temperature of studied tubes

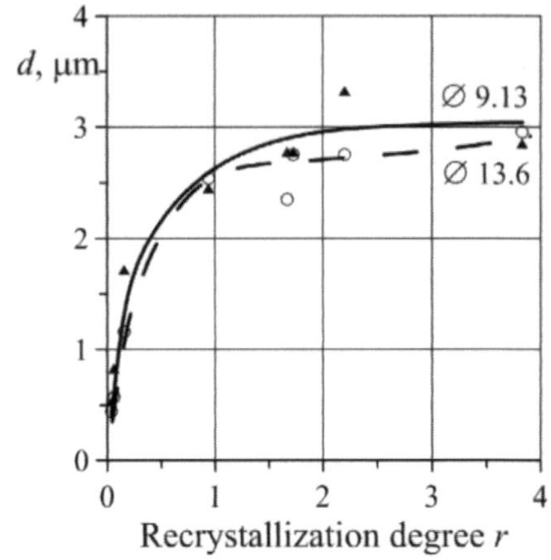

Figure 5. Correlation of grain size *d* from recrystallization degree *r* of studied tubes

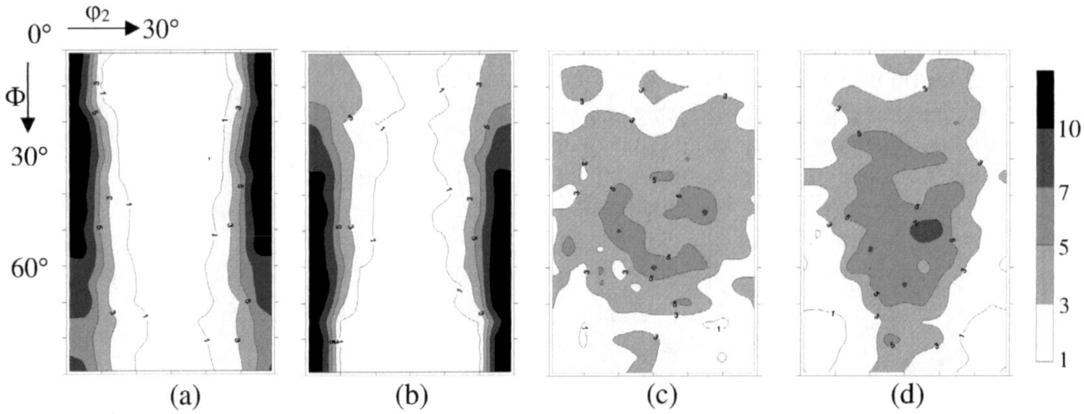

Figure 6. Change in ODF-section for φ_1=0 by increase of annealing temperature of cladding tubes: (a) rolled tube; (b) 480 °C – 3 h; (c) 540 °C – 3 h; (d) 600 °C – 3 h

4. Discussion of obtained results

The following discussion of X-ray data, presented in the given paler, is based on analysis of revealed regularities of structure and texture formation in zirconium alloys by recrystallization.

The first of all, it is the improvement of the crystalline structure of the deformed alloys, when the parameters of α-Zr unit cell attain the level, typical for their stable state. At that, the parameter *a* decreases, while the parameter *c* increases (figure 2), as well as the sharpening of X-ray peaks, coming down due to active diffusion processes, the drop of dislocation density, and, finally, the growth of grains with new orientations (figure 3).

Secondly, it is the weakening of the texture maximum in the ODF-section (compare figure 6-a and 6-b), corresponding to the orientation of basal normals along the radial direction. Grains with basal normals, oriented near the radial direction are localized within twinning regions of the deformed matrix [1, 3], which were formed during the final deformation of a tube with a high Q-factor. Analysis of texture formation at successive stages of tube manufacturing [22] allows us to state that before the last rolling the basal normals of grains, appeared near the radial direction were oriented near the tangential direction. Therefore, the misorientation of the twinned grains relative to neighboring regions of the deformed matrix can reach 85° and cause significant stresses during heating of tubes for heat treatment. Part of twins can decrease in size as a result of the arising high stresses caused by the anisotropy of the thermal variation of the linear grain sizes along the *a* and *c* directions. In this case, the twinned volume can return to its initial orientation, which we observed in figures 6-a and 6-b.

Namely, as a result of annealing at temperature of 480 ° C, i.e. when polygonization takes place, there is an increase in the pole density of the basal normals near the tangential direction and its decrease near radial direction.

In the case of intensive growth of new grains misoriented relative to the deformed matrix by rotation around the basal normals, a sharp changes in the ODF-section are observed (compare figures 6-a and 6-c or 6-d). At the same time, a sharp drop in the broadening of the X-ray line (11$\bar{2}$0) characterizing the grains of the new orientation is noticeable, while improvement of grains with the former orientation occurs at a lower rate of perfection (figure 3). It is likely that this is the reason for the subsequent absorption of distorted regions, in which only the process of structure polygonization occurs, by more perfect recrystallized grains.

It was shown in [13] that in the case of short-term annealings, conducted to establish mechanisms of recrystallization in Zr and Ti, at the initial stage, perfect regions formed in the deformed matrix, they maintain their orientations by means of the migration of defects to sinks or of polygonization. However, with an increase in the annealing time, a rotation of the crystalline lattice in the growing grains about basal normals is observed.

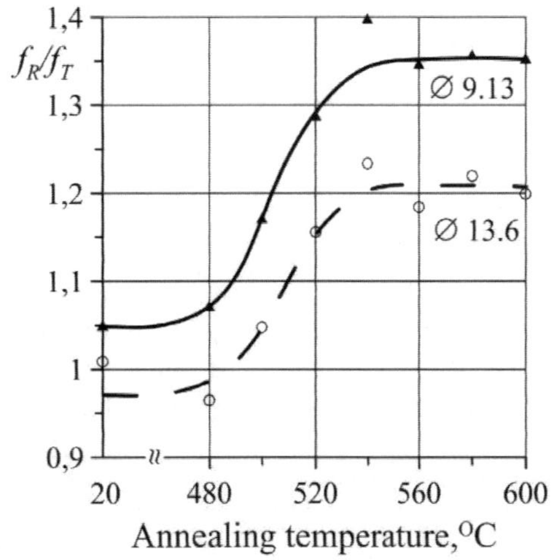

Figure 7. Changes of the ratio of integral texture parameters f_R/f_T of annealed tubes by increase of annealing temperature for different diameter tubes

According to thermodynamic concepts at the initial moment of heat treatment in the deformed material, the formation of grains of different orientations and by different mechanisms must take place, being determined by local conditions of diffusion processes in the deformed material. The presence of dislocations of various types, their orientations, microstresses that result from the anisotropy of the expansion of grains with different orientations in the heating process, predetermine the orientation of the primary nuclei of the future grains. Nuclei of new grains can appear both in the body of deformed grains and on their boundaries, characterized by greater stored energy. This point of view is shared by the authors of the theories of oriented nucleation and oriented growth of recrystallized grains [4, 22].

Evidence of the formation of new grains, disoriented with respect to the deformed matrix at an angle of 20-30° about basal normals, precisely in the boundary of deformed grains is the results of comparison of recrystallization in two-phase titanium alloys and in zirconium alloys with a two-component texture, where similar grains are absent [3, 23].

At the heat treatment of deformed materials, all possible thermal mechanisms operate simultaneously in the workpiece being processed: recovery and polygonization inside grains as well as the formation of new perfect grains at the boundaries of the initially deformed grains. The presence of two competing processes is observed throughout the investigated temperature range, but at 480 °C the nuclei of new grains, formed at the boundaries, can not grow because of the lack of speed for movement of the 30°-boundaries.

In connection with the above facts, the rate of heating of the sample has fundamental importance for activating of one or another mechanism of recrystallization. As it was shown in paper [1], in the case of accelerated heating (100°/s) of deformed samples from a Zr-based alloys, the growth of nuclei disoriented with respect to the deformed matrix by rotation about basal normals is suppressed.

5. Conclusions

1. In competing processes of crystal lattice perfection in cladding tubes at annealing temperatures over 540 °C, the oriented growth of grain nuclei, formed at grain boundaries and disoriented with respect to the deformed matrix by an angle of 20-30° about basal normals, is predominant.

2. After recrystallization of cladding tubes from E110 alloy, a significant change in their crystallographic texture of cold rolling was observed, consisting in the increase of integral texture parameter f_R and the decrease of texture parameter f_T.
3. The grains, formed at the stage of plastic deformation by twinning, prove to be unstable even at the stage of polygonization due to the mobility of highly disoriented boundaries under the conditions of anisotropic thermal expansion of neighboring regions.

References:

1. Isaenkova M, Perlovich Yu, Fesenko V et al 2014 Regularities of recrystallization of rolled single crystals and polycrystals of zirconium and alloy Zr-1% Nb. *The Physics of Metals and Metallography* **115**, **8** 756-64
2. Isaenkova M G, Perlovich Yu A, Fesenko V A et al 2015 Regularities of recrystallization of rolled Zr single crystals *IOP Conf. Series: Materials Science and Engineering* **82** 012038
3. Perlovich Yu, Isaenkova M et al 2015 *Recrystallization in materials processing* ed Glebovsky V (InTech, Rijeka, Croatia) ISBN 978-953-51-2196-1.
4. Kocks U F, Tome C N and Wenk H-R 1998 *Texture and Anisotropy. Preferred orientation in polycrystals and their effect on materials properties* (Cambridge: Cambridge University Press) p 676.
5. Coleman C E et al. 2009 Mechanical properties of Zr-2.5%Nb pressure tubes made from electrolytic powder. *Zirconium in the Nuclear Industry: 15th Int. Symp. ASTM STP* **1505** 699-723.
6. Lichter B D, Flanagan W F, Lee D N 2002 The role of texture in stress-corrosion cracking of metals and alloys *Materials Science Forum* **408-412** 991-8.
7. Perlovich Yu A, Isaenkova M G, Medvedev P N et al 2015 Mechanisms of the texture influence on the corrosion behavior of Zr-alloy cladding tubes. *Inorganic Materials: Applied Research* **6** (3) 259–66
8. Isaenkova M, Perlovich Yu, Fesenko V and Krymskaya O 2016 Change of Basal Normal Orientation during Recrystallization of Sheets and Tubes from Zr-Based Alloys. *In book: Proceedings of the 6th International Conference on Recrystallization and Grain Growth* 293-8
9. Perlovich Yu, Isaenkova M, Shmelyova T et al 1989 Texture change of tubes from Zr-2,5% Nb alloy at recrystallization *Soviet Atomic Energy* **67** (5) 327-31
10. Perlovich Yu and Isaenkova M 2002 Distribution of c- and a-dislocations in tubes of Zr alloys. *Metallurgical and Materials Transactions A* 867-74
11. Shikov A K, Bocharov O V, Arzhakova V M et al. 2003 Use of hafnium in control elements of nuclear reactors and power units. *Metal Science and Heat Treatment* **45** (7-8) 300-3.
12. Isaenkova M, Perlovich Yu 2005 Distribution of Dislocation Density in Tubes from Zr-Based Alloys by X-Ray Data. *Solid State Phenomena* **105** 89-94
13. Dewobroto N, Bozzolo N, Batberis P and Wagner F 2007 On the mechanisms governing the texture and microstructure evolution during static recrystallization and grain growth of low alloyed Zirconium sheets. *International Journal of Material Research* **97** 826-33
14. Isaenkova M and Perlovich Yu 2014 *Regularities of development of crystallographic texture and substructure inhomogeneity in zirconium alloys during deformation and heat treatment* (Moscow: NRNU MEPhI) p 528
15. Borodkina M.M. and Spector E.N. 1981 *X-ray analysis of texture of metals and alloys* (Moscow: Metallurgiya) p 272
16. Isaenkova M, Perlovich Yu and Fesenko V 2016 Modern methods of experimental construction of texture complete direct pole figures by using X-ray data. *IOP Conf. Series: Materials Science and Engineering* **130** 012055
17. Tempest P A 1980 Preferred orientation and its effect on bulk physical properties of hexagonal polycrystalline materials *J. Nuclear Materials* **92** 191-200
18. Pawlik K 1986 Determination of the orientation distribution function from pole figures in arbitrarily defined cells. *Phys. Stat. Sol. (b)* **134** K477-K483.
19. LaboTex v. 3.0 by LaboSoft (Krakow, Poland) on http://www.labosoft.com.pl
20. Krymskaya O, Isaenkova M and Perlovich Yu 2010 Determination of grain size for different texture components by statistical fluctuations of intensity, registered in the course of texture measurement *Solid State Phenomena* **160** 135-40.
21. Filippov V P, Bateev A V, Lauer Y A and Kargin N I 2013 Mossbauer spectroscopy of zirconium alloys. *Hyperfine Interactions* **217 (1-3)** 45-55
22. Vishnyakov Ya D, Babareko A A, Vladimirov S A et al 1979 *Theory of texture formation in metals and alloys* (Moscow: Nauka) p 343
23. Isaenkova M, Perlovich Yu, Fesenko V et al 2017 Behaviour of sheets from Ti-alloys by rolling and heat treatment *AIP Conference Proceedings* **1896** 190020; http://doi.org/10.1063/1.5008233

IOP Conf. Series: Journal of Physics: Conf. Series **1270** (2019) 012042 doi:10.1088/1742-6596/1270/1/012042

The effect of continuous deformation behaviors on texture formation in magnesium alloy

D Han[1] and K Kim[1,2]

[1] Marine Convergence Design Co-work, Pukyong National University, 45 Yongso-ro, Nam-gu, Busan 48513, Republic of Korea

[2] Department of Metallurgical Engineering, Pukyong National University, 45 Yongso-ro, Nam-gu, Busan 48513, Republic of Korea

E-mail: mrppeng@pknu.ac.kr

Abstract. In previous study, it was investigated texture formation behavior of high temperature compression test in various AZ magnesium alloy. These authors investigated AZ31, AZ61, AZ80 and AZ91 magnesium alloy of compression test at temperature 673K to 723K, a strain rate of $5.0x10^{-2}s^{-1}$, and a strain -0.4 to -1. It was found two characteristics of texture formation behaviour – 1) the basal texture was increase with solute concentration increasing, 2) the main texture component developed depending on deformation conditions. However, additional experiment was needed to clarify why the development of these texture occur. Therefore, this study investigated the mechanism of texture development behaviors in the AZ91 magnesium alloy with three different types initial textures during high-temperature plane strain compression deformation. This compression test conditions were temperature of 673K to 723K, and a strain rate of $5.0x10^{-4}s^{-1}$, with up to a strain -1.0. After compression test, the specimens were immediately quenched in water. The texture measurement was conducted by the Schulz reflection method using Cu Kα radiation and EBSD. As a result, in all specimens, fiber texture was formed at located at a position 25° away from (0001) texture. In addition, changes in the main component occurred depending on the temperature and the initial texture.

1. Introduction

The magnesium and its alloys is a lightweight material that has attracted attention because it has great synergy when used in transportation and portable devices. The AZ91 magnesium alloy, a type of magnesium alloy, has favourable mechanical properties that can be used as a structural material [1]. However, magnesium and its alloys has poor formability at room temperature, due to having the hexagonal crystal structure. Due to this limited slip system, there is a poor formability and high solute AZ91 alloy is mostly used as a casting material [2,7]. Therefore, it is necessary to understand the texture control in order to improve poor formability, and is important to understand the mechanism of texture formation in AZ91 magnesium alloy.

In previous study, these authors investigated the behavior formation of texture in AZ31, AZ61, AZ80, AZ91 magnesium alloys during high temperature compression test at various deformation conditions. It was found that two characteristic of texture formation behaviors occur. First, the basal texture was increasing with solute concentration increase. Second, the main texture component developed depending on deformation conditions. They reported that texture formation of magnesium alloy was affected by dynamic recrystallization [1-3]. However, there is no enough to clarify study on the texture formation behaviors of AZ91 alloy.

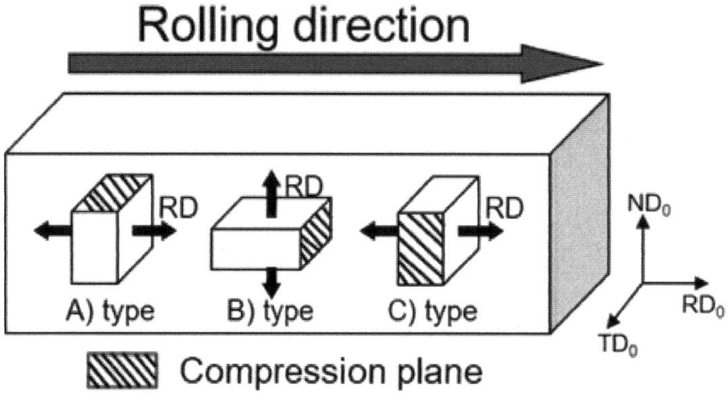

Figure 1. The geometry of specimens [2].

Therefore, in this study, texture formation behavior of AZ91 magnesium alloy was investigated by plane strain compression tests under the various initial texture and deformation conditions.

2. Experimental procedure

The base materials used a cast AZ91 ingot with Mg-8.77%Al-0.74%Zn wt% chemical composition. The ingot rolled at 673K with rolling reduction of 30%. Specimens were machined out three type of specimens as shown in Figure 1. Figure 1 shows the geometric arrangement of the specimens and RD means the direction in which the specimen extension in plane strain compression. Each specimen is rotated 90 degrees in TD or RD direction. After machining, the specimen is annealed at 723K for 1h for producing homogeneous microstructure. Figure 2 shows axis densities of three types of specimens before deformation. The pole densities are projected onto the compression plane. Mean axis density is used as a unit to draw the contour lines. In case of specimens A, the {0001} pole densities are distributed at the periphery of the pole figure. The (0001) pole densities in specimen B and C are distributed on the great circle 90° from TD and RD, respectively.

Plane strain compression tests were carried out at temperature 673K and 723K, and a strain rate of $5.0 \times 10^{-4} s^{-1}$ with an up to strain of -1.0. After compression test, the specimens were immediately quenched in water. The microstructure is observed the mid-plane after mechanical polishing and electropolishing. Texture measurements were conducted by the Schulz reflection method using Cu Kα

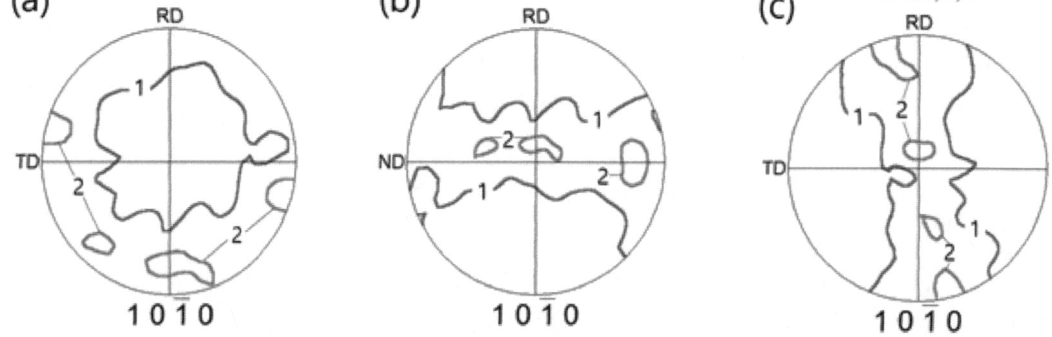

Figure 2. $\{10\bar{1}0\}$ pole figures showing the crystallographic characteristics of the three types of specimens before deformation. (a), (b) and (c) show the specimens of A, B and C, respectively. The pole densities are projected onto the compression plane. Mean axis density is used as a unit to draw the contour lines.

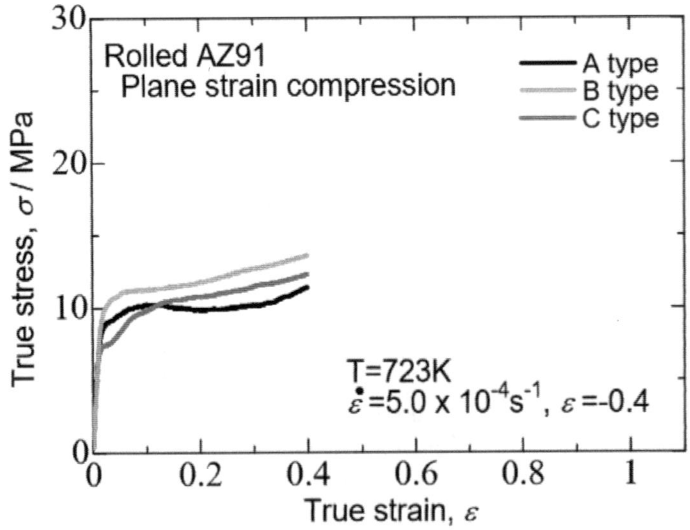

Figure 3. True stress – true strain curves for deformation at temperature of 723K with a strain rate of $5.0x10^{-4}s^{-1}$ up to a strain of -0.4.

radiation and EBSD. The texture evaluation through five pole figures of {0001}, $\{10\bar{1}0\}$, $\{\mathbf{10\bar{1}1}\}$, $\{\mathbf{10\bar{1}2}\}$, $\{\mathbf{11\bar{2}0}\}$. Based on these five pole figures the ODF (Crystal Orientation Distribution Function) was determined by the Dahms and Bunge method [4]. Electron Backscatter Diffraction (EBSD) measurements were performed after the electrolytic polishing.

3. Results and discussion

3.1. Behavior of deformation

Figure 3 shows true stress – true strain curves for deformation at temperature of 723K with a strain rate of $5.0x10^{-4}s^{-1}$ up to a strain of -0.4. The flow stress decrease with increasing of temperature.

Figure 4. Microstructure of specimen A observed by EBSD measurement (a) initial specimen, (b) deformation at 723K up to the strain of -1.0 with a strain rate of $5.0x10^{-4}s^{-1}$.

3.2. Behavior of microstructure and texture formation

Figure 4 shows the grain structure maps observed by EBSD measurements for specimen A before and after deformation – (a) before deformation and (b) at 723K with a strain rate of $5.0x10^{-2}s^{-1}$ up to a strains of -1.0. The light grey grains are within 15° from (0001) (compression plane) and the dark grey grains are range from 15° to 30° from (0001) (compression plane). Figure 4 (a) and (b) have the same magnification. Black lines express the grain boundaries with a misorientation higher than 15°.

Figure 4 (a) shows that the most of equiaxed grains formed, and these grains indicated light and dark grey. This means that the basal texture was formed before deformation as shown the pole figure of initial texture in Figure 2(a). Also, two types of grey grain show similar fraction. However, with increasing of the strain, the mean grain size gradually increase and the grains with dark grey continuously increase as shown in Figure 4. (b). Although the experiment results of different kinds of specimens are not shown here, specimen B and C were also has same texture formation behaviors.

Figure 5 is the inverse pole figure showing the density distribution of compression axis after deformation at strain rate of $5.0x10^{-4}s^{-1}$ – (a) 673K, ε = -1.0, (b) 673K, ε = -1.0, (c) 673K, ε = -1.0, (d) 723K, ε = -1.0 (e) 723K, ε = -1.0, (f) 723K, ε = -1.0, respectively. The mean axis density is used as a unit. The value of the maximum axis density and its position are shown at the bottom of the inverse pole figure. The position of maximum axis density is written in (α, β) form. α is the angle from (0001), and β is rotation angle around [0001] from the [0001]-[$10\bar{1}0$] line to the position of the maximum axis density.

In this study, three kinds of specimens show different texture before deformation. The values of maximum axis densities and its position change with increasing of strain and their positions are α angle 20~25°from (0001) (compression plane) at a true strain of 1.0. This result shows that basal texture was

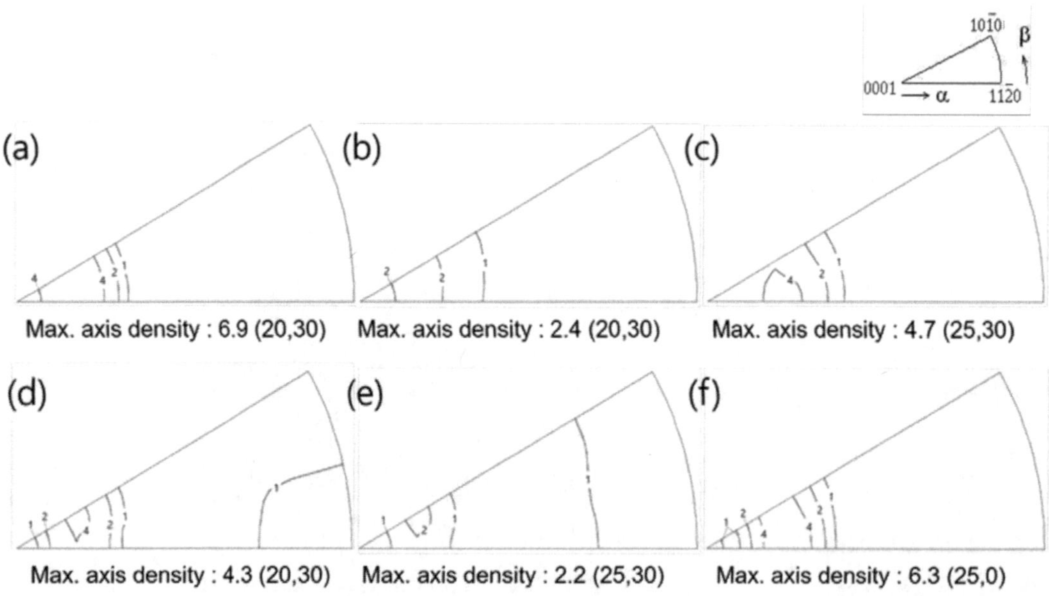

Figure 5. Inverse pole figure showing the density distribution of compression axis at strain rate of $5.0x10^{-4}s^{-1}$ – (a) 673K, ε = -1.0, (b) 673K, ε = -1.0, (c) 673K, ε = -1.0, (d) 723K, ε = -1.0 (e) 723K, ε = -1.0, (f) 723K, ε = -1.0, respectively. The mean axis density is used as a unit. (a), (d) initial specimen orientation A. (b), (e) initial specimen orientation B. (c), (f) initial specimen orientation C.

not developed at deformation condition of this study unlike previous study [2]. This means that texture formation behavior changes depending on the initial texure.

In order to investigate the quantitative variation of texture formation behaviour after deformation, the relationship between maximum axis density and misorientation from (0001) is shown in Figure 6. The horizontal axis in Figure 6 is α angle from inverse pole figure, and the vertical axis is maximum axis density of each β angles which means rotation angle around [0001] from the [0001]-[10$\bar{1}$0] line. With increasing strain, the position of maximum axis densities in α angle 20~25° gradually increases. After the deformation up to a strain of 1.0, maximum axis densities at α angle 20~25° show higher than any other α angle in specimens A and C. These results show that axis densities after deformation more developed when axis densities exist at α angle 20~25° before deformation. That is, this means that formation and development of texture after deformation is affected by initial texture. Also, it could confirmed that the maximum axis density in three kinds of specimens continuously moves into α angle 20~25° and the maximum axis density at α angle 20~25° highly increases with increasing of strain. From these results, it is considered that texture formation behavior is affected by continuous deformation and grain growth and the deformed texture is stable orientation for deformation of this study.

4. Conclusions

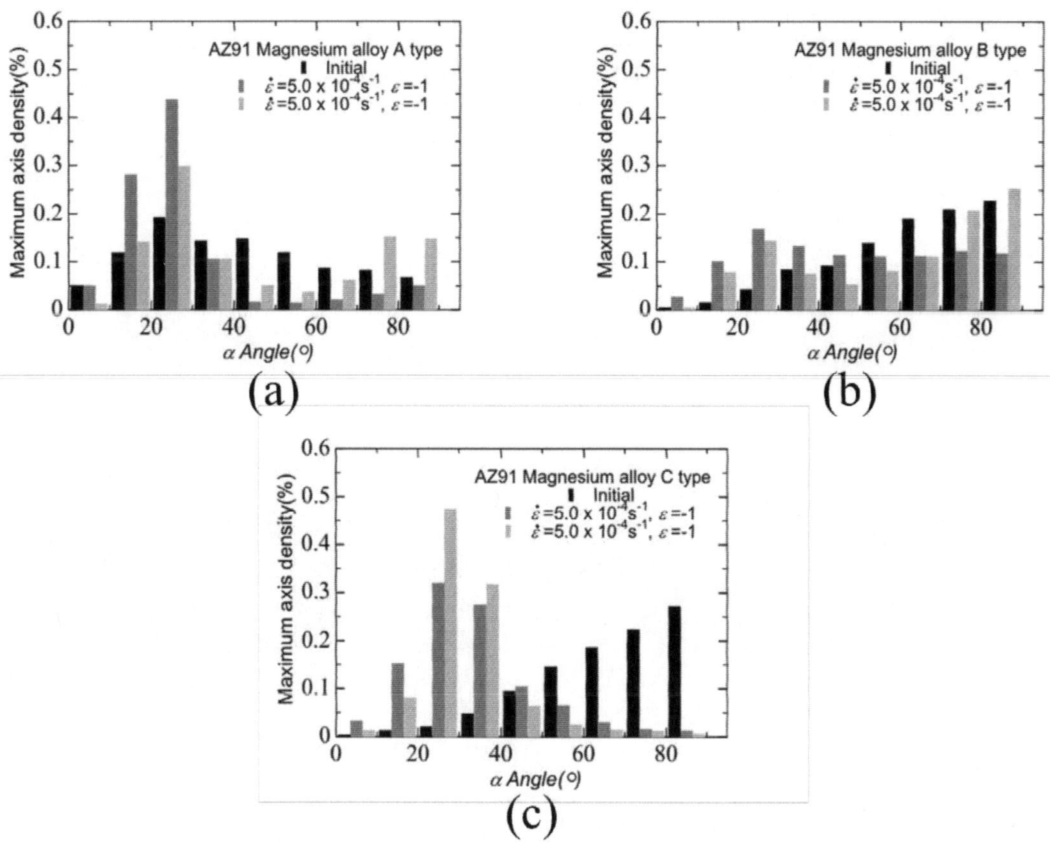

Figure 6. Compared main component α of Fraction A, B and C specimens to before and after deformation. The one bar means the 10 degree unit of average.

This study investigated the texture formation behavior of AZ91 magnesium alloy with different initial texture during plane strain compression at high temperature. The investigations for microstructure and texture gave the following results.

1) Regardless of initial texture, the fiber texture is formed at located at a position 20~25° away from (0001) after the deformation.
2) With increasing of strain, the fiber texture is formed and developed at a position 20~25° away from (0001). Texture formation behavior in this study is affected by continuous deformation and grain growth.

Acknowledgments
This work was supported by the Marine Convergence Design Co-work of Busan Korea 21 PLUS and the National Research Foundation of Korea grant funded by the Korea government (No.2017R1C1B5076690)

References
[1] Kim J, Okayasu K and Fukutomi H 2013 *Mater. Sci. Forum* **753** 493
[2] Park M, Okayasu K, Fukutomi H and Kim K 2016 *Met. Mater. Int.* **22** 1129
[3] Kim K, Okayasu K, and Fukutomi H 2012 *Mater. Trans.* **53** 1870
[4] Dahms M and Bunge H J 1989 *J. Appl. Crystallogr.* **22** 439
[5] Kim K, Okayasu K and Fukutomi H 2015 *Mater. Trans.* **56** 17
[6] Kim J, Okayasu K and Fukutomi H 2013 *Mater. Trans.* **54** 2
[7] Gottstein G and Samman T Al 2005 *Mate. Sci. Forum* **495** 623
[8] Horiuchi R and Otsuka M 1972 *Trans. JIM* **13** 284

Basal texture formation behavior of M1 magnesium alloy during high-temperature compression deformation

M Park[1] and K Kim[2,3]

[1] Institute of Innovative Research (IIR), Tokyo Institute of Technology, 4259 nagatsuta-cho, Midori-ku, Yokohama, Kanagawa 226-8503, Japan
[2] Department of Metallurgical Engineering, Pukyong National University, 45 Yongso-ro, Nam-gu, Busan 48513, Republic of Korea
[3] Marine Convergence Design Co-work, Pukyong National University, 45 Yongso-ro, Nam-gu, Busan 48513, Republic of Korea

mrppeng@pknu.ac.kr

Abstract. Understanding the formation behavior of basal texture is important for improving formability. Many researchers found that deformation condition was an important factor in order to texture control on AZ system magnesium alloy. In case of M1 magnesium alloy has been investigated in texture formation resulting from grain size manipulation by precipitation, but there seems no studies on the texture development during high temperature on M1 magnesium alloy by deformation condition. In this study, microstructure evolution and texture formation behavior is investigated under various deformation conditions in M1 magnesium alloy. In order to make the same initial conditions, M1 magnesium ingot was rolled at 673K with a rolling reduction of 30%. Uniaxial compression test specimens were machined in such a way that the compression plane was parallel to the rolling plane, and subsequently annealed at 823 K for 1h. Uniaxial compression tests were carried out at 673K and 723K under a strain rate of $5.0 \times 10^{-2} s^{-1}$ up to a strain of -1.0. EBSD measurement was performed for observation of crystal orientation distribution. As a result, occurrence of the dynamic recrystallization was confirmed. The position of maximum axis density of the texture and its intensity were varied depending on deformation conditions.

1. Introduction

The interest in lightweight materials of automotive industries has been recently growing due to reducing weight, and especially, magnesium and its alloy have been paid attention as lightweight materials because of low density and high specific strength. However, magnesium and magnesium alloy have generally poor room-temperature formability due to having a limited number of slip systems, and many researchers have been investigated on texture control of magnesium and magnesium alloy in order to improve their room-temperature formability [1-2].

In previous study, two authors were investigating on microstructure and texture formation of M1 magnesium alloy during high-temperature deformation at limited deformation condition. It was found that basal (0001) texture decreases and main texture component changes with increasing of strain. This results was affected by dynamic recrystallization and grain growth. However, there was not enough to understand the basal texture formation behavior of M1 magnesium alloy during high-temperature deformation [3]. Also, the basal texture formation behavior of AZ system magnesium alloy in previous study was strongly affected by deformation condition - especially deformation temperature [4-5].

Therefore, in this study, basal texture formation behavior of M1 magnesium alloy was investigated by uniaxial compression tests under two types of deformation temperature with strain rate of $5.0x10^{-2}s^{-1}$ up to a strain of -1.0.

2. Experimental procedure

The material used in this study is a commercial M1 (Mg - 1.2% Mn wt%) magnesium alloy. This cylindrical specimen was hot-rolled at 673 K with a rolling reduction of 30 % and then was machined out in such a way that the compression plane was parallel to rolling plane for uniaxial compression tests. The specimens were annealed at 823 K for 1h to produce a uniform grain structure. Figure 1 show the inverse pole figure showing the density distribution of compression axis. The mean axis density is used as a unit. The value of the maximum axis density and its position is expressed as (α, β) at the bottom of the figure. It was found that [0001] is the position of the maximum axis density, where the basal texture is being formed and the maximum axis density is 4.5 times the mean axis density.

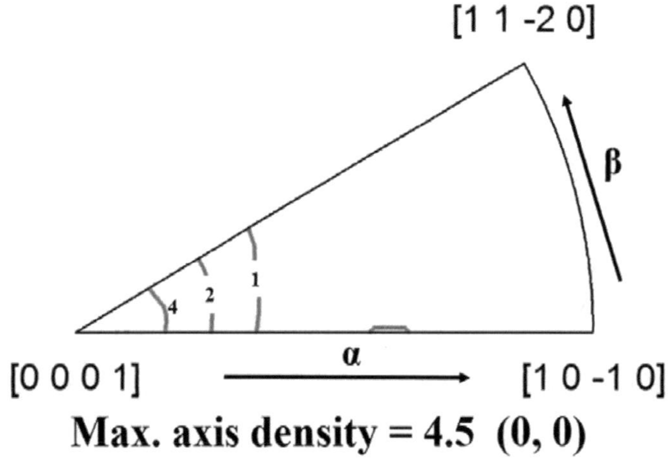

Figure 1. Inverse pole figure showing the density distribution of compression axis before deformation. Mean axis density is used as a unit.

Uniaxial compression tests were carried out at 673K and 723K under a strain rate of $5.0x10^{-2}s^{-1}$ up to a strain of -1.0. In order to prevent a change in the microstructure after compression test, the specimens were immediately quenched in oil after deformation. The mid-plane section of the specimen was prepared by polishing. The surfaces were finished using an Emery paper and a silica particle (0.04μm) suspension, and then the microstructure observation and texture measurement were performed. The texture measurements were carried out by the Schulz reflection method using Cu Kα radiation. Based on five pole figures, crystal orientation distribution function (ODF) was determined by the Dahms and Bunge method [6]. The main component and degree of texture development were determined by drawing the inverse pole figures based on ODF. In addition to the X-ray measurements, EBSD (Electron Back-Scattered Diffraction) measurements were performed. The EBSD measurements were carried out after electrolytic polishing. The region of 0.45mm × 1.33mm was scanned using the beam control option with a step of 4μm.

3. Results and discussion

3.1. Behavior of deformation

Figure 2 shows the true stress – true strain curves for the deformation at 673K and 723K under a strain rate of $5.0x10^{-2}s^{-1}$ up to a strain of -0.4.

The flow stresses increase with decreasing of temperature, and the work hardening occurs in all experimental deformation condition at the beginning of the true stress – true strain curve.

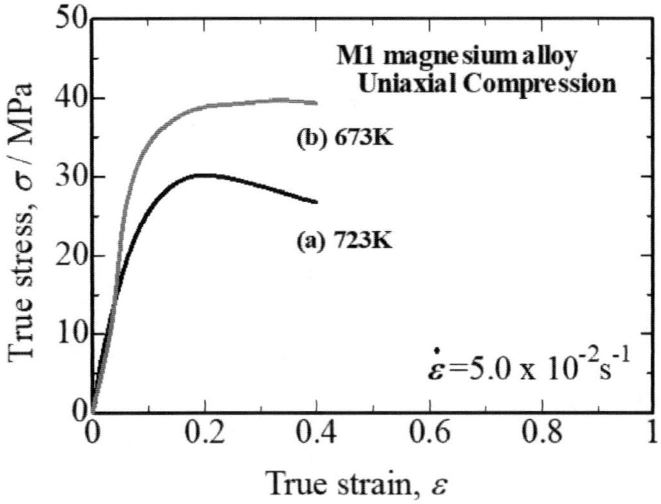

Figure 2. True stress – true strain curves for the deformation at $5.0x10^{-2}s^{-1}$ up to a strain of -0.4 with the temperature ranging from 673K to 723K.

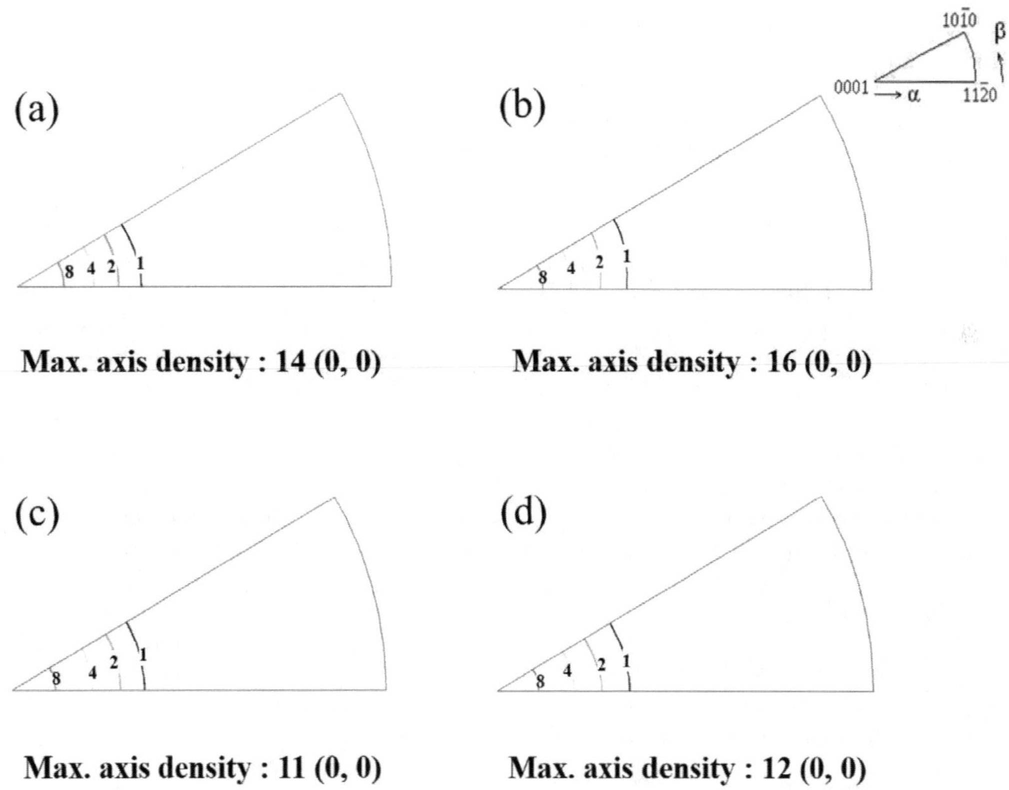

Figure 3. Inverse pole figure showing the density distribution of compression axis at strain rate of $5.0x10^{-2}s^{-1}$ – (a) 673K, ε = -0.4, (b) 673K, ε = -0.7, (c) 723K, ε = -0.4 and (d) 723K, ε = -0.7, respectively. The mean axis density is used as a unit.

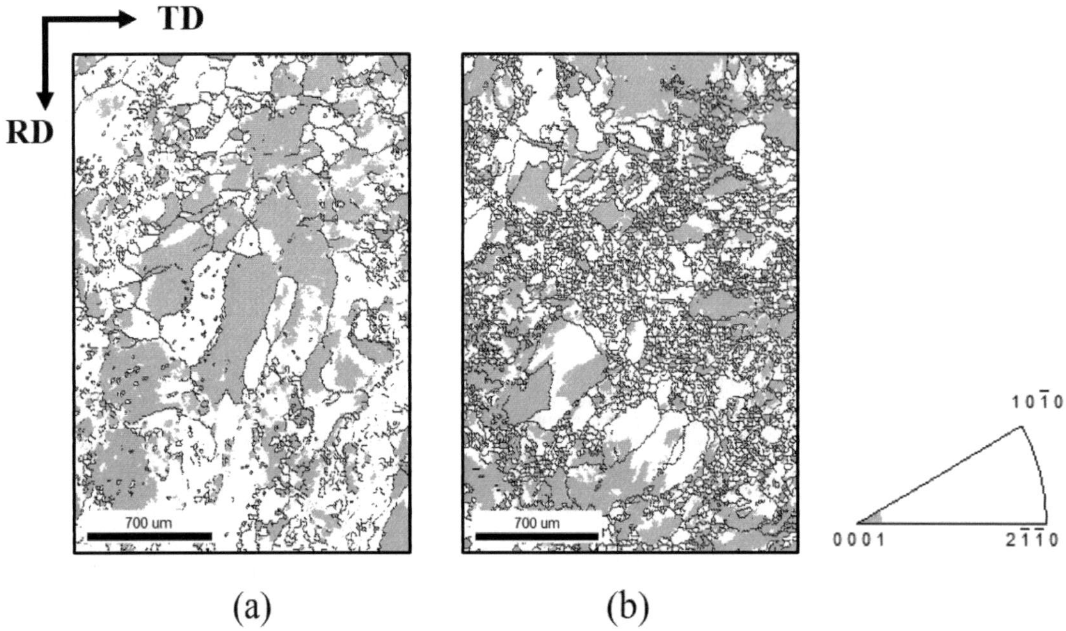

Figure 4. Grain structure map observed by EBSD measurements after deformation at 723K with a strain rate of $5.0\text{x}10^{-2}\text{s}^{-1}$ – up to strains of (a) -0.4 and (b) -0.7.

3.2. Behavior of texture formation

Figure 3 is the in inverse pole figure showing the density distribution of compression axis after deformation of this study - (a) 673K, ε = -0.4, (b) 673K, ε = -0.7, (c) 723K, ε = -0.4 and (d) 723K, ε = -0.7, respectively. The mean axis density is used as a unit. The value of the maximum axis density and its position are shown at the bottom of the inverse pole figure. The position of maximum axis density is written in (α, β) form. α is the angle from (0001), and β is rotation angle around [0001] from the [0001]-[10$\bar{1}$0] line to the position of the maximum axis density.

The values of maximum axis densities increase with an increase of true strain and with a decrease of deformation temperature. Also, the specimens underwent deformation up to a true strain of 1.0, but there was no significant change in their position of all specimens in experimental deformation conditions of this study. This means that (0001) (basal texture) is stable orientation for deformation of this study.

3.3. Grain structure

Figure 4 shows Grain structure map observed by EBSD measurements after deformation at 723K with a strain rate of $5.0\text{x}10^{-2}\text{s}^{-1}$ up to strains of (a) -0.4 and (b) -0.7. The gray grains are within 15° from (0001) (compression plane). As shown in Figure 4, the mean grain size decreases and volume fraction of grain with (0001) orientation component increases with increasing of the strain.

The researchers investigated deformation behavior and texture formation in AZ80 magnesium alloy during uniaxial compression deformation at high-temperature at various deformation conditions [4]. They reported that basal texture was formed under deformation conditions showing flow stress of more than 15-20MPa, and basal texture develops with the increase of flow stress due to the continuous dynamic recrystallization. As mentioned in section 3.2., it was found that basal texture is formed and developed with increasing of strain. Also, new grains with basal orientation in coarse grain is formed by development of subgrain boundary towards high angle boundary as shown in Figure 4 (a) and (b). This results are consistent with the previous study of AZ80 magnesium alloy, which reported the mechanism

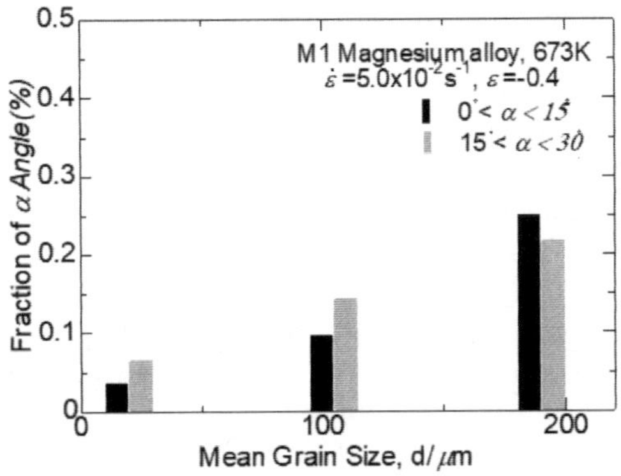

(a)

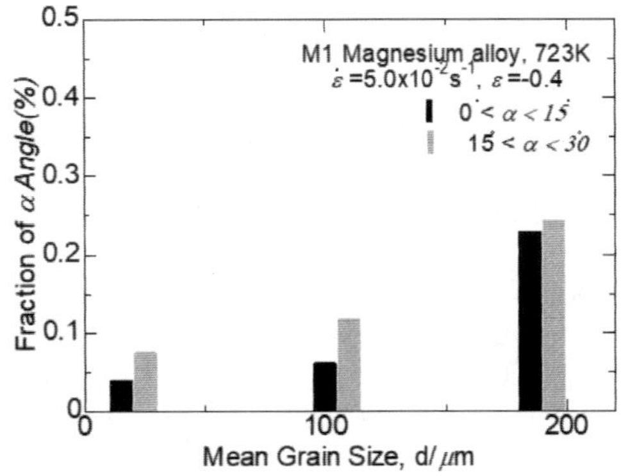

(b)

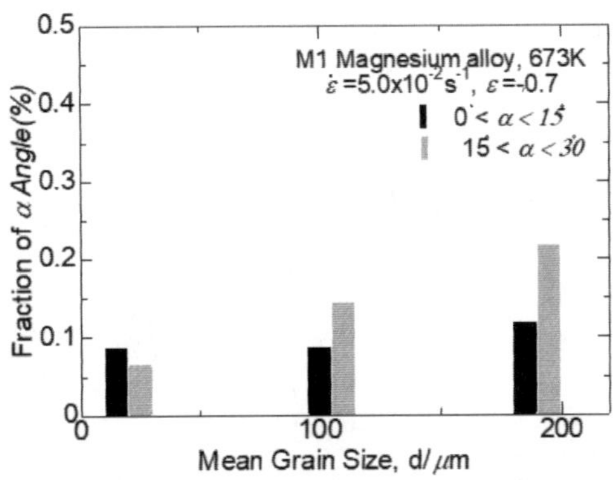

(c)

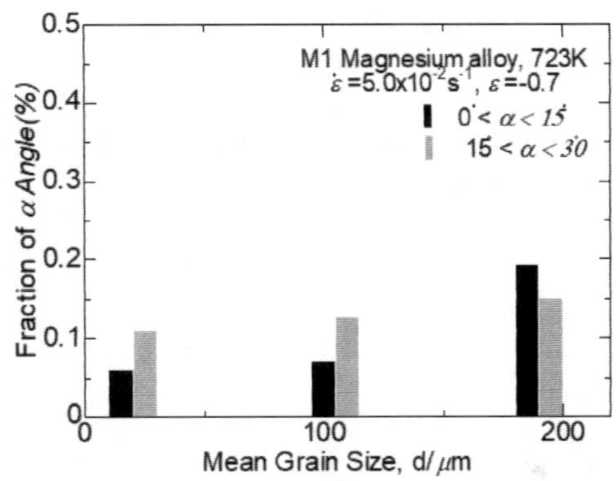

(d)

Figure 5. The relationship between basal texture component and grain size distribution after deformation at a strain rate of $5.0x10^{-2}s^{-1}$ – (a) 673K, ε = -0.4, (b) 723K, ε = -0.4, (c) 673K, ε = -0.7 and (d) 723 K, ε = -0.7, respectively.

of development of subgrain boundaries towards high angle boundaries. It was considered that basal texture is formed and developed by the continuous dynamic recrystallization during the temperature of 673K in this study.

Figure 5 shows the relationship between basal texture component and grain size distribution at a strain rate of $5.0x10^{-2}s^{-1}$ – (a) 673K, ε = -0.4, (b) 723K, ε = -0.4, (c) 673K, ε = -0.7 and (d) 723K, ε = -0.7, respectively. The Black bars are within 15° from (0001) orientation and the gray bars are from 15° to 30° from (0001) orientation. The grains having more than 200μm show high fraction of basal orientation and decrease with increasing of the strain. However, the fraction of small grains gradually increase with increasing of the strain. This means that small grains with basal orientation is formed by development of subgrain boundaries towards high angle boundaries in the coarse grains.

4. Conclusion

In order to clarify the basal texture formation behavior of M1 magnesium alloy during high-temperature deformation, M1 magnesium alloy was experimentally studied by uniaxial compression. The major results are as follows.

(1) Dynamic recrystallization occurs during plastic deformation.

(2) Basal texture is formed in all deformation condition of this study, and develops with increasing of strain.

(3) Development of basal texture is considered to the continuous dynamic recrystallization.

Acknowledgments
This work was supported by the Marine Convergence Design Co-work of Busan Korea 21 PLUS and the National Research Foundation of Korea grant funded by the Korea government (No.2017R1C1B5076690)

References

[1] Raynor G V 1959 *The Physical Metallurgy of Magnesium and Its Alloys* (London: Pergamon Press LTD)
[2] Al-Samman T, Li X and Chowdhury S G 2010 *Mater. Sci. Eng. A* 527 3450
[3] Lee K and Kim K 2019 *J. of the Korean Society for Heat Treatment.* 32 1
[4] Kim J, Okayasu K, and Fukutomi H 2013 *Mater. Trans.* 54 192
[5] Park M, Okayasu K, Fukutomi H and Kim K 2016 *Met. Mater. Int.* 22 1129
[6] M. Dahms and H. J. Bunge 1989 *J. Appl. Cryst.* 22 439

IOP Conf. Series: Journal of Physics: Conf. Series **1270** (2019) 012044 doi:10.1088/1742-6596/1270/1/012044

Effect of Ca concentration on Microstructure formation behaviors in AZ61 magnesium alloy

K Kim[1] D Han[2] and K Kim[1,2]

[1] Department of Metallurgical Engineering, Pukyong National University, 45 Yongso-ro, Nam-gu, Busan 48513, Republic of Korea
[2] Marine Convergence Design Co-work, Pukyong National University, 45 Yongso-ro, Nam-gu, Busan 48513, Republic of Korea

mrppeng@pknu.ac.kr

Abstract. Magnesium alloys have been paid attention as lightweight materials in various industrial fields. However, their use is limited by poor formability at room temperature which was caused by the limited number of slip systems. The texture control plays an important role in plastic workability of magnesium. Addition of Ca as alloying element is known to improve the general corrosion resistance and mechanical integrity of magnesium alloys in chloride environment. In order to investigate the Ca concentration on microstructure formation behaviors, Ca addition on AZ magnesium alloy were experimentally investigated by high-temperature uniaxial compression. Uniaxial compression test was conducted at 673K and 723K with a strain rate of $5.0\times10^{-2}s^{-1}$. The working softening was observed and the main component of texture and the sharpness of basal texture in two kinds of specimens varies depending on the deformation conditions.

1. Introduction

Magnesium alloys have been paid attention as lightweight materials in various industrial fields. However, their use is limited by room temperature poor formability, which was caused by the limited number of slip systems. In order to improve poor formability, it is important to understand the texture control in magnesium [1-2].

The researchers investigated the characteristics of texture formation behavior of AZ80 magnesium alloy during high-temperature deformation. They reported the main component and its sharpness of texture was varied depending on deformation condition [3]. Also, Park et al. have studied the effects of the solute element and its concentration on the texture formation behavior of AZ system magnesium alloys with different aluminum solute concentration. It was found that the main components and its sharpness of texture varied depending on deformation conditions and Al concentrations. Especially, basal texture was developed with an increasing of Al concentration and these texture were stable for deformation of that study [4].

Addition of Ca as alloying element has been known to improve the corrosion resistance and mechanical integrity of magnesium alloys in chloride environment [5-6]. The authors reported that AZ system magnesium alloys were formed second- phase by adding element Ca and these second-phases play an important role to improve their corrosion resistance and the mechanical properties. However, the study of the crystallographic microstructure change according to the solute content in AZ system magnesium alloy is insufficient.

IOP Conf. Series: Journal of Physics: Conf. Series **1270** (2019) 012044 doi:10.1088/1742-6596/1270/1/012044

In order to investigate the effect of Ca concentration on microstructure formation behaviors, AZ61 magnesium alloy and Ca addition on AZ61 magnesium alloy were experimentally investigated by high-temperature uniaxial compression. Uniaxial compression test is conducted at 673K and 723K with a strain rate of $5.0\times10^{-2}s^{-1}$ and the microstructure is observed by the optical microscope.

2. Experimental procedure

The materials used in this study are a commercial AZ61 (Mg-6.0%Al-1%Zn wt. %) and AZX611 (Mg-6.0%Al-1%Zn-1%Ca wt. %) magnesium alloys. Table 1 shows chemical composition of AZ61, AZX611.

Table 1. Chemical Composition of AZ61 and AZX611 magnesium alloy (wt.%)

	Al	Zn	Ca	Mn	Mg
AZ61	5.5-6.5	0.5-1.5	-	0.15-0.4	Bal.
AZX611	5.8-7.2	0.4-1.5	1.0	0.15-0.5	Bal.

These starting materials were hot-rolled at 673K with a rolling reduction of 30% and then machined out in such a way that the compression plane was parallel to rolling plane for uniaxial compression tests. The experimental specimens were annealed at 723K for 1h to produce a homogeneous microstructure. Figure 1 shows crystallographic microstructure in two kinds of specimens before high-temperature deformation. Figure 1 (a) shows initial microstructure of AZ61 and Figure 1 (b) shows initial microstructure of AZX611. As-cast Mg alloys formed second-phase particles such as Al-Ca compound and $Mg_{17}Al_{12}$ around the grain boundaries. Depending on the types of second-phase particles, these second-phase particles disappear or exist after annealing. Figure 1 (a) shows removed second-phase particles in microstructure after annealing. Whereas, Figure 1 (b) shows second-phase particles exist on grain boundary after annealing. A recent study showed that $Mg_{17}Al_{12}$ disappeared on annealing at temperature 673K to 723K. However, Al-Ca compound has high-melting point, which means that hardly dissolution. AZX611 shows uniform grain size and smaller microstructure size than AZ61 [8]. That means that microstructure formation was affected by second-phase particles.

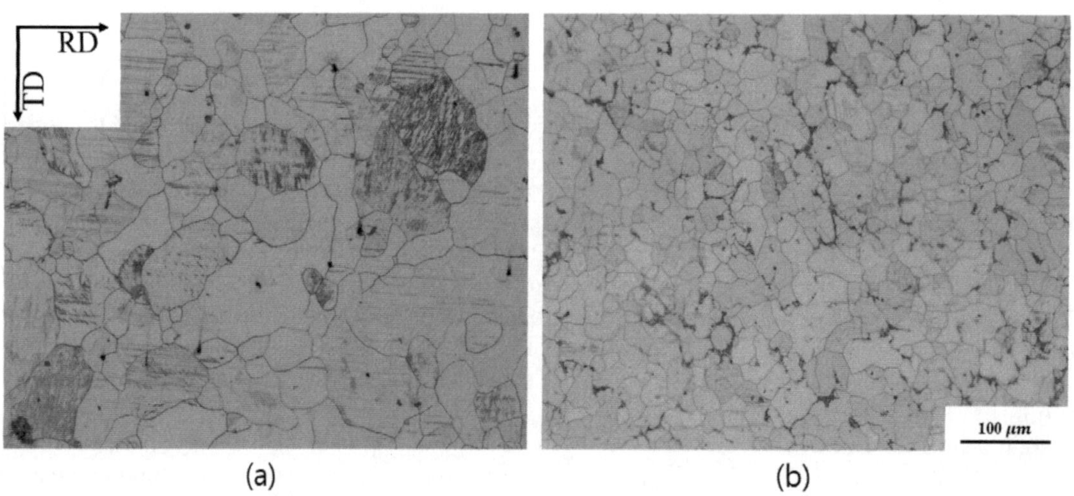

Figure 1. The optical micrographs in two kinds of specimens before deformation: (a) AZ61 magnesium alloy, (b) AZX611 magnesium alloy

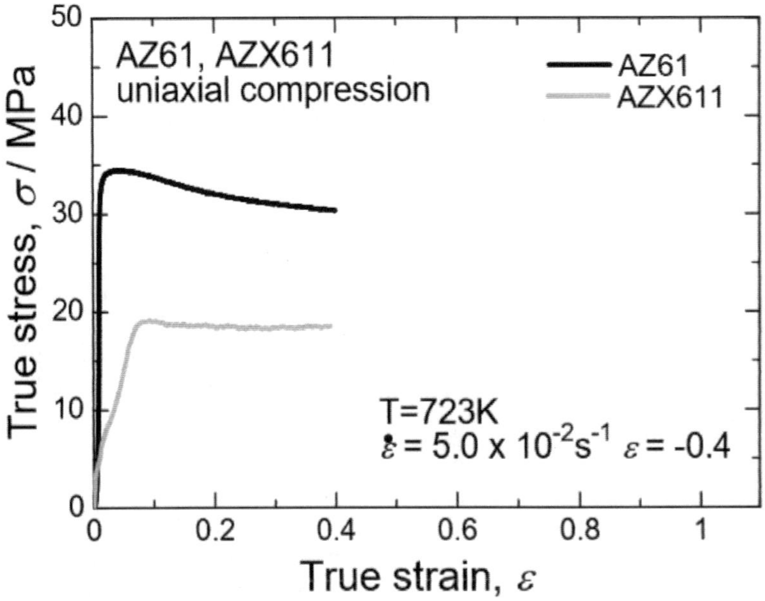

Figure 2. True stress - true strain curves for the deformation at 723K under a strain rate of $5.0 \times 10^{-2} s^{-1}$ up to a strain of -0.4.

Uniaxial compression tests were carried out at temperature 673K and 723K under a strain rate of $5.0 \times 10^{-2} s^{-1}$ up to a strain of -1.0. The direction of compression was marked after the rolling direction, and all specimens compressed same direction. These specimens were immediately quenched in water after deformation to prevent changes in the microstructure. The preparation of specimens is mid-plane by polishing. The surfaces were finished using as silica suspension. Moreover, electro-polishing carried out at 2% perchloric acid and 98% pure ethanol. After electropolishing immediately wash ethanol to prevent etching.

3. Result and discussion

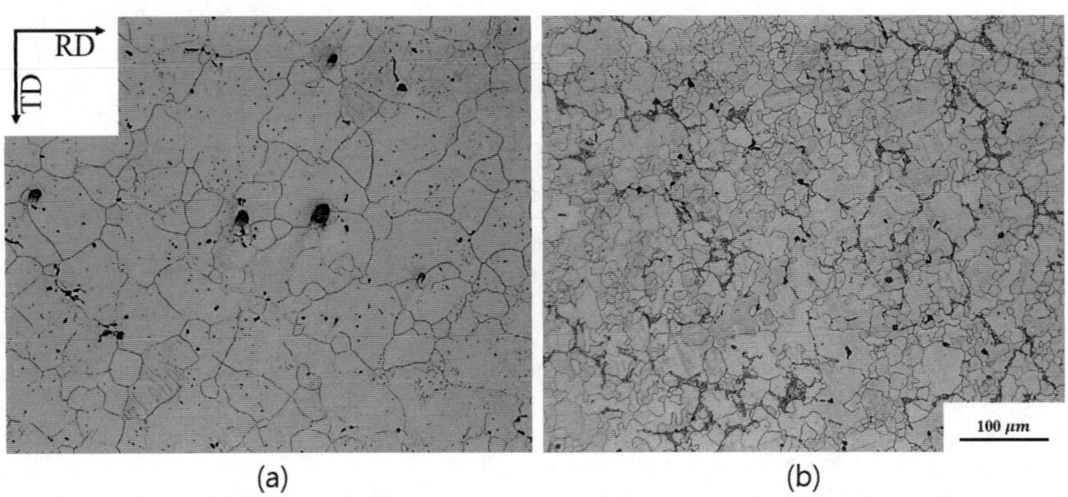

Figure 3. Microstructure of specimen observed by Optical Microscopy. The deformation condition was up to the strain of -0.4 with the strain rate of $5.0 \times 10^{-2} s^{-1}$ and temperature at 723K. (a) AZ61 magnesium alloy. (b) AZX611 magnesium alloy.

IOP Conf. Series: Journal of Physics: Conf. Series **1270** (2019) 012044 doi:10.1088/1742-6596/1270/1/012044

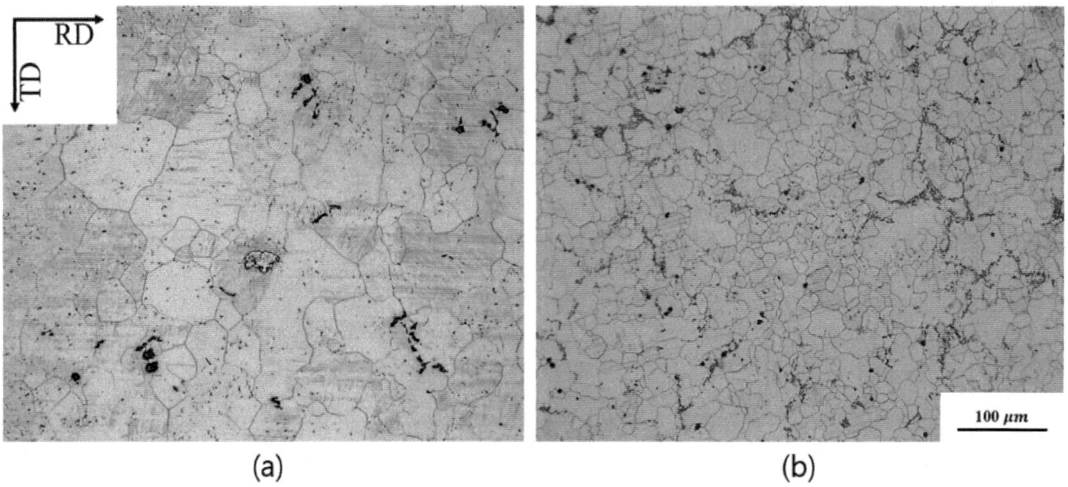

Figure 4. Microstructure of specimen observed by OM measurement. The deformation condition was up to the strain of -0.7 with the strain rate of $5.0x10^{-2}s^{-1}$ and temperature at 723K. (a) AZ61 magnesium alloy. (b) AZX611 magnesium alloy.

3.1. Behavior of deformation

Figure 2 shows the true stress - true strain curves for the deformation at 723K under a strain rate of $5.0x10^{-2}s^{-1}$ up to a strain of -0.4 in AZ61 and AZX611 magnesium alloys. The flow stress of AZX611 was lower than that of AZ61, due to initial grain size and composition. All specimen observed work softening such as the deformation behaviour of AZ80 reported by Kim et al [3].

3.2. Behavior of microstructure

Figure 3 (a) and (b) show the grain structure at compression plane after deformation at 723K under a strain rate of $5.0x10^{-2}s^{-1}$ up to the strain of -0.4. As reported on the deformation behavior of AZ magnesium alloys, small grains and coarse grains simultaneously existed as shown in Fig. 3 (a). However, Figure 3 (b) shows different microstructure formation behavior. It was found that the second-phase particles and fine grains are mostly distributed in the microstructure and second-phase particles are formed around fine grains.

Figure 4 (a) and (b) show the grain structure at compression plane after deformation at 723K under a strain rate of $5.0x10^{-2}s^{-1}$ up to the strain of -0.7. The microstructure formation behaviors in Fig.5 show similar formation behaviors as shown in Fig.4. Also, the grain size decrease with increasing of a strain in two kinds of specimens.

Figure 5 shows the grain structure at compression plane after deformation at 723K under a strain rate of $5.0x10^{-2}s^{-1}$ up to the strain of -1.0. After deformation, it shows different microstructure formation behaviors at figure 5 (a) and (b). Figure 5 (a) shows fine and uniform grain microstructure, and it was found confirmed that grain size of AZ61 magnesium alloy decreases with increasing of a strain. This means that dynamic recrystallization occurred as reported by Kim *et al* [3].

Figure 5 (b) shows mixed fine grain and large grain microstructure. Moreover, as compared with figure 5 (a), fine grain was smaller, and second-phase particles are surrounding fine grains As reported in recent study, second-phase particles stimulates nucleation and recrystallization [8]. That is, Ca addition affected on dynamic recrystallization of AZ magnesium alloys and different formation mechanisms.

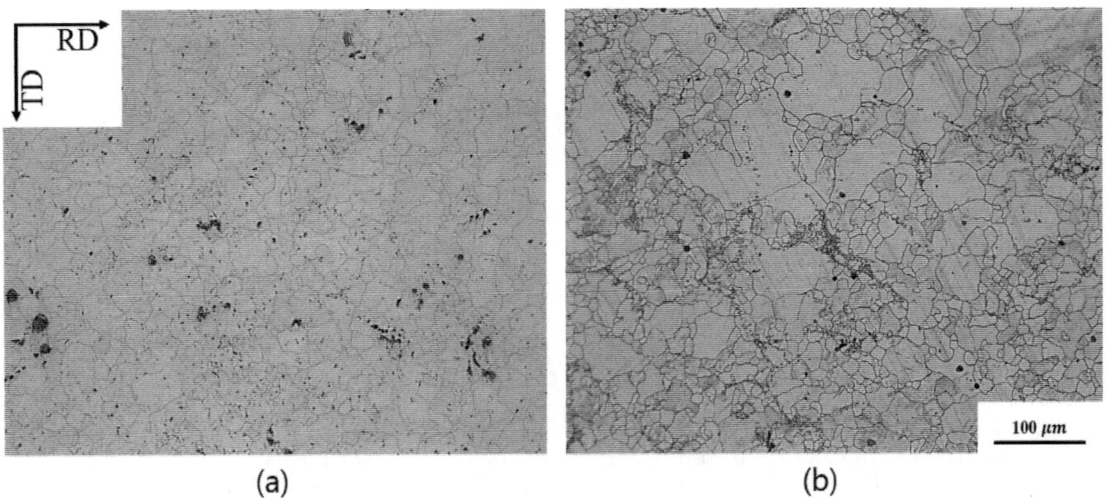

Figure 5. Microstructure of specimen observed by OM measurement. The deformation condition was up to the strain of -1.0 with the strain rate of $5.0x10^{-2}s^{-1}$ and temperature at 723K. (a) AZ61 magnesium alloy. (b) AZX611 magnesium alloy.

Consequently, microstructure formation behavior of AZ61 magnesium alloy is formed and developed by dynamic recrystallization as reported Park et al [4], whereas AZX611 magnesium alloys are affected by particle stimulated nucleation (PSN) in this study.

Figure 6 shows the variation of mean grain size with true strain in two kinds of specimens. White symbol is AZ61 magnesium alloy and black symbol is AZX611 magnesium alloy. The grain size was decreasing with true stain increase. That means dynamic recrystallization was occurred in AZ61 and AZX611 magnesium alloy. However, it is necessary to study that Al-Ca compounds affect dynamic recrystallization.

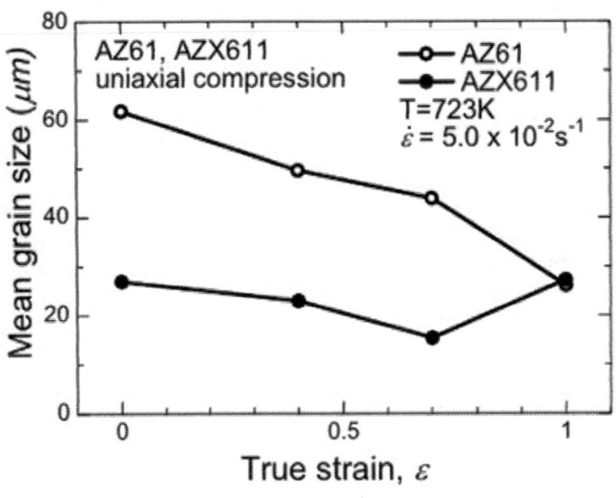

Figure 6. The variation of mean grain size with true strain in two kinds of specimens.

4. Conclusion

In order to investigate the effect of additional Ca on microstructure formation behaviour of AZ61 and AZX611 magnesium alloy during high-temperature deformation, AZ61 and AZX611 magnesium alloy were experimentally studied by uniaxial compression. The major results are as follows.

(1) Dynamic recrystallization occurs at AZ61 and AZX611 magnesium alloys in deformation conditions of this study.
(2) Two kinds of specimens shows different formation behaviors.

Acknowledgments
This work was supported by the Marine Convergence Design Co-work of Busan Korea 21 PLUS and the National Research Foundation of Korea grant funded by the Korea government (No.2017R1C1B5076690)

References

[1] Mordike B and Ebert T 2001 *Mater. Sci. Eng. A* **302** 37
[2] Roberts C S 1960 *Magnesium and its alloys* (New York: Wiley)
[3] Kim K, Okayasu K and Fukutomi H 2015 *Mater. Trans.* **56** 17
[4] Park M, Park H, Choi J and Kim K 2016 *Mater. Sci. Forum* **879** 1449
[5] Beak S, Kim H, Jeong H, Sohn S, Shin H, Choi K, Lee K, Lee J, Yim C, You H, Ha H and Park S 2016 *Corrosion Science* **112** 44
[6] Hakamada M, Watazu A, Saito N and Iwasaki H 2010 *Mater. Sci. Eng. A* **527** 7143
[7] Xie B, Zhang B, Ning Y and Fu Y W 2019 *Alloy and Compounds* **786** 25
[8] Xu S W, Kamado S and Honma T 2010 *Scripta. Met.* **63** 293

IOP Conf. Series: Journal of Physics: Conf. Series **1270** (2019) 012045 doi:10.1088/1742-6596/1270/1/012045

Influence of a pre-deformation on the growth of titanium multicrystals

D Chaubet[1], W Beucia[1], P Franciosi[2] and B Bacroix[2]

[1]Université Paris 13, LSPM CNRS, 93430 Villetaneuse, France
[2]LSPM CNRS UPR 3407, Université Paris13, 93430 Villetaneuse, France

brigitte.bacroix@univ-paris13.fr

Abstract.
In the present study, we have developed a new elaboration process which facilitates the growth of titanium single crystals by combining plastic strain and thermal cycling around the transition temperature. It consists in introducing a small plastic deformation into annealed specimens (like in the critical strain hardening method) before carrying out annealing cycles of only a few days under high vacuum. This method has been applied to titanium of commercial purity (CP-Ti). At the end of each cycle, samples are examined in SEM / EBSD for the characterization of grain sizes and shapes, crystallographic orientations and crystalline quality.
Our first results demonstrate that we can obtain grains bigger than one square centimetre (thickness 2mm) after a preliminary plastic deformation of 3% followed by two 2 pairs of annealing cycles (total time: 96 hours). This enhanced growth is thought to be due to the combination of two driving forces associated (i) to the energy stored within the grains during plastic strain and (ii) to the phase transformation, which acts alone in the transus cycling process. These single crystals are intended to be used to study their mechanical behaviour with the small tensile machines developed to perform in situ mechanical tests into microscopes (SEM, AFM) or diffractometers.

1. Introduction

Titanium alloys are widely used in many industrial sectors because of excellent mechanical properties, combined with a high corrosion resistance and lightness. However, the basic plasticity mechanisms still remain insufficiently characterized because of the large variety of possible mechanisms in hcp structures and a lack of basic studies on single crystals, difficult to produce. Studies reported in a fairly old literature [1] have reported the development of multi-centimetre titanium "multicrystals" by means of very long thermal cycles (i.e. about 8 days) around the hcp – bcc transition temperature, under ultra-vacuum conditions required for a satisfactory crystal quality. This very long duration of such cycling methods was the reason for which we formerly choosed another way to carry on preparation of single crystals of another hcp metal, zirconium, by a strain-annealing method [2]. In these two solid ways for single crystal preparation, there is a first step that brings energy to the metal, and a second one which use this energy to produce limited nucleation or grain growth in order to get very large grain sizes. For the two ways, this second step generally lasts few days, or more. In the present study, we intend to develop a faster procedure, mixing pre-deformation and thermal cycling with short thermal cycles.

2. Experimental material and procedures

The starting material was a 2.0 mm-thick sheet of commercially pure titanium (Ti grade 2 according to the American Society for Testing and Materials standards).The initial microstructure was in an annealed state with an equiaxed grain shape. The EQuivalent Average Diameter (EQAD) of the grains, as calculated from EBSD measurements, was about 47 µm (surface weighted). Samples were cut in a rectangular shape of 14.0 cm x 2.0 cm for the grain growth procedure. For EBSD, the sample preparation consisted first in mechanical grinding with silicon carbide paper down to grade 4000, followed by OPS (silica suspension) polishing. Then, a final electrochemical polishing was performed with A3 electrolyte of Struers, 35V for 30s.

The EBSD diagrams were performed within a ZEISS SUPRA40VP FEG- SEM equipped with the automatic Orientation Imaging Microscopy (OIM^{TM}) software from Tex-SEM laboratories and the processing of the raw data using the TSL OIM^{TM} Analysis7 software. The orientation maps covered a typical area of 2.00 mm x2.00 mm scanned with a step size varying from to 2µm to 3 µm. As most samples had β-transformed microstructures, we considered the colonies of α laths as grains. Such a choice led us to the following grain parameters for the analysis software: minimal grain boundaries misorientation of 2°, minimum grain size of 5 pixels, and multi-row shape. Grains of less than 5 pixels size were cleaned-up using the grain dilatation method provided by the software (with a 2° misorientation tolerance angle).

Tensile strain of the first step of thermo-mechanical treatment was performed with a Schenck/MTS testing machine at room temperature, with a strain rate of $10^{-4}s^{-1}$, tensile axis being parallel to the transverse direction (TD) of the as-received state. Gauge test length was 8.0 cm. Strain was measured with an extensometer, in order to obtain a final plastic strain of 3%.

High vacuum annealings were performed in a Nabertherm RHTC 80-710/15 high temperature tube furnace with a ceramic working chamber. Ramping of 5° per minute was used for heating and the beginning of cooling down to 550°C for each step of the procedure. High vacuum was obtain using turbomolecular pumping until a maximum gas pressure of $1.5.10^{-5}$ mbar was reached.

3. Experimental results and discussion

We will first present the thermo-mechanical procedure that led to the growth of large titanium crystals, and then the investigations on the influence of pre-deformation, comparing the microstructure evolution during the whole process of no pre-deformed (0%) and pre-deformed (3%) samples.

3.1. Thermomechanical procedure leading to large titanium grains

The five steps thermo-mechanical procedure used is sketched in figure 1. First step was a 3% tensile deformation, followed by two annealing cycles. Each cycle consisted of two annealing steps: the first one of 24 hours in the β phase at 920°C, and the second one of 24 hours in the α phase at 860°C.

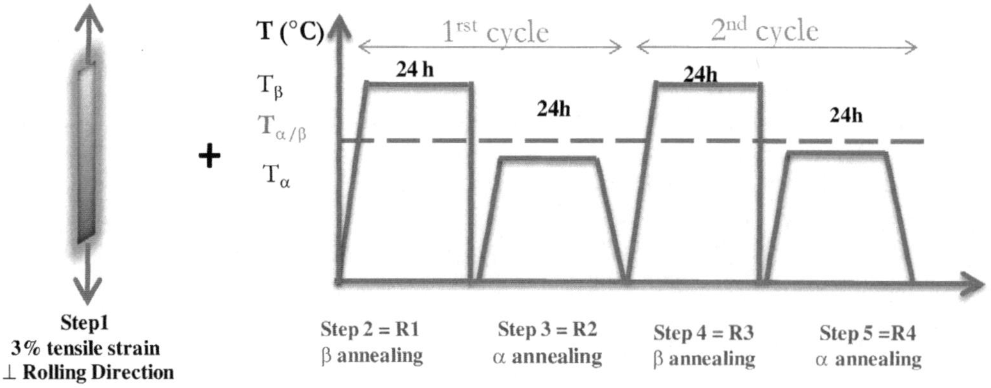

Figure 1. Schematic representation of the experimental thermo-mechanical procedure applied to produce large titanium crystals.

With this procedure some very large grains developed, so large that a complete view needed numerous backscattered SEM pictures to be constructed as can be seen in figure 2. For such large grains, IPF maps were plotted from EBSD measurement on both sides of the sample, showing the same uniform colour, which means that the orientation is the same on both sides, and that there is no substructure. We concluded that these grains had the thickness of the sheet and a good crystalline quality. As can be seen in figure 2, large grains are surrounded with much smaller grains, suggesting the occurrence of an Abnormal Grain Growth (AGG) mechanism.

Figure 2. Example of large titanium crystal obtained after 2 cycles (reconstructed from backcattered SEM pictures)

3.2. Comparison of microstructure evolution during the procedure for 0 % and 3 % pre-deformed samples.

As seen in former section, our procedure is less time-consuming than the usual thermal-cycling method used by Jourdan [1]. In order to understand how pre-strain could influence the grain growth, we compared the microstructures at the end of steps 2 to 5 for no pre-deformation and 3% pre-deformation (step 1 of the procedure). At the end of each of these steps, the microstructure was characterized through SEM/EBSD. From OIM analysis, IPF maps and boundaries misorientation distributions were also examined, and the average equivalent grain size (EQAD) was calculated (surface weighted).

3.2.1. First cycle and first 24h β annealing (step 2)

During this step, some growth is expected during heating in the α phase, particularly for the pre-deformed sample where Strain Induced Boundary Migration (SIBM) is likely to occur. Then, samples undergo complete phase transformation at 920°C: it has been checked that in some samples which were just heated to 920°C and immediately cooled down, the α grains have indeed completely lost their initial equiaxed grain shape. During the following 24 hours within the β phase, we assume that the large β grains could grow, but unfortunately, we could not observe them directly.

The microstructures of the 0% and the 3% pre-deformed samples that can be examined at the end of this step are the ones produced by the subsequent $\beta \rightarrow \alpha$ transformation during cooling. As can be seen in figure 3, the morphology is typical from transformation, and composed of colonies of laths. At the end of this step, the main difference between the two microstructures is a bigger size of the colonies (considered as "grains") within the pre-deformed sample, as shown by the EQAD values given under [001] IPF maps.

0%	3%

EQAD : 333μm	EQAD : 535 μm

Figure 3.[001] IPF map of microstructure after R1 (first β anneal, step 2) showing bigger colonies in the pre-deformed sample.

3.2.2. First cycle and first 24h α annealing (step 3)

In classical cycling procedure this step is the "growth" step, for which the driving force is to be found in the reduction of interfacial boundary energies for all kinds of boundaries produced by the β→α transformation, namely the inter-lath boundaries or inter-variants boundaries within the former β grains. Actually, it can be observed that the EQAD value increases for the 0% pre-deformed sample, while it does not seem to vary for the pre-deformed sample in the areas in which no abnormal grain growth occurs. Indeed, the most noteworthy event of this step is the occurrence of abnormal growth of few grains as can be seen in figures 4 and 5.

0%	3%
	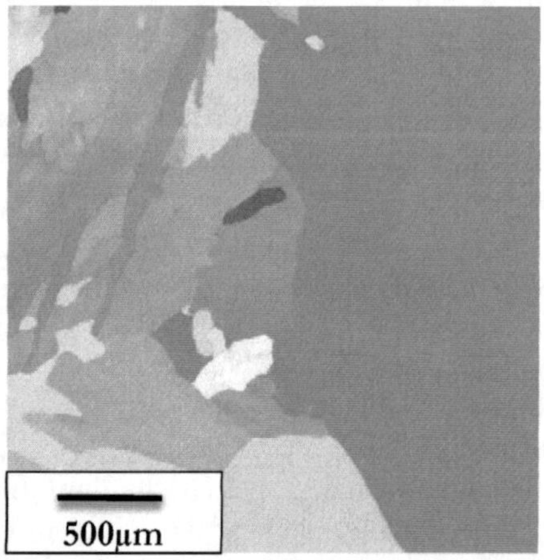
EQAD : 589 μm	EQAD : 533μm (in no AGG area)

Figure 4.[001] IPF map of microstructure after R2 (first α anneal, step3) showing abnormal grain growth in the pre-deformed sample

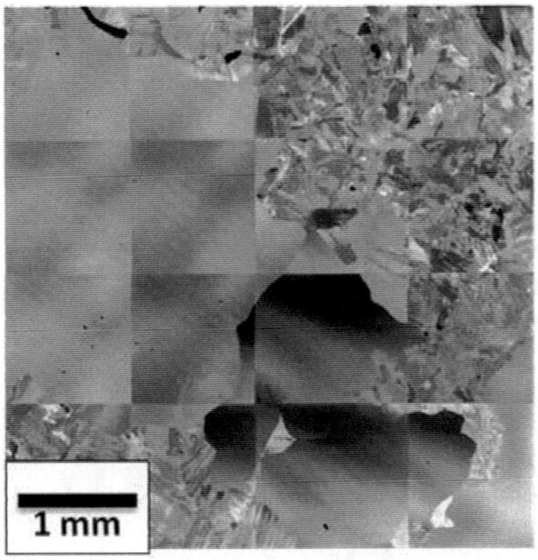

Figure 5. Reconstructed picture of abnormally grown grains for 3% pre-deformed Ti, at the end of the first cycle (backscattered SEM)

3.2.3. Second thermal cycle (steps 4 and 5)

As in the first step of the second cycle (step 4), the samples went through a new run of α→β→α transformations, it was surprising to observe that no needle or lath could be seen in the abnormally grown grains from the first cycle, as if no phase transformation occurred for them. So, apparently, these "metastable"α grains could take advantage of this step to continue their growth.

The comparison of the two microstructures at the end of the second α annealing is presented in figure 6. It is seen that the normal growth which took place in the 0% pre-deformed sample has led to grains with a diameter of few millimetres. At the same time, as the grains which have abnormally grown during the first cycle in the 3% deformed sample are only subjected to grain growth during the second cycle, their size can exceed 1 cm. No sub-structure could be detected on the maps within the very large grains (figure 6 right). Grains produced by this relatively short thermal process are thus of good crystalline quality. We can also notice that the orientations of different large grains are not all the same.

3.3. Why abnormal grain growth could be facilitated by pre-deformation?

It is likely that hardening due to the low plastic pre-deformation is eliminated during α heating (step 2), by a recrystallization mechanism implying limited nucleation and enhanced SIBM. As a consequence, we expected to see the influence of that prestrain simply on the average grain size after β annealing (step 2, figure 3), but not on the initiation of abnormal grain growth during further α annealing (step 3, figure 4).

In fact, it is likely that the SIBM mechanism could have also changed the impurities distribution owing to possible interactions either with fast moving boundaries, or with residual dislocations in the lower energy growing grains. If the impurities distribution in solid solution or as precipitates is still different at the end of the first β annealing step, it is then possible that some boundaries have become free of solute drag impurities or pinning particles, thus allowing the abnormal growth of some grains.

In this respect, it is also significant that abnormally grown grains seem to resist to phase transformation, as if no nucleation site was available within the large grains, or as if these grains contained sufficient solute elements capable of stabilizing the α phase.

IOP Conf. Series: Journal of Physics: Conf. Series **1270** (2019) 012045 doi:10.1088/1742-6596/1270/1/012045

0%	3%
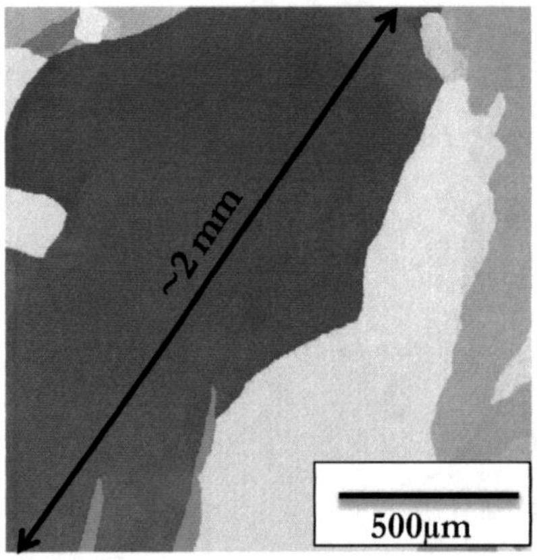	
Maximum grain size : few mm	Maximum grain size : >1 cm

Figure 6.[001] IPF map of microstructure at the end of the 2nd cycle

Otherwise, if we simply assume that bigger α grains produced by SIBM (because of the 3% strain) on heating transform into bigger β grains during β annealing, we understand why the α colonies that grow on cooling from the reverse phase transformation are larger in size that those observed in the 0% pre-deformed sample, as seen in figure 3.

4. Conclusion

We proposed here a new thermo-mechanical procedure allowing to produce titanium large crystals, of good crystalline quality and of sizes greater than one centimeter. These crystals may be valuable materials for fundamental investigations on single crystals with small testing machines.

The procedure consists of 3% pre-deformation followed by two annealing cycles under high vacuum. Each thermal cycle consists of a first step of 24 hours β annealing at 920°C followed by a second step of 24 hours α annealing.

We compared the microstructures obtained within the 0% and 3% predeformed samples at the end of each step of the procedure. A small pre-deformation allows to activate abnormal grain growth of few grains during the second step of the first thermal cycle. Experiments are under way in order to investigate the mechanism involved in such a rapid activation of abnormal growth.

References

[1] C Jourdan, D Rome-Talbot and J Gastaldi 1972 *Philosophical Magazine* **26:4** 1053-55

[2] D Chaubet, J P Fondere and B Bacroix 2001 *Materials Science and Engineering* A**30** 245-53

Simulations of the relative importance of initial size advantage and boundary energy anisotropy in abnormal grain growth

E J Payton[1,*], A R C Gerlt[1,2], A L Pilchak[1]

[1] Air Force Research Laboratory, Wright-Patterson AFB, OH 45433
[2] UES, Inc., Beavercreek, OH 45432

[*] corresponding author. E-mail: eric.payton@us.af.mil

Abstract. In textured materials, initial size advantage competes with boundary energy and mobility anisotropy as a driving force for grain coarsening. High local boundary anisotropy and large pre-existing size advantage encourage abnormal grain growth. In the present work, the importance of these driving forces relative to one another is explored for various initial grain size advantages and texture intensities using a Potts Monte Carlo approach with anisotropic grain boundary energy varying according to the Reed-Shockley model. A size greater than approximately twice the mean grain size was sufficient to virtually ensure a grain has a growth advantage regardless of texture intensity.

1. Introduction

The control of grain size during processing is of practical significance because almost every important property of a structural metallic material is a function of grain size. Accordingly, a number of deterministic and statistical models have been developed to predict changes in grain size during annealing. These models are largely based on the original model of Burke [1] and Burke and Turnbull [2], who assumed that the pressure acting on the grain boundary is due solely to its curvature. The authors' analysis revealed that grain growth kinetics could be described by the equation $D^n-D_o^n = kt$ where D_o is the initial mean grain size, D is the mean grain size at time t, n is the grain growth exponent, and k is a constant. The original analysis yielded $n = 2$ for ideal grain growth, though this is rarely observed experimentally. Deviation from ideality may be due to specimen size, texture, or the presence of pinning defects like second phases and inclusions [3]. While numerous rules of thumb and models have been developed to describe specimen thickness and pinning particle effects, respectively, incorporating texture effects into analytical grain growth models has received considerably less attention. Notable exceptions include the work of Abbruzzese and Lucke [4] and Eichelkraut et al. [5]. These authors introduced orientation-dependent grain boundary energy and mobility into Hillert's statistical model [6] and showed that texture had a strong effect on grain growth kinetics and the development of the grain size distribution. The theory also introduced the concept that the critical grain radius, above which grains will grow and below which grains will shrink, is an orientation-dependent quantity. Thus, there is a competition in the early stages of grain growth that is set up by the spatially stochastic distribution of grains present upon the completion of recrystallization. Due to the complexity of recrystallization process (see [Doherty et al. [7] for a comprehensive review), one can expect that some grains have a slight size advantage compared to the neighbors while others have a slight mobility advantage due to having a larger misorientation angle with neighboring grains. Some small fraction of grains may have both a size and mobility advantage. In this work, we explore this competition using Monte Carlo simulations that account for the effect of grain boundary misorientation on grain boundary mobility.

2. Methodology

Potts Monte Carlo simulations with were run using the AppPottsRS code [8] in the Stochastic Parallel PARticle Kinetic Simulator (SPPARKS) framework [9]. Within AppPottsRS, the grain boundary properties (specifically, the trans-boundary diffusivity, or the product of grain boundary mobility and interfacial energy) are assumed to vary with grain boundary misorientation according to the Reed-Shockley equation. Unimodal orientation distribution functions (ODFs) were generated using MTEX [10], then random samples were taken from the ODFs and assigned to 3x3 px square blocks in a 1500x1500 px 2D grid, creating 2.5E4 simulated grains. A 180° halfwidth was used to approximate a uniform orientation density function (i.e., a polycrystal with randomly oriented grains). Upon this grid, another evenly-spaced grid of 400 grains was superimposed, with each of the 400 grains given sized uniformly as either 4x4 px, 5x5 px, 6x6 px, or 9x9 px. The superimposed grid thus presents a set of grains with a pre-existing size advantage. The 400 grains in the superimposed grid were all assigned an orientation of the modal value of the ODF. By assigning all grains without an initial size advantage to have identical sizes, the present work deliberately avoids the additional complicating role of grain size distribution (which is known to have an effect on final grain size and grain growth rates [11]). Potts Monte Carlo simulations were then run for 40 Monte Carlo steps (MCS), since after this time grains with the initial size advantage impinged upon each other under some simulation conditions.

3. Results and Discussion

Figure 1 shows pole figures (top row) corresponding to the ODFs from which the grain orientations in the corresponding inverse pole figure maps (bottom row) were sampled. The bottom row shows inverse pole figure (IPF) maps of the results after 20 MCS of the simulations that initially contained 6x6 px "seed grains" with pre-existing size advantage of the 6x6 px simulations after 20 MCS. In the pole figures, the texture intensity decreases from left to right, with 1°, 5°, 10°, 15° and 180° halfwidths. Note in the IPF maps that the grid of "seed grains" with the initial size advantage become more apparent from the left to the right, such that nearly all of these grains with the size advantage persist to 20 MCS while their neighbors are eliminated.

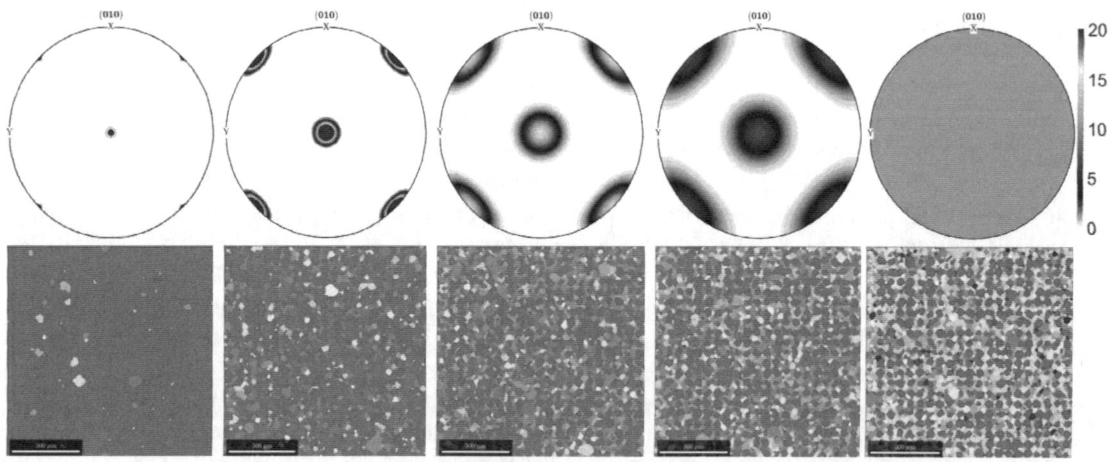

Figure 1: (001) pole figures for initial simulation conditions and corresponding normal-direction inverse pole figure maps after 20 MCS. Left to right: 1°, 5°, 10°, 15° and 180° halfwidths. The pole figure colorbar range is from zero to 20 in times random. Red in the inverse pole figure corresponds to the [001] crystal direction. Scale bar lengths correspond to 500 px.

IOP Conf. Series: Journal of Physics: Conf. Series **1270** (2019) 012046 doi:10.1088/1742-6596/1270/1/012046

The structure of the simulation is illustrated in figure 2, which displays snapshots of the results of simulated grain structure evolution for the case of the 180° halfwidth with 9x9 px square "seed" grains in a matrix of uniform 3x3 px square grains. The initial condition is shown on the left, while the center shows the results at 5 MCS and the right shows the simulated microstructure after 20 MCS. The whole microstructure is shown in the top row with each grain being plotted in a unique color, while the bottom row highlights the arrangement of the "seed" grains given an initial size advantage.

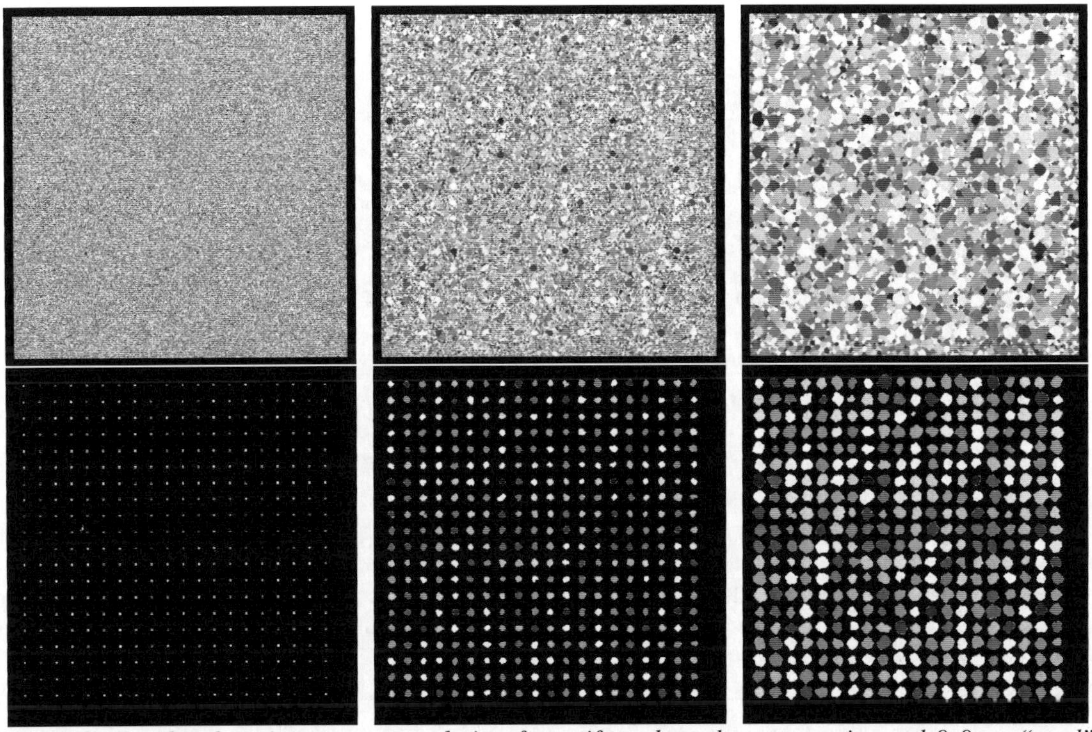

Figure 2: Simulated grain structure evolution for uniform boundary properties and 9x9 px "seed" grains in matrix grains uniform 3x3 px size. Colors indicate unique grain IDs. Left column: initial state; Middle column: 5 MCS; Right column: 20 MCS. Top row: overall structure; Bottom row: Only "seed" grains given a pre-existing size advantage shown.

Figure 3 shows the evolution of the mean circle equivalent grain size, <R>, in px with number of Monte Carlo steps (MCS) completed. The data has been split into four with each subplot displaying all texture intensity cases with the same initial seed grain size. The solid line (darkest color) corresponds to the case of a uniform ODF, the dashed line corresponds to the case of the 15° halfwidth, the dash-dot line to the 10° halfwidth, dotted line to 5°, and solid lightest color line to 1°. Since the grain growth rate varies significantly with the boundary energy on account of the texture, all subsequent analyses comparing the behavior in the different simulations are performed with respect to the mean grain size in the system as opposed to the Monte Carlo step, to facilitate comparisons between the different simulations.

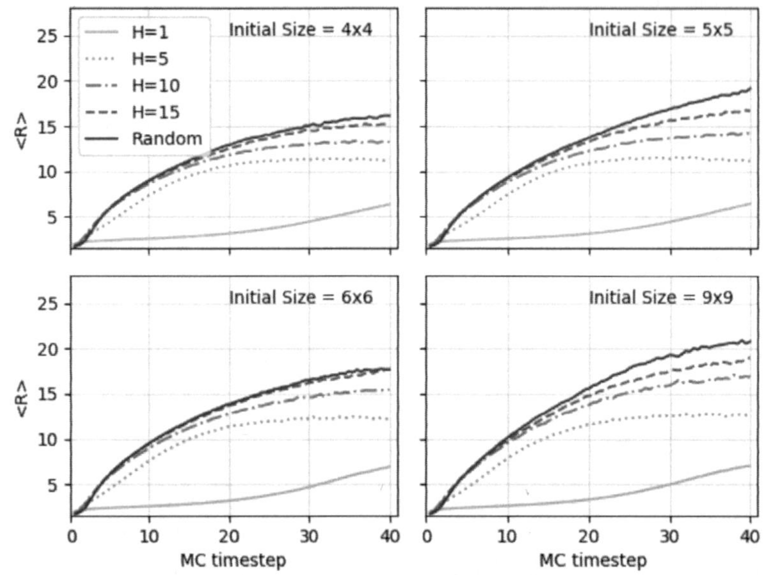

Figure 3: Evolution of mean grain size vs. number of Monte Carlo steps.

Figure 4 shows the number of retained "seed" grains vs mean grain size in the simulation. The linestyles are the same as those used for figure 3. Larger initial size results in a greater fraction of the "seed" grains persisting longer in the simulation. In the case of a size advantage of approximately two times the mean equivalent diameter, the probability of the neighborhood topology being such that the large grain is overcome by its neighbors becomes negligible for the 180° halfwidth ODF (approximating a random polycrystal, labeled "random" in figure legends). As texture intensity increases, the effect of pre-existing size advantage is reduced.

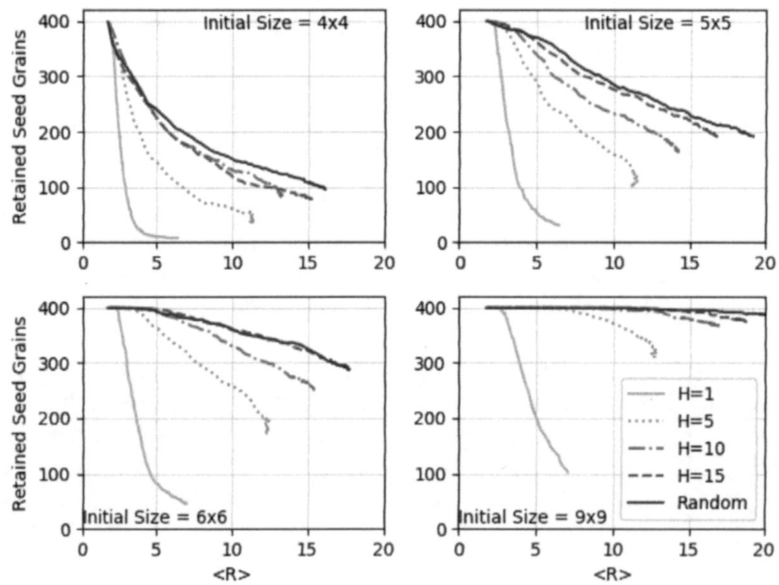

Figure 4: Number of retained seed grains vs. mean grain size during simulated grain coarsening. H=1 corresponds to a 1° halfwidth of the unimodal texture distribution, H=5 to 5° and so on. Line styles are the same as in previous figures.

The observation that pre-existing size advantage is mitigated by texture strength is reinforced in figure 5, which shows the evolution of the fraction of the grains in the system that were originally seed grains. If the seed grains exhibit no initial growth advantage, their line appears approximately horizontal, maintaining the initial ~0.02%. This only occurs for the highest texture strength, and is lost for the largest size advantage. For all other texture strengths investigated, even a small initial size advantage drastically increases the likelihood that a grain will be among the grains that remains at the end of 20 MCS.

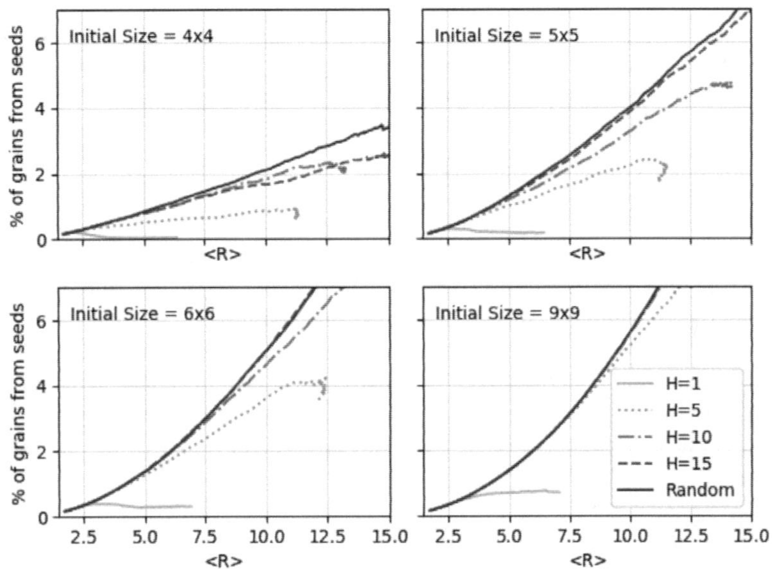

Figure 5: Evolution of the fraction of grains in the system that were "seed" grains given an initial size advantage. H=1 corresponds to a 1° halfwidth of the unimodal texture distribution, H=5 corresponds to 5°, and so on. Line styles are the same as in previous figures.

The misorientation angle distributions for the initial state and after 20 MCS are shown in figure 6. The shape of the 180° halfwidth distribution compares well with the expected Mackenzie distribution both initially and after 20 MCS. As evidenced by the increase in the peak of the 1° halfwidth distribution over the course of the 20 MCS (comparing left and right plots), the texture intensity increases during this simulation. This does not appear to be a strong effect in the other cases.

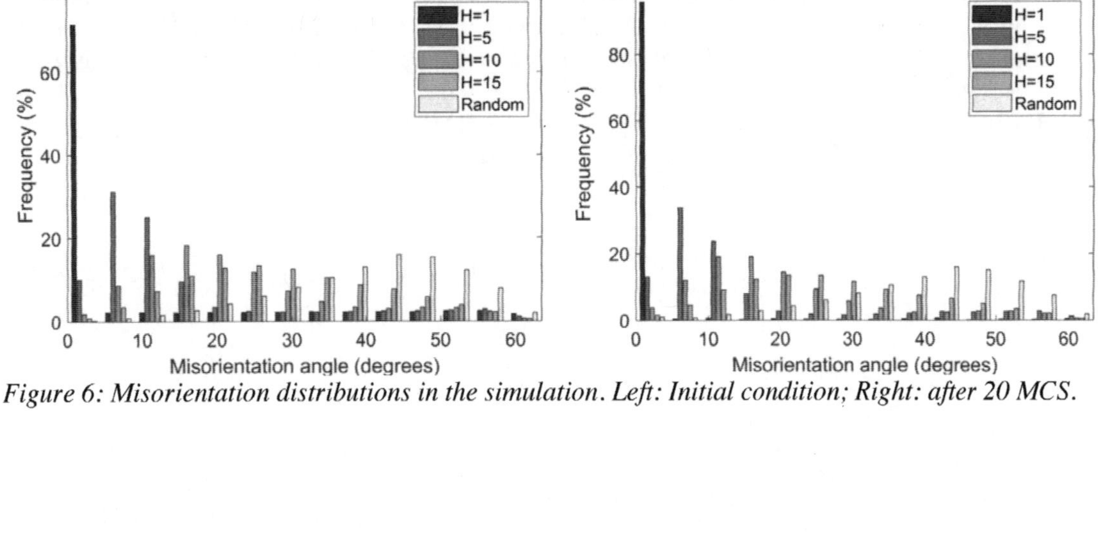

Figure 6: Misorientation distributions in the simulation. Left: Initial condition; Right: after 20 MCS.

The 10° halfwidth texture peaks around 15 times random. The present results suggest that, as a texture intensity reaches above 15 times random, the fraction of low angle boundaries increases to the point that grain neighborhood begins to matter to the extent that grains slightly larger than average do not have a significant advantage over other grains in the system which have randomly been assigned favorable neighborhoods. In light of this, a pressing question is how to characterize the frequency of neighborhoods that enable size advantages to develop in the first place, since obtaining a 2x the mean size advantage drastically increases the likelihood that a grain will persist for a long time during coarsening. Even in the absence of a size advantage, abnormal grain growth can occur when a single sharp texture component exists due to random fluctuations in local texture intensity. For mechanical processing routes that produce two or more texture components, an isolated grain that is highly misoriented with respect to its neighbors is likely to have higher volatility to its grain boundary mobility. These grains are either likely to rapidly disappear or become grains with an early size advantage, which may subsequently grow abnormally.

4. Conclusions

The present work analyzes a rather idealized case of grain coarsening in the presence of textures in order to gain insights into the relative importance of initial grain size on subsequent grain structure evolution in a textured polycrystal. Our results reveal the larger a grain is relative to the mean, the more likely it is to still exist at a snapshot of a later time during grain coarsening up to very strong texture intensities in excess of 15 times random. A size advantage of 2x the mean equivalent diameter appears to virtually guarantee that a grain has a sufficient size advantage over its neighbors to be among the grains that is not eliminated when less than 1% of the original grains remain.

References

[1] Burke J E 1949 *Trans. Metall. Soc. A.I.M.E.* **180** 73
[2] Burke J E and Turnbull D 1952 *Prog. Metal Physics* **3** 220
[3] Humphreys F J and Hatherly M 2004 *Recrystallization and Related Annealing Phenomena, 2nd ed.* (Oxford, UK: Elsevier) p 333-378
[4] Abbruzzese G and Lücke K 1986 *Acta Metall.* **34** 905
[5] Eichelkraut H, Abbruzzese G, and Lücke K 1988 *Acta Metall.* **36** 55
[6] Hillert M 1965 *Acta Metall.* **13** 227
[7] Doherty R D, Hughes D A, Humphreys F J, Jonas J J, Juul Jensen D, Kassner M E, King W E, McNelley T R, McQueen H J, and Rollett A D 1997 *Mater. Sci. Eng. A* **238** 219
[8] Hernández–Rivera E, Goins P E, Murdoch H A 2018 *Technical Report ARL-TR-8507: Anisotropic Grain Growth Modeling under the SPPARKS Framework* (Aberdeen Proving Ground, MD: Army Research Laboratory)
[9] Plimpton S, Battaile C, Chandross M, Holm L, Thompson A, Tikare V, Wagner G, Webb E, Zhou X, Garcia Cardona C, Slepoy A 2009 *Technical Report SAND2009-6226: Crossing the Mesoscale No-Man's Land via Parallel Kinetic Monte Carlo* (Albuquerque, NM: Sandia National Laboratories)
[10] Bachmann F, Hielscher R, Schaeben H 2010 *Solid State Phenomena* **160** 63
[11] Payton E J, Wang G, Mills M J, Wang Y 2013 *Acta Mater.* **61** 1316

IOP Conf. Series: Journal of Physics: Conf. Series **1270** (2019) 012047 doi:10.1088/1742-6596/1270/1/012047

The operative dynamic recrystallization mechanism of austenite during the transient deformation in a Ni-30%Fe model alloy

Wenxiong Chen [1,2], Chengwu Zheng[1,2*] and Dianzhong Li[1,2]

[1] Shenyang National Laboratory for Materials Science, Institute of Metal Research, CAS, Wenhua Road 72, Shenyang 110016, China

[2] School of Materials Science and Engineering, University of Science and Technology of China, Wenhua Road 72, Shenyang 110016, China

* E-mail: cwzheng@imr.ac.cn.

Abstract. Effect of strain rate and its discontinuous changes on dynamic recrystallization (DRX) behaviours of a Ni-30%Fe austenitic model alloy were investigated by hot compression tests. The results show that, during deformation at slow strain rates, DRX occurs with discontinuous nature, i.e. the DRX nuclei preferentially appear along initial grain boundaries followed by a quick growth through the strain-induced boundary migration (SIBM). Interestingly, during the further deformation after an abrupt increase in the strain rate, it can be found that the subgrains which developed in the first stage of deformation would gradually transform into new DRX grains. Therefore, in spite of the low stacking fault energy austenite, the continuous DRX (CDRX) is indeed possible to be activated during the transient deformation. The strain-induced sub-boundaries developed in the first deformation stages are believed to transform into higher-angle boundaries easily in further straining after abrupt increase of the strain rate. By this way, the CDRX mechanism in austenite is significantly promoted. Moreover, the local growth of new grains also plays an important role in this dynamic softening process.

1. Introduction

Dynamic recrystallization (DRX) process involves the formation of new strain-free grains on the deformation matrix during hot deformation, which plays a key role in tailoring the microstructure during the hot processing of steels. In general, there are two types of DRX that occur in metallic materials, which are closely related to the stacking fault energy (SFE) of the material [1]. The discontinuous DRX (DDRX) mechanism, usually takes place in metals of low to medium SFE, which has clear nucleation and growth stages. Continuous DRX (CDRX) mechanism, generally occurs in metals and alloys with high SFE, in which new DRXed grains with high angle boundaries (HABs) mainly formed by gradual coalescence of subgrains with minor boundary migration upon straining [2].

Up to now, a significant number of studies has been carried on to investigate the DRX mechanism in austenite under varying deformation conditions [3]. It recently has been proposed that the DRX mechanism is not only controlled by the deformation parameters, but also influenced by the initial grain structures. For instance, the fine initial grain size promoted CDRX in austenite [4]. However, limited attention has been given to the role of initial substructures on the DRX behaviors. In this study, the effect of strain rate as well as its discontinuous changes during the microstructural development in the austenite is considered. The aim of the present work is to investigate the evolution of pre-induced substructures under the transient change of deformation conditions.

2. Experiment

An austenitic model alloy with a composition of Ni-30.5Fe-0.02C-0.04Mn (wt.%) with an average grain size of about 300 μm has been used in the current study. This alloy has approximately the same SFE as the austenite phase in low carbon steels but does not undergo any phase transformation when quenched to the room temperature, which thus offers a facility to characterize microstructures of austenite from hot deformation [5]. Hot compression tests were performed on a Gleeble-3500 thermo-mechanical simulator. Cylindrical samples for compression tests have the length of 15 mm and the diameter of 8 mm. Uniaxial compression tests were performed both at constant deformation conditions with constant strain rates of 0.001 s^{-1} and 0.1 s^{-1} at 900 °C and at transient deformation conditions with an abrupt increase of the strain rate from 0.001 s^{-1} to 0.1 s^{-1} at the strain of 0.3. The samples were water quenched immediately after the deformation to preserve the microstructure formed after the hot deformations.

The microstructures were analyzed by the electron backscattered diffraction (EBSD) and the transmission electron microscopy (TEM) on middle sections perpendicular to the compression axis of samples. EBSD scans were carried out using a FEI Nova NanoSEM 430 field-emission scanning electron microscope operated at 20 kV. The HKL Channel 5 software was used for the EBSD data acquisition and post-processing. Samples for EBSD were prepared by mechanical polishing and finished with Ion milling to remove the surface deformation layer. TEM foils were mechanically grinded to reduce the thickness followed by twin-jet electro-polished with a solution containing 10% perchloric acid and 90% ethanol at the temperature of about -30 °C and the voltage of 15 V. The TEM analysis was performed on a FEI Tecnai G2 F20 field-emission microscope at 200 kV.

3. Results and Discussion

True flow stress curves for the Ni-30%Fe alloy at different strain rates are displayed in figure 1a. At the constant strain rate condition, the flow curve shows a stress peak followed by gradually softening to the steady-state stress regime during the deformation at a low strain rate of 0.001 s^{-1}. This type of flow behavior displays a typical feature of metals undergoing DRX. Increasing the strain rate to 0.1 s^{-1} leads to a broader plateau and a much high stress level on the flow curve, which implies the dynamic softening is hindered at this condition. In the case of the transient deformation condition with strain rate changed from 0.001 s^{-1} to 0.1 s^{-1}, the flow stress is adjusted immediately according to the imposed strain rate and finally approached the same stress level appearing in the constant deformation of 0.1 s^{-1}.

EBSD maps exhibit the microstructures of samples deformed to a strain of 0.8 at varying strain rates (as shown in figures 1b-d). These images revealed that the fraction and size of DRX grains are quite different under various deformation conditions. The deformed grains were completely replaced by newly coarse DRX grains in the sample deformed at a constant strain rate of 0.001 s^{-1}. On the contrary, only a few of fine DRX grains can be observed at the original boundaries at the higher strain rate of 0.1 s^{-1}. It is suggested that the DRX is suppressed by the increase of strain rate due to the shorter duration of deformation. However, it is noteworthy that, under the transient deformation condition, the DRX is accelerated significantly, which with a larger DRX fraction than that at the constant high strain rate deformations, but the average DRX grain size is approximately the same as that at the constant high strain rate condition. This implies that the DRX behavior has been changed due to the strain rate variation.

3.1. DRX behaviour at the constant low strain rate

Figure 2 shows a typical microstructure of samples deformed to strain of 0.3 with the strain rate of 0.001 s^{-1}. It can be seen that new DRX grains preferentially develop along initial grain boundaries, and the nucleation of DRX resulted from bulging of pre-existed grain boundaries associated with the formation of annealing twins or strain-induced sub-boundaries. Some Σ 3 annealing twin boundaries can be observed within new DRX grains, which usually indicates that the development of the DRX grain is assisted by the migration of grain boundaries over a large distance [3,6]. The presence of annealing twins within the new grains, therefore, is widely recognized as an indicator for the occurrence of DDRX.

The highly cumulative (point-to-origin) misorientations estimated by EBSD reveal that a large strain gradient developed within the deformed matrix, which would generate the driving force for the DRX grain growth by the migration of grain boundaries (figure 2b). TEM observation reveals that the substructures within the deformed grains displays a complex character with varieties of subgrain/cell morphologies (figure 2c). In spite of long duration of the test at low strain rate, the local angular misorientations across sub-boundaries are not exceed 3°, which implies that the substructure development is stunted. It then implies that the CDRX mechanism which new grains form through the gradually evolution of sub-boundaries can hardly operate in this austenitic alloy at the current deformation condition.

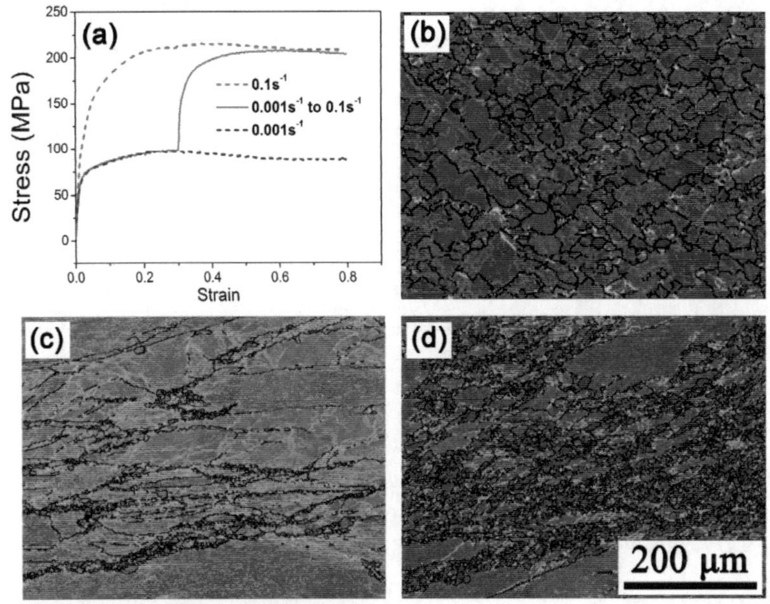

Figure 1. (a) Stress-strain curves of the Ni-30%Fe alloy obtained from the constant deformation tests (dash lines) and transient deformation test (solid line) at 900°C. Typical OIM maps of samples deformed to strain of 0.8 at (b) $0.001s^{-1}$; (c) $0.1s^{-1}$; (d) $0.001s^{-1}$ to $0.1s^{-1}$. The silver, green, black, red and magenta lines represent boundaries with $1° < \theta < 5°$, $5° < \theta < 15°$, $15° < \theta$, $\Sigma 3$ and $\Sigma 9$ twin boundaries respectively.

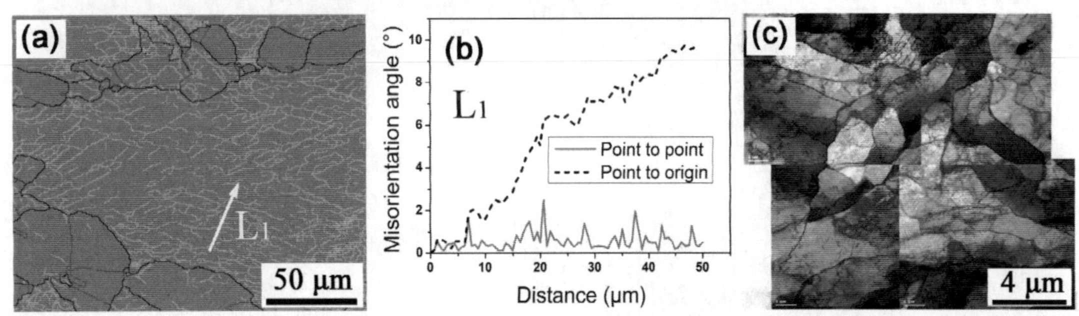

Figure 2. (a) OIM map of the Ni-30%Fe alloy deformed to a strain of 0.3 at a strain rate of $0.001s^{-1}$ at 900 °C; (b) misorientation profiles along the corresponding arrow L_1; (c) TEM image of deformed matrix corresponding to (a). The silver, green, blue, black, red and magenta lines represent boundaries with $0.8° < \theta < 2°$, $2° < \theta < 5°$, $5° < \theta < 15°$, $15° < \theta$, $\Sigma 3$ and $\Sigma 9$ twin boundaries respectively.

3.2. Microstructure evolution at the constant high strain rate

The microstructures of the sample deformed to a strain of 0.5 at the strain rate of 0.1 s^{-1} are displayed in figure 3. Similar to the deformation at low strain rate, new grains preferentially appear at the original grain boundaries due to bulging of the boundaries in this condition. It should be noted that an abundance of sub-boundaries is forming around the original grain boundaries (blue lines in figure 3a). The boundaries are impeded by the drag force derived from attachment of the various strain-induced sub-

IOP Conf. Series: Journal of Physics: Conf. Series **1270** (2019) 012047 doi:10.1088/1742-6596/1270/1/012047

boundaries, leading to the limited growth of new DRX grains. It indeed indicates that the DDRX process is inhibited at the high strain rate condition.

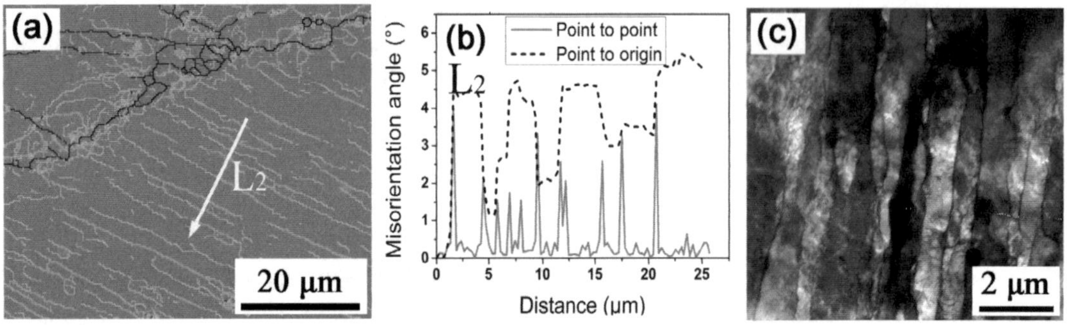

Figure 3. (a) OIM map of the Ni-30%Fe alloy deformed to a strain of 0.5 at a strain rate of $0.1s^{-1}$ at 900 °C; (b) misorientation profiles along the corresponding arrow L_2; (c) TEM image of deformed matrix corresponding to (a). The silver, green, blue, black, red and magenta lines represent boundaries with $0.8^o < \theta < 2^o$, $2^o < \theta < 5^o$, $5^o < \theta < 15^o$, $15^o < \theta$, $\Sigma 3$ and $\Sigma 9$ twin boundaries respectively.

The microstructure within the deformed grains mainly shows regular arrays of sub-boundaries delineated as elongated bands. The misorientation angles across the band boundaries measured by EBSD are almost below to 5°, and the cumulative misorientation across the band tends to systematically alternate across the consecutive but extended sub-boundaries (figure 3b). The TEM image reveals that the band boundaries could be considered as dense dislocation walls, which separate the deformed grains into parallel segments with relatively low dislocation densities (figure 3c). This kind of dislocation boundaries are always considered as micro-shear bands, as suggested by previous reports for the deformation of metals under high strain rates or at low temperatures [7]. It then indicates that some of slip systems can be activated by increasing flow stress under the applied high strain rate condition.

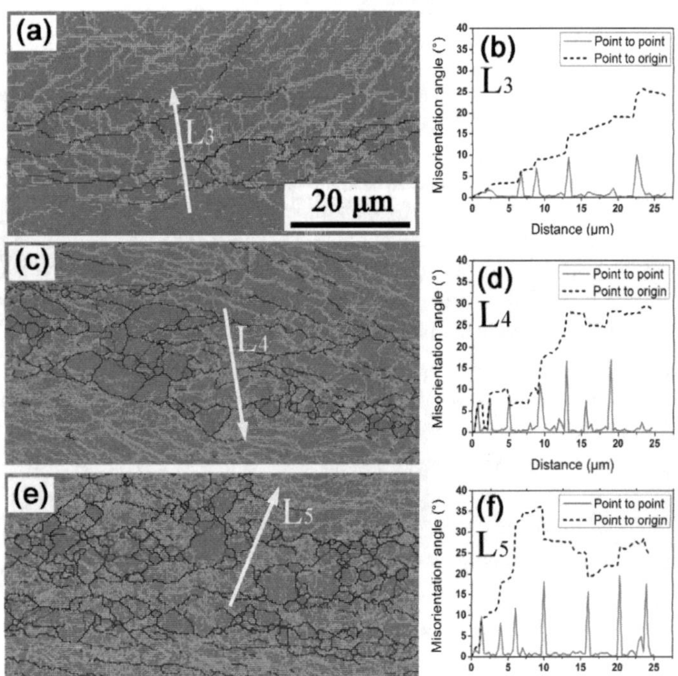

Figure 4. OIM maps of microstructures evolved in the transient deformations: deformed with strain rate jumping from $0.001s^{-1}$ to $0.1s^{-1}$ at strain of 0.3 and further straining to total strains of 0.4 (a), 0.5(c) and 0.8 (e). (b), (d) and (f): misorientation profiles along the corresponding arrows L_3, L_4, and L_5, respectively. The silver, green, blue, black, red and magenta lines represent boundaries with $0.8^o < \theta < 2^o$, $2^o < \theta < 5^o$, $5^o < \theta < 15^o$, $15^o < \theta$, $\Sigma 3$ and $\Sigma 9$ twin boundaries respectively.

IOP Conf. Series: Journal of Physics: Conf. Series **1270** (2019) 012047 doi:10.1088/1742-6596/1270/1/012047

3.3. Variation of DRX mechanisms under the transient deformation

During the transient deformation, after the strain rate changed from 0.001 s^{-1} to 0.1 s^{-1} at the strain of 0.3, the substructures formed in the first straining stage can be reasonably considered to do not alter immediately. The preformed microstructures have to evolve continuously to adapt to the imposed new strain rate in further deformations. Figure 4 shows the typical microstructures within the original grain interior after the strain rate jumps from 0.001 s^{-1} to 0.1 s^{-1}. Under continued straining to 0.4 after the strain rate changing, the microstructures evolved into an elongated subboundary network, these subboundaries generally with small to medium misorientation angles (figure 4a). The misorientations of some sub-boundaries are observed sharply increasing to about 7–15° (figure 4b). With the further straining to 0.5 and 0.8, the point-to-point misorientations of some segments in the subgrain boundary network are continuously increasing to above 15°, leading to the development of HABs during the deformation (figure 4d and 4f). The appearance of incomplete HABs confirms progressive transformation of sub-boundaries towards general grain boundaries during the deformation [8]. Furthermore, many fine DRX grains delimited partly by low angle boundaries (LABs) and partly by HABs are developed within the subgrain network (figure 4c and 4e). It thus provides a strong evidence to support the operation of the CDRX under the present transient deformation condition. The occurrence of the CDRX in austenite under current condition is in good agreement with the model suggested by Gourdet and Montheillet [2] who have proposed that during CDRX, LABs become increasingly transformed to higher-angle boundaries by absorbing dislocations.

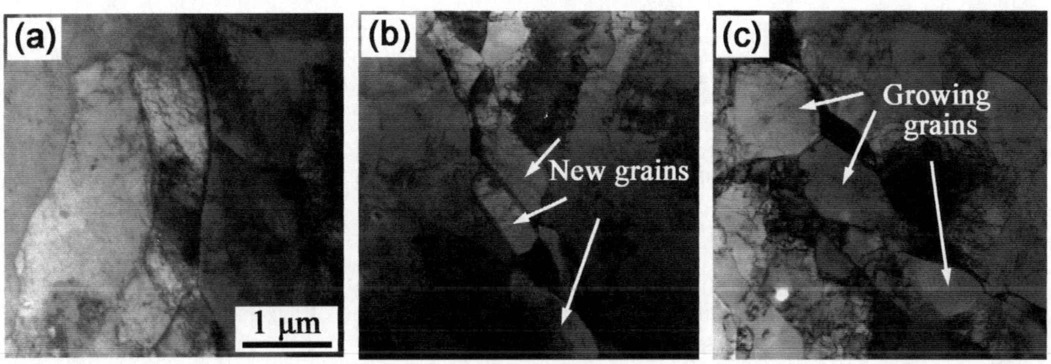

Figure 5. Typical TEM micrographs of (sub)structures evolved in the transient deformations: deformed with strain rate jumping from 0.001s^{-1} to 0.1s^{-1} at the strain of 0.3 and further straining to the total strain of 0.4 (a), 0.5 (b) and 0.8 (c).

TEM micrographs in Figure 5 show the evolution of sub-boundaries during the transient deformation. When the strain rate changes, the preformed substructures become elongated along the deformation direction. At the same time, abundant dislocations are generated and further accumulated around the sub-boundaries due to rapid strain hardening concomitantly with the instant strain rate jump (figure 5a). Furthermore, these newly formed dislocations will be effectively trapped and absorbed by the existing sub-boundaries or be rearranged to form new strain-induced sub-boundaries, leading to a rapid increase in their misorientations. Therefore, new CDRX grains will be preferentially developed from the existing sub-boundaries (figure 5b). It should be noted that the new grains formed by the CDRX generally have a relatively finer grain-size. Other than the CDRX grains in materials within high SFEs, these new grains in this austenitic alloy might be able to further grow by the migration of grain boundaries due to the high density of dislocations around them (figure 5c). The annealing twins can also be clearly observed inside many of new DRX grains (figure 4c and 4e), which also provides evidences for the occurrence of grain growth during this softening process, albeit most of new grains are forming by the CDRX mechanism.

In this study, it should be noted that the substructures formed during the transient deformation are quite different from the low-angle shear bands developed under the condition with constant high strain rates. It is easy to found that preserved substructures developed at former deformation stages will enhance the transformation of strain-induced subboundaries towards higher-angle boundaries in further straining after the abrupt jump of the strain rates. High density of preserved sub-boundaries can effectively absorb new generated dislocations. Meanwhile, some additional slip systems should be activated by the increased flow stress, which can induce rotations of individual crystallites towards different stable orientations during the deformation [9]. Both reasons can lead to quick increase in misorientations of the sub-boundaries, which can contribute to the occurrence of the CDRX in austenite [10]. It seems that the CDRX provides partially softening by the rearrangement of lattice dislocations to low-energy configurations (i.e. LABs to HABs) due to the absence of grain boundary migration inside the deformed coarse grains. The CDRX mechanism, however, cannot consume the lattice dislocation sufficiently because of the limited mobility of the dislocation in austenite. Hence, it can be expected that the newly formed CDRX grains would eliminate dislocations further by the local migration of new formed HABs. It can be reasonably speculated that the apparent acceleration of the DRX observed after the strain rate transition in Fig.1 is ascribed to the beneficial effect of the occurrence of CDRX and the subsequent grain growth.

4. Conclusions

In summary, although the austenitic alloy is of low SFE, two types of dynamic recrystallization might operate in the Ni-30%Fe austenitic model alloy, depending on the deformation conditions. At a slow strain rate, dynamic softening of the austenite takes place through a DDRX mechanism, whereas the DRX process was remarkably limited at the high strain rate condition. The transient deformation during which the strain rate jumps from low to high strain rates, however, can significantly promote occurrence of the CDRX in the low SFE austenitic alloy. The microstructure behaviours during the transient deformation are strongly affected by the deformation history. The preserved substructures are found significantly facilitate the strain-induced sub-boundaries increasingly transforming to higher-angle boundaries by continuing straining after the strain rate jumps, which promotes the occurrence of the CDRX in the low SFE austenitic alloy. Furthermore, the formation of new CDRX grains may further promote the dynamic softening of austenite by the subsequent grain growth.

Acknowledgements

The financial support from the National Natural Science Foundation of China (Grant Nos. 51771192, 51371169 and 51401214) is sincerely acknowledged.

References

[1] Humphreys FJ, Rohrer GS and Rollett A 2017 *Recrystallization and Related Annealing Phenomena (Third Edition)*, Elsevier, Amsterdam.
[2] Gourdet S and Montheillet F 2003 *Acta Mater.* **51** 2685–99.
[3] Sakai T, Belyakov A, Kaibyshev R, Miura H and Jonas JJ 2014 *Prog. Mater. Sci.* **60** 130–207.
[4] Manshadi AD and Hodgson PD 2008 *Metall. Mater. Trans. A* **39** 2830–240.
[5] Hurley PJ, Muddle BC and Hodgson PD 2001 *Metall. Mater. Trans. A* **32** 1507–17.
[6] Chen WX, Jia CN, Hu BJ, Zheng CW and Li DZ 2018 *Mater. Sci. Eng. A* **733** 419–428.
[7] Hansen N and Mehl RF, Medalist A 2001 *Metall. Mater. Trans. A* **32** 2917–35.
[8] Yanushkevich Z, Belyakov A and Kaibyshev R 2015 *Acta Mater.* **82** 244–254.
[9] Hughes DA and Hansen N 1997 *Acta Mater.* **45** 3871–86.
[10] Chen WX, Hu BJ, Jia CN, Zheng CW and Li DZ 2019 *Mater. Sci. Eng. A* **751** 10–14.

IOP Conf. Series: Journal of Physics: Conf. Series **1270** (2019) 012048 doi:10.1088/1742-6596/1270/1/012048

A continuous dynamic recrystallization model to describe the hot deformation behaviour of a Ti5553 alloy

R H Buzolin[1], F Krumphals[1], M Lasnik[2], A Krumphals[2], M C Poletti[1]
[1] Christian-Doppler Laboratory for Design of High Performance Alloys by Thermomechanical Processing, TU Graz, Kopernikusgasse 24/I. 8010 Graz, Austria
[2] voestalpine BÖHLER Aerospace GmbH & Co KG, Mariazeller Straße 25, 8605 Kapfenberg, Austria

ricardo.buzolin@tugraz.at

Abstract. A physical based model is developed to describe recrystallization phenomena of titanium alloys during hot working of the β phase. Continuous dynamic recrystallization is attributed as the restoration mechanism based on the progressive transformation of low angle boundaries (subgrains) into high angle boundaries (grains). The model describes both, the microstructure, and the flow stress evolutions during hot deformation with large strains. The microstructure is conceived as been formed by three different populations of dislocations, and high as well as low angle grain boundaries. Evolution rates of all microstructural features are determined based on the effects of generation, interaction and annihilation of dislocations during deformation due to dynamic recovery, continuous dynamic recrystallization and static recovery. Continuous dynamic recrystallization is modelled and is considered to occur homogeneously within the microstructure. The model is able to predict the formation of subgrains from a fully annealed microstructure and the progressive formation of high angle grain boundaries. The subgrain and grain sizes are also obtained as output of the model. The model was applied to describe the hot compression behaviour of a Ti5553 alloy deformed between 880°C to 920°C and strain rates from 0.001 s^{-1} up to 10 s^{-1}. The model is validated with flow curves and microstructural characterisation of hot deformed, and with microstructural information of non-deformed samples. The critical strain rate increases with increasing strain rate and decreasing temperature, similar to the discontinuous dynamic recrystallization phenomenon. The model can be implemented to simulate the microstructure and predict flow stresses of titanium alloys in industrial processes.

1. Introduction

The plastic deformation of high stacking fault energy (SFE) alloys at high temperatures involves dynamic recovery (DRV) [1] that can be followed by continuous dynamic recrystallization (CDRX) at larger strains [2]. The microstructures of hot deformed samples are not homogeneous with increasing the strain rate and the formation of boundaries in the interior of the grains are observed after some deformation. Moreover, after 50% deformation with low strain rate (0.01 s^{-1}) serrated grain boundaries are present in the microstructure and two complementary mechanisms have been proposed to explain this behaviour. The conventional lattice rotation leads to the development of geometrically necessary dislocations forming subgrains that migrate and evolve to high angle grain boundaries. Furthermore, a heterogeneous distribution of dislocations near the prior β grain boundaries results in the formation of substructures that are arranged in bands allocated in one side of β grain boundaries because of heterogeneous deformation. Progressive deformation can eventually cause the merging of such substructures, forming new high angle grain boundaries, parallel to prior β grain boundaries. The hot deformation behaviour of a similar alloy, the Ti-5Al-5Mo-5V-3Cr-1Zr, was also systematically investigated [3]–[5]. Warchomicka *et al.* [5] observed that near the prior β grain, CDRX occurs by progressive rotation of the lattice in the β phase at high strain rates. At low strain rates and large strains, geometric dynamic recrystallization (GDRX) is observed. The elongated prior β grain boundaries are pinched off leading to the formation of new small grains with similar size compared to the subgrains. The occurrence of GDRX during hot deformation of Ti-5Al-5Mo-5V-3Cr-1Zr was also observed by Dikovits *et al.* [4] at low strain rates. Furthermore, at high strain rates CDRX plays a role above β transus temperature. However, it has been observed that dynamic recovery (DRV) of the β phase is the dominant mechanism during deformation over a wide range of temperatures and strain rates. Instabilities have been attributed to the formation of deformation bands (DBs). A continuous misorientation change towards the grain boundaries has been attributed to be the major origin of the formation of the deformation bands at moderate and higher strain rates.

Physical models based on the evolution of internal variables such as the dislocation density have been developed for Ti alloys. Alabort *et al.* [6] proposed a mesoscale physical model to predict the hot tension behaviour of an equiaxed Ti-6Al-4V alloy during superplastic behaviour. The developed model is based on internal state variables and could predict the grain size, dislocation density and void fraction. A mesoscale model for hot deformation behaviour of a Ti-6Al-4V alloy was also developed [7]. The stress has been proposed to follow Taylor's assumption. The internal state variables to predict the flow stress and the microstructure evolution are the dislocation density and the vacancy concentration.

Although several attempts have been made to model both the flow stress, as well as microstructure evolution during deformation of Ti alloys, most of them consider discontinuous dynamic recrystallization as the deformation mechanism. In this work we developed a model to predict not only the flow stress but also the microstructure evolution during hot deformation of a cogged Ti5553 in β phase field based. The internal variables are three different populations of dislocations, high and low angle grain boundaries, and boundary misorientation angle distribution. The evolution rates of the microstructural features are determined based on dynamic recovery, continuous dynamic recrystallization and static recovery.

2. Experimental procedures

A cogged commercial Ti-5Al-5V-5Mo-3Cr with large prior β grains elongated in the cogging direction, and globular α phase was used in this investigation. The β transus temperature of this material lies around 850°C. Cylindrical samples of 10 mm diameter and 15 mm length were compressed using a Gleeble® 3800 device in the single β phase field at 880, 900 and 920°C at 5 strain rates between 0.001 and 10 s^{-1}, and up to a total strain of 0.9. The specimens were heated up at 5 K s^{-1}, held for 15 min before deformation, and subsequently in situ water quenched. The high cooling rate did not allow any static restoration of the β phase or diffusion-controlled phase transformation after deformation.

Specimens for metallographic characterization were ground with SiC abrasive papers from P320 to P2000, polished with 1 μm diamond suspension, followed by polishing with OPS suspension. Finally, vibro-polishing for 24 h using an acidic suspension of Alumina was performed. Scanning electron microscopy (SEM) was conducted using a Tescan Mira3 SEM equipped with a Hikari Electron Backscattered Electrons (EBSD) camera. EBSD measurements were performed for an area of 500 μm x 500 μm using a step size of 0.5 μm. The data treatment was performed using the OIM Analysis® software. A confidence index standardization was performed considering a grain with minimum size of 10 points and grain boundary misorientation tolerance angle of 12°. Finally, the neighbour confidence index correlation was utilized to re-index the data-points with confidence index lower than 0.75. A subgrain is defined as a region within grains where at least 2 points share a similar orientation with a boundary misorientation less than 15° but greater than 2°. In order to access the local microstructural features, kernel average misorientation (KAM) was calculated from the EBSD measurements [8], [9]. The density of geometrically necessary dislocations or GNDs was also calculated

[10]. A maximum misorientation of 5° is considered and the misorientation is calculated according to the first nearest neighbours.

3. Modelling description

The representative microstructure entity is considered as a grain. Within a grain, LAGBs are produced as result of dynamic and static recovery. The dominant recrystallization mechanism is assumed to be continuous dynamic recrystallization (CDRX) and experimental evidence is also shown. In CDRX, the progressive increase in misorientation of the LAGB occurs due to subgrain rotation or accumulation of dislocations along a subgrain boundary. The progressive increase in misorientation of the LAGBs gradually forms new HAGBs. When the steady state is reached, a fully recrystallized microstructure is obtained.

The dislocation populations are separated into three different ones: mobile, immobile and walls. The mobile dislocation density is considered as the average density of dislocations that are moving along the slip systems causing plastic deformation. The wall dislocations are spaced and organized so that a boundary misorientation is observed in the microstructure. The immobile dislocation density is considered as the average density of dislocations that do not glide and are located at the grain interiors, such as forest of dislocations, pinned dislocations within the slip systems, and dipoles.

The total stress is calculated as a sum of three components (Equation (1)): an athermal or internal stress (σ_{ath}) related to the strain field caused by the dislocations, a thermal or effective stress (σ_{th}) attributed to the activation energy for the movement of dislocations in the lattice, and a Hall-Petch component (σ_{HP}) to account for the effect of the subgrain and grain sizes on the total stress.

$$\sigma = \sigma_{ath} + \sigma_{th} + \sigma_{HP} \tag{1}$$

The athermal stress is a function of the different dislocation populations, equation (2).

$$\sigma_{ath} = \mathrm{M}\alpha Gb\sqrt{\rho_i + \rho_m + f\rho_w} \tag{2}$$

where M is the Taylor factor, α is the Taylor constant and considered 0.1, G is the shear modulus, b is the Burgers vector, ρ_i is the immobile dislocation density, ρ_m is the mobile dislocation density, ρ_w is the wall dislocation density, and f is a factor that varies from zero for well-organized LAGB (observed at low strain rates), to 1 for dislocations piled up at the boundaries due to localized deformation, progressive lattice rotation, etc. (usually the case at high strain rates).

The evolution of the microstructural components is modelled taking into account recovery and CDRX. The rate equations to account for the dislocation densities evolution as well as the equations used to compute the glide and climb velocities are based on the work of Ghoniem *et al.* [11].The description of continuous dynamic recrystallization was proposed by Montheillet *et al.* [12]. Finally, the dynamic as well as static restoration phenomena are considered for the evolution of the dislocation densities, equation (3).

$$\frac{\partial \rho_x}{\partial t} = \left(\frac{\partial \rho_x}{\partial t}\right)_{Dynamic} + \left(\frac{\partial \rho_x}{\partial t}\right)_{Static} \quad (x = m, i, w) \tag{3}$$

where *m*, *i* and *w* corresponds to mobile, immobile and wall dislocations densities, respectively.

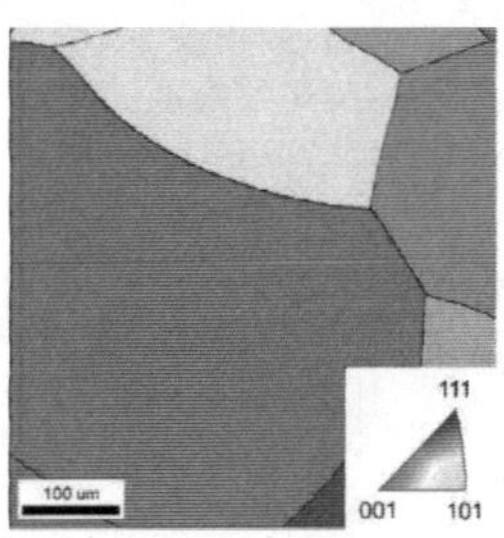

Figure 1: Inverse pole figure map of Ti5553 annealed at 920°C for 15 min.

4. Results and discussion

The microstructures of the samples before the hot compression exhibits a fully static recrystallized microstructure as shown in figure 1 for the heat treatment performed at 920°C for 15 min.

Figure 2 shows the inverse pole figures and the kernel average misorientation (KAM) maps of the Ti5553 deformed at 920°C at 0.001 s^{-1} (a and b, respectively) and 1 s^{-1} (c and d, respectively). The β phase shows notable substructure formation. Subgrains were surrounded by low and high angle boundaries, in agreement with works on similar alloys and the CDRX mechanism. The amount of LAGB formed is larger at 1 s^{-1}, as highlighted by the local misorientation in the KAM map in figure 2(d). The substructure was homogeneously distributed for the deformation at 0.001 s^{-1} (figure 2(a)), while some regions within the grains are observed with large misorientation accumulation along the grain and not visible subgrain formation at 1 s^{-1} (figure 2(c)). Despite of this localization of the plastic deformation inside a grain, no indication of discontinuous dynamic recrystallization was observed for the investigated Ti5553. The degree of recrystallization is very low for 920°C. In figure 2(a) the substructure formation seems to be completed for the microstructure but only a low amount new high angle boundaries is observed.

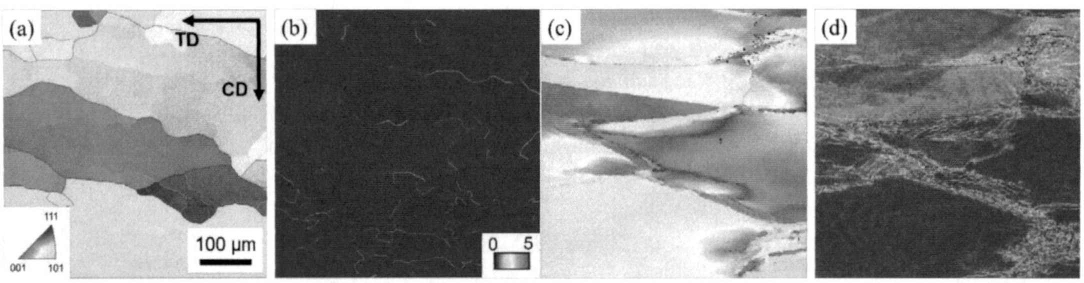

Figure 2: Inverse pole figure maps and kernel average misorientation (KAM) maps of the typical microstructure of the Ti5553 samples deformed at: (a),(b) 920°C and 0.001 s^{-1}; (c),(d) 920°C and 1 s^{-1}, respectively. Black lines indicate high angle grain boundaries (misorientation > 12°) and white lines indicate low angle grain boundaries (2°<misorientation < 12°); CD is the compression direction and TD is the transversal direction.

Figure 3 shows the results of the simulated and experimental flow curves for the investigated deformation conditions. A good agreement between the measured and simulated flow curves is found. The model is able to predict the nearly constant flow stress. The formation of HAGB from LAGB must lead to flow softening at much larger strains. The localization of the deformation, as shown in figure 2 (c), is attributed to the deviation between simulated and measured flow stress for the deformation at high strain rates (10 s^{-1}).

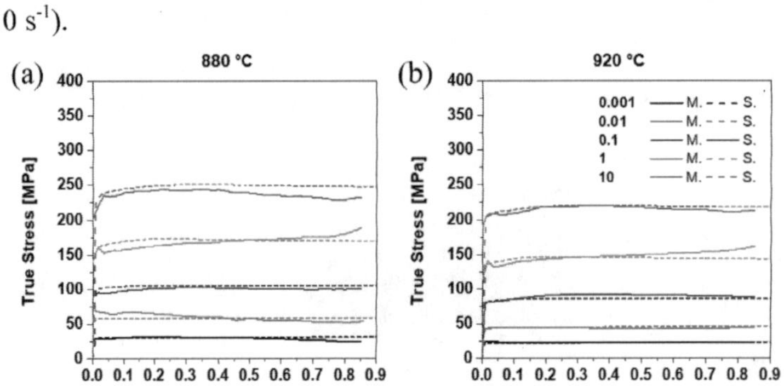

Figure 3: Measured (M.) and simulated (S.) flow curves: (a),(b) 880°C and 920°C, respectively.

Figure 4 exhibits the simulated and the measured results for wall dislocation density and subgrain size for different deformation conditions. The measured wall dislocation density is considered as the GND density obtained from the EBSD measurements. A large deviation is observed between the measured and simulated data when comparing the results of the low strain rate. The large initial grain size and the non-homogeneous deformation would require notably larger areas to be measured with EBSD to obtain satisfactory statistic data although the trend shows almost always an overestimation of the model.

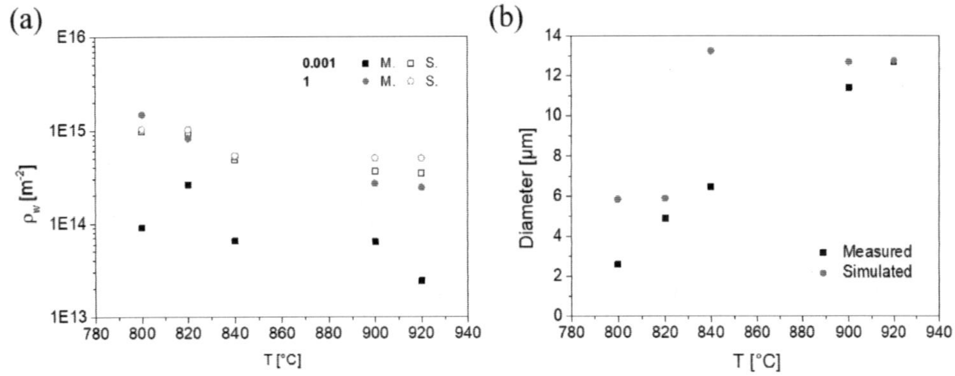

Figure 4: Comparison of: (a) measured (M.) and simulated (S.) wall dislocation density for different deformation temperatures and strain rates of 0.001 s^{-1} and 1 s^{-1}; (b) average subgrain diameter for the strain rate of 1 s^{-1} for different deformation temperatures.

To illustrate the capability of the proposed model to predict the microstructure evolution for larger strains, figure 5 shows the subgrain, grain sizes and fraction of HAGB evolution for strains up to 10 at 920°C. A decrease in subgrain size, which reaches a minimum, is predicted until the formation of new HAGB is pronounced. The steady state grain size corresponds to the grain size when CDRX is finished. The proposed model predicts larger steady state grain size for lower strain rates. The formation of the initial substructure from a fully recrystallized microstructure is shown by the decrease of HAGB until the minimum. After that, the onset of CDRX is pronounced leading to a recrystallized microstructure. The fraction of HAGB is considered as the fraction of recrystallization. The HAGB fraction (thus, the recrystallization fraction) is higher for lower strain rates for a given strain. Therefore, the onset of CDRX is delayed with increasing the strain rate for a given temperature.

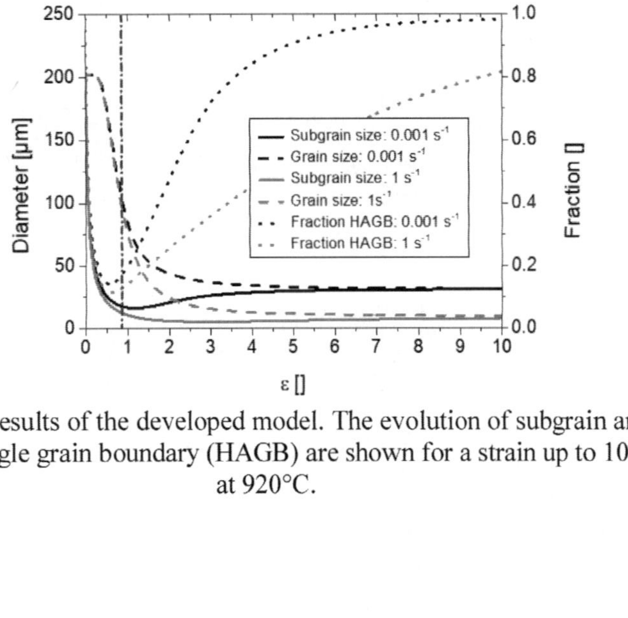

Figure 5: Simulation results of the developed model. The evolution of subgrain and grain sizes as well as fraction of high angle grain boundary (HAGB) are shown for a strain up to 10 and deformation at 920°C.

5. Summary and conclusions

A physically based model was developed for predicting the microstructure evolution and the flow stress of a Ti5553 alloy. The model is able to predict the grain size, subgrain size, boundary misorientation and dislocation densities evolution during deformation. The production, annihilation and conversion of one type of dislocation population into another are modelled. Moreover, glide and climb velocity can be determined. CDRX and DRV are considered as the major restoration mechanisms for the microstructure evolution. In this case, a progressive increase in boundary misorientation and the progressive formation of new subgrain boundaries are the proposed mechanisms for CDRX.

The proposed models were able to predict the investigated microstructure. Subgrain formation from a fully recrystallized microstructure is predicted and the subgrain size decreases with increasing strain rate. The investigated hot deformed microstructures show the presence of subgrains and new high angle boundaries in the vicinity of low angle boundaries, evidencing the role of CDRX. Smaller subgrains are formed for higher strain rates, while the size of the subgrains increases with increasing temperature due to higher HAGB mobility and faster CDRX. A more homogeneous microstructure is observed for low strain rates, while a notable degree of heterogeneity in the microstructure is found for high strain rates. Regions with intense misorientation spread are found for hot compression at high strain rates, which can be considered as regions with a notable pile-up or entanglement of dislocations. This localization of deformation is not considered in the model and assumed to be the reason for the observed deviations.

References

[1] Drury M R and Humphreys F J 1986 *Acta Metall.* **34.11** 2259–71.
[2] Gourdet S and Montheillet F 2000 *Mater. Sci. Eng. A* **283.1** 274–88.
[3] Poletti C, Germain L, Warchomicka F, Dikovits M and Mitsche S 2016 *Mater. Sci. Eng. A* **651** 280–90.
[4] Dikovits M, Poletti C and Warchomicka F 2014 *Metall. Mater. Trans. A* **45.3** 1586–96.
[5] Warchomicka F, Poletti C and Stockinger M 2011 *Mater. Sci. Eng. A* **528.28** 8277–85.
[6] Alabort E, Putman D and Reed R C 2015 *Acta Mater.* **95** 428–42.
[7] Picu R C and Majorell A 2002 *Mater. Sci. Eng. A* **326.2** 306–16.
[8] Gurao N P, Kumar P, Bhattacharya B, Haldar A and Suwas S 2012 *Metall. Mater. Trans. A* **43.13** 5193–01.
[9] Zhong Y, Yin F, Sakaguchi T, Nagai K and Yang K 2007 *Acta Mater.* **55.8** 2747–56.
[10] Field D P, Trivedi P B, Wright S I and Kumar M 2005 *Ultramicroscopy* **103.1** 33–39.
[11] Ghoniem N, Matthews J and Amodeo R 1990 *Res Mech.* **29** 197–19.
[12] Gourdet S and Montheillet F 2003 *Acta Mater.* **51.9** 2685–99.

IOP Conf. Series: Journal of Physics: Conf. Series **1270** (2019) 012049 doi:10.1088/1742-6596/1270/1/012049

Continuous dynamic recrystallization during hot torsion of an aluminum alloy.

M C Poletti[1], T Simonet-Fotso[1], D Halici[2], D Canelo-Yubero[1], F Montheillet[3], D Piot[3], Z Kovács[4], N Schell[5], D Tolnai[5]

1Institute of Materials Science Joining and Forming, Graz University of Technology, Graz, Austria;
2Andritz AG, Graz, Austria
3 Physique et Mécanique des Matériaux, Ecole des Mines Saint Etienne, Saint Etienne, France
4 Department of Materials Physics, Lorand Eötvös University. Budapest, Hungary
5. Institute of Materials Research, Helmholtz Zentrum Geesthacht, Geesthacht, Germany
e-mail address corresponding author: cecilia.poletti@tugraz.at

Abstract. Most commercial aluminum alloys are characterized by dynamic recrystallization at very large deformations in a continuous manner. The present study deals with the characterization and modeling of the evolution of the microstructure of an aluminum wrought alloy at large plastic deformations. Hot torsion tests of the AA6082 aluminum alloy are carried out using the thermomechanical simulator Gleeble®3800 in a wide range of temperatures and strain rates. The use of water quenching immediately after deformation avoids any static restoration during cooling. Microstructural investigations are carried out by means of electron back scattered diffraction using a scanning electron microscope to determine the grain and subgrain structures, as well as the misorientation distributions. In-situ synchrotron radiation tests during hot torsion are used to confirm the continuous dynamic recrystallization (CDRX) by the evidence of the conversion of low angle boundaries (LAGBs) into high-angle boundaries (HAGBs) and the formation of new texture. Experimental investigations show that CDRX starts with the formation of LAGBs at low strains (center of the sample). By subsequent straining (close to the surface of the sample), the accumulation of dislocations at the LAGBs causes an increase in their misorientation until a critical value is reached and LAGBs transforms into HAGBs. The developed model consists of a microstructural model, equation rates and constitutive equations. The microstructure is described by three internal variables. Their rates are evaluated using the Kocks-Mecking model. The modelled and experimental flow stresses show softening due to the consumption of dislocations and the continuous formation of new HAGBs.

1. Introduction

Dynamic recrystallization (DRX) often occurs when metallic parts undergo hot deformation. Discontinuous dynamic recrystallization (DDRX) is a well-known process that occurs in moderate and low stacking fault energy (SFE) materials by nucleation followed by grain growth [1,2]. When dealing with high SFE metals two different mechanisms have been proposed in the literature to explain the modification of the microstructure: 1) continuous dynamic recrystallization (CDRX) correlated with a change in texture [1-3], and 2) geometric dynamic recrystallization (GDRX) where texture remains unaltered [4,5]. The CDRX implies a smooth change in the microstructure by a progressive formation and rotation of subgrains as deformation proceeds. Thus, the amount of dislocations forming the walls of the subgrains increases, resulting in a total misorientation increment. During this process, subgrain boundaries or low angle grain boundaries (LAGB) are converted into high angle grain boundaries (HAGB) [3]. On the other hand, the GDRX is consistent with Taylor theory requirements [6] and involves formation of subgrains followed by migration and pinching off of original HAGBs [7]. Although the models are based on different mechanisms to explain the microstructure evolution, both mechanisms can take place simultaneously [8]. Finally, it still has to be clarified if the formed HAGBs are a consequence of the transformation from LAGB or by HAGB migration. [9,10].

This work aims at elucidating and modelling the physical processes occurring during plastic deformation up to large strains of a high SFE aluminum alloy. The evolution of the microstructure could be tracked using in-situ high energy synchrotron radiation diffraction (HEXRD) during a torsion test.

2. Experimental methods

Aluminum samples were hot deformed at very large strains by using torsion tests. The microstructure was characterized at the center and close outer surface of the torsion sample after deformation and water quenching. The alloy used in this study is an AA6082 (AlSi1MgMn) aluminum alloy produced by the company Zultner Metall GmbH by hot extrusion and aged to T6 condition. Torsion samples were prepared with a gauge length of 20 mm and a gauge diameter of 10 mm.

Hot torsion tests using a Gleeble ® 3800 were carried out to produce deformed samples and flow curves. Temperature control during deformation was done with a thermocouple welded at the fixed border of the gauge. The tests were carried out between 350 and 500°C, with effective strain rates of 0.01-1 s^{-1}. Equivalent flow stresses and strains were calculated from the torque and revolution data, respectively. All samples were water quenched after deformation to avoid any further restoration.

The microstructure of the deformed samples was observed across their gauge to obtain information with respect to the strain, this last varying from zero in the middle axe along the gauge to a maximum value at outer surface. Scanning electron microscopy (SEM) was conducted using a Tescan Mira3 SEM equipped with a Hikari Electron Backscattered Diffraction (EBSD) camera to characterize grain sizes, local misorientations given by Kernel Average Misorientation (KAM) and misorientation distributions.

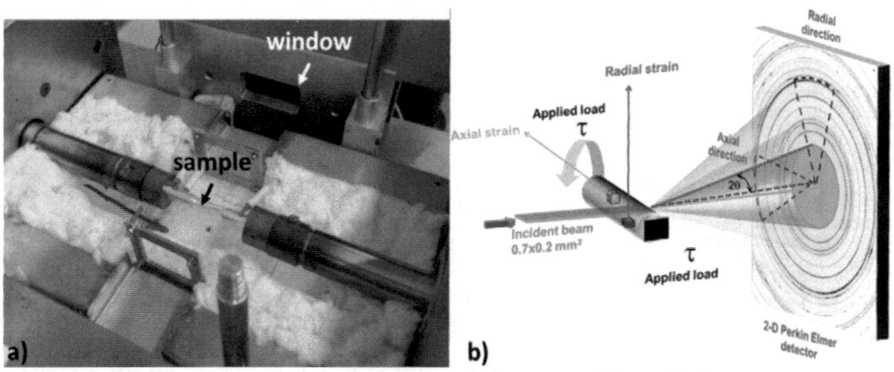

Figure 1 In-situ torsion test configuration a) furnace with a window transparent to X-rays, and b) drawing of the set up with incident beam close to surface of sample

X-Ray diffraction measurements using synchrotron radiation were done in a thin slice of material close to the outer surface while torsion was applied on the Al alloy at high temperatures (see Figure 1). The collected Debye-Scherrer rings were processed to produce the azimuthal angle (Ψ) - effective strain (AS) plots, showed together with the τ_{eff} - ε_{eff} curves in Figure 6.

3. Modelling

The model developed based on [11] consists of three main components: 1) internal variables that constitute the microstructure, 2) evolution of these internal variables, and 3) constitutive equations.

The microstructure was described as an aggregate of subgrains delimited by LAGBs and HAGBs. Three internal variables were used: the density of dislocations inside the subgrains (ρ_i), the average subgrain size (D), and the fraction of LAGBs (f_{LAGB}). During a strain increment $d\varepsilon$ dislocation density increases by $hd\varepsilon$ and $r\rho_i d\varepsilon$ annihilate or are incorporated in LAGBs by dynamic recovery. A part $\rho_i dV$ is consumed by the migration of HAGBs.

$$d\rho_i = hd\varepsilon - r\rho_i d\varepsilon - \rho_i S_V f_{HAGB} v_M d\varepsilon/\dot{\varepsilon} \quad (1)$$

f_{HAGB} = fraction of HAGBs, v_M = average velocity of grain boundaries, S_V is the surface of boundaries, and h and r are the hardening and recovery coefficients, respectively. Among the recovery part, a fraction β is annihilated, a fraction $(1-\beta)\alpha$ is consumed by the formation of new LAGB, and the rest $(1-\beta)(1-\alpha)$ is absorbed by the existent LAGBs and HAGBs. The incorporation of dislocations in LAGBs induces a misorientation increase ($d\theta$): (2)

$$d\theta = (1-\beta)(1-\alpha)\frac{b}{2n}D(r\rho_i)d\varepsilon$$

n = constant, b is the Burger's vector, . The evolution in total area of boundary per unit volume is given by $dSV = dSV^+ - dSV^-$, where dSV^+ corresponds to the area of LAGBs created during straining and dSV^- corresponds to the area of boundary eliminated by the HAGBs migration:

$$dS_v^+ = (1-\beta)\alpha\frac{b}{n\theta_0}(r\rho_i)d\varepsilon \text{ and } dS_v^- = S_V dV = \frac{S_V^2 f_{HAGB} v_M d\varepsilon}{\dot{\varepsilon}} \quad (3)$$

The fraction of LAGBs and the density of dislocations forming the LAGBs, ρ_w were calculated using the distribution function of the misorientations ($\varphi(\theta)$).

$$f_{LAGB} = \int_{\theta_0}^{\theta_c} \varphi(\bar{\theta},\varepsilon)d\theta \quad \text{and} \quad \rho_w = \left(\frac{2.n}{b.D}\right)\int_{\theta_0}^{\theta_c} \varphi(\bar{\theta},\varepsilon).\bar{\theta}d\theta \quad (4)$$

θ_0 = minimum misorientation angle of new created LAGB, $\bar{\theta}$ = mean misorientation, and θ_c = critical angle for the transformation of LAGB into HAGB. The flow stresses (Figure 2) were calculated using the Taylor factor M, the shear modulus G, and material constants α_1, A_1 and A_2.

$$\sigma = M\alpha_1 Gb\left(A_1\sqrt{\rho_i} + A_2\sqrt{\rho_w}\right) \quad (5)$$

4. Results and discussions

The experimental and calculated flow stresses are plotted in Figure 2 for different deformation conditions. The model could capture the temperature and strain rate influences on the stress. Furthermore, flow softening is observed and calculated for all the conditions.

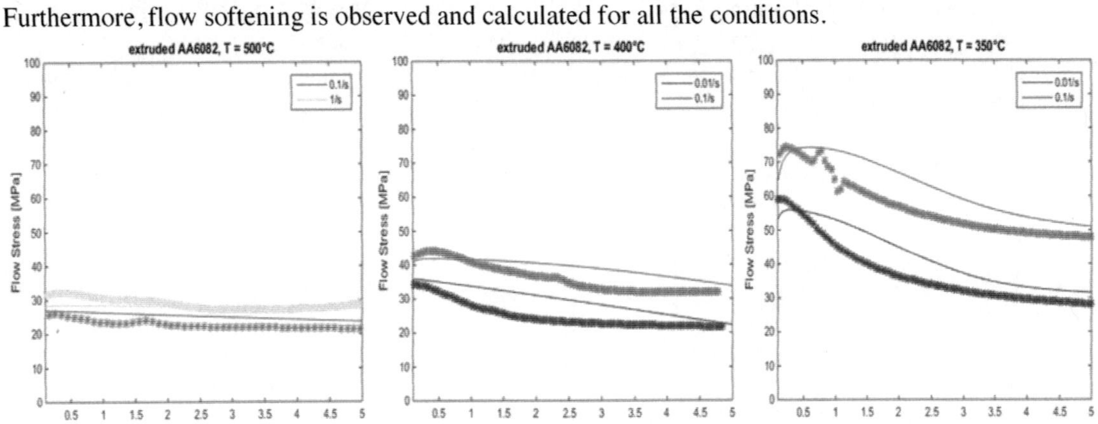

Figure 2. Experimental (symbols) and modelled (full line) flow stresses at different conditions

The effect of the strain on the microstructure can be followed from close to the centre (low strain) across the diameter to the edge of the sample (large strains) in Figure 3a) and Figure 3b) and c), respectively. By increasing the deformation, a continuous formation of new HAGB and grain refinement was observed in general. Furthermore, the recrystallized grain size as well as the grade of recrystallization for a given strain is shown to be dependent on the temperature and strain rate. In agreement, the flow steady state is achieved at large strains. The steady state is achieved faster the lower the strain rate and the higher the temperature are.

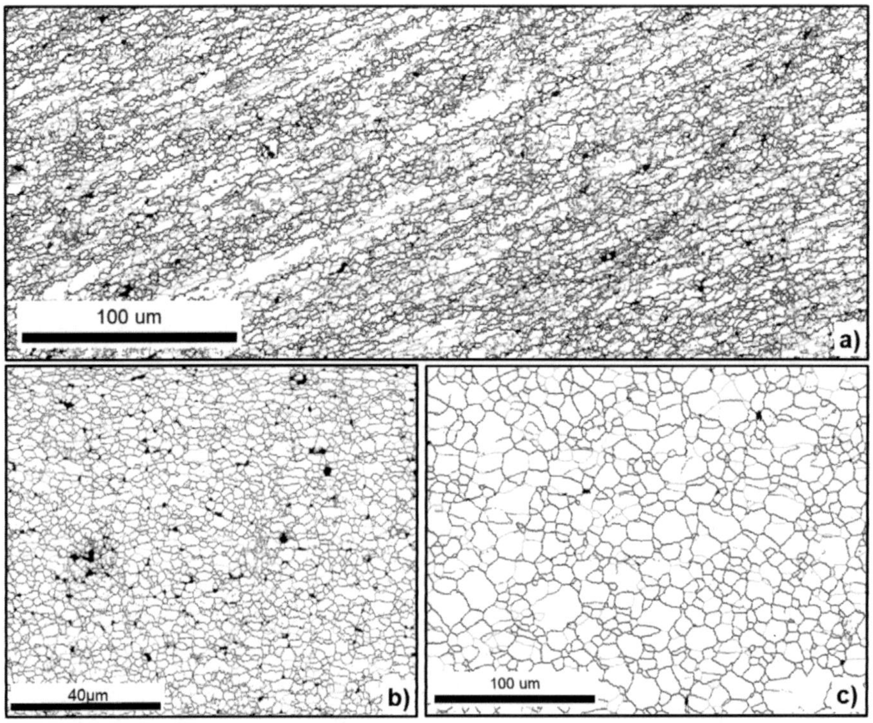

Figure 3. Map of boundaries: red between 2-5°, green between 5-11° and blue above 11°. Samples deformed at $0.1s^{-1}$ and a) 350°C, b) 350°C and c) 500°C. Pictures b) and c) are obtained from samples taken from the edge (strains close to 5), and picture a) shows less HAGB in the middle of the sample. Note the differences in the scales.

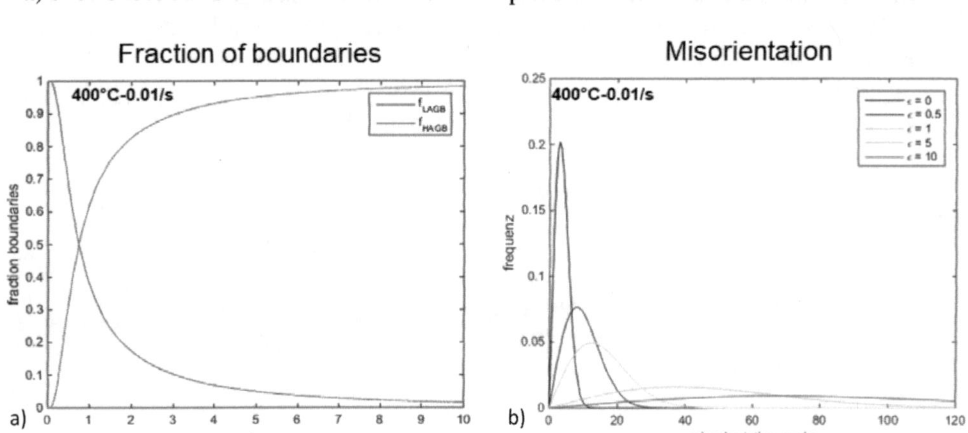

Figure 4. Microstructural outputs of the model: a) f_{HAGB} and f_{LAGB} with increasing strain, and b) evolution of the misorientation distributions with the strain.

The starting microstructure is modelled as being formed by one coarse grain. Therefore, LAGBs are firstly formed and then consumed during recovery and CDRX (Figure 4a). Furthermore, the evolution of the misorientation refers to what is formed during the deformation (Figure 4b), and it does not account for the HAGBs existing prior to deformation. The distributions of measured misorientations are shown for different conditions in Figure 5. LAGBs fraction decrease by increasing both, the strain and the strain rate. The distributions show large values of misorientations for the original grains, and very low values for the formation of LAGBs. The increasing values in the 15-40° range might correspond to the ones formed during CDRX, in agreement with the predictions of the model shown in Figure 4b).

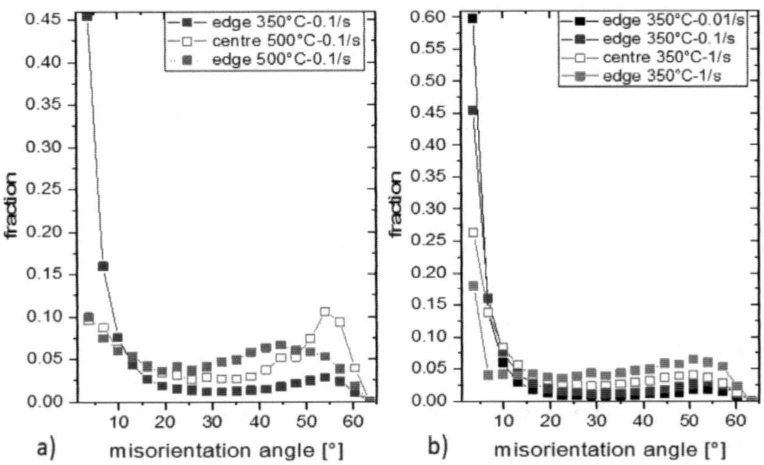

Figure 5. Misorientation angle distributions showing the influence of a) temperature and strain b) strain rate and strain.

The analysis of the in-situ data during hot torsion is shown in Figure 6 (AS-plots). It indicates a strong (111) starting texture and a less intense (200)-fibre texture due to the extrusion process of the initial condition. During the early stages of deformation, the high intensity regions broaden monotonically due to the formation of a substructure up to an effective strain of $\varepsilon_{eff}\sim$ 0.7. The resulting final texture is different from the starting one, being a strong proof of the CDRX process.

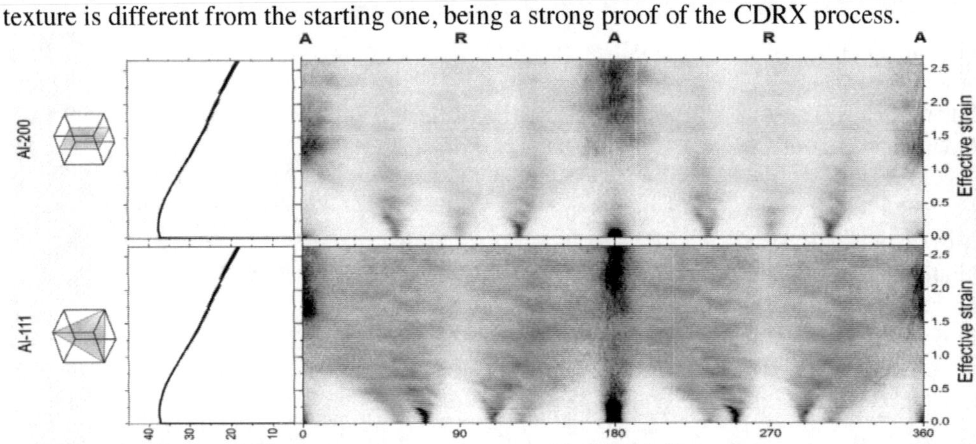

Figure 6. AS-plots for the Al-200 and Al-111 crystallographic planes.

Finally, the grain size was calculated as being decreased with the strain (**Figure 7**a). The driving force for CDRX is given by the pressure at the LAGBs as a result of the dislocation density, meaning that low temperatures and high strain rates are favorable. On the other hand, the mobility of the boundary increases exponentially with the temperature, therefore accelerating the CDRX and decreasing at the same time the driving force by the rapid consumption of dislocations. Measured grain sizes (Figure 7b) show in general larger values than the modelled ones.

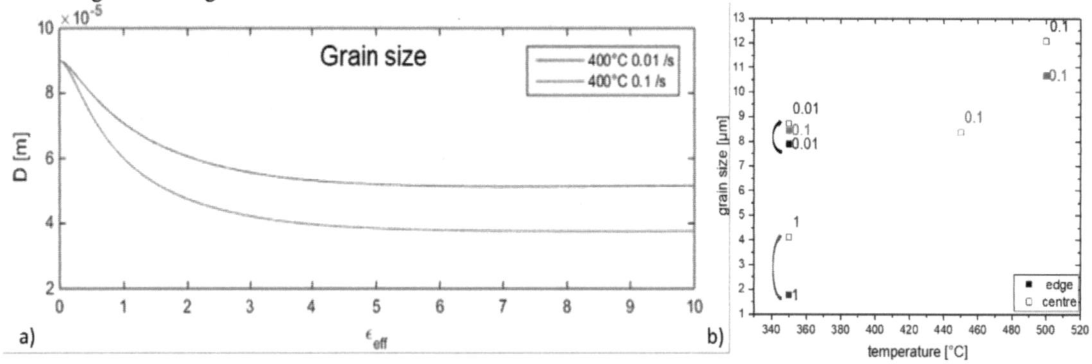

Figure 7. a) Calculated evolution of the grain size b) measured grain size at different temperatures, strains and strain rates (blue for $1s^{-1}$, red for $0.1s^{-1}$, and black for $0.01s^{-1}$).

5. Summary and conclusions

We could prove, quantify and model the microstructural and stress evolution of AA6082 aluminum alloy during large deformations at high temperatures. We found that the grain refinement and the decrement of dislocation densities within grains (KAM) can be described by CDRX. We described the continuous dynamic recrystallization as follows: during deformation at high temperatures, cell formation happens (wall dislocation density increases rapidly) by dynamic recovery. As the deformation increases, the wall dislocation density reaches a maximum value, then decreases due to the formation of new HAGBs. The softening is due to the absorption of dislocation by the existent and also the newly created boundaries.

6. Acknowledgments

The authors would like to thank to Ricardo Buzolin for the EBSD measurements. The work was supported by the Austrian Science Fund (FWF) within the stand alone project P 27471.

References

[1] Humphreys F J, Hatherly M 2004 *Recrystallization and related annealing phenomena* (Elsevier, Oxford).
[2] Sakai T, Belyakov A, Kaibyshev R, Miura H and Jonas J J 2014 *Prog. Mater. Sci.* **60** 130
[3] Gourdet S and Montheillet F 2000 *Mater. Sci. Eng. A* **283** 274
[4] McQueen H J, Knustand O, Ryum N, Solberg J K 1985 *Scripta Metall. Mater.* **19** 73
[5] Solberg J K, McQueen H J, Ryum N and Nes E 1989 *Part I, Philos. Mag. A* **60** 447
[6] McQueen H J, Kassner M E 2004 *Scripta Mater.* **51** 461
[7] Gholinia A, Pragnell P B and Markushev M V 2000 *Acta. Mater.* **48** 1115
[8] Pari Jr. L D and Misiolek V D 2008 *Acta Mater.* **56** 6174
[9] Driver J 2018 *Matter. Lett.* **222** 135
[10] Huang K and Logé R E 2016 *Mater. Design.* **111** 548
[11] Gourdet S, Montheillet F 2003 *Acta Mater.* **51-9** 2685

The dynamic recrystallization behavior in Al-Mg alloys

Yu Borisova, A Kalinenko, D Yuzbekova and A Mogucheva

Belgorod State University, Belgorod 308015, Russia

Corresponding author: borisova_yu@bsu.edu.ru

Abstract. The influence of phase composition on the behavior of dynamic recrystallization in alloys Al-3Mg (A1), Al-4.57Mg-0.35Mn-0.2Sc-0.09Zr (A2) and Al-5.4Mg-0.52Mn-0.1Zr (A3) (wt.%) during the equal-channel angular pressing (ECAP) was studied. The ECAP processing of alloy A1 resulted in a partially recrystallized microstructure with the mean size of coarse and fine grains of 20 and 1.4 μm, respectively In contrast, the formation of fully-recrystallized microstructure was observed in the alloys A2 and A3. The mean grain sizes in the alloys A2 and A3 were 1.1 and 1.5 μm, respectively. The hardness values in the alloys A2 and A3 after ECAP processing were increased by about 30% more than those in the alloy A1. This may be attributed to the presence of Al_6Mn and $Al_3(Sc,Zr)$ particles.

Keywords – **Aluminum alloys, Dynamic recrystallization, Equal-channel angular pressing**

1. Introduction

In recent decades, the refinement of the grain structure of aluminum alloys in order to obtain a fine-grained microstructure, i.e. microstructure with a grain size less than 1 micron, aroused a great attention. The great interest in fine-grained materials is primarily due to the possibility of improving their characteristics and achieving a new level of physical and mechanical properties [1-3]. Dynamic recrystallization (DRX) is an important mechanism controlling the formation of fine-grained microstructure for obtaining the required properties in various materials [4–7]. The most significant advantage of DRX is that the required microstructure can be formed directly during hot deformation [4,6,8].

The microstructures formed during DRX depend significantly on the temperature, strain and strain rate. In particular, the pressing technologies adhere to the strategy of applying large reductions, which lead to significant values of the strain. However, after reaching a certain strain, the final properties of the products are determined by the temperature and strain rate [9].

The formation of the fine-grained microstructure during hot deformation of aluminum alloys was fairly clarified [10]. This study dealt with the effect of deformation and heat treatment on the recrystallization and polygonization processes. However, little attention has been paid to the influence of the phase composition on the developing dynamic recrystallization in various aluminum alloys. The purpose of this work is to study the effect of phase composition on the DRX development in aluminum alloys.

2. Materials and method

In present work, the following aluminum alloys with different chemical composition were investigated. Alloy A1: Al-3Mg (wt.%); alloy A2: Al-4.57Mg-0.35Mn-0.2Sc-0.09Zr (wt.%); alloy A3: Al-5.4Mg-0.52Mn-0.1Zr (wt.%). Alloy A1 was produced by direct chill casting. The obtained ingots were homogenized at 500°C for 4 h, followed by slow cooling in the furnace. Then, billets with cross-section of 20 × 20 mm^2 and a length of 110 mm were subjected to a two-stage ECAP at a temperature of 300°C to a strain of ~ 8 and at a temperature of 200°C to a strain of ~ 4. Alloy A2 was obtained by continuous casting. The ingots of this alloy were subjected to homogenization annealing at 360–380°C for 12 h. For the formation of a fine-grained microstructure, ECAP was carried out at a temperature of 300°C to a strain of ~ 12. The initial microstructure of alloy A2 contained non-coherent Al_6Mn particles and coherent Al_3(Sc, Zr) particles. A3 alloy was obtained by continuous casting. The obtained ingots were subjected to homogenization annealing at 360°C for 6 h. The ECAP of the A3 alloy was carried out at 300°C to a strain of ~ 12. In this alloy, a fine-grained microstructure with non-coherent Al_6Mn particles was formed after ECAP processing.

The structural parameters were studied using the orientation imaging microscopy (OIM) with automated indexing of electron back scattering diffraction (EBSD) patterns using a Quanta 600 FEG scanning electron microscope (SEM). For the microstructure examination, the EBSD maps with a size of 100 × 100 μm^2 and a step of 0.2 μm were obtained. Due to the experimental error of the EBSD method, all low-angle boundaries with a misorientation of less than 2° were excluded from consideration using the TSL software. The misorientation of 15° was used as a criterion for distinguishing low- and high-angle boundaries (LAB and HAB, respectively).

3. Results and discussion

3.1. Initial microstructure

The initial microstructures of the studied alloys before ECAP represent a coarse-grained (CG) state. The initial microstructure of the alloy A1 consists of large grains with an average size of about 40 μm. No evidence of secondary phase particles in the alloy A1 in CG state is observed. The microstructure of alloy A2 consists of coarse elongated grains with a transverse size of ~ 20 μm and Al_3(Sc, Zr) particles with coherent interfaces and incoherent Al_6Mn particles distributed homogeneously throughout the matrix. The study of the microstructure of the alloy A3 in the CG state reveals the mean grain size of 22 μm. The uniform distribution of incoherent Al_6Mn particles is also observed inside the grains in this state. The dislocation densities in the studied alloys in CG state are approximately 10^{13} m^{-2}. Some parameters of the microstructures of studied alloys are presented in Table 1.

Table 1. Microstructure and mechanical properties of aluminum alloys.

Alloys	A1		A2		A3	
State	CG	FG	CG	FG	CG	FG
Grane size, μm	40	1.4 and 20	20	1.1	22	1.5
Fraction HAB's, %	84	48	35	78	93	62
Average Misorientation Angle, °	21	23	18	33	12	27
Microhardness, HV	55	95	97	129	94	128

3.2. Microstructure after ECAP

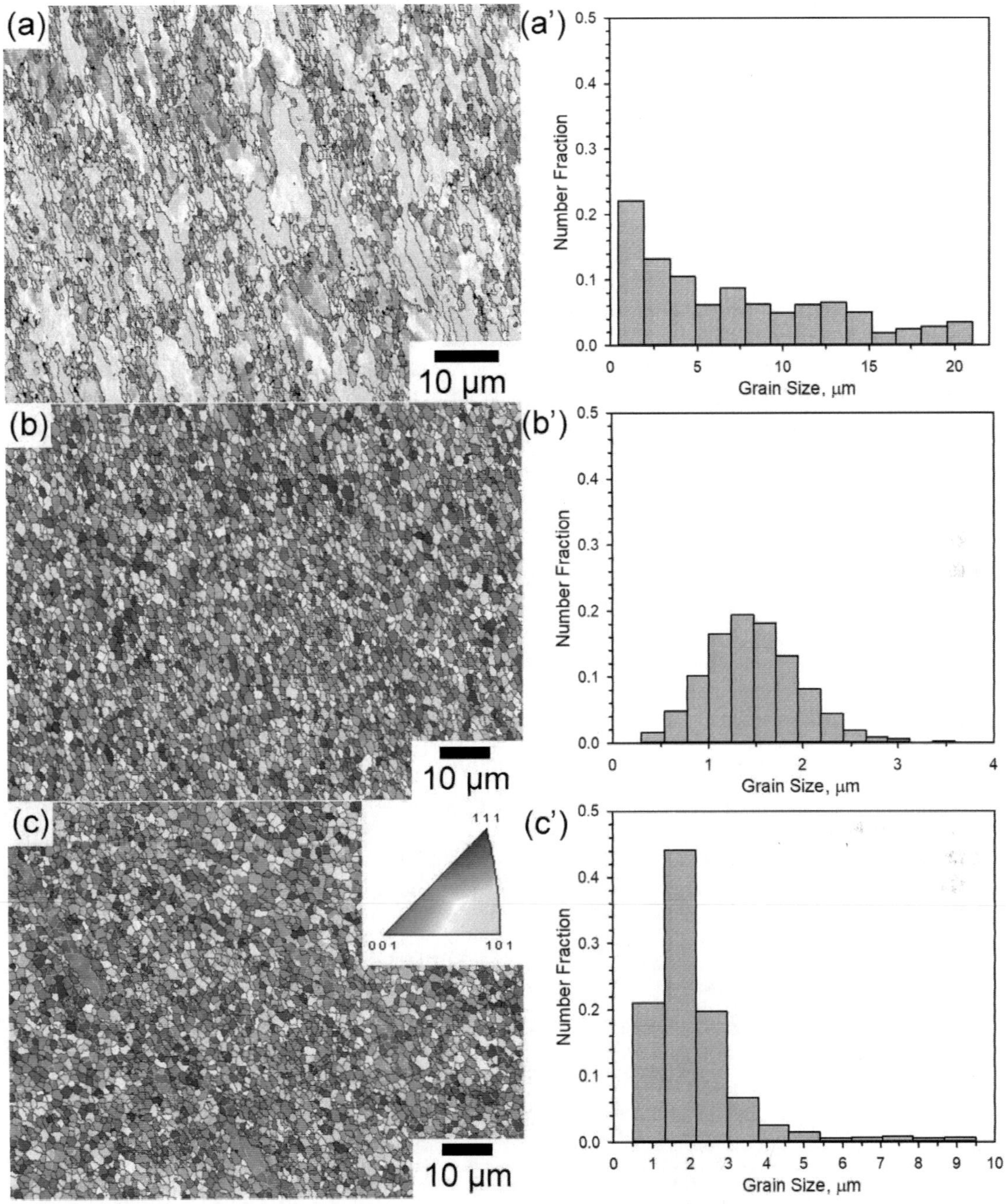

Figure 1. EBSD misorientation maps (a-c) and grain size distributions (a'-c') of the alloys A1 (a, a'), A2 (b,b') and A3 (c,c'). On the EBSD maps, the low-angle and high-angle boundaries are indicated by white and black lines, respectively.

A typical recrystallized microstructure, which develops during ECAP at 300°C, is shown in figure 1 and table 1. The bimodal grain size distribution is observed in alloy A1, where the partially-recrystallized microstructure consists in both the large and fine-grains (FG). It can be seen in figure 1a that the coarse remnants of original grains are surrounded by numerous fine recrystallized grains after 12 passes of ECAP. Thus, in the partially recrystallized microstructure shown in figure 1a, so-called microstructure of the necklace develops. The mean size of the coarse and fine grains in this state are 20 and 1.4 μm respectively. For all studied alloys, the fraction of recrystallized grains increases with an increase in the number of ECAP passes. The formation of new recrystallized grains occurs along the initial grain boundaries [11-13]. It is clearly seen in figures 1b and 1c that the homogeneous fine-grained microstructures are completely formed after 12 passes of ECAP in alloys A2 and A3. The average grain sizes in the alloys A2 and A3 are 1.1 and 1.5 μm respectively.

It should be noted that the grain sizes in the alloys A2 and A3 subjected to ECAP, are smaller than in A1, subjected to ECAP at the same temperature [14]. It is known [14] that the several alloying elements lead to substantial solid-solution hardening and slow down the recovery process. Such elements also lead to more pronounced grain refinement during severe plastic deformation. It is known that dissolved zirconium significantly slows down the diffusion processes in the aluminum matrix [15], preventing recovery and inhibiting the migration of grain boundaries. As a result, the size of recrystallized grains is reduced. Thus, in the alloys A2 and A3, the combined effect of pinning pressure from the Al_6Mn particles and dissolved zirconium results in much lower grain sizes as compared to that in the alloy A1 [14].

The fine-grained microstructures of the studied alloys are characterized by large fractions of LAB. In the alloy A1, the largest fraction of LAB and the higher value of the average misorientation angle θav as compared to other studied alloys are observed (figure 2). The average misorientation angles in the alloys A2 and A3 increase by 15° after ECAP processing. Note that a gradual increase in the average misorientation angle in these alloys is associated with a gradual increase in the fraction of the high-angle boundaries during deformation.

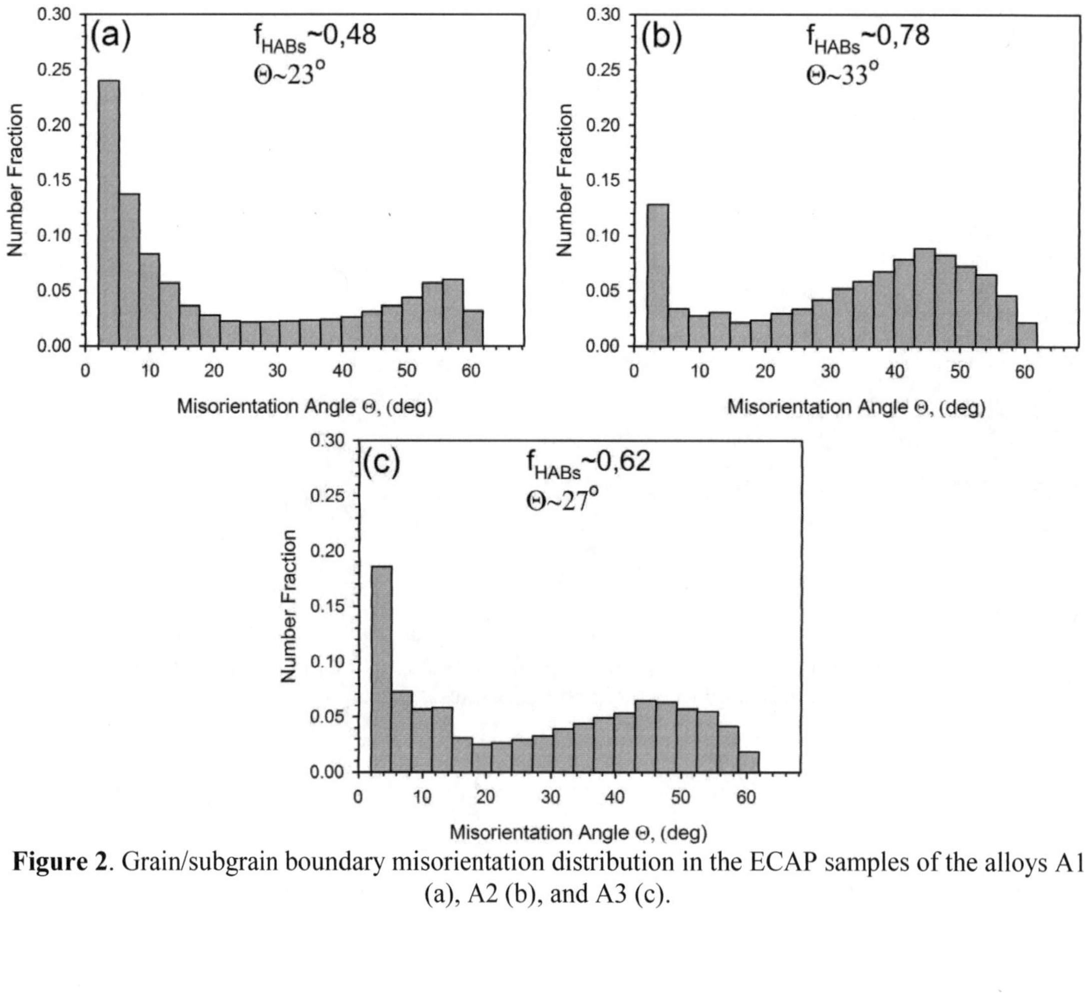

Figure 2. Grain/subgrain boundary misorientation distribution in the ECAP samples of the alloys A1 (a), A2 (b), and A3 (c).

An average lattice dislocation density was evaluated from the EBSD data from the lattice curvature using the kernel average misorientation option [16] via Frank's equation:

$$\theta = 2\sin\frac{\theta}{2} = \frac{Nb}{h} \tag{1}$$

where θ is the misorientation created by a wall consisting of N dislocations of height h, and b is the Burgers vector. In EBSD observations, distance h corresponds to the step size of the scanning. The dislocation density ρ is then given by the ratio of the dislocation number per surface area, Eq. (2):

$$\rho = \frac{N}{S} \tag{2}$$

where the surface area of the hexagon S (scanning geometry) is $S = \sqrt{3}h^2/2$. It follows that the dislocation density can be estimated from the following relationship:

$$\rho = \frac{2\theta}{\sqrt{3}hb} \tag{3}$$

The values used for dislocation density calculations are given in table 2.

Table 2. Values of physical parameters in Eq. (3)

Alloys	h, μm	θ/h, m^{-1}	b, m	ρ, m^{-2}
A1	0.2	1.15×10^5	2.86×10^{-9}	2×10^{14}
A2	0.2	1.65×10^5	2.86×10^{-9}	1.5×10^{14}
A3	0.2	1.35×10^5	2.86×10^{-9}	1.4×10^{14}

For the alloy A1 the ECAP processing leads to 70% increase in hardness as compared to the CG state. The increase in hardness in alloys A2 and A3 is less pronounced. However, the hardness values in these alloys after ECAP is even higher than that in the A1 alloy, which may be due to the influence of Al_6Mn particles [17].

4. Summary and conclusions

In the present work, the influence of the phase composition on the DRX process in three aluminum alloys was investigated. The main results are as follows:

1. ECAP processing leads to the partially recrystallized microstructure in alloy A1. Such microstructure can be referred to as necklace microstructure. In the A2 and A3 alloys, fully recrystallized fine-grained microstructures are observed. The completion of DRX process in these alloys may be associated with the presence of $Al_3(Sc,Zr)$ and Al_6Mn particles.
2. The higher fraction of LABs and the smaller value of the average misorientation angle are observed in alloy A1 as compared to alloys A2 and A3. ECAP leads to an increase of the average misorientation angle by 15° in the A2 and A3 alloys.
3. The hardness of the alloys A2 and A3 is higher by approximately 30% than that in the A1 alloy. This may be due to the presence of incoherent Al_6Mn particles and coherent $Al_3(Sc,Zr)$ particles in the alloys A2 and A3.

Acknowledgement

The study was financially supported by the Russian Science Foundation, Belgorod State University project No. 17-72-20239. The main results were obtained by using the equipment of the Joint Research Center, «Technology and Materials», Belgorod State University.

References

[1] Markushev M, Vinogradov A 2006 *Severe Plastic Deformation: Towards Bulk Production of Nanostructured Materials* (Burnhanettin A. USA: Nova Science Publishers) P. 233–247.
[2] Dao V, Lu L, Azaro R et al. 2007 *Acta Materialia* vol 55 (Amsterdam: Elsevier) p 4041.
[3] Sitdikov O, Avtokratova E, Sakai T, Tsuzaki K 2013 *Metallurgical and Materials Transactions A* vol. 44 (Berlin, Germany: Springer) p 1087.
[4] Sakai T, Belyakov A, Kaibyshev R, Miura H, Jonas J.J 2014 Progress in Materials Science vol 60 (Amsterdam: Elsevier) p 130.
[5] Humphreys F, Hatherly M 2004 *Recrystallization and Related Annealing Phenomena* (Elsevier Science: New York, NY, USA).
[6] Huang K, Logé R 2016 Materials & Design vol 111 (Amsterdam: Elsevier) p 548.
[7] Zhang D, Yang X, Fang Z, Xiao Z, Yang Y, Sakai T 2016 *Materials Science and Engineering: A* vol. 675 (Amsterdam: Elsevier) p 128.
[8] Zhaoa P, Wanga Y, Niezgodaab S R 2018 International Journal of Plasticity vol 100 (Elsevier) p 52.
[9] Loginov Yu, Antonenko L 2010 *Zagotovitel'nyye proizvodstva v mashinostroyenii* [Preparation production in mechanical engineering] vol. 7 (Moscow: Innovative Engineering Press) p 14
[10] Trifonov V, Bobruk E 2004 *The Physics Of Metals And Metallography* vol 97 (Road Town: Pleiades Publishing, Ltd) p 188.
[11] Sakai T, Jonas J 1984 *Acta Metallurgica* vol 32 (Elsevier Science Publishing Company) p 189.
[12] Humphreys F, Hatherly M 2004 *Recrystallization and Related Annealing Phenomena* (Elsevier Science: New York, NY, USA).
[13] Dudova N, Belyakov A, Sakai T, Kaibyshev R 2010 *Acta Materialia* vol 58 (Amsterdam: Elsevier) p 3624.
[14] Sitdikov O, Sakai T, Avtokratova E, Kaibyshev R, Tsuzaki K, Watanabe Y 2008 *Acta Materialia* vol 56 (Amsterdam: Elsevier) p 821.
[15] Doherty R, Hughes D, Humphreys F. Jonas J, Jensen D, Kassner M, King W, T.R. McNelley, McQueen H, Rollett A 1997 *Materials Science and Engineering: A* vol. 238 (Amsterdam: Elsevier) p 219.
[16] Zhilyaeva A, Shakhova I, Morozova A, Belyakov A, Kaibyshev R 2016 *Materials Science and Engineering: A* vol. 654 (Amsterdam: Elsevier) p 131.
[17] Sato Y, Kokawa H, Enmoto M, Jogan S 1999 *Metallurgical and Materials Transactions A* vol. 30 (Berlin, Germany: Springer) p 2429.

Consideration of recrystallization modeling of γ/γ′-strengthened Ni-based superalloys at sub-solvus temperatures

Takashi Nishimoto[1]*, Takuma Okajima[1], Qiaofu Zhang[2], Jiadong Gong[2] and Greg Olson[2]

[1] Corporate R&D Center, Daido Steel Co., Ltd., 2-30 Daido-cho, Minami-ku, Nagoya, Aichi, Japan

[2] QuesTek Innovations, 1820 Ridge Avenue; Evanston, IL, 60201, USA

* t-nishimoto@ac.daido.co.jp

Abstract. New Ni-based superalloys and methods for controlling their microstructures have recently received much investigation to further improve their high-temperature properties, sometimes via control of precipitates such as γ′ (Ni_3(Al,Ti,Nb,Ta)) and γ″ (Ni_3Nb). In Ni-based superalloys, fine grains favor deformability and mechanical properties; therefore, control of grain size, achieved by recrystallization, is also important. Some new studies have reported the pinning effects of coarse γ′. However, few studies have investigated modeling methods for recrystallization behaviors. Difficulties in grain size and phase distribution control have hindered fabrication by open-die forging utilizing the pinning effect of the γ′ phase. Therefore, this study examines the recrystallization behaviors of Ni-based superalloys containing coarse primary γ′ precipitates and considers applicable modeling methods. The recrystallization behaviors at sub-solvus temperatures are investigated in a new Ni-based superalloy with abundant coarse γ′ precipitates. Grain growth is limited by the pinning effect of coarse γ′ and the recrystallized grain size corresponds to increasing applied strain. Dynamic and static recrystallization is promoted by high strains and high temperatures. The kinetics of the recrystallized area fraction and recrystallization grain growth influenced by the pinning effect of coarse γ′ is modeled by Zurob's method.

1. Background

Ni-based superalloys have excellent high-temperature strengths. Recently, in order to further improve their high-temperature mechanical properties, the abundance of intermetallic compounds such as γ′ (Ni_3(Ti,Al,Nb,Ta)) and γ″ (Ni_3Nb) has been increased. However, control of grain size and precipitation morphology is also important in superalloys. Generally, Ni-based superalloys require fine grain structures to attain good hot workability [1] and high-temperature mechanical properties. Creep properties, which are of great importance, are enhanced by coarse grains; however, other properties such as fatigue properties, yield stress, and tensile stress are improved by fine grains. Therefore, obtaining structures with grains that promote both types of properties is necessary. Ni-based superalloys comprise only γ phases; refinement by phase transformation cannot be applied and recrystallization is the only method available for grain refinement. Therefore, microstructural control is achieved by hot rolling and hot forging. These refinement methods have received significant study. Grain refinement can be achieved by the pinning effects of precipitations such as carbides [2] and spheroidized δ phases [3]. A forging method applying the pinning effect of coarse primary γ′ was also reported recently [4-6]. Studies

have suggested that the recrystallization behaviors of precipitated coarse primary γ′ are very complex. The mechanism of recrystallization behavior has been extensively investigated.
Daido Steel Co. has developed the Daido Numerical Engineering System (DAINUS) to optimize the manufacturing process [7]. In previous studies, the manufacturing processes for Ni-based superalloys 706 and 718 have been optimized by applying DAINUS [8]. However, methods for modeling recrystallization behaviors with coarse primary γ′ have received little attention. Therefore, DAINUS cannot be directly used for microstructural predictions regarding coarse γ′. It is therefore difficult to optimize the manufacturing process of γ′-strengthened Ni-based superalloys with abundant γ′ precipitates. In this study, the recrystallization behaviors with precipitated coarse primary γ′ are investigated and modeling methods for this behavior are considered for microstructural predictions using DAINUS.

2. Experimental procedure

2.1. Materials and initial heat treatment

In order to investigate the recrystallization behaviors at sub-solvus temperatures, a new γ′-strengthened Ni-based superalloy Alloy A was used in this study. The chemical composition of Alloy A is shown in Table 1. The γ′-solvus temperature of Alloy A is measured by differential thermal analysis (DTA) as 1078 °C. Alloy A was melted by vacuum induction melting–electric slag remelting–vacuum arc remelting. Afterward, homogenization heat treatment and forging at super-solvus temperatures were performed. In order to precipitate coarse primary γ′, over-aging heat treatment was applied after forging. The average grain size after heat treatment is approximately 200 μm, with petal-shaped coarse primary γ′ precipitates. The average size of γ′ precipitates is approximately 1 μm.

Table 1. Chemical composition of Alloy A (mass %).

	Ni	Cr	Co	Mo	W	Al	Ti	Nb	C
Alloy A	Bal.	16.0	20.0	3.0	2.5	3.3	1.8	2.7	0.01

2.2. Experimental condition of recrystallization behavior

An investigation of the recrystallization behavior was performed by Rastegaev compression testing [9] and reheating. Figure. 1 shows the experimental heat pattern; the detailed conditions of the forging and reheating tests are shown in Tables 2 and 3, respectively. All experimental temperatures were below the γ′-solvus temperature of 1078 °C to check the recrystallization behavior of the precipitated coarse primary γ′. All specimens were cooled by water to freeze the microstructures. To observe the microstructures and coarse primary γ′, all specimens were etched observed by optical micrography and scanning electron microscopy (SEM). In order to observe optical micrograph, cupric chloride solution (50ml ethanol + 50ml HCl + 5g cupric chloride) is used. To observe γ' phase, 10% nitric acid is used and electrolyzed at 0.6mA for 2seconds. To analyze the details of the recrystallization behavior, electron backscatter diffraction (EBSD) was used.

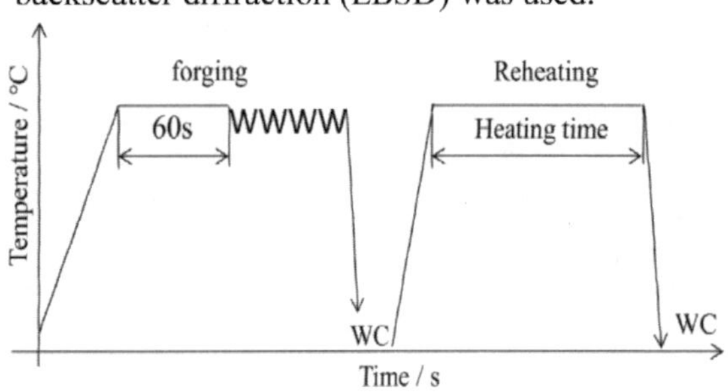

Figure 1. Heating applied for compression and reheating tests. WC: water cooling.

Table 2. Experimental Condition of compression test

Strain	0.2, 0.5, 0.8
Strain rate	0.5 s^{-1}
Forging temperature	1020, 1040 °C

Table 3. Experimental Condition of reheating test

Holding time	10800–72000 s
Heating temperature (Compression at 1020 °C)	1020, 1040, 1060 °C
Heating temperature (Compression at 1040 °C)	1040 °C

IOP Conf. Series: Journal of Physics: Conf. Series **1270** (2019) 012051 doi:10.1088/1742-6596/1270/1/012051

3. Experimental results

3.1. Microstructure as observed by optical micrography

Optical micrography results at the applied strains of 0.2 and 0.8 are shown in Figure 2. All microstructures follow the coarse prior γ grains; new recrystallized grains have originated from the previous grain boundaries. After reheating, the sizes of the prior γ grains are reduced to promote static recrystallization under all heating conditions. Recrystallization is promoted by the addition of high strains and heating at high temperatures for long periods. These results indicate features similar to those of the recrystallization of single-phase γ.After the reheating test, recrystallized grains grow from 10 to 20 μm. Compared with the grain growth behaviors of Waspaloy and Alloy 718 [10], grain growth is more restrained in Alloy A. One of the reasons for this behavior is the pinning effect of coarse primary γ′.

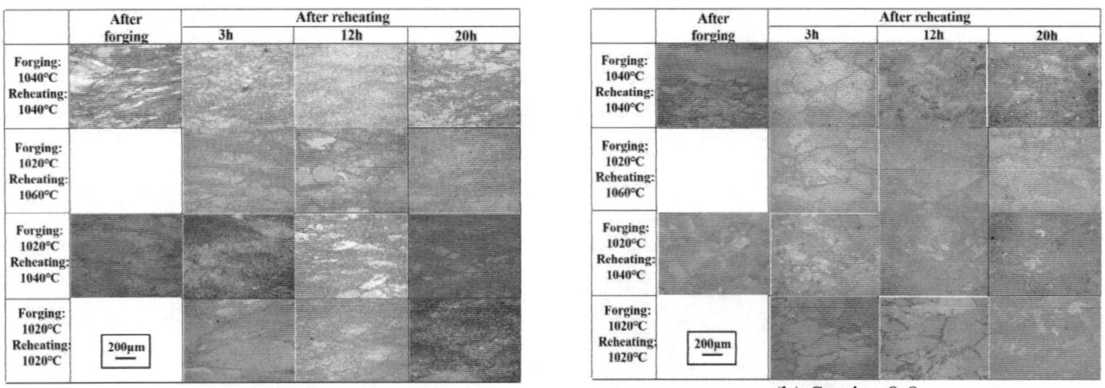

(a) Strain: 0.2 (b) Strain: 0.8

Figure 2. Optical micrograph after the experiments

3.2. Measurement results of coarse primary γ′

From the SEM images, the area fractions and diameters of coarse primary γ′ phases are measured by image analysis. The measured results at the strain level of 0.2 are shown in Figure 3. The area fractions of γ′ are increasing as the experimental temperature decreases. Compared to the results at different forging temperatures, the amount of γ′ differs. The amount of precipitated γ′ increases with increasing forging temperature. However, the amount of primary γ′ remains constant after reheating test. This indicates that the amount of coarse primary γ′ after reheating is less sensitive to the forging temperature. In addition, the reheating time has little influence on the amount of γ′.

Regarding the grain growth behavior of coarse primary γ′, the grain size is increased with increasing reheating temperature and reheating time.

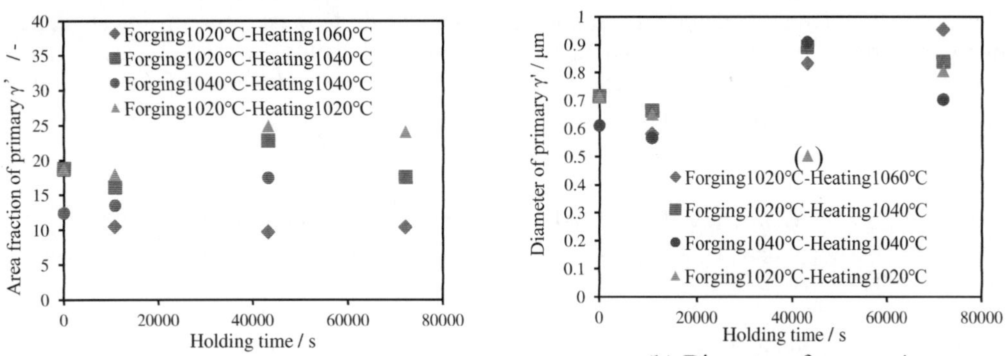

(a) Area fraction of coarse γ′ (b) Diameter of coarse γ′

Figure 3. Precipitation kinetics of gamma' phase at various temperature

3.3. Kernel average misorientation map analyzed by EBSD

In order to investigate the driving force of recrystallization, the dislocation density is measured by kernel average misorientation (KAM) analysis. KAM is a measurement of local orientation that relates to dislocation density. One report indicates that the dislocation density is correspondingly increased with increasing KAM values [11]. Figure 4 shows the KAM map obtained by EBSD at the strain of 0.2, as well as the relationship between temperature and holding time for the KAM values, which are defined by

$$K = \frac{\sum_{i=1}^{6n} \alpha_i}{6n}, \tag{1}$$

where K is the KAM value, n is the number of measurement points, and α_i is the misorientation angle. From the results of optical micrograph and KAM maps, the recrystallized grain area has generally low misorientation, while the high-misorientation areas do not experience recrystallization. Therefore, strain is recovered with grain recrystallization, while non-recrystallized grains restrain the dislocation density. In comparing the results after forging, the KAM value is increased with decreased forging temperatures. This indicates that the driving force for recrystallization corresponds with the forging temperature. After heating, the misorientation is reduced by strain recovery. The KAM values are decreased to a greater extent with higher-temperature and longer-duration heating.

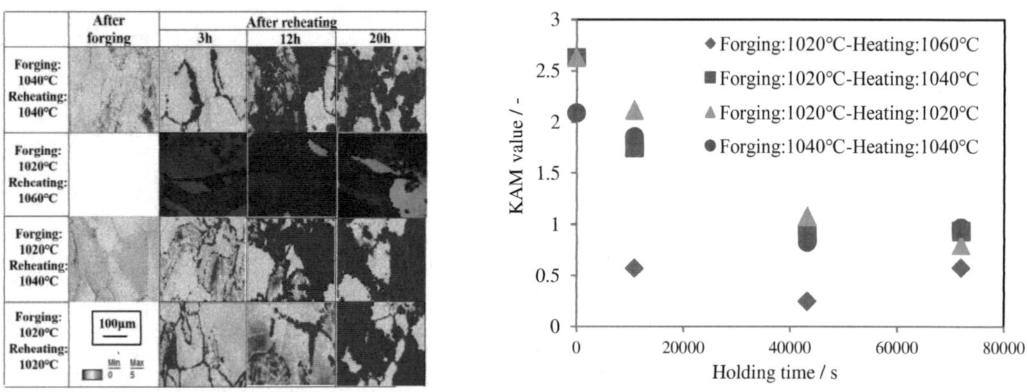

(a)Results of KAM map (b)Results of KAM value

Figure 4. Reslt of KAM maps from EBSD analysis after experiments

4. Consideration of modeling method of recrystallization behavior

4.1. Definition of static recrystallized area fraction

In order to consider modeling methods for this alloy at sub-solvus temperatures, the recrystallized area fraction and recrystallized grain size are measured by the grain size distribution results obtained via EBSD analysis. In this study, modeling methods for dynamic and static recrystallization are considered separately. For dynamic recrystallization, the recrystallized area fraction and grain sizes are measured using the forging test results. The grain size of static recrystallization is obtained using the reheating test results; the area fraction of static recrystallization is defined by:

$$X_{st} = \frac{X_{total} - X_{dyn}}{1 - X_{dyn}} \tag{2}$$

where X_{total} is the total area fraction of static and dynamic recrystallization from the reheating test results, X_{dyn} is the area fraction of dynamic recrystallization, and X_{st} is the area fraction of static recrystallization. X_{st} is calculated by equation (2).

4.2. Modeling of area fraction of dynamic recrystallization

Modeling of the dynamic recrystallization is considered using the X_{dyn} results. In this study, the area fraction of dynamic recrystallization is determined by equation (3):

$$X_{\mathrm{dyn}} = 1 - \exp\left\{-\ln 2 \times \left(\frac{\varepsilon}{\varepsilon_{0.5}}\right)^n\right\} \tag{3}$$

where ε is the strain, $\varepsilon_{0.5}$ is the strain at which X_{dyn} is 0.5, and n is a coefficient. $\varepsilon_{0.5}$ is a function of temperature. A comparison of the measured and modeled results is shown in Figure 5. The modeled results have good precision compared to the measured results.

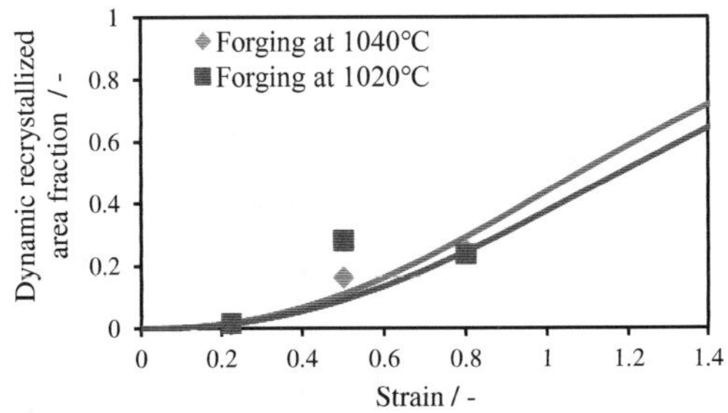

Figure 5. Modeling result of dynamic recrystallized area fraction.

4.3. Modeling of area fraction of static recrystallization

Modeling of static recrystallization is considered using the X_{st} results calculated by equation (2). Generally, the recrystallized area fraction is quantified by an Avrami-type equation:

$$X_{\mathrm{st}} = 1 - \exp\left\{-\ln 2 \times \left(\frac{t}{t_{0.5}}\right)^n\right\} \tag{4}$$

where t is the holding time, $t_{0.5}$ is the time when X_{st} is 0.5, and n is a coefficient. However, in this case, it is difficult to apply because it does not consider the effect of driving force and pinning force. Therefore, the model reported by Zurob et al., which does consider these effects, is applied in this study [12]:

$$X_{\mathrm{st}} = 1 - \exp(-X_{\mathrm{ext}}) \tag{5}$$

$$X_{\mathrm{ext}} = \frac{A_0 \exp\left(\frac{Q}{RT}\right)}{RT} \int_0^t \left[B(\varepsilon) - 1.2\frac{\gamma_{gb} F_\gamma^{2/3}}{r}\right] dt \tag{6}$$

where X_{ext} is the extended recrystallized fraction, Q is the activation energy, R is the gas coefficient, T is heating temperature, $B(\varepsilon)$ is the driving force of recrystallization, γ_{gb} is the grain-boundary energy of γ/γ', r is the radius of coarse primary γ', and F_γ is the area fraction of γ'. From the KAM map obtained after forging, the KAM increases with low-temperature forging, indicating that the driving force depends on the forging temperature. However, the gap in X_{st} for forging at different temperatures is slight. Therefore, $B(\varepsilon)$ is a function of strain in this study. F_v and r are obtained from the measurement results. The modeling results of static recrystallization are shown in Figure 6, which shows high strain. The precision of the modeling result increases with increasing strain.

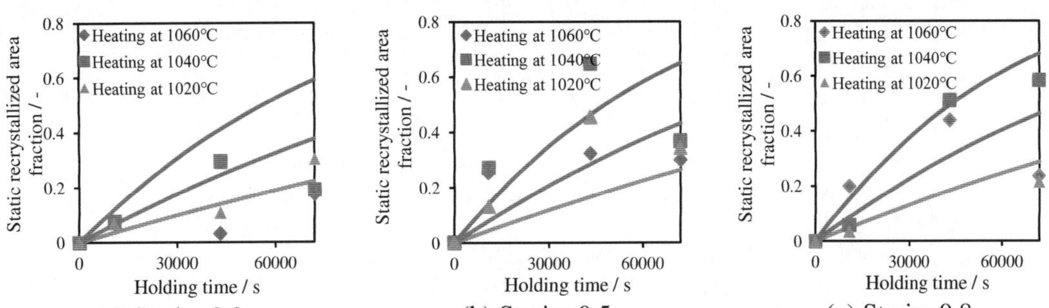

Figure 6. Static recrystallization modeling result of Alloy A

4.4. Modeling results for integrated dynamic and static recrystallization

The modeling results for integrated dynamic and static recrystallization are shown in Figure 7. The modeling results fit very well with the experimental results, especially at high-strain conditions. The pinning force for coarse primary γ′ and the driving force of recrystallization fluctuate continuously during forging and reheating. Therefore, microstructural prediction via the applied Avrami-type equation is difficult. A new modeling method for predicting the recrystallized area fraction is thus established by applying the modified model developed in this study.

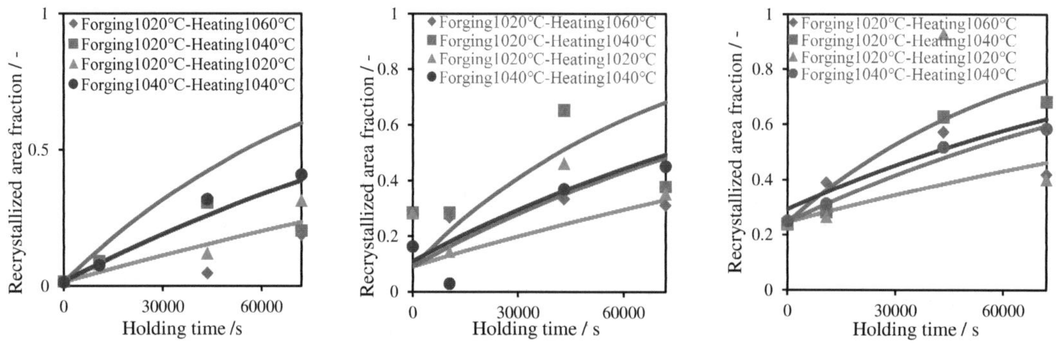

(a) Strain: 0.2 (b) Strain: 0.5 (c) Strain: 0.8

Figure 7. Recrystallization modeling result of Alloy A

5. Conclusion

(1) Recrystallization occurred in Alloy A from the grain boundaries of prior γ with coarse primary γ′. In addition, recrystallization is promoted by heating at high temperature, applying high strain, and maintaining heating for a long duration. Grain growth is restricted by the pinning effect of coarse γ′.

(2) The area fraction of coarse primary γ′ is changed by the heating temperature; the grain size of γ′ is increased with heating. The KAM value correspondingly increases with increases in the forging temperature. These trends indicate that the pinning force and driving force of recrystallization fluctuate continuously during forging and reheating.

(3) The microstructure is difficult to predict by applying Avrami-type equations because they cannot consider the fluctuation of these effects. The proposed modified model may accommodate these effects and shows good microstructural prediction for Alloy A with coarse primary γ′.

References

[1] Bi Z, Lv X and Zhang J 2014 *EUROSUPERALLOYS 2014 – 2nd European Symp. on Superalloys and their Applications* **14** 07002

[2] Ruan J J, Ueshima N and Oikawa K 2018 *J. Alloys Compd* **737** 83–91

[3] Masanuga A, Furuse T, Yoshida H, Okajima T, Hatta T and Isogawa S 2006 *IFM2006: 16th International Forgemasters Meeting* **16** p. 571

[4] Collins D M, Conduit B D, Stone H J, Hardy M C, Conduit G J and Mitchell R J 2013 *Acta Mater* **61** 3378.

[5] Charpagne M A, Billot T, Franchet J M and Bozzolo N (2016) *Superalloy2016 Proc.* p. 417.

[6] Thebault J, Solas D, Rey C, Baudin T, Fandeur O and Clavel M (2008) *Superalloys 2008 Proc.* p. 985

[7] Yoshida H, Hatta T, Hironaka T, Isogawa S and Sekiguchi H 2007 *NUMIFORM Proc.* p. 987

[8] Hatta T, Hironaka T, Yoshida H, Isogaiwa S and Sekiguchi H 2008 *Int. Conf. IFM2008* p. 472.

[9] P. Christiansen,P. A. F. Martins and N. Bay 2016 *Experimental Mechanism* 56 1271-1279

[10] Okajima T, Sakurai S, Fujiwara M 2014 *Int. Conf. IFM2014* p. 198.

[11] Umezaki S, Murata Y, Nomura K and Kubushiro K 2014, *J. Jpn. Inst. Met. Mater.* **78** 218–24

[12] Zurob H S, Hutchinson C R, Brechet Y and Purdy G 2002 *Acta Mater.* **50** 3075–92

IOP Conf. Series: Journal of Physics: Conf. Series **1270** (2019) 012052 doi:10.1088/1742-6596/1270/1/012052

Effect of hot deformation on grain refinement in a Re containing nickel-based superalloy

Sh Kh Mukhtarov[1], V M Imayev[1], A A Ganeev[1], R V Shakhov[1] and A V Logunov[2]

[1] Institute for Metals Superplasticity Problems, Russian Academy of Sciences 39 Stepan Khalturin str, Ufa 450001, Russia
[2] PJSC "UEC-Saturn", 163 Lenin Ave, Rybinsk 152903, Russia

E-mail: shamil@anrb.ru

Abstract. Hot deformation experiments have been performed for a novel polycrystalline superalloy with a nominal composition Ni-12.5(Al,Ti,Nb,Ta)-37(Cr,Co,W,Mo,Re)-0.17(C,La,Y,Ce,B) (in wt. %) intended for disc applications in gas turbine engines. The as-cast superalloy was characterized by a higher amount of the γ' phase, a coarse γ grain size and a high level of dendritic segregation. The as-cast superalloy was subjected to long-term annealing to homogenize and heterogenize the material prior to hot deformation. The hot working experiments were performed in a different manner. First, isothermal compression experiments at temperatures below and above the γ' solvus temperature were performed. These experiments allowed us to determine the temperature-strain rate conditions providing improved workability. Second, unidirectional forging experiments were performed under quasi-isothermal conditions using a specially designed can made of a stainless steel. The use of the optimal temperature-strain conditions and the can provided reasonable hot workability and delocalized occurrence of recrystallization processes during hot forging. EBSD analysis of the forged workpieces revealed that the grain refinement resulted from continuous dynamic recrystallization. For the first time hot deformation processing was successfully developed for the rhenium containing ingot-metallurgy nickel-based superalloy.

1. Introduction

The development of advanced nickel-based superalloys for gas turbine engines follows towards increasing the content of refractory alloying elements to increase the high-temperature capability of superalloys [1,2]. For example, single-crystal superalloys of the second and third generations contain about 3 and 5-6 wt.% of Re, respectively [3-7]. No doubt that heavy alloying with refractory metals makes more difficult hot deformation processing of as-cast ingots [8]. The fact is that heavy alloying can lead not only to a higher content of the γ' phase but also to hardening of the matrix phase. At the same time, the use of hot deformation of as-cast ingots is mandatory requirement for high-loaded parts like discs in gas turbine engines. The hot working processing was successfully developed for different wrought superalloys with a volume fraction of the γ' phase around 50% [9]. The general approach includes the use of annealing followed by slow cooling before hot working, which leads to coagulation of the γ' phase and a partial loss of coherence between γ and γ' phases. Subsequent hot deformation is typically performed at subsolvus temperatures under isothermal or near isothermal conditions in order to impede precipitating the γ' phase during hot working. In the present work, the same approach was applied with respect to heavily alloyed rhenium containing superalloy with a higher content of the γ' phase. The purpose of the present work was to analyse the microstructure evolution during annealing and following hot working in different conditions. Finally, the work is aimed to develop robust hot working processing, which might be used for manufacturing of large-scale superalloy billets.

2. Experimental

The chemical composition of the nickel-based superalloy used in this study was Ni-12.5(Al,Ti,Nb,Ta)-37(Cr,Co,W,Mo,Re)-0.17(C,La,Y,Ce,B) (in wt. %). The superalloy contains 2 wt. % of rhenium and is similar to some of the second generation single-crystal superalloys. The initial as-cast ingot was produced by induction melting. To improve chemical homogeneity, small pieces were cut out of the initial ingot and remelted in a laboratory arc-melting furnace under argon atmosphere. The remelted ingots had a size of 45 mm in a diameter and 15 mm in a height. The solvus temperature T_s was measured via water quenching experiments from temperatures around T_s. The minimal temperature corresponding to a full dissolution of the γ' phase was assumed as the γ' solvus temperature. This temperature was found to be T_s=1220±5°C. The remelted superalloy ingots were subjected to long-term homogenization and heterogenization annealing in the temperature range of (T_s-120)÷(T_s+20). The heterogenization annealing was followed by slow cooling with a rate of 25°C/min. The hot compression tests were carried out using a Schenck machine and a furnace with operating temperature up to 1250°C. Compression tests of samples Ø6 mm × 8 mm were carried out under isothermal conditions in the temperature range of 1100-1225°C with an initial strain rate of $\acute{\varepsilon}=10^{-3}$ s^{-1} [10]. The tests were conducted in air. At the temperature providing highest workability, the samples were compressed to an engineering strain of ε=25, 50 and 70% with an initial strain rate of $\acute{\varepsilon}=10^{-3}$ and to a strain of ε=70% with the strain rate of $\acute{\varepsilon}=10^{-3}$-$10^{-2}$ s^{-1}. All compressed samples were removed from the furnace during not more than 30 sec and then cooled in air. The results of compression tests were further used for development of canned unidirectional forging under quasi-isothermal conditions. The canned hot forging was fulfilled in two steps with intermediate recrystallization annealing using a specially designed thick-walled can made of a stainless steel. To do it, the canned workpieces were heated up to 1175°C and forged with the strain rate of $\acute{\varepsilon}\approx10^{-2}$ s^{-1} to a total engineering strain of 75% using the die tool preheated to T=930°C. Intermediate recrystallization annealing was performed at 1150°C (2 h).The forged workpieces were removed from the furnace during not more than 30 sec and then cooled in air and aged. As a result, sound forgings were obtained.

The microstructure examination was carried out in the cross sections of the deformed samples and workpieces using scanning electron microscopy (SEM) in backscattering electron (BSE) mode and transmission electron microscopy (TEM). The volume fractions of the γ' phase and the carbides were quantitatively evaluated by the systematic point count method using mechanically polished samples and BSE images at high magnifications. Electron backscatter diffraction (EBSD) analysis was performed with a scan-step size of 1 μm from central parts of the deformed samples and workpieces. The grain boundaries with misorientation more than 15 were assumed as high-angle ones. The grain boundaries having misorientation angle less than 2° were excluded from consideration.

3. Results and discussion

3.1. Microstructure examination of the superalloy in as-cast condition and after annealing

Figure 1 represents BSE images of the superalloy in the as-cast condition. It was characterized by a coarse grained structure with a pronounced dendritic segregation. Non-equilibrium eutectic colonies typical of as-cast heavily alloyed nickel-based superalloys were detected (figure 1 a). In the matrix γ phase there were about 68 vol.% of the γ' phase and about 2.5 vol.% of carbides. The γ' phase consisted of fine precipitates with a size of 0.1-0.25 μm (figure 1 b); there were also coarse coagulated γ' particles with a size of 1-2 μm. In order to improve the hot workability, the as-cast superalloy was subjected to long-term homogenization and heterogenization annealing [9]. One can see that the annealing led to dissolution of nonequilibrium eutectics and coagulation of the γ′ phase. TEM observations showed that many γ/γ′ boundaries contained misfit dislocations (figure 2). This suggests that the annealing also resulted in a partial loss of coherence between the γ′ phase and the γ matrix.

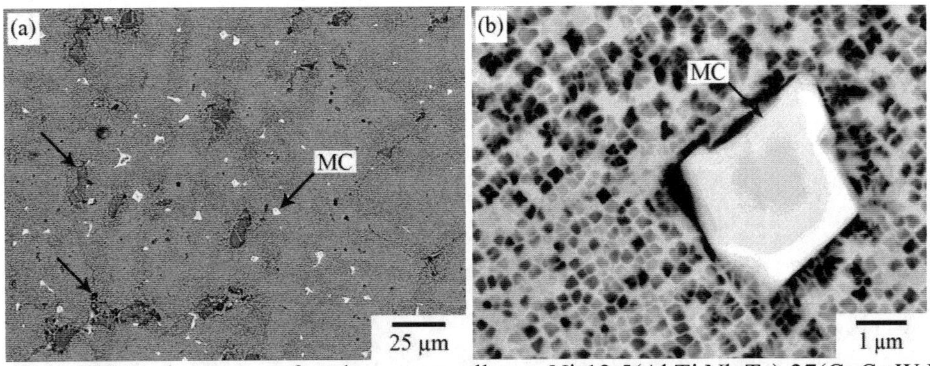

Figure 1. BSE images of the superalloy Ni-12.5(Al,Ti,Nb,Ta)-37(Cr,Co,W,Mo,Re)-0.17(C,La,Y,Ce,B) in as-cast condition at different magnifications: (a) arrows show nonequilibrium eutectic colonies; (a, b) fine γ′ precipitates were formed during cooling of ingot, white carbides (MC) are arrowed.

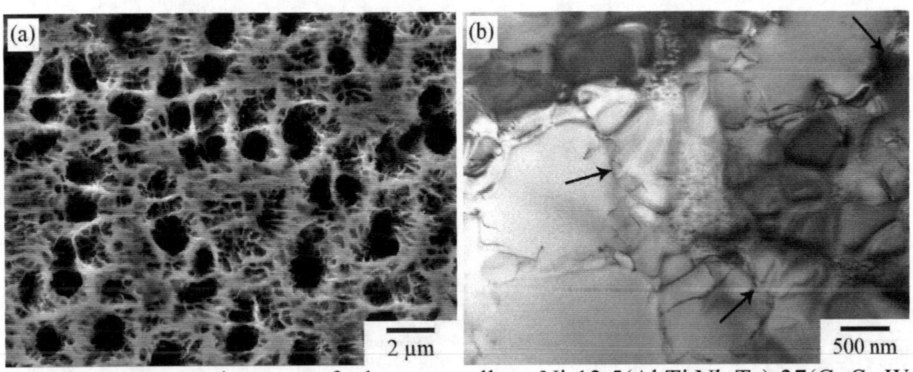

Figure 2. Microstructure images of the superalloy Ni-12.5(Al,Ti,Nb,Ta)-37(Cr,Co,W,Mo,Re)-0.17(C,La,Y,Ce,B) in the cast condition subjected to long-term annealing: (a) coarse γ′ precipitates resulted from coagulation of the γ′ phase during annealing and slow furnace cooling (BSE); (b) arrows show incoherent γ/γ′ boundaries (TEM).

3.2. *Effect of hot working on the microstructure*

The compression tests in the cast condition and after long-term annealing before compression showed that the annealing led to appreciable decreasing the flow stresses during compression at subsolvus temperatures (1100-1175°C) and improving the hot workability. Evidently, it was a result of the γ′ phase coagulation and the partial loss of coherence between the γ and γ′ phases. Figure 3 shows BSE images obtained from central parts of the compressed samples of the superalloy. One can see that recrystallization processes occurred extensively leading to refinement of the microstructure after straining at subsolvus temperatures. The average size of recrystallized γ grains increased from $d \approx 3$ μm to $d \approx 20$ μm with increasing the test temperature from 1100 to 1200°C (figures 3 a-c). After compression at 1225°C the γ′ phase was completely dissolved and dynamic grain growth during deformation was limited only by carbides (figure 3 d). The average size of recrystallized γ grains after compression at 1225°C was around 100 μm. The largest volume of refined recrystallized γ grains was found to be after compression at temperature 1175°C (figure 3b) that corresponded to the highest hot workability of the superalloy. Therefore, this deformation temperature was further chosen as the optimal one. Microstructure examination of samples compressed at 1175°C with the strain rate ranging from 10^{-3} to 10^{-2} s^{-1} showed that the volume fraction of recrystallized γ grains decreased with increasing the strain rate. After compression in the optimal temperature-strain rate conditions (T=1175°C, $\dot{\varepsilon}=10^{-3}$ s^{-1}) the average size of recrystallized γ grains was around 10 μm. In the microstructure there were both coarse γ′ particles with a size of 1-5 μm, which were not dissolved during compression, and dispersed γ′ precipitates with a size of 0.15-0.25 μm (figure 4 a). TEM observations suggest that coherent γ′ phase particles precipitated in the interior of γ grains during cooling of the compressed sample (figure 4 b).

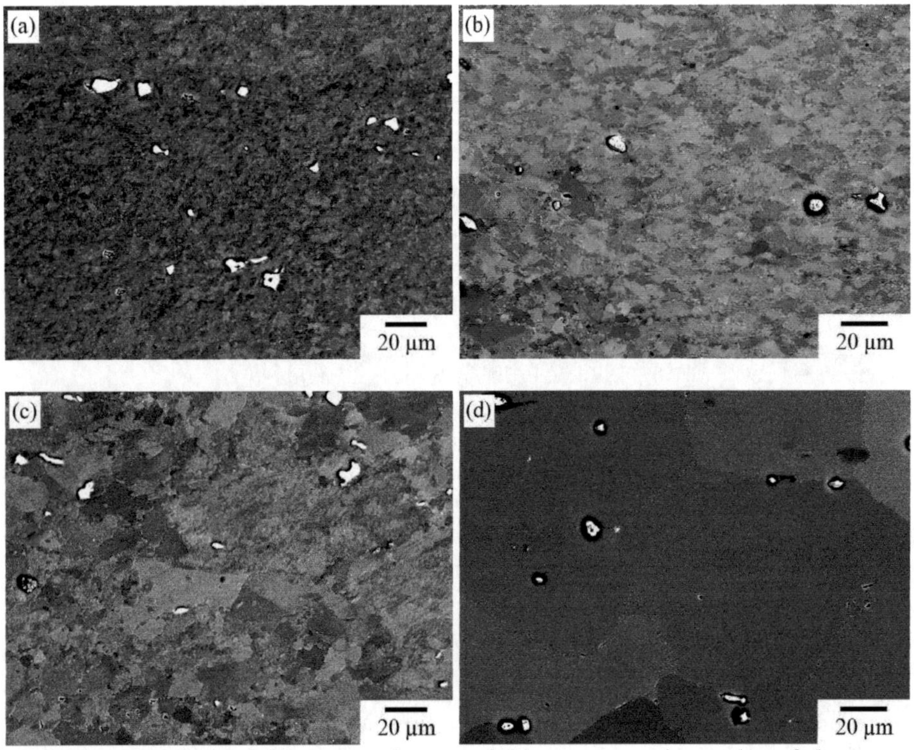

Figure 3. BSE images obtained from central parts of the compressed samples of the superalloy Ni-12.5(Al,Ti,Nb,Ta)-37(Cr,Co,W,Mo,Re)-0.17(C,La,Y,Ce,B) ($\dot{\varepsilon}=10^{-3}$ s^{-1}, ε=70%: (a) 1150°C; (b) 1175°C; (c) 1200°C; (d) 1225°C. The samples were subjected to long-term annealing prior to compression tests.

EBSD analysis was performed in central parts of samples compressed in the optimal temperature-strain rate conditions to different strain values (ε=25, 50 and 70%) (figure 5). One can see that compression to a strain of 25% led to formation of developed substructure, which consisted of subgrains with a size of 1-10 μm divided by low-angle grain boundaries. Only separate recrystallized γ′ grains were observed after compression to a strain of 25% near the carbides, which promoted formation of recrystallization nuclei (figure 5 a). An increase of the strain value up to 50% led to extensive occurrence of recrystallization processes first of all along the initial γ grain boundaries (figure 5 b). After compression to a strain of 70%, the microstructure was found to be near completely recrystallized (figure 5 c). The size of recrystallized γ grains was in the range of 5-100 μm. Some large grains contained low-angle boundaries. The fraction of high-angle grain boundaries in the selected area (figure 5 c) was defined as 63%. Taking into account a short removal time of the compressed samples from the furnace, one can conclude that the main mechanism responsible for the microstructure refinement during compression at 1175°C was continuous dynamic recrystallization.

Thus, the novel heavily alloyed superalloy showed quite reasonable hot workability at subsolvus temperatures and its recrystallization behaviour was generally similar to that of other nickel-based superalloys with a lower content of the γ' phase.

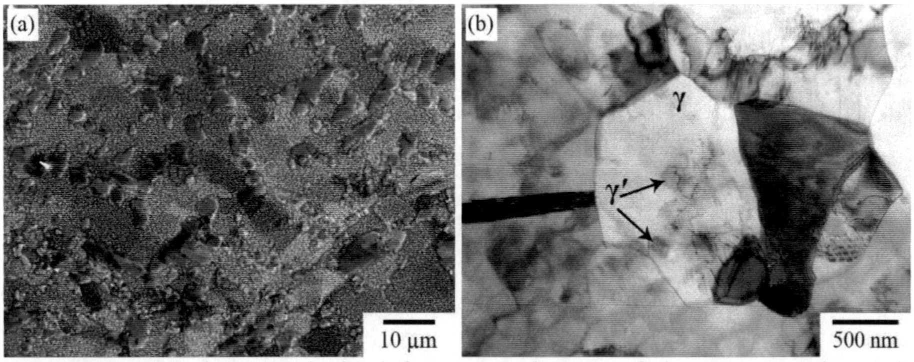

Figure 4. Microstructure images obtained from central parts of the compressed samples of the superalloy Ni-12.5(Al,Ti,Nb,Ta)-37(Cr,Co,W,Mo,Re)-0.17(C,La,Y,Ce,B) subjected to long-term annealing and compression in the optimal temperature strain-rate conditions (T=1175°C, $\dot{\varepsilon}=10^{-3}$ s^{-1}, ε=70%): (a) coarse γ' particles were not dissolved during compression (BSE); (b) fine γ' precipitates were formed in the interior of γ grains during cooling of the compressed sample (TEM).

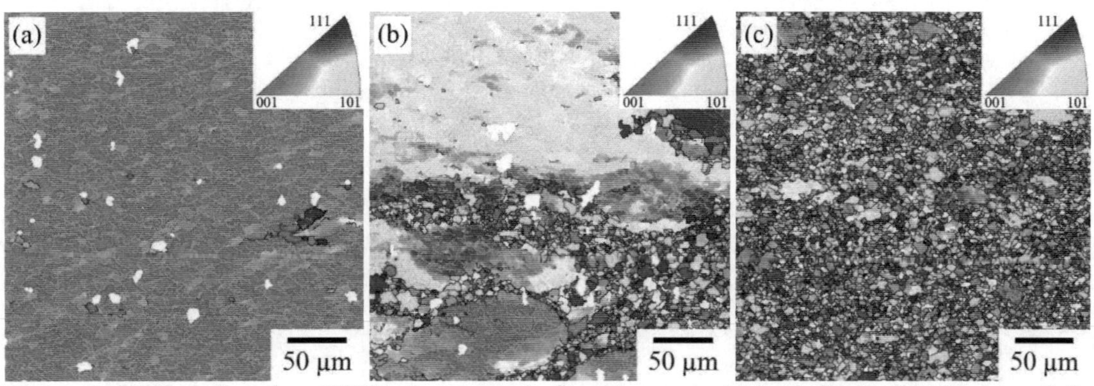

Figure 5. Normal-direction EBSD (inverse-pole-figure) maps obtained from central parts of the compressed samples of the superalloy Ni-12.5(Al,Ti,Nb,Ta)-37(Cr,Co,W,Mo,Re)-0.17(C,La,Y,Ce,B) subjected to long-term annealing and compression in the optimal temperature strain-rate conditions (T=1175°C, $\dot{\varepsilon}=10^{-3}$ s^{-1}): (a) ε=25%; (b) ε=50%; (c) ε=70%. The compression axis is vertical.

To produce large-scale forged semifinished products made of the studied superalloy, canned hot forging under quasi-isothermal conditions was fulfilled in two steps with intermediate recrystallization annealing. In this case, the strain rate was higher than the optimal one to avoid cooling of the workpiece during forging procedure. Sound forgings near free of cracks were produced by this technique.

Figure 6 shows EBSD map obtained from the central part of the forged workpiece of the superalloy subjected to long-term annealing, two-step canned forging with intermediate recrystallization annealing and ageing. In spite of the use of a higher strain rate, near fully recrystallized microstructure with predominantly high-angle grain boundaries was obtained in the forged workpiece. A slightly finer size of recrystallized γ grains was obtained in the workpiece as compared with the sample compressed with $\dot{\varepsilon}=10^{-3}$ s^{-1}. This can be ascribed to a higher strain rate used for canned forging. As in the case of isothermal compression of the small samples, continuous dynamic recrystallization was the main process providing refinement of the microstructure during the forging procedure.

Figure 6. Normal-direction EBSD (inverse-pole-figure) map obtained from the central part of the forged workpiece of the superalloy Ni-12.5(Al,Ti,Nb,Ta)-37(Cr,Co,W,Mo,Re)-0.17(C,La,Y,Ce,B) subjected to long-term annealing, two-step canned forging with intermediate recrystallization annealing and ageing (T=1175°C, $\dot{\varepsilon}=10^{-2}$ s^{-1}, ε=75%). High-angle and low-angle grain boundaries are shown as black and white lines, respectively. The forging direction is vertical.

4. Conclusions

For the first time hot working processing was successfully developed for the rhenium containing ingot-metallurgy nickel-based superalloy Ni-12.5(Al,Ti,Nb,Ta)-37(Cr,Co,W,Mo,Re)-0.17(C,La,Y,Ce,B) (wt.%). It was demonstrated that forged semifinished products can be produced by canned forging at subsolvus temperatures. Microstructure examination of the forged samples and workpieces showed that hot forging led to extensive occurrence of recrystallization processes, which provided formation of the refined microstructure with predominantly high-angle grain boundaries. Continues dynamic recrystallization was found to be the main process responsible for refining the cast structure of the superalloy. The novel superalloy showed quite reasonable hot workability at subsolvus temperatures and its recrystallization behaviour was generally similar to that of other nickel-based superalloys with a lower fraction of the γ' phase.

Acknowledgments

The present work was funded by the Russian Foundation for Basic Research according to the research Project № 18-08-00997-a. The work was performed using the facilities of the shared services center «Structural and Physical-Mechanical Studies of Materials» at the Institute for Metals Superplasticity Problems of Russian Academy of Sciences.

References

[1] Pollock T M 2006 *Journal of propulsion and power* **22** 361
[2] Reed R C 2006 *The superalloys: Fundamentals and Applications* (New York: Cambridge University Press)
[3] Walston W S, O'Hara K S, Ross E W, Pollock T M and Murphy W H 1996 *Proc. of the 8th Int. Symp. Superalloys 1996* (Warrendale, PA: TMS) p 27
[4] Li J R, Zhong Z G, Tang D Z, Liu S Z, Wei P, Wei P Y, Wu Z T, Huang D and Han M 2000 *Proc. of the 9th Int. Symp. Superalloys 2000* (Warrendale, PA: TMS) p 777
[5] Sato A, Harada H, Yeh A-C, Kawagishi K, Kobayashi T, Koizumi Y, Yokokawa T and Zhang J-X. 2008 *Proc. of the 11th Int. Symp. Superalloys 2008* (Warrendale, PA: TMS) p 131
[6] Kawagishi K, Yeh A-C, Yokokawa T, Kobayashi T, Koizumi Y and Harada H 2012 *Proc. of the 12th Int. Symp. Superalloys 2012* (Hoboken, NJ: Wiley) p 189
[7] Mottura A and Reed R C 2014 What is the role of rhenium in single crystal superalloys? *MATEC Web of Conferences* 14(2014)01001
[8] McDevitta E T 2014 Vacuum induction melting and vacuum arc remelting of Co-Al-W-X gamma-prime superalloys *MATEC Web of Conferences* 14(2014)02001
[9] Kaibyshev O A and Utyashev F Z 2005 *Superplasticity: Microstructural Refinement and Superplastic Roll Forming* (Arlington: Futurepast)
[10] Mukhtarov Sh, Utyashev F Z and Shakhov R 2018 *Defect and Diffusion Forum* **385** 424

Kinetics of recrystallization and grain growth in an ultra-fine grained CoCrFeNiMn-type high-entropy alloy

M V Klimova[1], D G Shaysultanov[1], R S Chernichenko[1], V N Sanin[2], S V Zherebtsov[1] and N D Stepanov[1,*]

[1] Laboratory of Bulk Nanostructured Materials, Belgorod National Research University, Belgorod 308015, Russia
[2] Merzhanov Institute of Structural Macrokinetics and Materials Science, Russian Academy of Sciences, Chernogolovka, Moscow, 142432 Russia

Corresponding author: stepanov@bsu.edu.ru

Abstract. A novel high entropy alloy based on the CoCrFeMnNi system with substantial amounts of Al and C was studied. After cold rolling and annealing at 973-1273 K a duplex ultra-fine grained structure composed of the recrystallized fcc grains and $M_{23}C6$ and B2 particles was produced. Analysis of the coarsening behavior of grains and particles growth suggested that kinetics of both was controlled by volume diffusion. The apparent activation energy of structure coarsening during recrystallization was evaluated.

1. Introduction

The so-called high entropy alloys (HEAs), multicomponent alloys of 5 or more principal elements taken in nearly equiatomic concentration (5-35 at.%), have become a very attractive research field in materials science [1,2]. The well-known HEAs family is based on 3d transition metals like Cr, Mn, Fe, Co and Ni [3]. A typical and well-investigated representative of this family is the equiatomic CoCrFeMnNi alloy, also known as the Cantor alloy [4–7]. This alloy has a single disordered face-centered cubic (fcc) structure at temperatures >900°C [6,8–13] and is widely considered as a "model" single phase HEA. In addition, the alloy has attractive mechanical properties; namely very high ductility and fracture toughness at room temperature [5,7], which yet increases even more under cryogenic condition. Nevertheless, the strength of the alloy is quite low [5,11].

Numerous efforts have been undertaken to increase the strength of the CoCrFeMnNi and similar alloys. Precipitation hardening was found to be a particularly effective [14–16]. Compound-forming elements like Al and Ti are often used to produce strengthening precipitates. For example, after proper heat treatment, a CoCrFeNi alloy, containing 4 at.% of Al and 2 at.% of Ti, demonstrates the ultimate tensile strength of ~1100 MPa with elongation of ~40% due to strong precipitation strengthening by $L1_2$ phase particles [14]. Similarly, strong precipitation strengthening can be achieved in the alloys which are doped with carbon thanks to precipitation of fine carbide particles [16,17].

Another approach to increase the strength of the CoCrFeNiMn alloys is to refine the fcc grain size. It is well established that the CoCrFeMnNi alloy can have very high Hall-Petch coefficient, about ~0.5 GPa×$\mu m^{-0.5}$ [5]. Indeed, some earlier reports demonstrated promising mechanical properties of the equiatomic CoCrFeMnNi alloy with the grain size of d≈0.5 μm [18,19]. In these works, cold working followed by annealing at temperatures ≥650°C was used to produce the ultra-fine grained (UFG) structure. Because of relatively fast kinetics of recrystallization and grain growth in the single-phase CoCrFeNiMn alloy [20,21], very precise control over the annealing condition is needed for preserving the fine grain size.

A different approach to produce UFG structure was used in [22,23]. Here, the studied alloys contained substantial amounts of Al or Al and C. Presence of additional elements resulted in the precipitation of second phase particles during post-rolling annealing, that impeded the growth of the recrystallized fcc grains. As a result, duplex UFG structure providing a good balance between strength and ductility was formed. However, the stability of such a structure requires additional studies. In present work, the type and the kinetics of coarsening of second-phase precipitates and grain growth during annealing at 973-1273 K for up to 10 hours in the previously reported Al, C containing CoCrFeMnNi-type alloy [22] were determined.

2. Materials and methods

In this work, the studied alloy was fabricated via the self-propagating high-temperature synthesis (SHS)-casting technique. The initial products were taken in the form of a mixture of powders including oxides of the target elements (Co, Cr, Fe, Mn, Ni), Al (both as the metal reducer and alloying element), and pure C (graphite powder). Prior to the synthesis, the powders were mechanically mixed for 20 min. The mixture was placed in centrifuge with a graphite mold of 80 mm diameter under artificial gravity of 50-60 G.

The obtained ingot had the diameter of 40 mm and the length of 70 mm. The chemical composition of the obtained ingot is given in Table 1. Samples for the thermomechanical processing having the dimensions of 4×5×22 mm^3 were cut from the as-cast ingot by an electric discharge machine were. These samples were cold rolled to the final thickness of ≈0.32 mm corresponding to a thickness reduction of 92% with reduction per pass of 5-10%. Cold-rolled samples were annealed at 973-1273 K for 0.5, 1, 5, and 10 hours. For the annealing, the samples were placed in a preheated oven and held for the desired durations of time; cooling was carried by air.

Table 1. Chemical composition of the studied alloy (in at.%).

Co	Cr	Fe	Mn	Ni	Al	C
21.55	18.73	21.44	10.01	21.44	5.31	1.52

Microstructures of the annealed alloy were studied by scanning (SEM) electron microscopy mainly. The samples for scanning electron microscopy were prepared by careful mechanical polishing. The SEM studies were performed using a FEI Quanta 600 FEG microscope equipped with backscattered electron (BSE) and energy dispersive spectrometry (EDS) detectors. The grain/particle size was measured using SEM-BSE images per standard linear interception method. At least 300 grains/particles per each condition were analyzed. The fraction of the second phase(s) was measured from SEM-BSE images using a commercial Digimizer Image Analysis software and a binarization procedure. To identify the crystallographic structure of different phases, transmission electron microscopy (TEM) was employed. Details on TEM procedures can be found elsewhere [22].

3. Results

Cold rolling and annealing in the interval of 973-1273 K resulted in the formation of a duplex ultrafine-grained structure composed of the recrystallized fcc grains and $M_{23}C_6$ type carbides and B2 particles (see more details in [22]). The measured chemical composition of the fcc phase has retained rather close to the nominal one (Table 1), however it was depleted of Al and enriched in Co and Fe (~24-26 at.% each). The carbides were primarily composed of Cr (≈63 at.%), and depleted of other metallic elements. The B2 phase, in turn, was enriched with Ni (~35-37 at.%), Mn (~14-17 at.%), and Al (~14-21 at.%), and depleted of the rest of the elements. Since the clear distinction between the morphology of the particles was not possible, the size and the fraction of the $M_{23}C_6$ and B2 second phase's particles were measured jointly.

The effect of annealing temperature or time on the hardness of cold rolled samples is shown in Figure 1. Annealing at 973 K for 0.5 or 1 hour leads to a slight increase in hardness from 478 HV (in the cold-rolled state) to 513 or 495 HV. Such hardening may be due to the formation of second phase particles in a non-recrystallized or partially recrystallized state [13]. However, a further increase in the annealing temperature or time resulted in a remarkable decrease in the hardness in all interval of annealing times which associated with primary static recrystallization.

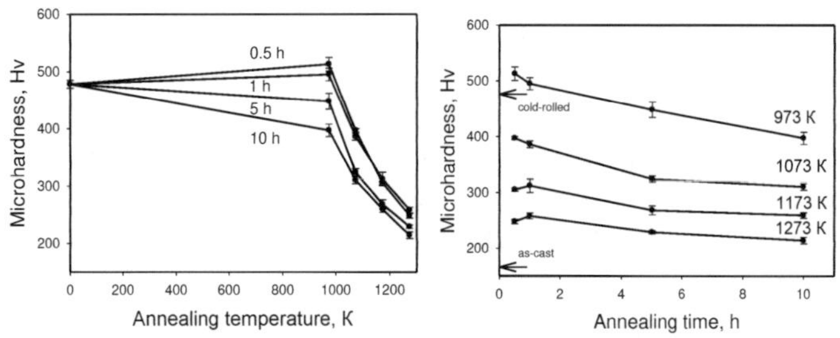

(a) (b)

Figure 1. Effect of annealing temperature (a) and time (b) on the microhardness of the Al, C-containing CoCrFeMnNi-type alloy.

Typical microstructures, which formed in the cold rolled samples after annealing for 0.5 or 10 hours at temperatures of 973-1273 K, are shown in Figure 2. Annealing of the rolled alloy during 0.5 h at temperatures 973 - 1073 K resulted in partial recrystallization of the fcc matrix phase (Figure 2 a, b). The corresponding microstructures consisted of deformed areas and equiaxed recrystallized grains. An increase in the annealing temperature to 1173-1273 K gave rise to complete recrystallization (Figure 2 c, d). After annealing at 973 K the recrystallized fraction and the size of the recrystallized grains were 0.52 and 0.57±0.23 μm, respectively (Table 2). An increase in the annealing temperature to 1273 K resulted in an increase in grain size to 1.75±0.85 μm.

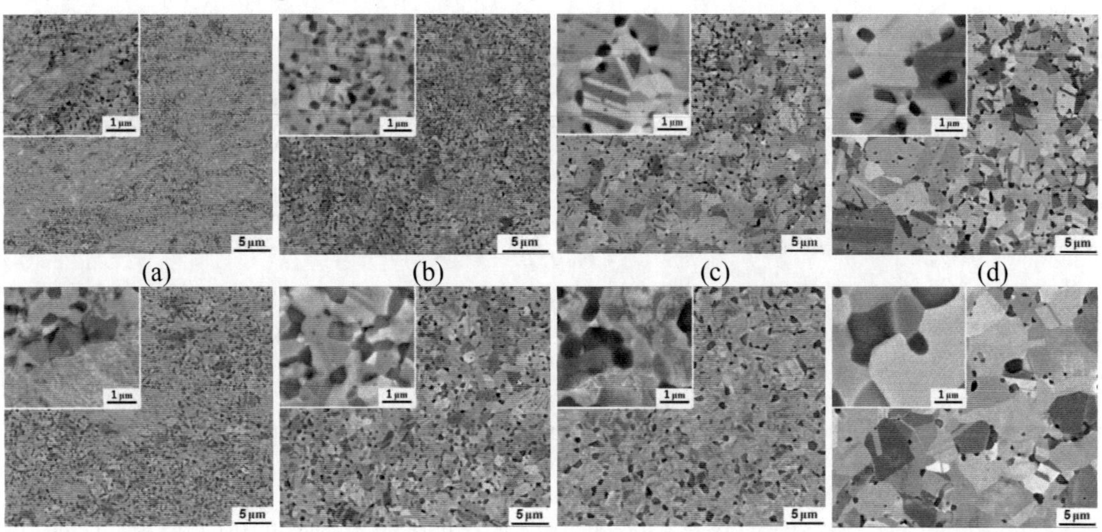

Figure 2. SEM-BSE images of the alloy after cold rolling to 92% thickness reduction and subsequent annealing for 0.5 h (a-d) and 10 h (e-h) at (a, e) – 973 K, (b, f) – 1073 K, (c, g) – 1173 K, (d, h) – 1273 K.

It is worth noting that an increase in the annealing time to 10 hours at the temperature of 973 K does not lead to the formation of a fully recrystallized structure (Figure 2e). The average grain size in the recrystallized microstructures increased with an increase in the annealing time (Table 2). A number of equilibrium triple junctions of grain boundaries (120° between boundaries in triple junctions) indicates normal grain growth as the main mechanism of microstructure evolution. The growing grains are characterized by numerous annealing twins, which are typical of low SFE metals and alloys during recrystallization followed by normal grain growth.

The grain growth takes place concurrently with the carbide and B2 particles coarsening. In the partially recrystallized condition the second phases were predominantly found in the recrystallized regions. The particles located at the grain boundaries were generally coarser than the particles inside recrystallized grains. The estimated fraction of the second phases decreased with increasing of annealing time and temperature at 1073-1273 K. A nonlinear dependence of the fraction of the particles at low annealing temperature can be associated with the development of primary recrystallization, when particle formation occurs in recrystallized regions. On the other hand, the average size of the second phase's particles increased from 0.14±0.05 μm to 0.53±0.27 μm when temperature increased from 973 K to 1273 K.

Table 2. Microstructure parameters of the studied alloy after cold rolling to 92% thickness reduction and subsequent annealing at 973 K, 1073 K, 1173 K, and 1273 K for 0.5-10 hours.

Temperature	973 K				1073 K				1173 K				1273 K			
Time, h	0.5	1	5	10	0.5	1	5	10	0.5	1	5	10	0.5	1	5	10
Grain size, μm	0.57 ± 0.23	0.42 ± 0.16	0.84 ± 0.31	1.10 ± 0.52	0.55 ± 0.22	0.88 ± 0.42	1.00 ± 0.36	1.41 ± 0.65	1.03 ± 0.35	0.90 ± 0.31	1.81 ± 0.65	1.85 ± 0.44	1.75 ± 1.15	2.28 ± 1.05	3.10 ± 1.36	4.69 ± 2.40
Particle size, nm	136 ± 52	165 ± 66	246 ± 122	294 ± 114	240 ± 100	258 ± 117	314 ± 183	416 ± 212	299 ± 160	356 ± 156	459 ± 268	542 ± 231	530 ± 270	465 ± 217	967 ± 435	935 ± 459
Particles fraction, %	5.9	9.6	20.2	10.6	19.9	15.2	13.7	9.4	17.3	12.9	7.85	6.5	14.1	8.16	6.0	3.5

The kinetics of the grain/precipitate coarsening was well described by the power-law dependence:

$$\boldsymbol{d^n - d_0^n = Kt} \quad (1)$$

where n is the grain/particle growth exponents, d_o is grain/particle size, and K is the rate constant at a given temperature. The exponent n =3 suggests that coarsening was controlled by volume diffusion (Figure 3 a, b).

The temperature dependence of the grain size is given by the dependence of rate constant K on temperature, Eq. (1), which has an Arrhenius form:

$$\boldsymbol{K = K_0 \exp(-Q/RT)} \quad (2)$$

Here, T is the absolute temperature, Q is the apparent activation energy for grain/particle growth, R is the gas constant, and K_o is a material constant. The apparent activation energy was calculated from the linear dependence of ln(K) on 1/T (Figure 3c). In particular, the present analysis showed that Q = 137 kJ/mol for grain growth and Q = 125 kJ/mol for particles coarsening. The obtained values are considerably lower than that were reported for bulk self-diffusion of constitutive elements of the $CoCrFeMn_{0.5}Ni$ alloy (288.4-317.5 kJ/mol) [24]. Note that the information on diffusion behavior of Al and C in the fcc HEAs are currently lacking, yet the self-diffusion energies of these species in γ (fcc) Fe are ~300 kJ/mol [25,26]. Therefore, additional work is required to correlate the obtained values of activation energies of grain growth/particles coarsening with the responsible mechanisms. This work will be performed in future studies.

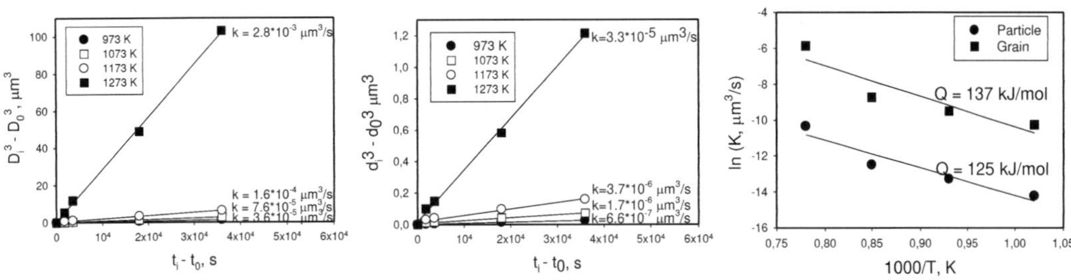

(a) (b) (c)

Figure 3. Dependence of (a) the fcc grain and (b) particles size on annealing time during annealing at 973-1273 K and (c) Arrhenius plots describing the apparent activation energy for grain or particle growth.

4. Conclusions

In the present work, recrystallization and grain growth of the CoCrFeMnNi-type high entropy alloy with subsequent annealing at 973-1273 K were studied. Following conclusions were drawn:

1) The annealed alloy is composed of the fcc matrix phase with the presence of B2 and M23C6 particles. Fraction and size of recrystallized fcc grains and size of the second phase particles increased with the increase of the annealing temperature, while the fraction of second phase particles exhibited complex dependence on annealing conditions.
2) Power law-type dependence between the grain/particle size and annealing time was observed. The exponent value of n=3 suggested that coarsening was controlled by volume diffusion. The activation energy was Q = 137 kJ/mol for grain growth and Q = 125 kJ/mol for particles coarsening.

References

[1] Yeh J W, Chen S K, Lin S J, Gan J Y, Chin T S, Shun T T, Tsau C H and Chang S Y 2004 Nanostructured high-entropy alloys with multiple principal elements: Novel alloy design concepts and outcomes *Adv. Eng. Mater.* **6** 299–303+274

[2] Senkov O N, Scott J M, Senkova S V., Miracle D B and Woodward C F 2011 Microstructure and room temperature properties of a high-entropy TaNbHfZrTi alloy *J. Alloys Compd.* **509** 6043–8

[3] Miracle D B and Senkov O N 2017 A critical review of high entropy alloys and related concepts *Acta Mater.* **122** 448–511

[4] Cantor B, Chang I T H, Knight P and Vincent A J B 2004 Microstructural development in equiatomic multicomponent alloys *Mater. Sci. Eng. A* **375** 213–8

[5] Otto F, Dlouhý A, Somsen C, Bei H, Eggeler G and George E P 2013 The influences of temperature and microstructure on the tensile properties of a CoCrFeMnNi high-entropy alloy *Acta Mater.* **61** 5743–55

[6] Otto F, Yang Y, Bei H and George E P P 2013 Relative effects of enthalpy and entropy on the phase stability of equiatomic high-entropy alloys *Acta Mater.* **61** 2628–38

[7] Gludovatz B, Hohenwarter A, Catoor D, Chang E H, George E P and Ritchie R O 2014 A fracture-resistant high-entropy alloy for cryogenic applications *Science (80-.).* **345** 1153–8

[8] Schuh B, Mendez-Martin F, Völker B, George E P, Clemens H, Pippan R and Hohenwarter A 2015 Mechanical properties, microstructure and thermal stability of a nanocrystalline CoCrFeMnNi high-entropy alloy after severe plastic deformation *Acta Mater.* **96** 258–68

[9] Otto F, Dlouhý A, Pradeep K G, Kuběnová M, Raabe D, Eggeler G and George E P 2016 Decomposition of the single-phase high-entropy alloy CrMnFeCoNi after prolonged anneals at intermediate temperatures *Acta Mater.* **112**

[10] Stepanov N D, Shaysultanov D G, Ozerov M S, Zherebtsov S V. and Salishchev G A 2016

Second phase formation in the CoCrFeNiMn high entropy alloy after recrystallization annealing *Mater. Lett.* **185** 1–4

[11] Salishchev G A, Tikhonovsky M A, Shaysultanov D G, Stepanov N D, Kuznetsov A V, Kolodiy I V, Tortika A S and Senkov O N 2014 Effect of Mn and v on structure and mechanical properties of high-entropy alloys based on CoCrFeNi system *J. Alloys Compd.* **591** 11–21

[12] Laurent-Brocq M, Akhatova A, Perrière L, Chebini S, Sauvage X, Leroy E and Champion Y 2015 Insights into the phase diagram of the CrMnFeCoNi high entropy alloy *Acta Mater.* **88**

[13] Klimova M V, Shaysultanov D G, Zherebtsov S V and Stepanov N D 2019 Effect of second phase particles on mechanical properties and grain growth in a CoCrFeMnNi high entropy alloy *Mater. Sci. Eng. A* **748** 228–35

[14] He J Y, Wang H, Huang H L, Xu X D, Chen M W, Wu Y, Liu X J, Nieh T G, An K and Lu Z P 2016 A precipitation-hardened high-entropy alloy with outstanding tensile properties *Acta Mater.* **102** 187–96

[15] Gwalani B, Soni V, Lee M, Mantri S, Ren Y and Banerjee R 2017 Optimizing the coupled effects of Hall-Petch and precipitation strengthening in a Al0.3CoCrFeNi high entropy alloy *Mater. Des.* **121** 254–60

[16] Klimova M V., Shaysultanov D G, Chernichenko R S, Sanin V N, Stepanov N D, Zherebtsov S V. and Belyakov A N 2019 Recrystallized microstructures and mechanical properties of a C-containing CoCrFeNiMn-type high-entropy alloy *Mater. Sci. Eng. A* **740–741** 201–10

[17] Stepanov N D, Yurchenko N Y, Tikhonovsky M A and Salishchev G A 2016 Effect of carbon content and annealing on structure and hardness of the CoCrFeNiMn-based high entropy alloys *J. Alloys Compd.* **687** 59–71

[18] Sun S J, Tian Y Z, Lin H R, Dong X G, Wang Y H, Zhang Z J and Zhang Z F 2017 Enhanced strength and ductility of bulk CoCrFeMnNi high entropy alloy having fully recrystallized ultrafine-grained structure *Mater. Des.* **133** 122–7

[19] Wang Y H, Zhang Z F, Sun S J, Yang H J, Tian Y Z, Lin H R and Dong X G 2017 Transition of twinning behavior in CoCrFeMnNi high entropy alloy with grain refinement *Mater. Sci. Eng. A* **712** 603–7

[20] Otto F, Hanold N L and George E P 2014 Microstructural evolution after thermomechanical processing in an equiatomic, single-phase CoCrFeMnNi high-entropy alloy with special focus on twin boundaries *Intermetallics* **54** 39–48

[21] Liu W H, Wu Y, He J Y, Nieh T G and Lu Z P 2013 Grain growth and the Hall-Petch relationship in a high-entropy FeCrNiCoMn alloy *Scr. Mater.* **68** 526–9

[22] Stepanov N D, Shaysultanov D G, Chernichenko R S, Ikornikov D M, Sanin V N and Zherebtsov S V. 2018 Mechanical properties of a new high entropy alloy with a duplex ultra-fine grained structure *Mater. Sci. Eng. A* **728** 54–62

[23] Yasuda H Y, Miyamoto H, Cho K and Nagase T 2017 Formation of ultrafine-grained microstructure in Al0.3CoCrFeNi high entropy alloys with grain boundary precipitates *Mater. Lett.* **199** 120–3

[24] Tsai K-Y, Tsai M-H and Yeh J-W 2013 Sluggish diffusion in Co–Cr–Fe–Mn–Ni high-entropy alloys *Acta Mater.* **61** 4887–97

[25] Vasilyev A A, Sokolov S F, Kolbasnikov N G and Sokolov D F 2011 Effect of alloying on the self-diffusion activation energy in γ-iron *Phys. Solid State* **53** 2194–200

[26] Zhang B 2014 Calculation of self-diffusion coefficients in iron *AIP Adv.* **4** 017128

Acknowledgments

The authors gratefully acknowledge the financial support from the Russian Science Foundation Grant No. 18-19-00003. The authors are grateful to the personnel of the Joint Research Center, «Technology and Materials», Belgorod State University, for their assistance.

IOP Conf. Series: Journal of Physics: Conf. Series **1270** (2019) 012054 doi:10.1088/1742-6596/1270/1/012054

Microstructure and texture of a severely warm-rolled and annealed $AlCoCrFeNi_{2.1}$ eutectic high entropy alloy

S R Reddy [1, a], U Sunkari [1], A Lozinko [2], S Guo [2], P P Bhattacharjee [1, b]
[1] Department of Materials Science and Metallurgical Engineering, IIT Hyderabad, Kandi, Sangareddy, 502285, Telangana, India

[2] Industrial and Materials Science, Chalmers University of Technology, 41296 Gothenburg, Sweden

E-mail: ([a] ms14m16p000001@iith.ac.in, [b] pinakib@iith.ac.in)

Abstract. The effect of warm-rolling and subsequent annealing on the evolution of microstructure and texture was systematically investigated in a lamellar ($L1_2$+B2) $AlCoCrFeNi_{2.1}$ eutectic high entropy alloy (EHEA). The EHEA was warm-rolled at 400°C up to 90% reduction in thickness and annealed at different temperatures. Ultra-fine lamellar microstructure evolved after heavy deformation with the characteristic textures of the two phases. Annealing treatments resulted in the gradual development of an ultrafine micro-duplex structure with increasing annealing temperature, even though the alloy showed significant resistance to grain growth. The annealing process was featured by concomitant microstructural and textural changes.

1. Introduction

High entropy alloys (HEAs) are a new class of multicomponent alloys containing a large number of alloying elements (≥5) in equiatomic or near equiatomic proportions [1]. HEAs may show simple phases owing to their large configurational entropy [1, 2]. HEAs have shown tremendous potential due to their intriguing and attractive mechanical properties.

Eutectic HEAs (EHEAs) are a special class of HEAs developed due to their excellent castability and enhanced properties [3]. $AlCoCrFeNi_{2.1}$ EHEA is found to show superior mechanical properties due to the presence of ordered $L1_2$ (soft) and B2 (hard) phases [3, 4]. Thermo-mechanical processing comprising of heavy deformation (cold/cryo) and annealing [4-7] led to the development of tailored microstructures exhibiting superior strength-ductility synergy in these alloys. However, the effect of warm-rolling on the microstructure of these EHEAs has not been studied intensively, which constitutes the major objective of the present research.

2. Experimental procedure

The $AlCoCrFeNi_{2.1}$ EHEA used in the present work was prepared by arc melting from high purity elements followed by suction casting. Specimens with dimensions 20mm*15mm*3mm were extracted from the as-cast EHEA followed by isothermal warm-rolling at 400°C up to 90% reduction in thickness (~0.3mm) using a laboratory scale two-high rolling machine (SPX Precision Instruments, Fenn, USA). The 90% warm-rolled samples were annealed at different temperatures ranging between 800°C to 1200°C for 1 h.

The microstructure and texture of the deformed and annealed samples were investigated using an electron backscatter diffraction (EBSD) system (Oxford Instruments, UK) attached to a Field Emission Gun-Scanning Electron Microscope (FEG-SEM) (SUPRA 40, Carl-Zeiss, Germany). The samples for SEM-EBSD were prepared by mechanical polishing up to colloidal silica followed by electro-polishing using an electrolyte mixture of 90% ethanol + 10% perchloric acid. Several EBSD scans were acquired and analyzed by the TSL-OIM™ (EDAX, USA) software. The orientation distribution functions (ODFs) were calculated using a harmonic series expansion method (series rank 22) using orthotropic sample symmetry. The hardness measurements of the deformed and annealed samples were performed using micro Vickers Hardness Tester (DuraScan 20, Emco-Test, Austria).

3. Results and discussion

Figure 1 shows the EBSD image quality (IQ) maps of the as-cast EHEA and those following warm-rolling to different thickness reductions. The as-cast EHEA reveals the eutectic mixture (65:35) of $L1_2$ /ordered FCC (white) and B2/ordered BCC (dark) phases as shown in figure 1 (a). The average lamellar spacing of the as-cast EHEA is 0.36 μm with a hardness of 330 Hv. The EHEA shows progressive alignment of the lamellae along the rollig direction with increasing thickness reduction as shown by the IQ maps of the 20% (figure 1(b)) and 50% (figure 1(c)) reduction in thickness. The development of ultra-fine lamellar microstructure can be observed after 90% deformation (figure 1(d)). Presence of shear bands are also easily seen in the microstructure.

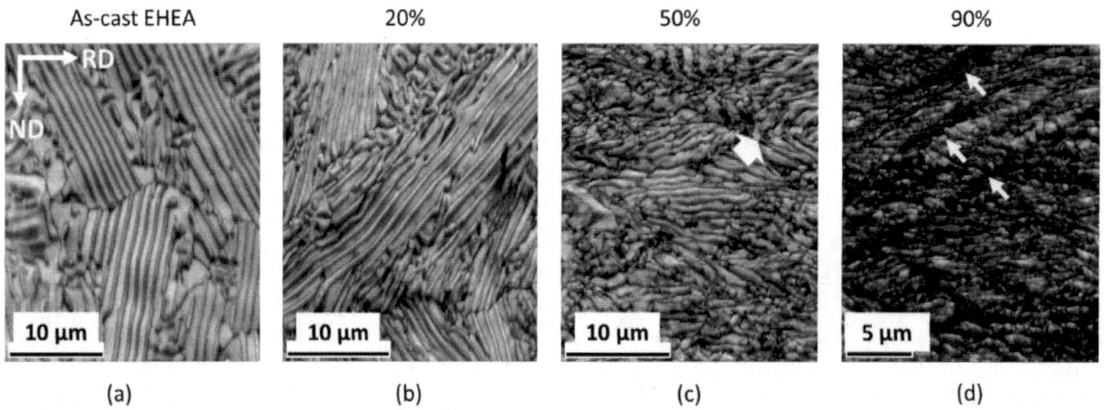

Figure 1. EBSD image quality (IQ) maps showing microstructure of (a) as-cast, (b) 20%, (c) 50% and (d) 90% warm rolled EHEA. The IQ maps confirms the $L1_2$/FCC (white) and B2 (dark) phase.

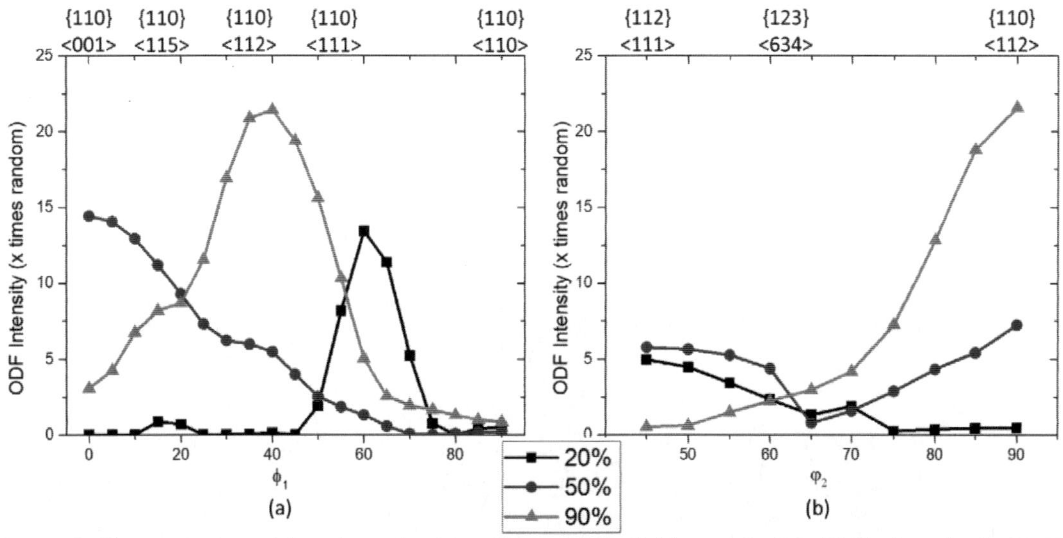

Figure 2. Change in intensities along the (a) α-fiber and (b) β-fiber of the $L1_2$/FCC phase in the EHEA with increasing deformation. The locations of important texture components in $L1_2$/FCC phase are indicated.

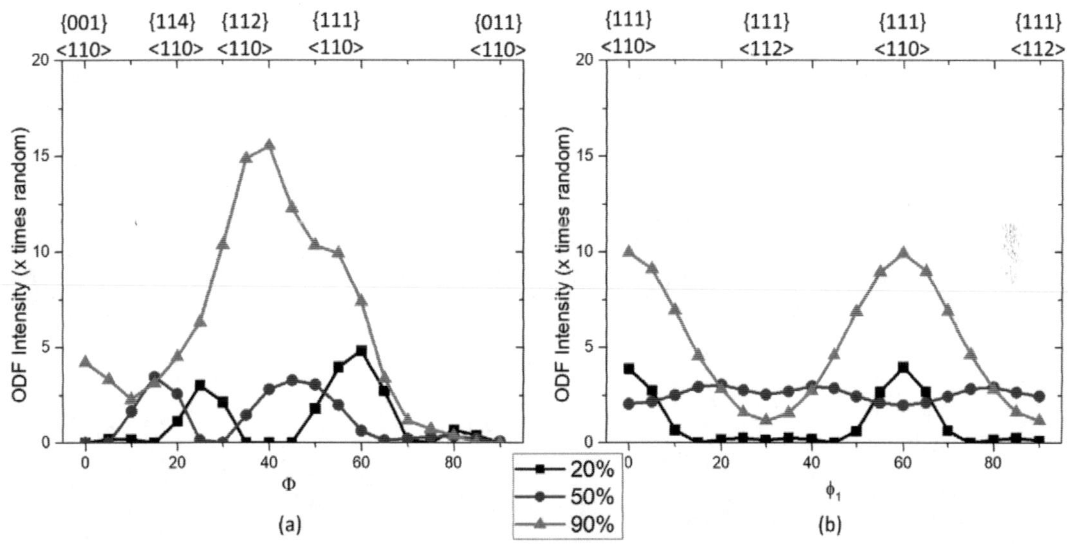

Figure 3. Change in intensities along the (a) RD-fiber (//<110>) and (b) ND-fiber (//<111>) of the B2/BCC phase in the EHEA with increasing deformation. The locations of important texture components in B2/BCC phase are indicated.

In order to understand the texture development during deformation, relevant texture fiber plots of the $L1_2$/FCC phase are represented in figure 2. The α-fiber plot (extending from Goss {011}<001> to Brass orientation {110}<112>) (figure 2(a)) confirms the development of strong Brass ({110}<112>) after 90% deformation. The β-fiber plot (extending from Copper {112}<111> to S {123}<634> to Brass orientation) (figure 2(b)) [8] also confirms the strong intensity of the Brass component. Evidently, the $L1_2$/FCC phase develops predominantly Brass type deformation texture.

Figure 3 shows relevant texture fiber charts (a) RD-fiber (//<110>) (extending from {001}<110> to {111}<110>) and (b) ND-fiber (//<111>) (extending from {111}<110> to {111}<112>) of the B2 phase of the warm rolled materials. The increase in deformation up to 90% led to the development of typical RD and ND-fiber with strong intensities of the RD-fiber centered on {112}<110> while the ND-fiber centered on {111}<110>. Therefore, the B2 phase shows typical RD and ND-fiber components.

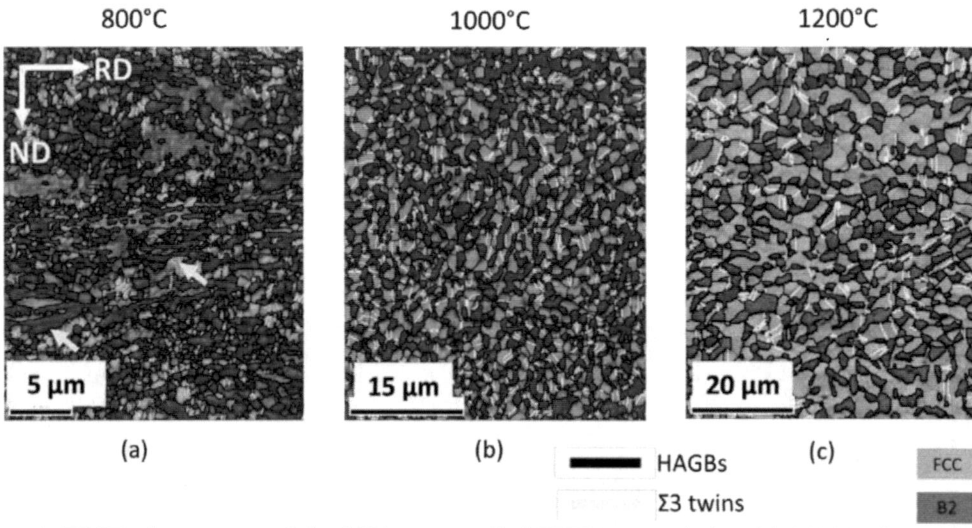

Figure 4. EBSD phase maps of the 90% warm-rolled EHEA annealed at (a) 800°C, (b) 1000°C and (c) 1200°C for 1 hr.

Table 1: Summary of key microstructural parameters of EHEA in deformed and annealed conditions

Material Condition	Phase Fraction (FCC:BCC)	Grain size (μm)	CSL Fraction	Hardness (Hv)
90% warm-rolled	65:35	0.12 ± 0.01	-	648 ± 19
Annealed at 800°C	48:52	0.42 ± 0.02	0.16	446 ± 16
Annealed at 1000°C	55:45	0.84 ± 0.01	0.27	362 ± 11
Annealed at 1200°C	63:37	1.59 ± 0.10	0.28	325 ± 11

Figure 4 shows the EBSD phase maps of 90% warm-rolled material after annealing at 800°C (figure 4(a)), 1000°C (figure 4(b)) and 1200°C (figure 4(c)) respectively. The phase maps show the development of micro-duplex structure consisting of $L1_2$/FCC and B2 phases in 800°C with remnant extended bands as indicated by the white arrow. The material shows an ultrafine grain size of 0.4 μm after annealing at 800°C. Annealing at 1000°C and 1200°C results in the development of the duplex structure with a systematic decrease in the B2 phase fraction. The increase in the coincidence site lattice (CSL) fraction with increasing annealing temperature is clearly observed. The EHEA shows significant resistance to the grain growth with an average grain size of ~1.6 μm at 1200°C. Systematic decrease in hardness with increasing annealing temperature is established. Table 1 summarizes the key microstructural parameters after annealing.

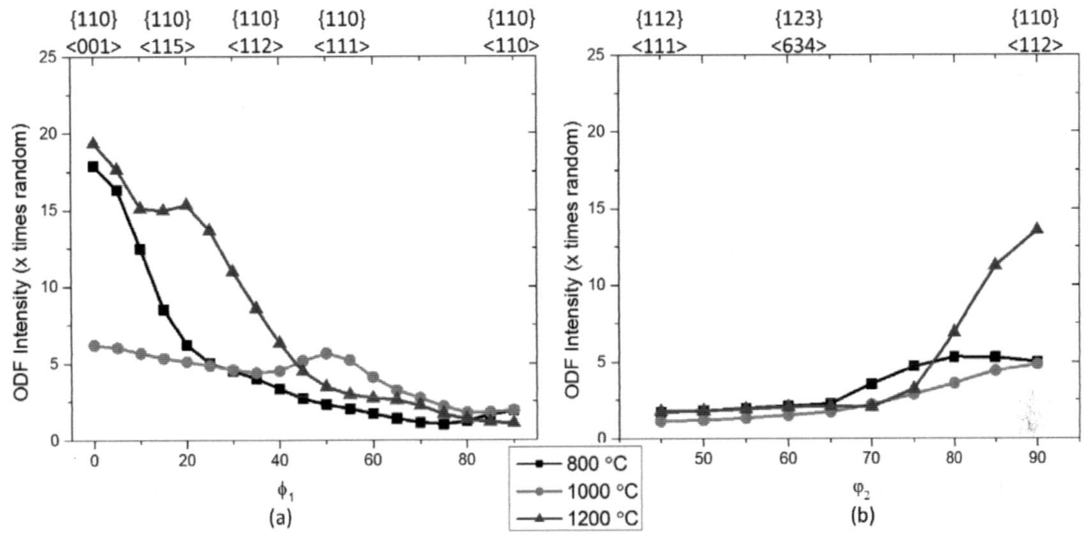

Figure 5. Intensities along the (a) α-fiber and (b) β-fiber of the $L1_2$/FCC phase in the 90% warm rolled EHEA annealed at different temperatures. The locations of important texture components in $L1_2$/FCC phase are indicated.

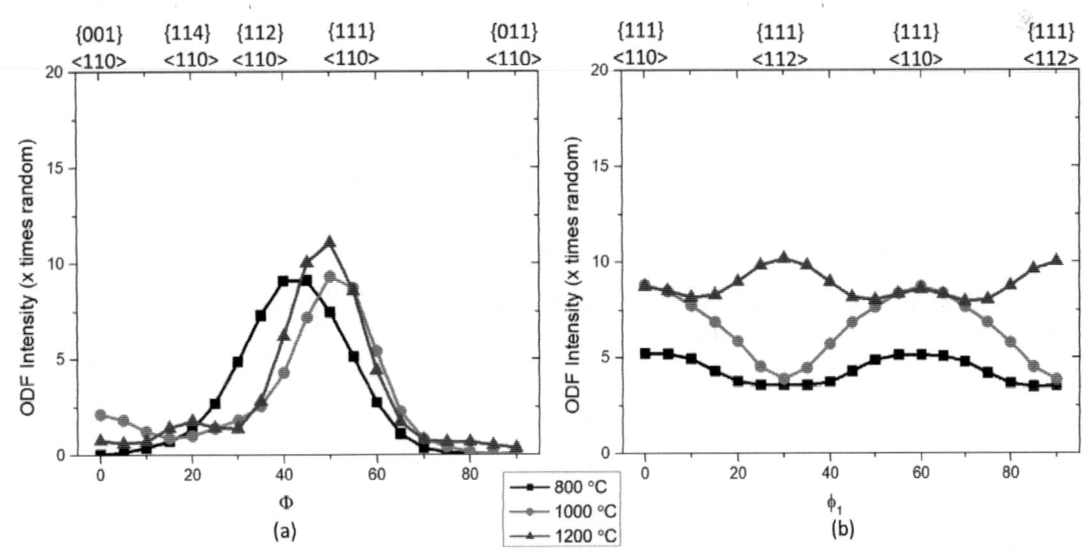

Figure 6. Intensities along the (a) RD-fiber (//<110>) and (b) ND-fiber (//<111>) of the B2/BCC phase in 90% warm rolled EHEA annealed at different temperatures. The locations of important texture components in B2/BCC phase are indicated.

The annealing texture of $L1_2$ phase is shown in figure 5. The α-fiber plot (figure 5(a)) shows strong intensities in between the Goss ({110}<001>) and Brass component. The retention of the Brass component can be confirmed from the β-fiber plot (figure 5(b)). Therefore, the deformation texture is mostly retained.

The texture evolution of the B2 phase as represented by the RD-fiber (figure 6 (a)) and ND-fiber (figure 6(b)) plots. The RD-fiber plot shows the presence of the strongest texture development at {112}<110> after 800°C, which confirms the retention of deformation texture. However, there is a shift to {111}<110> texture component with increasing annealing temperature. Stronger ND-fiber texture is confirmed upon further annealing up to 1200°C.

4. Conclusions

- Severe warm-rolling up to 90% reduction in thickness results in a gradual alignment of the lamellae along the RD.
- The $L1_2$ phase developed α-fiber (ND//<110>) texture, while the B2 phase developed the typical RD (//<110>) and ND (//<111>) fiber texture.
- Annealing resulted in the development of ultrafine duplex structure .
- The $L1_2$ phase retained the α-fiber texture, while the B2 phase displayed stronger ND-fiber texture throughout the annealing temperatures.

Acknowledgment

This work was supported by grants from DST, India (grant no. EMR/2016/002215) and JICA-CKP; and Junior Researcher Grant (grant no. 2015-04087) from Swedish Research Council, Sweden.

References

[1] Yeh J W, Chen S K, Lin S J, Gan J Y, Chin T S, Shun T T, Tsau C H, Chang S Y, 2004 *Adv. Eng. Mater.* **6(5)** 299-303.
[2] Murty B S, Yeh J W, Ranganathan S, Bhattacharjee P P, 2019 *High-entropy alloys* (London: Butterworth-Heinemann, Elsevier).
[3] Lu Y, Dong Y, Guo S, Jiang L, Kang H, Wang T, Wen B, Wang Z, Jie J, Cao Z, 2014 *Sci. Rep.* **4** 6200.
[4] Wani I S, Bhattacharjee T, Sheikh S, Lu Y P, Chatterjee S, Bhattacharjee P P, Guo S, Tsuji N, 2016 *Mater. Res. Lett.* **4(3)** 174-179.
[5] Bhattacharjee T, Wani I S, Sheikh S, Clark I T, Okawa T, Guo S, Bhattacharjee P P, Tsuji N, 2018 *Sci. Rep.* **8(1)** 3276.
[6] Patel A, Wani I S, Reddy S R, Narayanaswamy S, Lozinko A, Saha R, Guo S, Bhattacharjee P P, 2018 *Intermetallics* **97** 12-21.
[7] Wani I S, Bhattacharjee T, Sheikh S, Bhattacharjee P P, Guo S, Tsuji N, 2016 *Mater. Sci. Eng. A* **675** 99-109.
[8] Sidor J J, Kestens L A I, 2013 *Scr. Mater.* **68(5)** 273-276.

AUTHOR INDEX